Solutions Manual and Transparency Masters to Accompany

Engineering Fluid Mechanics

Sixth Edition

John A. Roberson
WASHINGTON STATE UNIVERSITY, PULLMAN

Clayton T. Crowe
WASHINGTON STATE UNIVERSITY, PULLMAN

John Wiley & Sons, Inc.
NEW YORK • CHICHESTER • BRISBANE
TORONTO • SINGAPORE • WEINHEIM

ISBN 0-471-17306-1

Printed in the United States of America

10 9 8 7 6 5 4 3 2 1

Printed and bound by Malloy Lithographing, Inc.

CONTENTS

NOTES TO INSTRUCTORS

Instructors often feel the need to exchange ideas about course content and novel ways of teaching it. For this reason we thought it may be of interest to you to know about how we have taught fluid mechanics at Washington State University (WSU).

Course Procedure

Currently fluid mechanics is taught as a traditional three-lecture-a-week course both through the School of Mechanical Engineering (MME) and the Department of Civil and Environmental Engineering (CEE). Students in the Department of Biological Systems Engineering (formerly Agricultural Engineering) take the CEE course. The students in MME are subdivided into groups of three for purposes of discussion of the open-ended problems and a design problem is assigned toward the end of the course.

Course Content

The majority of the material is covered in one semester with the exception of Chapters 12 (compressible flow), Chapter 15 (open-channel flow) and Chapter 16 (numerical methods in fluid mechanics). Depending on the instructor and the student progress, elements of these chapters may be addressed.

Ways in Which Chapter 16 Might Be Used

We envision that Chapter 16 could be used as a regular part of a fluid mechanics course or for special situations. For example, it might be used in the following ways:

A. Some colleges offer fluid mechanics as a two quarter sequence (3 hours each quarter). For such a program, the chapter could be offered on a regular basis in the second course.

B. For instructors who wish to introduce potential flow theory, Chapter 16 covers this topic along with an introduction to computational procedures for solving potential flow problems.

C. Utilize it in a special self-study type of minicourse for gifted students. Students might earn a semester's hour credit for such a study.

D. Include the chapter as part of a more advanced course in fluid mechanics. It would be especially appropriate for an "intermediate fluid mechanics" course.

E. Some students will be exposed to computational fluid mechanics in their jobs after graduation. Self study of Chapter 16 would be very valuable to them.

Use of Films

We often use films or videotapes to reveal the physical aspects of certain types of flow phenomena. The ones that we have found to be very educational are:

Surface tension in fluid mechanics;
Flow visualization:
Cavitation;
Turbulence;
Form drag, lift, and propulsion.

The first four were produced under the direction of the National Committee for Fluid Mechanics Films (NCFMF). Videotapes of these are available from

Enclopaedia Britannica Educational Corp.
310 S. Michigan Avenue
Chicago, IL 60604-9839

The last one was produced at the University of Iowa and a videotape of it is available from

University of Iowa
Audio-Visual Center
C215 Seashore Hall
Iowa City, Iowa 52242

Other tapes on fluid mechanics are also available from both of these sources.

Errors/Feedback

We would sincerely appreciate receiving information about any errors in this manual. Also, we would like to know about your opinions concerning the text material in general. Your comments and suggestions would be appreciated.

Use of Software

For the most part, the student requires only a hand-held calculator to carry out the solution procedures for problems in the text. Some software packages may be useful to the student for problems which require integration or plotting of results. MATHCAD® would be recommended as a general purpose package for executing mathematical manipulations. Spreadsheet programs, such as Excel® are useful for problems involving tables of numbers or numerical integrations.

Software illustrating various aspects of fluid mechanics accompanies the book titled *Fluid Mechanics - Programs* for the IBM® PC by Daniel Olfe (McGraw-Hill, 1987). Programs in BASIC are available for such problems as flow between rotating cylinders, trajectory of a golf ball, flow past an airfoil section, and compressible flow through a deLaval nozzle.

Software for Chapter 16

Instructors using Chapter 16 may obtain the software programs for that chapter by sending a blank 3.5" high density, double-sided floppy disk to Professor Clayton T. Crowe, Department of Mechanical and Materials Engineering, Washington State University, Pullman, WA 99164-2920.

CHAPTER TWO

2.1 $$\rho = \frac{P}{RT} = \frac{140{,}000}{R(30 + 273)} = \frac{462.0}{R}\ \text{kg/m}^3$$

From Table A-2: $R_{air} = 287$ J/kgK

$R_{He} = 2{,}077$

$R_{CO_2} = 189$

Then $$\rho_{air} = \frac{462.0}{287} = \underline{\underline{1.61\ \text{kg/m}^3}}$$

$$\rho_{He} = \frac{462.0}{2{,}077} = \underline{\underline{0.222\ \text{kg/m}^3}}$$

$$\rho_{CO_2} = \frac{462.0}{189} = \underline{\underline{2.44\ \text{kg/m}^3}}$$

2.2 $$\rho_{CO_2} = \frac{P}{RT} = \frac{300{,}000}{189(60 + 273)} = 4.767\ \text{kg/m}^3$$

Then $\gamma_{CO_2} = \rho_{CO_2} \times g = 4.767 \times 9.81 = \underline{\underline{46.764\ \text{N/m}^3}}$

2.3 $$\rho_{He} = \frac{P}{RT} = \frac{500{,}000}{2{,}077(60 + 273)} = 0.723\ \text{kg/m}^3$$

Then $\gamma_{He} = \rho_{He} \times g = 0.723 \times 9.81 = \underline{\underline{7.09\ \text{N/m}^3}}$

2.4 $$M = \rho V\!\!\!\!- = (p/RT)V\!\!\!\!-$$

Then $M_2/M_1 = (p_2/p_1)$ where p is the absolute pressure.

$M_2/M_1 = 300\ \text{kPa}/200\ \text{kPa} = \underline{\underline{1.5}}$

2.5 $$\rho_{air} = \frac{P}{RT} = \frac{103{,}000}{287(40 + 273)} = 1.147\ \text{kg/m}^3$$

$\rho_{water} = 1{,}000\ \text{kg/m}^3$ Then $\dfrac{\rho_{water}}{\rho_{air}} = \dfrac{1{,}000}{1.147} = \underline{\underline{872}}$

2.6 $\rho = p/RT$

$p_{abs.} = 200$ psia x 144 psf/psi = 28,800 psf

$R = 1{,}555$ ft-lbf/(slug - °R)

$T - 460 + 50 = 510°R$

$\rho = 28{,}800/(1{,}555 \times 510) = 0.0363$ slugs/ft^3

or $\gamma = \rho g = 0.0363 \times 32.2 = 1.69$ lbf/ft^3

then $W_{air} = 1.169$ lbf/ft^3 x 4 ft^3 = 4.68 lbf

W_{total} = 104.68 lbf

2.7 $\rho_{air} = \frac{P}{RT} = \frac{445{,}000}{287(38 + 273)} = 4{,}986$ kg/m^3

$\gamma_{air} = \rho_{air} \times g = 3{,}865 \times 9.81 = 43.91$ N/m^3

2.8 Assume average pressure based on elevation = (5,280/2) ft.

Then $P \approx 14.7 - \frac{5{,}280}{2} \times 0.00242 \times 32.2/144 = 13.3$ psia

Assume T = 50°F

Then $\rho = \frac{P}{RT} = \frac{13.3 \times 144}{1{,}716(50 + 460)} = 0.00219$ slugs/ft^3

Mass = $\rho \forall = 0.00219 \times (5{,}280)^3 = 3.22 \times 10^8$ slugs

$3.22 \times 10^8 \times 32.2 = 1.04 \times 10^{10}$ lbm

$1.04 \times 10^{10} \times 0.4536 = 4.72 \times 10^9$ kg

2.9 $\rho = p/RT$

$R_{CO_2} = 189$ J/kgK (from Table A-2)

$p = 400$ kN/m^2

$T = 20 + 273 = 293$ R

Thus $\rho = 400 \times 10^3/(189 \times 293) = 7.22$ kg/m^3

$\gamma = \rho g = 7.22$ kg/m^3 x 9.81 m/s^2 = 70.86 N/m^3

$\mu = 1.7 \times 10^{-5}$ N·s/m^2 (from Fig. A-2)

$\nu = \mu/\rho = 1.7 \times 10^{-5}/(7.22) = 2.35 \times 10^{-6}$ m^2/s

---

2.10 Mass and Weight are extensive properties; the remaining properties are intensive.

2.11 $c_p/c_v = k \qquad c_p - c_v = R$

$c_p/c_v - c_v/c_v = R/c_v$

$k - 1 = R/c_v; \quad c_v = \underline{\underline{R/(k-1)}}$

$c_p = R + c_v = R + R/(k-1) = \underline{\underline{kR/(k-1)}}$

2.12 Water: $\mu_{70} = 4.04 \times 10^{-4} N \cdot s/m^2$, $\mu_{10} = 1.31 \times 10^{-3} N \cdot s/m^2$ } from Table A-5

$\Delta\mu = \mu_{70} - \mu_{10} = \underline{\underline{-9.06 \times 10^{-4}\ N \cdot s/m^2}}$

$\rho_{70} = 978\ kg/m^3$, $\rho_{10} = 1{,}000\ kg/m^3$ } from Table A-5

$\Delta\rho = \rho_{70} - \rho_{10} = \underline{\underline{-22\ kg/m^3}}$

Air: $\mu_{70} = 2.04 \times 10^{-5}\ N \cdot s/m^2$, $\mu_{10} = 1.76 \times 10^{-5} N \cdot s/m^2$ } from Table A-3

$\Delta\mu = \underline{\underline{+0.28 \times 10^{-5} N \cdot s/m^2}}$

$\rho_{70} = 1.03\ kg/m^3$, $\rho_{10} = 1.25\ kg/m^3$ } from Table A-3

$\Delta\rho = \underline{\underline{-0.22\ kg/m^3}}$

2.13 $\Delta\nu_{air,10\to60} = (1.89 - 1.41) \times 10^{-5} = \underline{\underline{4.8 \times 10^{-6} m^2/s}}$

2.14

	Oil (SAE 10W)	kerosene	water
$\mu (N \cdot s/m^2)$	$\underline{\underline{3.6 \times 10^{-2}}}$	$\underline{\underline{1.4 \times 10^{-3}}}$ (Fig. A-2)	$\underline{\underline{6.8 \times 10^{-4}}}$
$\rho (kg/m^2)$	870		993
$\nu (m^2/s)$	$\underline{\underline{4.1 \times 10^{-5}}}$	$\underline{\underline{1.7 \times 10^{-6}}}$ (Fig.A-2)	$\underline{\underline{6.8 \times 10^{-7}}}$

2.15 $\mu_{air,20°C} = 1.81 \times 10^{-5} N\cdot s/m^2$; $\nu = 1.51 \times 10^{-5} m^2/s$

$\mu_{water,20°C} = 1.00 \times 10^{-3} N\cdot s/m^2$; $\nu = 1.00 \times 10^{-6} m^2/s$

$\mu_{air}/\mu_{water} = \underline{\underline{1.81 \times 10^{-2}}}$; $\nu_{air}/\nu_{water} = \underline{\underline{15.1}}$

2.16 $du/dy = 10/((1/4)/12) s^{-1}$; $\mu = 5.2 \times 10^{-4} lb\text{-}s/ft^2$

Then $\tau = \mu du/dy = 5.2 \times 10^{-4} \times 10 \times 48 = \underline{\underline{0.250 \ lb/ft^2}}$

2.17 $\mu_{air} = 1.76 \times 10^{-5} N\cdot s/m^2$ (Table A-3)

$\mu_{water} = 1.31 \times 10^{-3} N\cdot s/m^2$ (Table A-5)

$\rho_{air} = p/RT = 103,000/(287 \times 283) = 1.268 \ kg/m^3$

$\rho_{water} = 1,000 \ kg/m^3$

Then $\nu_{air} = \mu_{air}/\rho_{air} = (1.76 \times 10^{-5})/1.27 = \underline{\underline{1.39 \times 10^{-5} m^2/s}}$

$\nu_{water} = (1.31 \times 10^{-3}/1,000) = \underline{\underline{1.31 \times 10^{-6} m^2/s}}$

2.18 $\tau = \mu du/dy$ where $\mu = 10^{-3} N\cdot s/m^2$

$du/dy = (1/6)(10)(y)^{-5/6} s^{-1}$

$= (1/6)(10)(.002)^{-5/6} s^{-1}$

$= (10/6)(177.5) s^{-1}$

Then $\tau = (10^{-3} N\cdot s/m^2)(10/6)(177.5) s^{-1} = \underline{\underline{0.296 \ N/m^2}}$

2.19 $u = 100y(0.1 - y) = 10y - 100y^2$

$du/dy = 10 - 200y$

$(du/dy)_{y=0} = 10 s^{-1}$ $(du/dy)_{y=0.1} = -10 s^{-1}$

$\tau_0 = \mu du/dy = (8 \times 10^{-5}) \times 10 = \underline{\underline{8 \times 10^{-4} lb/ft^2}}$

$\tau_{0.1} = \underline{\underline{-8 \times 10^{-4} lb/ft^2}}$

y	u
0	0
0.02	0.16
0.04	0.24
0.06	0.24
0.08	0.16
0.10	0

2.20 $\tau = \mu\, dV/dy$

$\tau_{max} \approx \mu(\Delta V/\Delta y)_{\text{next to wall}}$

$\tau_{max} = (10^{-3}\ N{\cdot}s/m^2)(1\ m/s/0.001\ m) = \underline{1.0\ N/m^2}$

The minimum shear stress will be zero, midway between the two walls, where the velocity gradient is zero.

2.21 $V = -((dp/dx)/2\mu)(By - y^2)$; $B = 0.05m$

$dp/dx = -1{,}600 N/m^2$; $y = 0.012$; $\mu = 6.2 \times 10^{-1} N{\cdot}s/m^2$

Then $V_{12mm} = (1{,}600/(2 \times 0.62))(0.05 \times 0.012 - (0.012)^2)$

$V_{12mm} = \underline{0.588\ m/s}$

Shear stress: $\tau = \mu dV/dy$ where $dV/dy = -(1/2\mu)(dp/dx)(B - 2y)$

$\tau_0 = (1{,}600/2)(0.05) = \underline{40 N/m^2}$

$\tau_{12} = (1{,}600/2)(0.05 - 2 \times 0.012) = \underline{20.8 N/m^2}$

2.22 $\tau = \mu du/dy$

$\mu du/dy = -\mu(1/2\mu)(dp/ds)(H-2y) + u_t\mu/H$

Evaluate τ at $y = H$:

$\tau_H = -(1/2)(dp/ds)(H-2H) + u_t\mu/H$

$= (1/2)(dp/ds)H + u_t\mu/H$

Evaluate τ at $y = 0$

$\tau_0 = -(1/2)(dp/ds)\ H + u_t\mu/H$

Observation of the velocity gradient lets one conclude that the pressure gradient dp/ds is negative. Also u_t is negative.

Therefore $|\tau_H| > |\tau_0|$

The maximum shear stress occurs at $\underline{y = H}$

2.23 From solution to P2.22

$\tau = -(1/2)(dp/ds)(H-2y) + u_t\mu/H$

Set $\tau = 0$ and solve for y

$0 = -(1/2)(dp/ds)(H-2y) + u_t\mu/H$

$\underline{y = (H/2) - (\mu u_t/(Hdp/ds))}$

2.24 $\tau = \mu du/dy = 0$ at $y = 0$

$du/dy = -(1/2\mu)(dp/ds)(H-2y) + u_t/H$

Then, at $y = 0$: $du/dy = 0 = -(1/2\mu)(dp/ds) H + u_t/H$

Solve for u_t: $\underline{u_t = (1/2\mu)(dp/ds)H^2}$

Note: Because $dp/ds < 0$, $u_t < 0$.

2.25 Given $\omega = 10\text{rpm} = 10\text{x}(2\pi/60) = 1.047$ rad/s

$D = 100\text{mm} = 0.10\text{m}$; $R = 0.050\text{m}$

Spacing $= 1\text{mm} = 1\times10^{-3}\text{m} = \Delta R$

Temp. $= 38°C$; $\mu = 3.6\times10^{-2}\text{N-s/m}^2$ (From Table A.4)

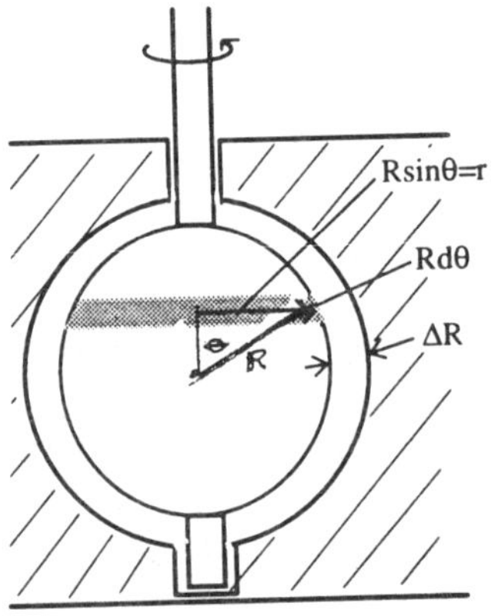

Rsinθ=r

Rdθ

ΔR

R

$$dT = rdF$$

$$dT = r\tau dA$$

where $\tau = \mu(dV/dy) = \mu(\Delta V/\Delta R)$

$$= \mu(\omega R\sin\theta/\Delta R)$$

$$= (3.6\times10^{-2}\text{N}\cdot\text{s}/\text{m}^2)(10\times2\pi/60)\text{rad}/\text{s}(0.05\text{m}\sin\theta/10^{-3}\text{m})$$

$$= 1.885\sin\theta\ \text{N}/\text{m}^2$$

$$dA = 2\pi R\sin\theta R d\theta$$

$$= 2\pi R^2\sin\theta d\theta$$

$$r = R\sin\theta$$

Then $dT = R\sin\theta(1.885\sin\theta)(2\pi R^2\sin\theta d\theta)$

$$dT = 11.84R^3\sin^3\theta d\theta$$

$$T = 11.84R^3\int_0^\pi \sin^3\theta d\theta$$

2.25 (continued)

$$= 11.84(0.05)^3[-(1/3)\cos\theta(\sin^2\theta + 2)]_0^\pi$$

$$= 11.84(0.05)^3[-(1/3)(-1)(2) - (-1/3)(1)(2)]$$

Torque $= \underline{\underline{1.97 \times 10^{-3}\text{Nm}}}$

2.26 Because the viscosity of gases increases with temperature $\mu_{100}/\mu_{50} > 1$. Correct choice is (c).

2.27 $\tau = \mu dV/dy$

$W/(\pi d\ell) = \mu V_{fall}/[(D-d)/2]$

$V_{fall} = W(D-d)\ /(2\pi d\ell\mu)$

$V_{fall} = 20(0.5 \times 10^{-3})/(2\pi \times 0.1 \times 0.2 \times 3.5 \times 10^{-1}) = \underline{\underline{0.23 \text{ m/s}}}$

2.28 $\Sigma F_z = 0$

$-W + F_\tau = ma$

$-W + \pi d\ell\mu V/[(D-d)/2] = W/g\ a$

$-W + (\pi \times 0.1 \times 0.2 \times 3.5 \times 10^{-1}V)/(0.5 \times 10^{-3}/2) = Wa/9.81$

Substituting V = 0.5 m/s and a = 14 m/s^2 and solving yields $\underline{\underline{W = 18.1 \text{ N}}}$

2.29 Assume linear velocity distribution: $dV/dy = V/y = \omega r/y$

$\tau = \mu dV/dy = \mu\omega r/y$

$\tau_2/\tau_3 = (\mu \times 1 \times 2/y)/(\mu \times 1 \times 3/y) = 2/3 = \underline{\underline{0.667}}$

$V = \omega r = 2 \times 0.03 = \underline{\underline{0.06 \text{ m/s}}}$

$\tau = \mu dV/dy = 0.01 \times 0.06/0.002 = \underline{\underline{0.30\text{N/m}^2}}$

2.30 $\tau = \mu\ dV/dy$

$\tau = \mu\omega r/y$

$= (0.01 \text{ N}\cdot\text{s/m}^2) \times (2 \text{ rad/s}) \times r/(0.002 \text{ m}) = 10\ r\text{N/m}^2$

d Torque $= r\tau dA$

2.30 (continued)

$$= r(10r)2\pi r dr = 20\pi r^3 dr$$

$$\text{Torque} = \int_0^{0.04} 20\pi r^3 dr = 20\pi\, r^4/4 \Big|_0^{0.04}$$

Torque = $\underline{\underline{4.02 \times 10^{-5} \text{N}\cdot\text{m}}}$

2.31 $\tau = \mu\, dV/dy$

$\tau = \mu\, r\omega/s$

On an elemental strip of area of radius r the differential shear force will be $\tau\, dA$ or $\tau(2\pi r dr)$. The differential torque will be the product of the differential shear force and the radius r

$$\text{or } dT_{\text{one side}} = r[\tau(2\pi r dr)]$$

$$= r[(\mu r\omega/s)(2\pi r dr)]$$

$$= (2\pi\mu\omega/s)r^3 dr$$

$$dT_{\text{both sides}} = (4\pi\mu\omega/s)r^3 dr$$

$$T = \int_0^{D/2} (4\pi\mu\omega/s)r^3 dr = \underline{\underline{(1/16)\pi\mu\omega D^4/s}}$$

2.32 One possible design solution is given below.

Assumptions:

1. Motor oil is SAE 10W-30; therefore, μ will vary from about 2×10^{-4}lbf-s/ft^2 to 8×10^{-3}lbf - s/ft^2 (Fig. A.2)
2. Assume the only significant shear stress develops between the rotating cylinder and the fixed cylinder.
3. Assume we want the maximum, rate of rotation ω, to be 3 rad/s.

Arbitrary decisions:

1. Let h = 4.0 in. = 0.333 ft
2. Let I.D. of fixed cylinder = 9.00 in. = 0.7500 ft.
3. Let O.D. of rotating cylinder = 8.900 in. = .7417 ft.

Let the applied torque which drives the rotating cylinder be produced by a force from a thread or small diameter monofilament line acting at a radial distance r_s. Here r_s is the radius of a spool on which the thread or line is wound. The applied force is produced by a weight and pulley system shown in the sketch below.

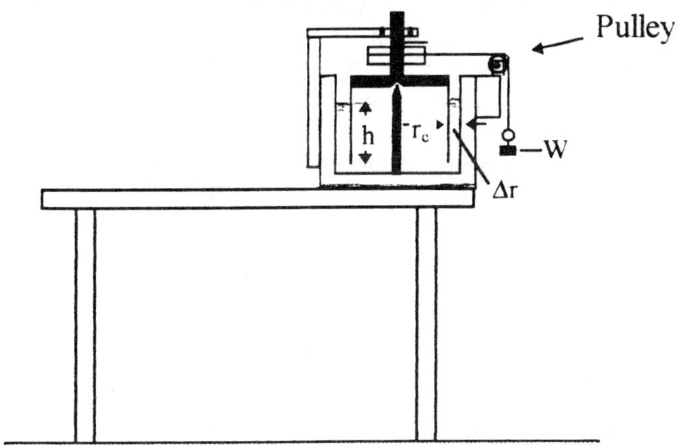

The relationship between μ, r_s, Δr, ω, h, and W is now developed:

$$T = r_c F_s \tag{1}$$

where T = applied torque
r_c = outer radius of rotating cylinder

F_s = shearing force developed at the outer radius of the rotating cylinder
but $F_s = \tau A_s$ where A_s = area in shear = $2\pi r_c h$

$\tau = \mu dV/dy \approx \mu \Delta V/\Delta r$ where $\Delta V = r_c\omega$ and Δr = spacing

$$\text{Then } T = r_c(\mu\Delta V/\Delta r)(2\pi r_c h)$$

$$= r_c\mu(r_c\omega/\Delta r)(2\pi r_c h) \tag{2}$$

But the applied torque $T = Wr_s$ so Eq. (2) becomes

$$Wr_s = r_c^3\mu\omega(2\pi)h/\Delta r$$

$$\text{or } \mu = (Wr_s\Delta r)/(2\pi\omega h r_c^3) \tag{3}$$

The weight W will be arbitrarily chosen (say 2 or 3 oz.) and ω will be determined by measuring the time it takes the weight to travel a given distance. So $r_s\omega = V_{fall}$ or $\omega = V_{fall}/r_s$. Equation (3) then becomes

$$\mu = (W/V_f)(r_s^2/r_c^3)(\Delta r/(2\pi h))$$

In our design let $r_s = 2\text{in.} = 0.1667$ ft.

$$\mu = (W / F_f)(0.1667^2/.3708^3)(0.004167 / (2\pi \times .3333)$$

Then
$$\mu = (W / V_f)(.02779/.05098)$$
$$\mu = (W / V_f)(1.085 \times 10^{-3})\text{lbf} - \text{s} / \text{ft}^2$$

Example: If W = 2 oz. = 0.125lbf. and V_f is measured to be 0.24 ft/s then $\mu = (0.125/0.24)(1.085\times10^{-3}), = 0.564\times10^{-3} = 5.65\times10^{-4}\text{lbf-s/ft}^2$

Other things that could be noted or considered in the design:

1. Specify dimensions of all parts of the instrument.
2. Neglect friction in bearings of pulley and on shaft of cylinder.
3. Neglect weight of thread or monofilament line.
4. Consider degree of accuracy.
5. Estimate cost of the instrument.

2.33 $E = -\Delta p/(\Delta \mathcal{V}/\mathcal{V})$

$2.2 \times 10^9 = -2 \times 10^6/(\Delta \mathcal{V}/1{,}000)$

Therefore, $\Delta \mathcal{V} = -(2 \times 10^6 \times 1{,}000)/(2.2 \times 10^9) = -0.909$ cc

Volume after pressure applied = $\mathcal{V} - \Delta \mathcal{V} = \underline{999.09 \text{ cc}}$

2.34 $E = -\Delta p/(\Delta \mathcal{V}/\mathcal{V})$

$2.2 \times 10^9 = -\Delta p/(-1/100)$

$\Delta p = 2.2 \times 10^7 \text{N/m}^2 = \underline{22 \text{ MN/m}^2 \text{ (increase)}}$

2.35 Refer to Fig 2-6(a). The surface tension force, $2\pi r\sigma$, will be resisted by the pressure force acting on the cut section of the spherical droplet or

$p(\pi r2) = 2\pi r\sigma$

$p = 2\sigma/r = \underline{4\sigma/d}$

2.36 $\Sigma F = 0$

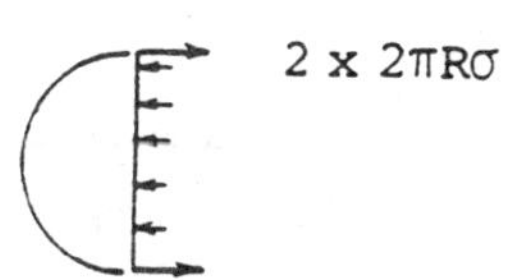

$\Delta p \pi R^2 - 2(2\pi R\sigma) = 0$

$\Delta p = \underline{\underline{4\sigma/R}}$; $\Delta p_{4mm\ rad.} = 4 \times 7.3 \times 10^{-2} N/m / 0.004\ m = \underline{\underline{73.0\ N/m^2}}$

Note: Effect of thickness, t, is assumed negligible.

2.37 $\Sigma F_z = 0$

+ surface tension force supporting bug - weight of bug = 0

Consider cross section of bug leg:

$F_T = (2/\text{leg})(6\ \text{legs})\sigma \ell$

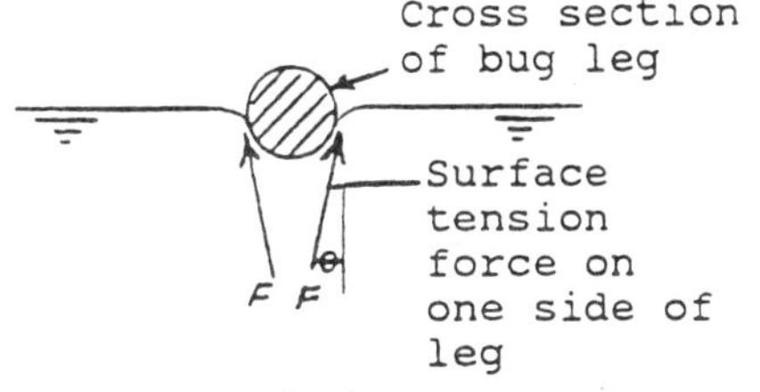

$= 12\sigma\ell$

$= 12(0.073\ N/m)(0.005m)$

$= 0.00438N$

Assume θ is small
Then cos θ = 1; F cos θ = F

$\therefore F_T - mg = 0$

$m = (0.00438/9.81) \times 1{,}000 = \underline{\underline{0.446\ \text{grams}}}$

2.38 $\Delta h = 4\sigma/(\gamma d) = 4 \times 0.005/(62.4 \times d) = 3.21 \times 10^{-4}/d$ ft

$d = 1/4$ in. $= 1/48$ ft; $\Delta h = 3.21 \times 10^{-4}/(1/48) = 0.0154$ ft $= \underline{\underline{0.185\ \text{in.}}}$

$d = 1/8$ in. $= 1/96$ ft; $\Delta h = 3.21 \times 10^{-4}/(1/96) = 0.0308$ ft $= \underline{\underline{0.369\ \text{in.}}}$

$d = 1/32$ in. $= 1/384$ ft; $\Delta h = 3.21 \times 10^{-4}/(1/384) = 0.123$ ft $= \underline{\underline{1.48\ \text{in.}}}$

2.39 $\Sigma F_y = 0$

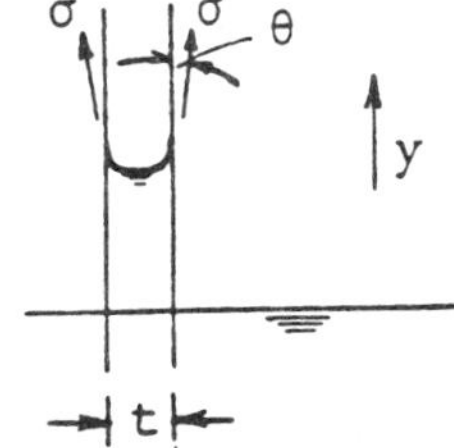

$2\sigma\ell - h\ell t\gamma = 0$

$h = 2\sigma/\gamma t$

$\sigma = 7.3 \times 10^{-2} N/m$

$h = 2 \times 7.3 \times 10^{-2}/(0.0010 \times 9{,}810) = 0.0149m = \underline{\underline{14.9\ mm}}$

2.40 Solution is similar to that for P2-36 except that only one surface exists here.

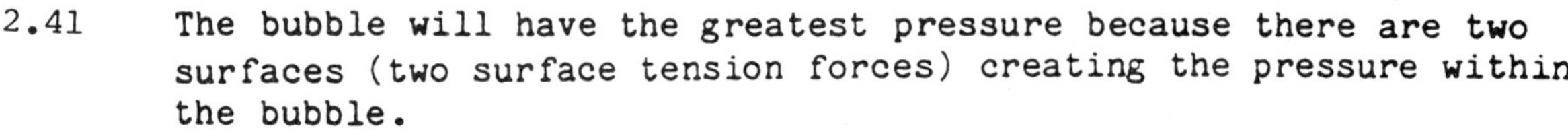

$\Delta p \pi R^2 - 2\pi R\sigma = 0$

$\Delta p = 2\sigma/R$

$\Delta p = 2 \times 7.3 \times 10^{-2}/(0.5 \times 10^{-3}) = \underline{\underline{292\ N/m^2}}$

2.41 The bubble will have the greatest pressure because there are two surfaces (two surface tension forces) creating the pressure within the bubble.

2.42 $100 - (101 - 69)/3.1 = \underline{\underline{89.7°C}}$

3.1 $p = 20/(\pi/4) \times (1)^2 = 25.46$ psi $= 175.4$ kPa

% gage error = $(26 - 25.46) \times 100/25.46 = \underline{2.10\%}$

3.2 $F = \Delta pA$

$= (100 \text{ kPa} - 30 \text{ kPa})\ \pi\ r^2$

$= (70{,}000 \text{ N/m}^2)\ \pi\ (.15\text{m})^2$

$= \underline{4.95} \text{ kN}$

3.3 This is an open-ended problem. The area of contact between each tire and pavement is a function of the weight of the car and the air pressure in the tires. The solution to the second part is a function of the air pressure and the diameter of the pump.

3.4 F per bolt at A-A = $p(\pi/4)D^2/20$

Assume same force per bolt at B-B

$p(\pi/4)D^2/20 = p(\pi/4)d^2/n$

$n = 20 \times (d/D)^2 = 20 \times (1/2)^2 = \underline{5}$

3.5 $p_{atm} = 100\text{kPa}$; $\rho_{water} = 1000 \text{ kg/m}^3$; $\gamma_{water} = 9810 \text{ N/m}^3$

Assume the water wets the glass and consider forces acting at the liquid surface inside the glass tube:

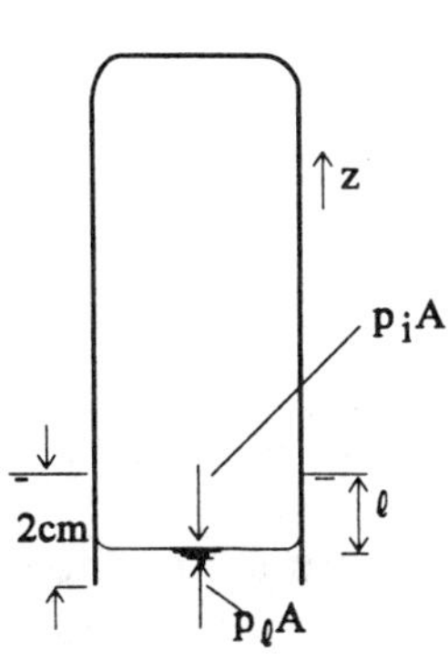

$$\sum F_z = 0$$

$$-p_iA + p_\ell A + \sigma\pi d = 0 \quad (1)$$

where p_i = pressure inside the tube

p_ℓ = pressure in the water at depth ℓ.

Also $p_i\forall_i = p_{atm}\forall_{tube}$

$$p_i = p_{atm}(\forall_{tube}/\forall_i) \quad (2)$$

$$= p_{atm}(0.10A_{tube}/((.08+\ell)(A_{tube}))$$

3.5 (Cont'd)

$$p_i = p_{atm}(0.10 / (.08 + \ell)) \quad (3)$$

$$p_\ell = p_{atm} + \gamma\ell \quad (4)$$

Solving for ℓ with Eqs. (1), (3) and (4) yields

$$\ell = 0.0192\text{m} = \underline{\underline{1.92\text{cm}}}$$

3.6 correct graph is (b)

3.7 (a) The water surface level in the left tube will be higher because of greater surface tension effects for that tube.

3.8

$$p_1 - \gamma\Delta z = p_2$$

or $p_2 = 200\text{N/A} - 0.85 \times 9{,}810\ \text{N/m}^3 \times 2\ \text{m}$

$= [200\text{N}/((\pi/4) \times (0.05)^2)] - 16{,}677\ \text{N/m}^2$

$= 85.18$ kPa

$F_2 = p_2 A_2 = 85{,}180 \times (\pi/4) \times (0.10)^2$

$= \underline{\underline{669\text{N}}}$

3.9

$p = \gamma\Delta z = 9{,}790 \times 50 = 489{,}500\ \text{N/m}^2 = \underline{\underline{489.5\ \text{kPa}}}$

$p_{40}/p_{atm} = (489.5 + 101.3)/101.3 = \underline{\underline{5.83}}$

3.10

Assume $\gamma = 9{,}810\ \text{N/m}^3$

The gage pressure at the 10 m depth = 10γ

$= 10\text{m} \times 9{,}810\ \text{N/m}^3 = 98.1$ kPa

The absolute pressure at the 10 m depth

$= 98.1$ kPa + 98 kPa = 2 x 98 kPa

The absolute pressure at the 10 m depth is $\underline{\underline{\text{2 times}}}$ that at the surface.

3.11 $p = (\gamma h)_{water} + (\gamma h)_{kerosene}$

$= 9,790 \times 80 \times 10^{-2} + 8,010 \times 1 = 16,821 \text{ N/m}^2$

$= \underline{15.84 \text{ kPa}}$

3.12 $p = 12 \times 8,630 = 103,560 \text{ Pa} = \underline{103.6 \text{ kPa}}$

3.13 $p = \gamma \times t$ where t = thickness of the atmosphere

$(14.7 \text{ lbs/in}^2)(144 \text{ in}^2/\text{ft}^2) = \gamma t$ where $\gamma = \rho g$

$14.7 \times 144 \text{ lb/ft}^2 = (0.00247 \text{ lbf}\cdot\text{s}^2/\text{ft}^4)(32.2 \text{ ft/s}^2)t$

$t = \underline{26,615 \text{ ft}}$

3.14 $\gamma = p/h = 73.6 \times 10^3/6 = 12.27 \text{ kN/m}^3$

Sp. gr. $= 12.27/9.81 = \underline{1.25}$

3.15

$\rho = \rho_{water}(1 + 0.01d)$

or $\gamma = \gamma_{water}(1 + 0.01d)$

$dp/dz = - dp/dd = - \gamma$

Then $dp/dd = \gamma_{water}(1 + 0.01d)$

Integrating: $p = \gamma_{water}(d + 0.01d^2/2) + C$

$p_{gage} = 0$ when $d = 0$; $C = 0$

Then $p_{d=10m} = \gamma_{water}(10 + 0.01 \times 10^2/2)$

$= \underline{103.0 \text{ kPa}}$ for $\gamma_{water} = 9,810 \text{N/m}^3$

3.16 From solution to Prob. 3.15 we have

$p = \gamma_{water}(d + 0.01 d^2/2)$

$50,000 \text{ N/m}^2 = (9,810 \text{ N/m}^3)(d + .005 d^2)$

Solving the above equation for d yields $d = \underline{4.973\text{m}}$

3.17

$$
\begin{aligned}
dp/dz &= -\gamma \\
&= -(50 - 0.1z) \\
p &= \int_0^{-20}(50 - 0.1z)dz \\
&= -50z + 0.1z^2/2\Big|_0^{-20} \\
&= 1000 + 0.1 \times 400/2 \\
&= \underline{\underline{1020\text{psfg}}}
\end{aligned}
$$

3.18 Given: $T = 20°C \therefore \gamma_w = 9790$ N/m^3 (Table A.5). Assume the air is an ideal gas; therefore,

$$p/\rho = RT(\text{Eq.}2.4).\ \text{But}\ \rho = M/\forall$$

so

$$p/(M/\forall) = RT; p\forall = MRT$$

Because M, R and T are constants for air in the tube we have

$$
\begin{aligned}
p_1\forall_1 &= p_2\forall_2 \\
p_1 &= 100{,}000\text{N}/\text{m}^2\text{abs.} \\
\forall_1 &= 1\text{m} \times A_{tube} \\
p_2 &= 100{,}000\text{N}/\text{m}^2 + \gamma_w(1\text{m} - \Delta\ell) \qquad (1) \\
\text{Also } p_2 &= p_1\forall_1/\forall_2 \\
&= (100{,}000\text{N}/\text{m}^2)(1\text{m} \times A_{tube})/[(1\text{m} - \Delta\ell)A_{tube}] \\
&= (100{,}000\text{N}/\text{m}^2)[1/(1 - \Delta\ell)] \qquad (2)
\end{aligned}
$$

Eliminate p_2 between Eqs. (1) and (2)

$$
\begin{aligned}
100{,}000 + \gamma_w(1 - \Delta\ell) &= (100{,}000)[1/(1 - \Delta\ell)] \\
\Delta\ell^2 - 12.214 + 1 &= 0
\end{aligned}
$$

Solving for $\Delta\ell$ yields $\Delta\ell = 0.082\text{m}$

3.19 The distance Δh is the height of the liquid surface in the manometer above the interface between the water and the heavier liquid. Thus

$$\gamma_{H_2O} \times 0.1\text{m} = 3\gamma_{H_2O} \times \Delta h$$
$$\Delta h = 0.1\text{m}/3$$
$$= 0.0333\text{m}$$
$$= 3.33\text{cm}$$

3.20 $0 + 4\ _{H_2O} + 3 \times 3\gamma_{H_2O} = p_{max}$

$p_{max} = 13 \times 9{,}810 = 127{,}530\ \text{N/m}^2 = \underline{127.5\ \text{kPa}}$

Maximum pressure will be at the bottom of the liquid with a S of 3.

$F_{CD} = pA = (127{,}530 - 1 \times 3 \times 9{,}810) \times 1\ \text{m}^2 = \underline{98.1\ \text{kN}}$

3.21 $\Delta p = \gamma h = 10{,}070 \times 6 \times 10^3$

$E_v = \Delta p/(d\rho/\rho)$

$(d\rho/\rho) = \Delta p/E_v = (10{,}070 \times 6 \times 10^3)/(2.2 \times 10^9) = 27.46 \times 10^{-3} = \underline{2.75\%}$

3.22 Volume added is shown in the figure below.
First get pressure at bottom of piston:

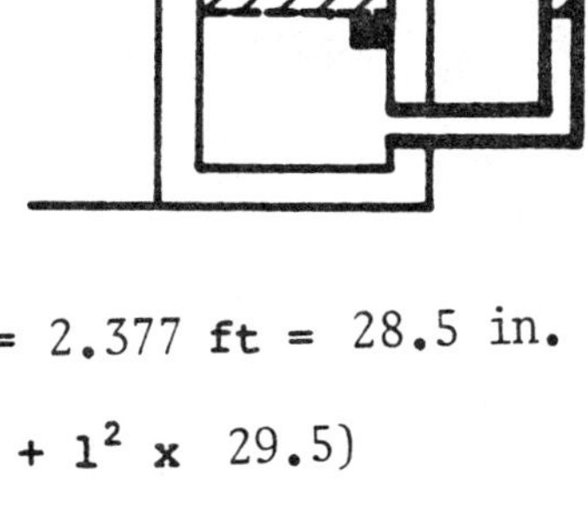

$p_p A_p = 11\ \text{lbf}$

$p_p = 11/A_p$

$= 11/((\pi/4) \times 4^2)$

$= 0.875\ \text{psig} = 126.1\ \text{psfg}$

Then $126.1\ \text{psfg} = \gamma_{oil} h$

or $h = 126.1/(62.4 \times 0.85) = 2.377\ \text{ft} = 28.5\ \text{in.}$

Finally $V_{added} = (\pi/4)(4^2 \times 1 + 1^2 \times 29.5)$

$= \underline{35.7\ \text{in.}^3}$

3.23 $$\rho = p/RT$$

$$\rho_{10} = p_{10}/RT \; ; \; \rho_{20} = p_{20}/RT \quad ; \quad \frac{\rho_{20}}{\rho_{10}} = \frac{p_{20}}{p_{10}}$$

where p = absolute pressure

$$p_{10} = 14.7 \text{ lb/in}^2 \times 144 \text{ in}^2/\text{ft}^2 + 10 \text{ ft} \times 62.4 \text{ lb/ft}^3 = 2741 \text{ lb/ft}^2$$

$$p_{20} = 14.7 \text{ lb/in}^2 \times 144 \text{ in}^2/\text{ft}^2 + 20 \text{ ft} \times 62.4 \text{ lb/ft}^2 = 3365 \text{ lb/ft}^2$$

$$\rho_{20}/\rho_{10} = p_{20}/p_{10} = 3365/2741 = \underline{\underline{1.23}}$$

3.24 $$p_v + \gamma_{Hg} h = p_{atm}$$

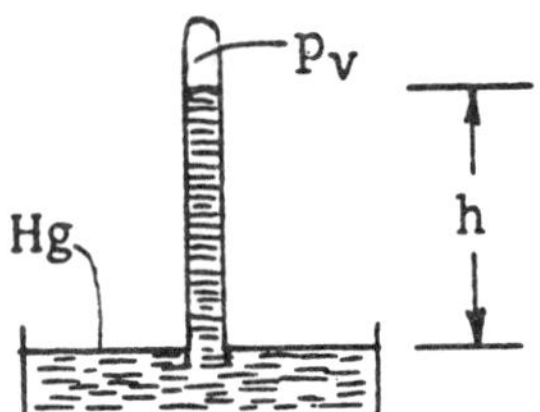

$$h = (p_{atm} - p_v)/\gamma_{Hg}$$

Assume $p_{v_{Hg}} = 0$

$$h = (99{,}000 - 0)/133{,}000 = 0.744 \text{ m} = \underline{\underline{744 \text{ mm}}}$$

3.25 Write manometer equation from the pipe to the open end of the manometer

$$p_{pipe} + (0.5 \text{ ft})(62.4 \text{ lbf/ft}^3) + (1 \text{ ft})(2 \times 62.4 \text{ lbf/ft}^3) - (2.5 \text{ ft})(62.4 \text{ lbf/ft}^3) = 0$$

$$p_{pipe} = (2.5 - 2 - 0.5) \text{ ft } (62.4 \text{ lbf/ft}^3) = \underline{\underline{0}}$$

3.26 $$p_A + (2.0 \text{ ft})(62.3 \text{ lbf/ft}^3) - (2/12 \text{ ft})(847 \text{ lbf/ft}^3) = 0$$

$$p_A = -124.6 \text{ lbf/ft}^2 + 141.2 \text{ lbf/ft}^2 = \underline{+16.6 \text{ lbf/ft}^2}$$

$$= \underline{+0.12 \text{ psi}}$$

3.27 $$p_A = 9{,}810 \, (1 \times 13.55 - 1.5 + 1.3 \times 0.9) = 129{,}700 \text{ N/m}^2 = \underline{\underline{129.7 \text{ kPa}}}$$

3.28 Write the manometer equation from pipe A to the top of the mercury column. Also assume S_{Hg}=13.55 and $\gamma_{H_2O} = 62.4 \text{lbf/ft}^3$.

$$p_A - 1.3\gamma_{oil} + 1.5\gamma_{H_2O} - 1\gamma_{Hg} = 0$$

$$\text{where } \gamma_{oil} = 0.9\gamma_{H_2O}$$

$$\gamma_{Hg} = 13.55\gamma_{H_2O}$$

$$\text{Then } p_A - 1.3(0.9)\gamma_{H_2O} + 1.5\gamma_{H_2O} - 13.55\gamma_{H_2O} = 0$$

$$p_A = \gamma_{H_2O}(13.55 + 1.17 - 1.5)$$

$$= (62.4 \text{lbf/ft}^3)(13.22 \text{ft})$$

$$= \underline{\underline{825 \text{psfg}}} = \underline{\underline{5.73 \text{psig}}}$$

3.29 $\Delta h_{\text{surface tension}} = 4\sigma/(\gamma d) = (4 \times 7.3 \times 10^{-2})/(9,810 \times 1 \times 10^{-3})$

$= 0.0298 \text{ m} = 2.98 \text{ cm}$

$p_A = \gamma h = 9,810 \ (10 - 2.98) \times 10^{-2} = \underline{\underline{689 \text{ Pa}}}$

3.30 $p_B = 50 \times (3/5) \times 10^{-2} \times 20 \times 10^3 - 10 \times 10^{-2} \times 20 \times 10^3 - 50 \times 10^{-2} \times 10 \times 10^3$

$p_B = -1,000 \text{ Pa} = \underline{\underline{-1.00 \text{ kPa}}}$

3.31 $(\pi/4) \ D^2_{tube} \times \ell = (\pi/4) \ D^2_{cistern} \times (\Delta h)_{cistern}$

$(\Delta h)_{cistern} = (1/8)^2 \times 50 = 0.781 \text{ cm}$

$p_{cistern} = (\ell \sin 10° + \Delta h)\rho g$

$= (50 \sin 10° + 0.781) \times 10^{-2} \times 800 \times 9.81 = \underline{\underline{743 \text{ Pa}}}$

3.32 $\Delta h = (1/10)^2 \times 3 = 0.03 \text{ ft}$

$p_{cistern} = (2 \sin 10° + 0.03) \ 50 = \underline{\underline{18.86 \text{ psf}}}$

3.33 One design possibility might be to use a piston and manometer as shown. The design should be such that the diameter of the piston and specific weight of the liquid will yield a reasonable deflection on the manometer (one that can be read by the person being weighed). The manometer could be marked to indicate weight in pounds or Newtons. Also, the designer may want to consider the seal on the piston.

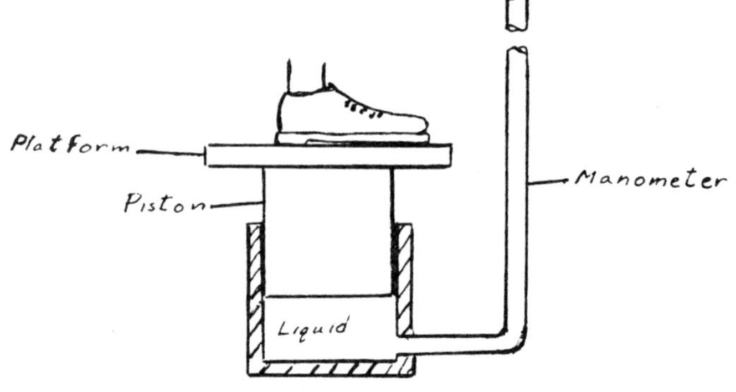

3.34 $p_A = 1.31 \times 847 - 4.59 \times 62.4 = \underline{\underline{823.2 \text{ psf}}} = \underline{\underline{5.72 \text{ psi}}}$

$p_A = 0.40 \times 1.33 \times 10^5 - 1.40 \times 9{,}810 = \underline{\underline{39.5 \text{ kPa}}}$

3.35 Volume of unknown liquid $= \forall = (\pi/4)d^2\ell = 2 \text{ cm}^3$

$\forall = (\pi/4)(0.5)^2\ell = 2$

$\ell = 10.186 \text{ cm}$

Write manometer equation from water surface in left leg to liquid surface in right leg:

$0 + (10.186 \text{ cm} - 5 \text{ cm})(10^{-2} \text{ m/cm})(9{,}810 \text{ N/m}^3)$

$- (10.186 \text{ cm})(10^{-2} \text{ m/cm}) \ \gamma_{liq.} = 0$

$508.7 \text{ Pa} - 0.10186 \ \gamma_{liq.} = 0$

$\underline{\gamma_{liq.} = 4{,}995 \text{ N/m}^3}$

3.36 (1): $y_L + y_R = 1/3$ ft

$$0 + (1.5 \times 62.4) + (y_L \times 847) - (y_R \times 847) = 0$$

(2): $y_L - y_R = -0.1105$ ft

(1) + (2): $2y_L = 0.333 - 0.1105$ $\quad y_L = \underline{\underline{0.111 \text{ ft}}}$

$$y_R = 0.333 - y_L = 0.333 - 0.111 = \underline{\underline{0.222 \text{ ft}}}$$

$p_{max} = 0.222 \times 847 = \underline{\underline{188 \text{ psf}}}$

3.37 First assume that the water will all stay in the left leg (it will not occupy any of the horizontal part of the tube). Then the resulting configuration of water and carbon tetrachloride will appear as shown below:

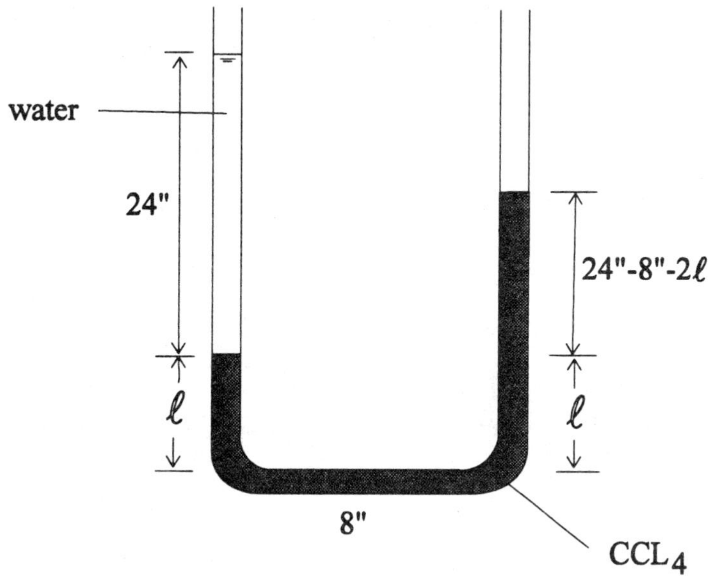

Write the manometer equation from the water surface to the carbon tetrachloride surface:

$$0 + (24/12)\gamma_{H_2O} + (\ell - \ell)\gamma_{CCL_4} - ((24 - 8 - 2\ell)/12)\gamma_{CCL_4} = 0$$

$$24\gamma_{H_20} - (24 - 8 - 2\ell)(1.59\gamma_{H_2O}) = 0$$

$$(24 - 1.59 \times 16 + 3.18\ell) = 0$$

$$3.18\ell = 1.44$$

$$\ell = 0.453\text{in.}$$

Therefore, the water surface will be $\underline{24.453}$ inches above the bottom of the tube and the carbon tetrachloride surface will be $\underline{15.547}$ inches above the bottom.

3.38 Assume the water and oil take a configuration in the tube as shown below. Also neglect the volume of tube at the bends of the tube.

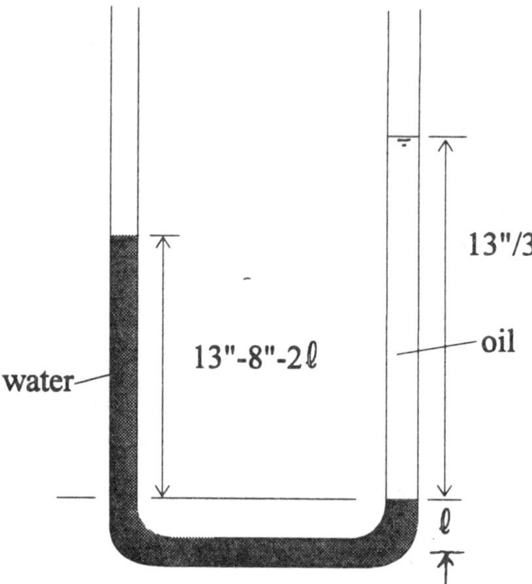

The pressure at the bottom of the column of oil is the same as the pressure at a depth of 13″ - 8″ -2 ℓ″ in the column of water. Equate these pressures and solve for ℓ.

$$((13-8-2\ell)/12)\gamma_w = ((13/3)/12)\gamma_{oil}$$
$$(5-2\ell)\gamma_w = (13/3)(0.87\gamma_w)$$
$$\ell = 0.615\text{in.}$$

Thus the interface between the oil and water is 0.615 in. above the centerline of the horizontal part of tube. The surface of the oil is 4.948in. above the centerline of the horizontal part of the tube and the surface of the water is 4.385in. above the centerline of the horizontal part of the tube.

3.39 $(34 - 10) \times 10^{-2} \times 9{,}810 = 30 \times 10^{-2} \times 9{,}810 \times S$

$\underline{S = 0.80}$

3.40

Δh

γ_m

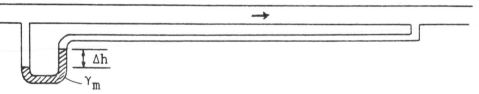

Use a manometer fluid heavier than water. The specific weight of the manometer fluid is identified as γ_m.

Then $\Delta h_{max} = \Delta p_{max}/(\gamma_m - \gamma_{H_2O})$.

If the manometer fluid is carbon-tetrachloride ($\gamma_m = 15{,}600$),

$\Delta h_{max} = 60 \times 10^3/(15{,}600 - 9{,}810) = 10.36$ m ---(too large).

3.40 (continued)

If the manometer fluid is mercury (γ_m = 133,000),

Δh_{max} = 60 x 10^3/(133,000 - 9,810) = 0.487 m ---(O.K.).

Assume the manometer can be read to ± 2 mm.

Then % error = ± 2/487 = ± 0.004 = ± 0.4%

The probable accuracy for full deflection (0.5m) is about 99.6%. For smaller pressure differencs the possible degree of error would vary inversely with the manometer deflection. For example, if the deflection were 10 cm = 0.1m, then the possible degree of error would be ±2% and the expected degree of accuracy would be about 98%.

Note: Error analysis is much more sophisticated than presented above; however, this simple treatment should be enough to let the student have an appreciation for the subject.

3.41 p_A + (4 + 2) 62.4 x 0.8 + 3 x 62.4 - (3 + 2)62.4 x 0.8 = p_B

$p_A - p_B$ = -237 psf = -1.65 psi

3.42 p_A + (3 + 1)9,810 x 0.8 + 2 x 9,790 - (2 + 1) 9,810 x 0.8 = p_B

$p_A - p_B$ = -27,430 Pa = -27.43 kPa

3.43 (1 + 3)51 + z x 180 - (z + 3)62.37 = 2 x 144

z = 2.31 ft

3.44 (0 + 3)51 + z x 847 -(z + 3)62.4 = 3 x 144

z = 0.594 ft

3.45 One possible apparatus might be a simple glass U-tube. Have each leg of the U-tube equipped with a scale so that liquid levels in the tube could be read. The procedure might be as described in steps below:

1. Pour water into the tube so that each leg is filled up to a given level (for example to 15in. level).
2. Pour liquid with unknown specific weight into the right leg until the water in the left leg rises to a given level (for example to 27 in. level).
3. Measure the elevation of the liquid surface and interface between the two liquids in the right tube. Let the distance between the surface and interface be ℓ ft.
4. The hydrostatic relationship will be $\gamma_{H_2O}(2') = \gamma_\ell \ell$ or $\gamma_\ell = 2\gamma_{H_2O}/\ell$
5. To accommodate the range of γ specified the tube would have to be about 3 or 4 ft. high.

The errors that might result could be due to:
1. error in reading liquid level
2. error due to different surface tension
   a. different surface tension because of different liquids in each leg
   b. one leg may have slightly different diameter than other one; therefore, creating different surface tension effect.

Sophisticated error analysis is not expected from the student. However, the student should sense that an error in reading a surface level in the manometer will produce an error in calculation of specific weight. For example, assume that in one test the true value of ℓ were 0.28 ft. but it was actually read as 0.29 ft. Then just by plugging in the formula one would find the true value of γ would be 7.14 γ_{H_2O} but the value obtained by using the erroneous reading would be found to be 6.90 γ_{H_2O}. Thus the manometer reading produced a -3.4% error in calculated value of γ. In this particular example the focus of attention was on the measurement of ℓ. However, the setting of the water surface in the left leg of the manometer would also involve a possible reading error, etc.

Other things that could be considered in the design are:
1. Diameter of tubing
2. Means of support
3. Cost
4. How to empty and clean tube after test is made

3.46

$$p_f = (0.02)(62.4)(50/(1/12))(V^2/64.4)$$

$$= 11.63\ V^2\ \text{lbf/ft}^2$$

Greatest change: $p_f = 11.63 \times 15^2 = 2617\ \text{lbf/ft}^2$

$= 41.9$ ft of head of water

Smallest change: $p_f = 11.63 \times 3^2 = 104.6\ \text{lbf/ft}^2$

$= 1.68$ ft of head of water

3.46 (Continued)

Since the pressure drop is quite large, we should use a manometer with a fluid that has a large specific weight. Use a mercury-water manometer. Then for a p_f of 41.9 ft of water the deflection can be calculated as given below:

$$\Delta \times 12.6 = 41.9 \text{ ft}$$

$$\Delta = 3.325 \text{ ft}$$

If one could read the deflection to within $\pm$ 1/8 in. then the degree of possible error in percent would be

$$(\pm 1/8/(3.325 \times 12)) \times 100 = 0.31\%$$

If one were to use the same manometer to measure the smallest pressure change, then the possible degree of error would be

$$(\pm 1/8/(0.133 \times 12)) \times 100 = 7.81\% \text{ (pretty high)}$$

Other things that could be considered in the design:

1. Cost
2. Kind of tubing and connections between pipe and manometer.
3. How to be sure there are no air bubbles in the tubing.
4. How to support the manometer.

3.47 $p_A = (0.9 + 0.6 \times 13.6 - 1.8 \times 0.8 + 1.5)\ 9{,}810 = 89{,}467$

$p_a = \underline{89.47 \text{ kPa}}$

3.48 $p_A = (90 + 60 \times 13.6 - 180 \times 0.8 + 150) \times (1/12) \times 62.4 = 4{,}742$

$p_A = \underline{32.93 \text{ psig}}$

3.49 $p_A - 1 \times 0.85 \times 9,810 + 0.5 \times 0.85 \times 9,810 = p_B$

$p_A - p_B = 4,169 \text{ Pa} = \underline{\underline{4.169 \text{ kPa}}}$

$(p_A/\gamma + z_A) - (p_B/\gamma + z_B) = (4,169/0.85 \times 9,810) - 1 = \underline{\underline{-0.50 \text{ m}}}$

3.50 Initial condition:

$$150,000 \text{ N/m}^2 - \gamma_{Hg}h = 100,000$$

$$\gamma_{Hg}h = 50,000 \text{ N/m}^2 \qquad (1)$$

Final condition:

$$300,000 \text{ N/m}^2 - K\gamma_{Hg}h = 100,000 \text{ N/m}^2 \qquad (2)$$

Solving Eqs. (1) and (2) for K yields K = 4.0

The final deflection will be $\underline{\underline{4.0 \text{ h}}}$

3.51 $\gamma_{oil} = (0.85)(62.4 \text{ lbf/ft}^3) = 53.04 \text{ lbf/ft}^3$

$\gamma_{Hg} = 847 \text{ lbf/ft}^3$ (from table A.4)

$p_A + (18/12)\text{ft }(\gamma_{oil}) + (2/12)\text{ft}\gamma_{oil} + (3/12)\text{ft}\gamma_{oil}$

$- (3/12)\text{ft}\gamma_{Hg} - (2/12)\text{ft}\gamma_{oil} = p_B$

$p_A - p_B = (-1.75 \text{ ft})(53.04 \text{ lbf/ft}^3) + (0.25 \text{ ft})(847 \text{ lbf/ft}^3)$

$= \underline{\underline{118.9 \text{ lbf/ft}^2}}$

$h = (p/\gamma) + z$ then $h_A - h_B = (p_A - p_B)/\gamma_{oil} + z_A - z_B$

$h_A - h_B = (118.9 \text{ lbf/ft})/(53.04 \text{ lbf/ft}^3) + (1.5 - 0)$

$= \underline{\underline{3.74 \text{ ft}}}$

3.52 $50,000 \text{ N/m}^2 + \gamma_{oil} \times 1\text{m} = 58,530 \text{ N/m}^2$

$$\gamma_{oil} = 8,530 \text{ N/m}^2$$

Therefore $S = (8,530 \text{ N/m}^2)/(9,810 \text{ N/m}^2) = \underline{\underline{0.87}}$

$$p_C = 58,530 + \gamma_{oil} \times 0.5 + \gamma_{water} \times 1$$

$$= 58,530 + 8,530 \times 0.5 + 9,810$$

$$= 72,605 \text{ N/m}^2 = \underline{\underline{72.6 \text{ kPa}}}$$

3.53 Neglect the change of pressure due to the column of air in the tube.

Then $p_{gage} - (d - 1)\gamma_{liquid} = 0$

$$20,000 - ((d - 1) \times 0.85 \times 9,810) = 0$$

$$d = (20,000/(0.85 \times 9,810)) + 1 = \underline{\underline{3.40m}}$$

3.54 $dp/dz = -\gamma$

Because γ becomes smaller with an increase in elevation the ratio of (dp/dz)'s will have a value greater than 1.

3.55 Let the horizontal gate dimension be given as b and the vertical dimension, h.

$$T_A = F(y_{cp} - \bar{y})$$

where F = the hydrostatic force acting on the gate and $(y_{cp} - \bar{y})$ is the distance between the center of pressure and the centroid of the gate.

Thus $T_A = \gamma(H - (h/2))(bh)(\bar{I}/\bar{y}A)$

$$= \gamma(H - (h/2))(bh)(bh^3/12)/((H - (h/2))(bh))$$

$$T_A = \gamma bh^3/12$$

Therefore, T_A does not change with H.

$$T_B = F((h/2) + y_{cp} - \bar{y})$$

$$= \gamma(H - (h/2))(bh)((h/2) + y_{cp} - \bar{y})$$

$$= \gamma(H - (h/2))(bh)((h/2) + \bar{I}/(\bar{y}A))$$

$$= \gamma(H - (h/2))(bh)[(h/2) + (bh^3/12)/((H - (h/2))bh)]$$

$$= \gamma(H - (h/2))bh^2/2 + \gamma bh^3/12$$

Thus, T_A is constant but T_B increases with H. Case (c) is the correct choice.

3.56 The correct answers obtained by looking at the solution to problem 3.55 are that a, b, and e are valid statements.

3.57 For standard atmosphere $T_{sea\ level} = 288\ K = 15°C$

$p = p_0[(T_0 - \alpha(z-z_0))/T_0]^{g/\alpha R} = 101.3[(288 - 6.5\ (z-z_0))/288]^{g/\alpha R}$

where $g/\alpha R = 9.81/(6.5 \times 10^{-3} \times 287) = 5.26$

Then $p_{1,500} = 101.3[(288 - 6.5(1.5))/288]^{5.26} = 84.5$ kPa

And $p_{3,000} = 101.3[(288 - 6.5(3.0))/288]^{5.26} = 70.1$ kPa

From Table A-5, $T_{boiling,\ 1,500\,m} \approx \underline{\underline{95°C}}$ (interpolated);

$T_{boiling,\ 3,000\,m} \approx \underline{\underline{90°C}}$

3.58 $p_0 = 101.3$ kPa

$$p_B = p_0\left(\frac{T_0 - \alpha(z-z_0)}{T_0}\right)^{g/\alpha R}$$

$$= 101.3\left(\frac{(273+15) - 6.5 \times 10^{-3}(4,000-0)}{(273+15)}\right)^{\frac{9.81}{6.5 \times 10^{-3} \times 287}}$$

$= 61.59$ kPa

$p_c = 101.3\ [(288 - 6.5 \times 10^{-3}(2,000 - 0))/288]^{5.259}$

$= \underline{\underline{79.46\ \text{kPa}}}$

$p_A = 101.3 + 9.810 \times 10 = \underline{\underline{199.4\ \text{kPa}}}$

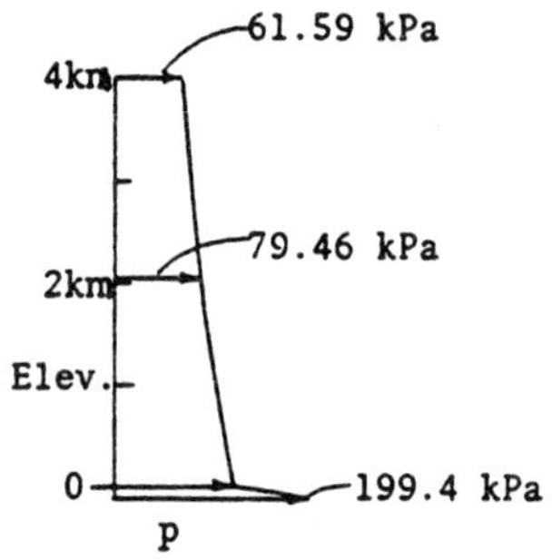

3.59 $p = p_0[(T_0 - \alpha(z-z_0))/T_0]^{g/\alpha R}$

$= 14.7[(520 - 3.566 \times 10^{-3}(20,000-0))/520]^{32.2/(3.566 \times 10^{-3} \times 1,715)}$

$= 6.76$ psia

$p_a = 101[(288 - 6.5 \times 10^{-3}(6,096-0))/288]^{9.81/(6.5 \times 10^{-3} \times 287)}$

$p_a = \underline{\underline{46.4\ \text{kPa}}}$ abs.

3.60 Assume $b\forall\rho$ = constant where b = breath rate, $\forall$ = volume per breath, and ρ = mass density of air. Assume point 1 is sea level and point 2 is 18,000 ft elevation.

Then $b_1\forall_1\rho_1 = b_2\forall_2\rho_2$

$$b_2 = b_1(\forall_1/\forall_2)(\rho_1/\rho_2)$$

Assuming $\forall_1 = \forall_2$, then $b_2 = b_1(\rho_1/\rho_2)$ but $\rho = (p/RT)$

Thus, $b_2 = b_1(p_1/p_2)(T_2/T_1)$

$p_2 = p_1(T_2/T_1)^{g/\alpha R}$; $p_1/p_2 = (T_2/T_1)^{-g/\alpha R}$

Then $b_2 = b_1(T_2/T_1)^{1-g/\alpha R}$

when b_1 = 16 breaths per minute and T_1 = 59°F = 519°R

$T_2 = T_1 - \alpha(z_2-z_1) = 519 - 3.566 \times 10^{-3}(18,000-0) = 454.8°R$

$b_2 = 16(454.8/519)^{1-32.2/(3.566 \times 10^{-3} \times 1,715)} =$ <u>28 breaths per minute</u>

3.61 $p = p_0[(T_0-\alpha(z-z_0))/T_0]^{g/\alpha R}$

$75 = 95[(283-6.5(z-1))/283]^{9.81/(6.5 \times 10^{-3} \times 287)}$

z = 2.91 km

$T = T_0 - \alpha(z-z_0) = 10 - 6.5(2.91-1) =$ <u>-2.41°C</u>

3.62 $p = p_0[(T_0 - \alpha(z-z_0))/T_0]^{g/\alpha R}$

$10 = 13.6[((70+460) - 3.566 \times 10^{-3}(z-2,000))/(70+460)]^{32.2/(3.566 \times 10^{-3} \times 1,715)}$

z = <u>10,430 ft</u>

3.63 $T = T_0 - \alpha(z-z_0) = 519 - 3.566 \times 10^{-3}(5,280 - 0) =$ <u>500°R</u>

$= 288 - 6.5 \times 10^{-3}(1,609 - 0) =$ <u>278°K</u>

$p = p_0(T/T_0)^{g/\alpha R} = 14.7(500/519)^{5.261} =$ <u>12.1 psia</u>

$p_a = 101.3(278/288)^{9.81/(6.5 \times 10^{-3} \times 287)} =$ <u>83.4 kPa</u>

$\rho = p/RT = (12.1 \times 144)/(1,715 \times 500) =$ <u>0.00203 slugs/ft³</u>

$\rho = 83,400/(287 \times 278) =$ <u>1.05 kg/m³</u>

3.64 Force of slurry on gate = $\bar{p}_s A$ and it acts to the right

Force of water on gate = $\bar{p}_w A$ and it acts to the left

$F_{net} = (\bar{p}_s - \bar{p}_w)A$

$= (8\gamma_s - 8\gamma_w)A$

$= (8\text{ ft})(16\text{ ft}^2)(150\text{ lbf/ft}^3 - 60\text{ lbf/ft}^3) = \underline{11{,}520\text{ lbf}}$

Because the pressure is uniform along any horizontal line the moment on the gate is zero; therefore, the moment required to keep the gate closed is zero.

3.65 The center of pressure for the ellipse is lower than the center of pressure for the circle. However, the hydrostatic force on the circle and ellipse are the same. Therefore, the moment on the ellipse will be greater and since it is counterclockwise it will cause the handle to move to the left.

3.66

$F = \bar{p}A = 10 \times 9{,}810 \times 4 \times 4 = 1{,}569{,}600\text{ N} = 32.8 \times 62.4 \times 13.1 \times 13.1 = 351{,}238\text{ lbf}$

$y_{cp} - \bar{y} = \bar{I}/\bar{y}A = (4 \times 4^3/12)/(10 \times 4 \times 4) = 0.133\text{ m}$

$= (13.1 \times 13.1^3/12)/(32.8 \times 13.1 \times 13.1) = 0.436\text{ ft}$

$F_{block} = 1{,}569{,}600 \times 0.133/2 = \underline{104{,}378\text{ N}}$

$= 351{,}238 \times 0.436/6.55 = \underline{23{,}380\text{ lbf}}$

3.67

$$F = \bar{p}A$$
$$= \gamma_{H_2O}\bar{y}A$$
$$= (62.4\text{lbf / ft}^3)(10\text{ft})(16\text{ft}^2)$$
$$= 9{,}984\text{lbf}$$
$$y_{cp} - \bar{y} = \bar{I}/(\bar{y}A)$$
$$= (4 \times 4^3/12)/(10 \times 4^2)$$
$$= 0.1333\text{ft.}$$

3.67 (Continued)

Sum moments about pivot:

$$\sum M_{pivot} = 0$$
$$9{,}984 \times 0.1333 - 2F_{block} = 0$$
$$F_{block} = 666 lbf$$

Force of gate on block is 666lbf acting to right

3.68 $F = \bar{p}A = 4.5 \times 150 \times (9 \times 1) = 6{,}075$ lbf

$y_{cp} = \bar{y} + \bar{I}/\bar{y}A = 4.5 + (1 \times 9^3)/(12 \times 4.5 \times 9) = 6.00$ ft

$F_{TIE} = 2 \times F \times y_{cp}/h = 2 \times 6{,}075 \times 6.00/9 = 8{,}100$ lbf

---

3.69 Force acting on gate:

$$F_G = \bar{p}A$$
$$\bar{p} = \gamma_{H_2O} \times 2$$
$$A = 40 ft^2$$

Then $F_G = 62.4 \times 2 \times 40 = 4{,}992$ lbf. F_G acts at 2/3 depth or 8/3 ft below water surface. Take moments about hinge to solve for F.

$$\sum M = 0$$
$$(F_G \times 8/3) - 4F = 0$$
$$F = 3{,}328 lbf \text{ to the left}$$

3.70 $F = \bar{p}A = (3 + 4.5)9{,}810 \times 9 \times 9 = 5{,}960{,}000$N

$y_{cp} = \bar{y} + \bar{I}/\bar{y}A = 7.5 + 9 \times 9^3/(12 \times 7.5 \times 9 \times 9) = 8.40$m

$F_{hinge} = F(d - y_{cp})/h = 5{,}960{,}000(12 - 8.40)/9 = 2{,}384{,}000N = 2{,}384$ kN

3.71 $F = \bar{p}A$

$= (3\text{m} + 3\text{m}\cos 30°)(9{,}810\text{N/m}^3) \times 24\text{m}^2$

$F = 1{,}318{,}000\ \text{N}$

$y_{cp} - \bar{y} = \bar{I}/\bar{y}A$ where $\bar{y} = 3\text{m} + 3\text{m}/\cos 30° = 6.464\text{m}$

$= (4 \times 6^3/12)\text{m}^4/(6.464\text{m} \times 24\text{m}^2)$

$= 0.4641\text{m}$

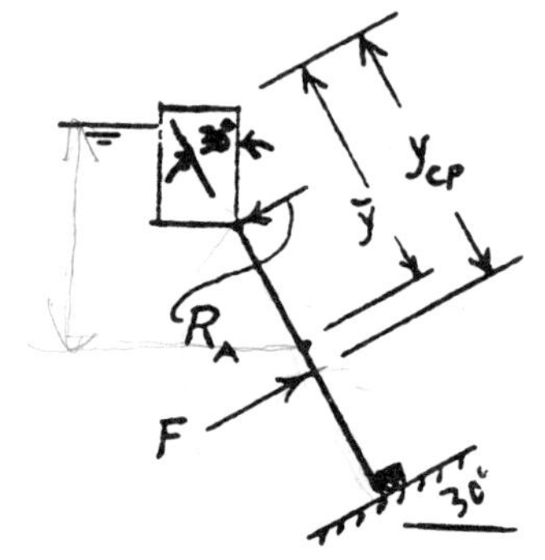

Take moments about the stop:

$\Sigma M_{stop} = 0$

$6R_A - (3 - 0.464)(1{,}318{,}000) = 0$

$R_A = 557\ \text{kN}\ \angle 30°$

3.72 Hydrostatic force on gate is: $F_G = \bar{p}A$ where

$$\bar{p} = (3\text{ft} + 3\text{ft}\cos 30°)\gamma_w$$
$$= 5.60\text{ft} \times 62.4\text{lbf/ft}^3$$
$$= 349\text{lbf/ft}^2$$
$$A = 6 \times 5 = 30\text{ft}^2$$
$$\text{Then } F_G = 349 \times 30 = 10{,}470\text{lbf}$$
$$y_{cp} - \bar{y} = \bar{I}/\bar{y}A$$
$$\text{where } \bar{I} = bh^3/12 = (5 \times 6^3/12)\text{ft}^4$$
$$\bar{y} = 3\text{ft} + 3\text{ft}/\cos 30° = 6.46\text{ft}$$
$$A = 30\text{ft}^2$$
$$\text{then } y_{cp} - \bar{y} = (5 \times 6^3/12)/[(6.46)(30)]$$
$$y_{cp} - \bar{y} = 0.464\text{ft}$$

Thus, F_G acts 3.464ft from hinge or 2.536 ft from stop. Get reaction at A by summing moments about stop.

$$\sum M_{stop} = 0$$
$$F_G \times 2.536 - 6R_A = 0$$
$$10{,}470 \times 2.536 - 6R_A = 0$$
$$R_A = 4{,}425\text{lbf}$$

3.73 $F = (7 + 2.5)62.4 \times 10 \times 6 = 35{,}568$ lbf

$y_{cp} - \bar{y} = (6 \times 10^3)/(12 \times 19 \times 10 \times 6) = 0.439$ ft

$F_A = (35{,}568 \times 5.439)/(10 \cos 30°) = \underline{\underline{22{,}338 \text{ lbf}}}$

3.74 $F = \bar{p}A = (0.4 + 0.4)9{,}810 \times 0.8 \times 0.8 \times 1.2 = 6{,}027$ N

$y_{cp} - \bar{y} = \bar{I}/\bar{y}A = 1.2 \times 0.8^3/(12 \times 0.8 \times 0.8 \times 1.2) = 0.067$ m

$M = 6{,}027 \times (0.4 - 0.067) = \underline{\underline{2{,}007 \text{ N}\cdot\text{m}}}$

3.75 $F = \bar{p}A = (5 + 2.5)9{,}810 \times 3 \times 5/\sin 60° = 1{,}274.4$ kN

$y_{cp} - \bar{y} = 3 \times (5/\sin 60°)^3/(12 \times (7.5/\sin 60°)(3 \times 5/\sin 60°)) = 0.321$ m

$T = 0.321 \text{ m} \times 1{,}274.4 \text{ kN} = \underline{\underline{409 \text{ kN}\cdot\text{m}}}$

3.76 $F = \bar{p}A = (12 + 6)62.4 \times 6 \times 12/\sin 60° = 93{,}381$ lbf

$y_{cp} - \bar{y} = \bar{I}/\bar{y}A = 6 \times (12/\sin 60°)^3/(12 \times (18/\sin 60°)(6 \times 12/\sin 60°))$

$= 0.770$ ft

$T = 0.770 \times 93{,}381 = \underline{\underline{71{,}903 \text{ ft-lbf}}}$

3.77 The length of gate is $\sqrt{4^2 + 3^2} = 5\text{m}$

$$\begin{aligned}
\bar{y} &= 2.5\text{m} + 5/4\text{m} \\
&= 3.75\text{m} \\
F &= \bar{p}A \\
&= (4/5)(3.75)(9.810)(2 \times 5) \\
&= 294{,}300\text{N} \\
&= 294.3\text{kN} \\
y_{cp} - \bar{y} &= \bar{I}/(\bar{y}A) \\
&= 2 \times (5^3/12)/(3.75 \times 2 \times 5) \\
&= 0.555\text{m}
\end{aligned}$$

Sum moments about hinge:

$$\sum M_{hinge} = 0$$

$$F \times (2.5 + .555) + 3P = 0$$

$$294.3(3.055) + 3P = 0$$

$$\underline{\underline{P = 299.7\text{kN}}}$$

3.78 Either a vertical plane gate or a tainter gate could be used. In any case, the horizontal component of hydrostatic force acting on the gate would be at least this much:

$$F_{horiz.} = \bar{p}A$$

$$= 10 \times 62.4 \times 20 \times 30 = \underline{\underline{374{,}400 \text{ lbf}}}$$

Many design details such as location, lift mechanism etc, depending upon what is desired by the instructor.

3.79 $y_{cp} - \bar{y} = 0.60\ell - 0.5\ell = 0.10\ell$

$0.10 = \bar{I}/\bar{y}A = \ell \times \ell^3/(12 \times (h + \ell/2)\ell^2)$

$h = \underline{\underline{0.333\ell}}$

3.80 Hydrostatic force on gate: $F_G = \bar{p}A$ where

$$\bar{p} = 12\text{m} \times \gamma_{water}$$
$$= 12 \times 9{,}810$$
$$= 117{,}720\text{Pa}$$
$$A = 12 \times 12 = 144\text{m}^2$$

Then $F_G = (117{,}720\text{N}/\text{m}^2) \times (144\text{m}^2) = 16.952\text{MN}$

$$y_{cp} - \bar{y} = \bar{I}/\bar{y}A$$
$$= (12 \times 12^3/12)/(12 \times 12^2) = 1\text{m}$$

$\therefore F_G$ acts 6m+1m below the hinge

Force of water in tank: $F_{tank} = W = 6 \times 12 \times L \times \gamma_{water}$
$$= 7.063 \times 10^5 \times L$$

Moment arm of $F_{tank} = L/2$

Sum moments about hinge:

$$\sum M_{hinge} = 0$$
$$(7.063 \times 10^5 L) \cdot (L/2) = (16.952 \times 10^6) \times (7) = 0$$
$$L^2 = 336m^2; \underline{\underline{L = 18.3m}}$$

3.81 $F = \bar{p}A$

$= (30 \text{ ft} \times 62.4 \text{ lb/ft}^3)((\pi \times D^2/4) \text{ ft}^2)$

$= (30 \times 62.4 \times \pi \times 10^2/4)\text{lb} = 147{,}027 \text{ lb}$

$y_{cp} - \bar{y} = \bar{I}/\bar{y}A = (\pi r^4/4)/(\bar{y}\pi r^2) = (5^2/4)/(30/.866)$

$= 0.1804 \text{ ft}$

Torque $= \underline{\underline{0.1804 \text{ ft} \times 147{,}027 \text{ lb} = 26{,}520 \text{ ft-lb}}}$

3.82 $F = \overline{p}A = (1 + 1.5)9{,}810 \times 1 \times 3\sqrt{2} = 104{,}050$

$y_{cp} - \overline{y} = \overline{I}/\overline{y}A = 1 \times (3\sqrt{2})^3/(12 \times (2.5 \times \sqrt{2})(1 \times 3\sqrt{2})) = 0.424 \text{ m}$

Overturning moment $M_1 = 90{,}000 \times 1.5 = 135{,}000 \text{ N}\cdot\text{m}$

Restoring moment $M_2 = 104{,}050 \times (3\sqrt{2}/2 - 0.424) = 176{,}606 \text{ N}\cdot\text{m} > M_1$

So the <u>gate will stay</u>.

3.83 $F = (4 + 3.535)62.4 \times (3 \times 7.07\sqrt{2}) = 14{,}103 \text{ lbf}$

$y_{cp} - \overline{y} = 3 \times (7.07\sqrt{2})^3/(12 \times 7.535\sqrt{2} \times 3 \times 7.07\sqrt{2}) = 0.782 \text{ ft}$

Overturning moment $M_1 = 18{,}000 \times 7.07/2 = 63{,}630 \text{ N}\cdot\text{m}$

Restoring moment $M_2 = 14{,}103(7.07\sqrt{2}/2 - 0.782) = 59{,}476 \text{ N}\cdot\text{m} < M_1$

So the <u>gate will fall</u>.

3.84 $F = \overline{p}A = (h + 2h/3)\gamma(Wh/\sin 60°)/2 = 5\gamma Wh^2/3\sqrt{3}$

$y_{cp} - \overline{y} = \overline{I}/\overline{y}A = W(h/\sin 60°)^3/(36 \times (5h/(3 \sin 60°)) \times (Wh/2 \sin 60°))$

$= h/(15\sqrt{3})$; $\Sigma M = 0$

$R_T h/\sin 60° = F[(h/(3 \sin 60°)) - (h/15\sqrt{3})]$

$R_T/F = \underline{\underline{3/10}}$

3.85 $F = \overline{p}A = (1 + 6)9{,}810 \times 0.5 \times 4 \times 9 = 1.236 \text{ MN}$

$y_{cp} - \overline{y} = \overline{I}/\overline{y}A = (4 \times 9^3)/(36 \times 7 \times 0.5 \times 4 \times 9) = 0.643 \text{ m}$

$P = 1{,}236{,}060 \times (3 - 0.643)/9 = \underline{\underline{323.7 \text{ kN}}}$

3.86 $dF = pdA = \gamma_0(1 + kd/d_0) d\, d(d) W$

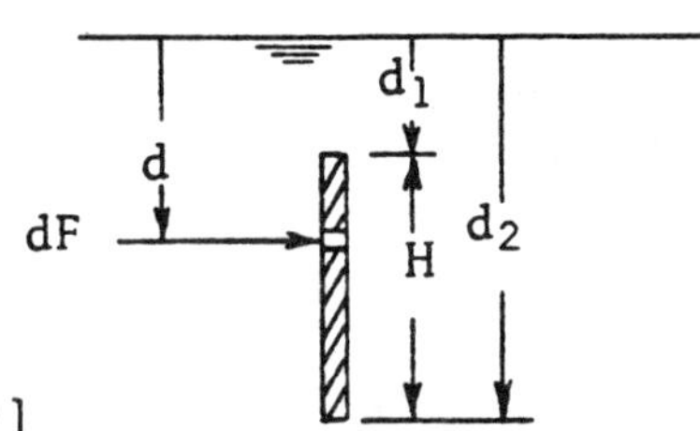

$$F = \gamma_0 W \int_{d_1}^{d_2} d(1 + kd/d_0) d(d)$$

$$F = \gamma_0 W[1/2(H^2 + 2d_1 H) + (k/3d_0)(H^3 + 3d_1 d_2 H)]$$

or $\underline{F = \gamma_0 W[1/2(d_2^2 - d_1^2) + (k/3d_0)(d_2^3 - d_1^3]}$

When $d_1 = 0$ $\quad F = \gamma_0 W(H^2/2 + kH^3/3d_0)$

Since the specific weight increases with the increase in depth, the location of the center of pressure will be located below that for constant density liquid.

3.87 a) $F_{Hydr} = \bar{p}A = (0.25\ell + 0.5\ell \times 0.707) \times \gamma x W\ell = 0.6036\gamma W\ell^2$

$y_{cp} - \bar{y} = \bar{I}/\bar{y}A = (W\ell^3/12)/(((0.25\ell/0.707) + 0.5\ell) \times W\ell)$

$y_{cp} - \bar{y} = 0.0976\ell; \quad \Sigma M_{hinge} = 0$

Then $-0.707R_A\ell + (0.5\ell + 0.0976\ell) \times 0.6036\gamma W\ell^2 = 0$

$\underline{R_A = 0.510\gamma W\ell^2}$

b) The reaction here will be less because if one thinks of the applied hydrostatic force in terms of vertical and horizontal components, the horizontal component will be the same in both cases, but the vertical component will be less because there is less volume of liquid above the curved gate.

3.88 Equivalent depth of liquid for 3 psi = (3 x 144)/(0.8 x 62.4) = 8.65 ft

$F = \bar{p}A = (8.65 + 2 + 5)(62.4 \times 0.8)(6 \times 10) = 46{,}875$ psf

$y_{cp} - \bar{y} = \bar{I}/\bar{y}A = (6 \times 10^3)/(12 \times 15.65 \times 6 \times 10) = 0.532$ ft

$P = 46{,}875 \times (5 + 0.532)/10 = \underline{25{,}931 \text{ lbf}}$

3.89 Equivalent depth of liquid for 40 kPa = 40,000/(0.8 x 9,810) = 5.10m

$F = (5.10 + 1 + 1.5)(0.8 \times 9{,}810)(3 \times 2) = 357{,}870$N

$y_{cp} - \bar{y} = \bar{I}/\bar{y}A = (2 \times 3^3)/(12 \times 7.60 \times 3 \times 2) = 0.099$ m

$P = 357{,}870(1.5 + 0.099)/3 = \underline{190{,}745 \text{ N}}$

3.90 $F = \bar{p}A = (1.5/2)24{,}000 \times (1.5/\sin 60°) = 31{,}177$ N

$y_{cp} - \bar{y} = \bar{I}/\bar{y}A = 1 \times (1.5/\sin 60°)^3/(12 \times (1.5/2 \sin 60°) \times (1.5/\sin 60°))$

$= 0.2887$ m

$M = 31{,}177 \times (1.5/2 \sin 60° - 0.2887) = 18{,}000$ N•m/m $=$ <u>18 kN•m/m</u>

3.91 A simple check shows that d will have to be less than 4 m. Thus

$F_{Hydrostatic} = \bar{p}A = 1/2\ d \times 9{,}810 \times 2\ d = 9{,}810\ d^2$N

The hydrostatic force will act 2/3 d below water surface; therefore the momentum will be (4 - (1/3)d) below the hinge.

$\Sigma M_{Hinge} = 0$

$5 \times 60{,}000 - (4 - (1/3)d)(9{,}810\ d^2) = 0$

Solving for d yields <u>d = 3.23m</u>

3.92 The hydrostatic force acting on the gate will be:

$F = \bar{p}A = (3 \times 9{,}810) \times (2 \times 4) = 235{,}440$ N

$y_p - \bar{y} = \bar{I}/\bar{y}A$

$= (2 \times 4^3/12)/(3 \times 2 \times 4) = 0.444$m

$\Sigma M_{Hinge} = 0$

$W \times 5 - 235{,}440 \times 2.444 = 0$

W = <u>115,100 N</u>

3.93 $y_p = (2/3) \times (8/\cos 45°) = 7.54$m

Point B is (8/cos 45°)m - 3.5 m = 7.81 m along the gate from the water surface; therefore, the gate is <u>unstable.</u>

3.94 $F_{AB,hydrostatic} = \bar{p}_{AB}A_{AB} = (h/2)\gamma h = \gamma h^2/2$

$F_{BC,hydrostatic} = \bar{p}_{BC}A_{BC} = \gamma h \times 4$ ft

$\Sigma M_B = 0$

$-(\gamma h^2/2)(h/3) + \gamma h \times 4 \text{ ft} \times 2 \text{ ft} = 0$

h = <u>6.93 ft</u>

3.95 $F = \bar{p}A = 1 \times 9{,}810 \times 2 \times 1 = 19{,}620$ N

$y_{cp} - \bar{y} = \bar{I}/\bar{y}A = (1 \times 2^3)/(12 \times 1 \times 2 \times 1) = 0.33$ m

Weight $W = 19{,}620 \times (1 - 0.33)/2.5 = 5{,}258$ N

Volume $\forall = 5{,}258/(23{,}600 - 9{,}810) = \underline{\underline{0.381 \text{ m}^3}}$

3.96 $F = 2.5 \times 62.4 \times 2 \times 5 = 1{,}560$ lbf

$y_{cp} - \bar{y} = (2 \times 5^3)/(12 \times 2.5 \times 2 \times 5) = 0.833$ ft

$W = 1{,}560(2.5 - 0.833)/6.25 = 416$ lbf

$\forall = 416/(150 - 62.4) = \underline{\underline{4.74 \text{ ft}^3}}$

3.97 Hydrostatic force F_H:

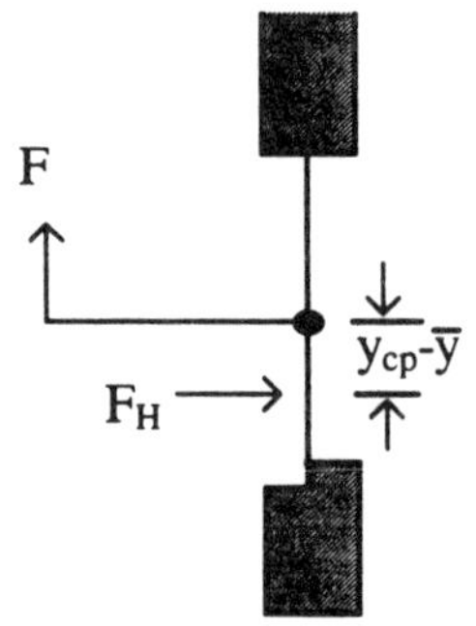

$$F_H = \bar{p}A$$
$$= 10 \times 9{,}810 \times \pi D^2/4$$
$$= 98{,}100 \times \pi(1^2/4)$$
$$= 77{,}048\text{N}$$
$$y_{cp} - \bar{y} = \bar{I}/(\bar{y}A)$$
$$= (\pi r^4/4)/(10 \times \pi D^2/4)$$
$$y_{cp} - \bar{y} = r^2/40 = 0.00625\text{m}$$

$$\sum M_{Hinge} = 0$$
$$F_H \times (0.00625\text{m}) - 1 \times F = 0$$
$$\text{But } F = F_{buoy} - \text{Wgt}$$
$$= A(10\text{m} - \ell)\gamma_{H_2O} - 200$$
$$= (\pi/4)(.25^2)(10 - \ell)(9{,}810) - 200$$
$$= 4815.5\text{N} - 481.5\ell\text{N} - 200\text{N}$$
$$= (4615.5 - 481.5\ell)\text{N}$$

where ℓ = length of chain

$$77{,}048 \times 0.00625 - 1 \times (4615.5 - 481.5\ell) = 0$$
$$481.55 - 4615.5 + 481.5\ell = 0$$
$$\underline{\underline{\ell = 8.59\text{m}}}$$

3.98 The horizontal component of force acting on the walls is the same for each wall. However, walls A-A' and C-C' have vertical components that will require greater resisting moments than the wall B-B'. If one thinks of the vertical component as a force resulting from buoyancy, it can be easily shown that there is a greater "buoyant" force acting on wall A-A' than on C'C'. Thus, wall A-A' will require the greatest resisting moment.

3.99 $W_{\text{in air}} = 650\text{ N} = \forall\gamma_{\text{block}}$ (1)

$W_{\text{in water}} = 500\text{ N} = \forall(\gamma_{\text{block}} - \gamma_{\text{water}})$ (2)

$\gamma_{\text{water}} = 9{,}810\text{ N/m}^3$ (3)

Solving Eqs. (1),(2), and (3) yield:

$$\forall = 0.0153\text{ m}^3$$

$$\gamma_{\text{block}} = 42{,}510\text{ N/m}^3$$

3.100

$$dF_z = -\sin\theta \times p \times 2\pi R^2 \cos\theta d\theta$$

but $p = \gamma(d - R\sin\theta)$

$$\therefore dF_z = -\sin\theta(\gamma d - \gamma R\sin\theta)(2\pi R^2\cos\theta d\theta)$$

$$F_z = \int_{-\pi/2}^{\pi/2} -\gamma d 2\pi R^2 \cos\theta\sin\theta d\theta + \int_{-\pi/2}^{\pi/2} \gamma 2\pi R^3 \cos\theta\sin^2\theta d\theta$$

$$F_z = -\gamma d 2\pi R^2 \int_{-\pi/2}^{\pi/2} \cos\theta\sin\theta d\theta + \gamma 2\pi R^3 \int_{-\pi/2}^{\pi/2} \cos\theta\sin^2\theta d\theta$$

$$= -\gamma d 2\pi R\left[(1/2)\sin^2\theta\right]_{-\pi/2}^{\pi/2} + \gamma 2\pi R^3\left[(1/3)\sin^3\theta\right]_{-\pi/2}^{\pi/2}$$

$$= -\gamma d 2\pi R(0) + \gamma 2\pi R^3[(1/3)(1+1)]$$

$$= \gamma 2\pi R^3(2/3) = (4/3\pi R^3\gamma)$$

3.101 $\forall(\gamma - \gamma_0) = \forall(\gamma - 9{,}810 \times 0.8) = 55$

$\forall(\gamma - \gamma_{Hg}) = \forall(\gamma - 133{,}000) = -45$

Equating $\forall$'s, $\gamma = 76{,}682\text{ N/m}^2$

$\forall = 7.99 \times 10^{-4}\text{m}^3$

$W = 61.3\text{ N}$

sp.gr. = 7.82

3.102 $\forall_0 = (\pi/6)\ D_0^3$

$= (\pi/6)\ m^3 = 0.524\ m^3$

$T_0 = 288K$

$p_{0,He} = p_{atm.} + 10{,}000\ P_a = 111{,}300\ Pa$

$\rho_{0,He} = (p_{0,He}/R_{He}T_0) = 111{,}300/((2077)(288)) = 0.186\ kg/m^3$

Conservation of mass

$m_0 = m_{alt.}$

$\forall_0\rho_{0,He} = \forall_{alt.}\rho_{He}\ ;\ \forall_{alt.} = \forall_0\rho_{0,He}/\rho_{He}$ (1)

Newton's second law:

$\Sigma F_z = ma = 0$

$F_{buoy.} - Wgt = 0$

$\forall_{alt.}\rho_{air}g - (mg + W_{He}) = 0$

Eliminate $\forall_{alt.}$ with Eq. (1)

$(\forall_0\rho_0/\rho_{He})\ \rho_{air}g = (mg + \forall_0\rho_{0,He}g)$

Eliminate ρ's with Eq. of state:

$$\frac{(V_0\rho_0)\ (p_{alt.}/R_{air}T)\ g}{(p_{alt.} + 10{,}000)/(R_{He}T)} = (mg + V_0\rho_0 g)$$

$$\frac{(0.524)(0.186)(9.81)(2077)p_{alt.}}{(p_{alt.} + 10{,}000)(287)} =$$

$(0.1)(9.81) + (0.524)(0.186)(9.81)$

Solve: $p_{alt.} = 3888\ Pa$

Check to see if $p_{alt.}$ is in the troposphere or stratosphere. Using Eq. (3.15) solve for pressure at top of traposphere.

$$p = p_0\left[\frac{t_0 - \alpha(z - z_0)}{T_0}\right]^{g/\alpha R}$$

$= 101{,}300\ [(288 - 6.5 \times 10^{-3})(10{,}769)/(288)]^{5.259}$

$= 23{,}422\ Pa$

Because $p_{alt.} < p_{at\ top\ of\ troposphere}$ we know that $p_{alt.}$ occurs in the stratosphere. Therefore, use Eq. (3.16) to solve for the altitude for the equilibrium condition.

3.102 (Cont'd)

$$p = p_0 e^{-(z-z_0)g/RT}$$

where z_0 = 10,769 m (beginning of stratosphere)

$R = R_{Air}$ = 287 J/kgK

T = Temperature in stratosphere = -55°C = 218 K

p = 3888 Pa

Then solving for z yields z = 22,100 m = 22.1 km

3.103

$\forall\gamma = 918$ N

$\forall(\gamma - 9{,}810) = 609$ N

$\forall = (918 - 609)/9{,}810 = 0.0315\ m^3$

3.104

Rod weight $= (2LA\rho_w + LA(2\rho_w))g = 4LA\rho_w g = 4LA\gamma_w$

Buoyant Force $= \forall\gamma_{Liq} = 3LA\gamma_{Liq}$

Rod weight = Buoyant force

$4LA\gamma_w = 3LA\gamma_{Liq}$

$\gamma_{Liq} = (4/3)\gamma_w$

The liquid is more dense than water.

3.105

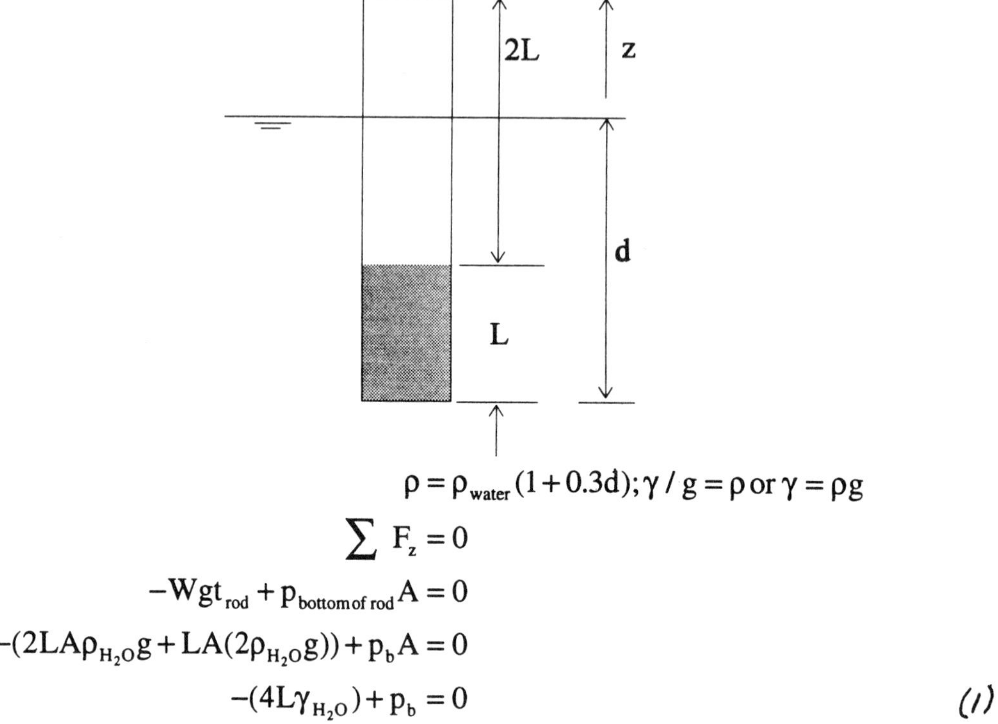

$$\rho = \rho_{water}(1+0.3d); \gamma / g = \rho \text{ or } \gamma = \rho g$$

$$\sum F_z = 0$$

$$-Wgt_{rod} + p_{bottom\,of\,rod} A = 0$$

$$-(2LA\rho_{H_2O}g + LA(2\rho_{H_2O}g)) + p_b A = 0$$

$$-(4L\gamma_{H_2O}) + p_b = 0 \qquad (1)$$

From given condition of the liquid

$$\gamma = \gamma_{H_2O}(1+0.3d)$$

$$\gamma = \gamma_{H_2O}(1-0.3z)$$

$$\frac{dp}{dz} = -\gamma$$

$$dp = -\gamma_{H_2O}(1-0.3z)dz$$

$$p = \gamma_{H_2O}\int(-1+0.3z)dz$$

$$p_b = \gamma_{H_2O}(-z+0.3z^2/2) + C \qquad (2)$$

when $z = 0$ $p = 0$ gage so $C = 0$

Substitute p_b of Eq. (2) into Eq. (1)

3.105 (continued)

$$-4L\gamma_{H_2O} + \gamma_{H_2O}(-z + 0.3z^2/2) = 0$$

But L = 1ft so

$$0.15z^2 - z - 4 = 0$$

$$z^2 - 6.667z - 26.67 = 0$$

Solve for z by the quadratic equation

$$z = (+6.667 \pm \sqrt{44.4 + 106.68})/2$$

$z = -2.812$ ft or depth d at which rod will float is -z or 2.812ft

---

3.106 $\text{Weight}_{anchor} = 0.50 \text{ ft}^3 \times (2.2 \times 62.4 \text{ lb/ft}^3) = 68.65 \text{ lb}$

The water displaced by boat due to weight of anchor

$$= 68.65 \text{ lb}/(62.4 \text{ lb/ft}^3) = 1.100 \text{ ft}^3$$

Therefore, when the anchor is removed from the boat, the boat will rise and the water level in the pond will drop:

$$\Delta h = 1.10 \text{ ft}^3/500 \text{ ft}^2 = 0.0022 \text{ ft}$$

However, as the anchor is dropped into the pond, the pond will rise because of the volume taken up by the anchor. This change in water level in the pond will be:

$$\Delta h = 0.500 \text{ ft}^3/500 \text{ ft}^2 = .001 \text{ ft}$$

Net change = -.0022 ft + .001 ft = -.0012 ft = -.0144 in.

The pond level will drop .0144 inches.

3.107 $S = 0.5 \Rightarrow \gamma_{block} = 0.5\gamma_{water}$

Wgt. of displaced water = weight of block

$$V_w\gamma_w = V_b\gamma_b$$

$$V_w = (\gamma_b/\gamma_w)V_b$$

$$V_w = .5V_b = 100 \text{ cm}^3$$

3.107 (continued)

Then the total volume below water surface when block is floating in water $= V_{W,orig.} + 100\ cm^3$

$V_{W,orig.} = (\pi/3)(10\ cm)^3$

$= 1047.2\ cm^3$

$V_{final} = 1047.2\ cm^3 + 100\ cm^3$

$(\pi/3)h^3_{final} = 1147.2\ cm^3$

$h_{final} = 10.309\ cm$

$\Delta h = \underline{0.309\ cm}$

3.108 The same relative volume will be unsubmerged whatever the orientation; therefore,

$\frac{V_{u.s.}}{V_s} = \frac{hA}{LA} = \frac{LA_{u.s.}}{LA}$

or $h/L = A_{u.s.}/A$

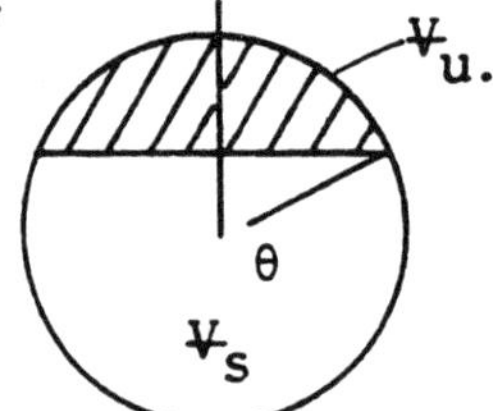

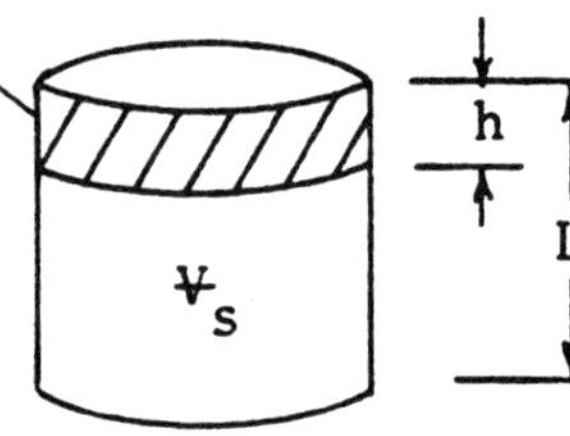

Also, $\cos\theta = 5'/10' = 0.50$

$\theta = 60°$ and $2\theta = 120°$

So $A_{u.s.} = (1/3)\pi R^2 - R\cos 60° R \sin 60°$

Therefore $h/L = R^2\left[((1/3)\pi) - \sin 60° \cos 60°\right]/\pi R^2 = \underline{\underline{0.195}}$; $\underline{\underline{= 7.80\ m}}$

3.109 The block will displace water equal to the weight of the block:

$\Delta V_W \gamma_W = Wgt_{block}$

$\Delta V_W = 2\ lb/(62.4\ lb/ft^3) = 0.03205\ ft^3$

$\therefore \Delta h A_T = \Delta V_W$

$\Delta h = \Delta V_W/A_T = 0.03205\ ft^3/((\pi/4)(1^2)\ ft^2)$

$\Delta h = 0.0408\ ft$

Water in tank $\underline{\underline{\text{will rise 0.0408 ft.}}}$

3.110 $\Sigma F_y = 0$

$- 30{,}000 - 4 \times 1{,}000L + 4 \times (\pi/4) \times 1^2 \times (10{,}000(L-1)) = 0$

$\underline{\underline{L = 2.24 \text{ m}}}$

3.111 Assume the block will sink a distance y into the fluid with S = 1.2.

$\Sigma F_y = 0$

$- Wgt + pA = 0$

$- (6L)^2 \times 3L \times 0.8\gamma_{water} + (L \times \gamma_{water} + y \times 1.2\gamma_w)\ 36L^2 = 0$

$y = 1.167L$

so $\underline{\underline{d = 2.167L}}$

3.112 When the ice is put in the tank the water will rise consistent with the weight of water displaced by the ice:

$W_{ice} = \forall_w \gamma_w$

but $\forall_w = 5 \text{ lb}/(62.4 \text{ lb/ft}^3) = 0.0801 \text{ ft}^3$

The amount of rise of water in tank $= 0.0801 \text{ ft}^3/(A_{cyl})$

$\Delta h = 0.0801 \text{ ft}^3/((\pi/4)(2 \text{ ft})^2)$

$= \underline{\underline{0.0255 \text{ ft}}}$

When the ice melts the melted water will simply occupy the same volume of water that the ice originally displaced; therefore, there will be no change in water surface level in the tank when the ice melts.

3.113 $\Sigma M_A = 0$

$- Wgt_{wood} \times (0.5L \cos 30°) + F_{buoy.} \times (5/6)L \cos 30° = 0$

$-\gamma_{wood} \times AL \times (0.5L \cos 30°) + ((1/3)AL\gamma_{H_2O}) \times (5/6)L \cos 30° = 0$

$\gamma_{wood} = (10/18)\gamma_{H_2O}$

$= \underline{\underline{5{,}450 \text{ N/m}^3}}$

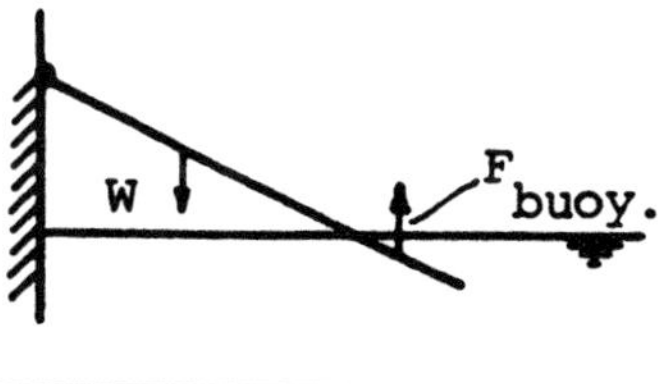

3.114 Take summation of moments about A to see if pole will rise or fall. The forces producing moments about A will be the weight of the pole and the buoyant force.

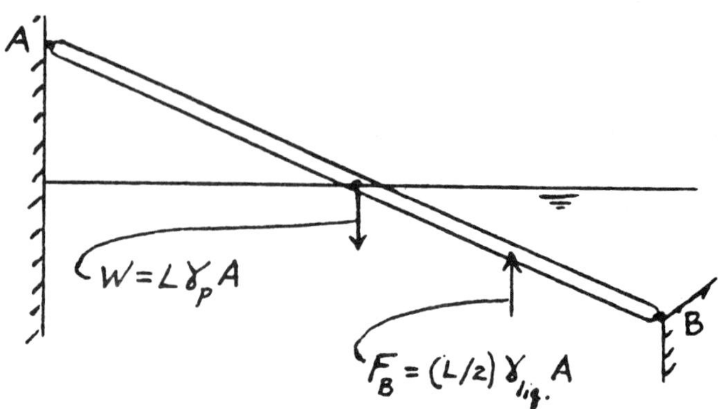

$$\Sigma M_A = -(1/2)(L\cos\alpha)(L\gamma_p A) + (3/4)(L\cos\alpha)(L/2)\gamma_{liq}A$$

$$= L^2 A\cos\alpha[-(1/2)\gamma_p + (3/8)\gamma_{liq}]$$

$$= K\ (-80 + 75)$$

A negative moment acts on the pole; therefore, it will fall.

3.115 Draft = $(40{,}000 \times 2{,}000)/40{,}000\,\gamma = (2{,}000/\gamma)$ ft

Since γ of salt water is greater than γ of fresh water, the ship will take a greater draft in fresh water.

$(2{,}000/62.4) - (2{,}000/64.1) = $ 0.85 ft

3.116 $\Sigma F_V = 0;\quad F_B - F_s - F_w - F_c = 0$

$$F_s = F_B - F_w - F_c$$

$$= (4/3)\pi(0.6)^3 \times 10{,}070 - 1{,}600 - 4{,}500$$

$$= 3{,}011 \text{ N of scrap}$$

3.117 Assuming standard atmospheric temperature condition:

$$T = 519 - 3.566 \times 10^{-3} \times 15{,}000 = 465.5°R$$

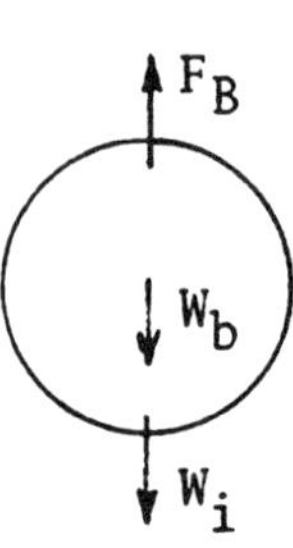

$$\rho_{air} = (8.3 \times 144)/(1{,}715 \times 465.5)$$

$$= 0.001497 \text{ slugs/ft}^3$$

$$\rho_{He} = (8.3 \times 144)/(12{,}429 \times 465.5)$$

$$= 0.000207 \text{ slugs/ft}^3$$

3.117 (Cont'd)

$$\Sigma F = 0 = F_L - F_b - F_i$$

$$= (1/6)\pi D^3 g(\rho_{air} - \rho_{He}) - \pi D^2(0.01) - 10$$

$$= D^3 \times 16.88(14.97 - 2.07)10^{-4} - D^2 \times 3.14 \times 10^{-2} - 10$$

$$D = \underline{\underline{8.22 \text{ ft}}}$$

3.118

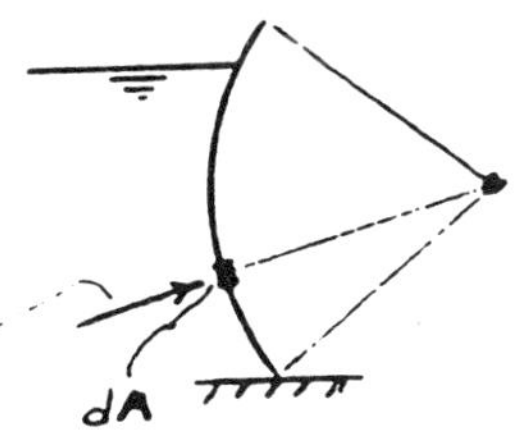

Consider all the differential pressure forces acting on the radial gate as shown. Because each differential pressure force acts normal to the differential area, then each differential pressure force must act through the center of curvature of the gate. Because all the differential pressure forces will be acting through the center of curvature (the pin), the resultant must also pass through this same point (the pin).

3.119

$$F_V = 1 \times 9{,}810 \times 1 \times 1 + (1/4)\pi \times (1)^2 \times 1 \times 9{,}810 = \underline{\underline{17{,}515 \text{ N}}}$$

$$x = M_0/F_V$$

$$= 1 \times 1 \times 1 \times 9{,}810 \times 0.5 + 1 \times 9{,}810 \times \int_0^1 \sqrt{1 - x^2}\, x dx/17{,}515$$

$$= \underline{\underline{0.467 \text{ m}}}$$

$$F_H = \bar{p}A = (1 + 0.5)9{,}810 \times 1 \times 1 = \underline{\underline{14{,}715 \text{ N}}}$$

$$y_{cp} = \bar{y} + \bar{I}/\bar{y}A$$

$$= 1.5 + (1 \times 1^3)/(12 \times 1.5 \times 1 \times 1) = \underline{\underline{1.555 \text{ m}}}$$

$$F_R = \sqrt{(14{,}715)^2 + (17{,}515)^2} = \underline{\underline{22{,}876 \text{ N}}}$$

$$\tan\theta = 14{,}715/17{,}515; \quad \theta = \underline{\underline{40°2'}}$$

3.120 From the reasoning given in the solution to problem 3.112, we know the resultant must pass through the center of curvature of the gate. The horizontal component of hydrostatic force acting on the gate will be the hydrostatic force acting on the vertical projection of the gate or:

$$F_H = \bar{p}A$$

$$= 25 \text{ ft} \times 62.4 \text{ lb/ft3} \times 40 \text{ ft} \times 50 \text{ ft}$$

$$F_H = 3{,}120{,}000 \text{ lb}$$

3.120 (Cont'd)

The vertical component of hydrostatic force will be the buoyant force acting on the radial gate. It will be equal in magnitude to the weight of the displaced liquid (the weight of water shown by the cross-hatched volume in the above Fig.).

Thus, $F_V = \gamma \forall$

where $\forall = [(60/360)\pi \times 50^2 ft^2 - (1/2)50 \times 50 \cos 30° ft^2] \times 40$ ft

$= 226.5$ ft$^2 \times 40$ ft

$= 9060$ ft^3

Then $F_V = (62.4$ lb/ft$^3)(9060$ ft$^3) = 565{,}344$ lbs

$F_{result} = 3{,}120{,}000\,\mathbf{i} + 565{,}344\,\mathbf{j}$ lb acting through the center of curvature of the gate.

3.121 $F_H = \bar{p}A = -2.5 \times 50 \times (3 \times 1) = \underline{-375 \text{ lbf}}$
(force acts to the right)

$F_V = \forall\gamma = (1 \times 3 + \pi 3^2 \times \frac{1}{4})50 = 503.4$ lbf (downward)

$y_{cp} = 2.5 + 1 \times 3^3/(12 \times 2.5 \times 1 \times 3)$

$= \underline{2.8 \text{ ft above the water surface}}$

3.122 $\Sigma F_y = 0$

$F_B - W_{H_2O} - 600 + pA = 0$

$F_B = 600 + 1.2 \times 62.4\pi/4(2 \times 4 + 8 \times 0.25) - 1.2 \times 62.4 \times 10\pi \times 4/4$

$= \underline{-1{,}164 \text{ lbf}}$

3.123 $W_{H_2O} = (2/3)\pi 4^3 \times 62.4 + 8 \times (\pi/4) \times (2/4)^2 \times 62.4$

$= \underline{8{,}462 \text{ lbf}}$

$W_{dome} = 1{,}300$ lbf

$F_{Pressure} = 12 \times 62.4 \times \pi \times (4)^2 = 37{,}639$ lbf

$F_{bolt} = F_{pressure} - W_{H_2O} - W_{dome}$

$= 37{,}639 - 8{,}462 - 1{,}300 = \underline{27{,}877 \text{ lb downward}}$

3.124 $\Sigma F_z = 0$

$p_{bottom}A_{bottom} + F_{bolts} - W_{H_2O} - W_{dome} = 0$

where $p_{bottom}A_{bottom} = 4.8 \times 9{,}810 \times \pi \times 1.6^2 = 378.7$ KN

$W_{H_2O} = 9{,}810(3.2 \times (\pi/4) \times 0.2^2 + (2/3)\pi \times 1.6^3)$

$= 85.1$ kN

Then $F_{bolts} = -378.7 + 85.4 + 6 = \underline{\underline{-287.6 \text{ kN}}}$

3.125 $p_b = 10 \times 144 - 3 \times 1.5 \times 62.4 \times 1.5 = \underline{\underline{1{,}019 \text{ psf}}}$

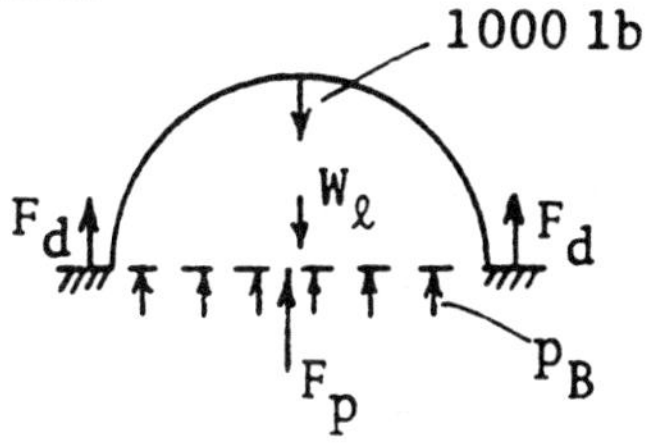

$F_p = p_bA = 1{,}019\pi \times 3^2 = 28{,}811$ lbf

$W_\ell = (2\pi/3)3^3 \times 62.4 \times 1.5 = 5{,}293$ lbf

$\Sigma F_V = 0; \quad F_d + F_p - W_\ell - 1{,}000 = 0$

$F_d + 28{,}810 - 5{,}293 - 1{,}000 = 0; \quad F_d = \underline{\underline{-22{,}518 \text{ lb}}}$

3.126 $F_H = (1+1)9{,}810 \times \pi \times (1)^2 = 61{,}640 \text{ N} = 61.64$ kN

61.64 kN force will act horizontally to the left to hold the dome in place.

$(y_{cp} - \bar{y}) = \bar{I}/\bar{y}A = (\pi \times 1^4/4)/(2\pi \times 1^2) = 0.125$ m

$F_V = (1/2)(4\pi \times 1^3/3)9{,}810 = 20{,}550 \text{ N} = \underline{20.55 \text{ kN}}$

To be applied downward to hold the dome in place.

3.127 The horizontal component of the hydrostatic force acting on the dome will be the hydrostatic force acting on the vertical projection of the bottom half of the dome or:

$F_H = \bar{p}A$

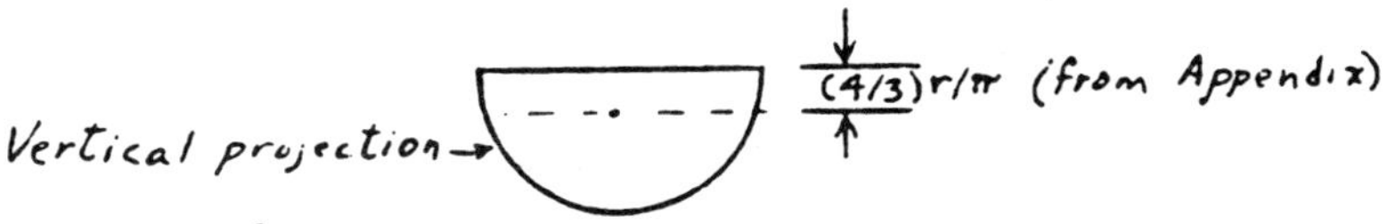

$\bar{p} = (4/3)(5/\pi)\text{ft } (62.4 \text{ lb/ft}^3)$

$= 132.4 \text{ lb/ft}^2$

$F_H = (132.4 \text{ lb/ft}^2)(\pi/8)(10^2)\text{ft}^2 = 5199$ lb

The vertical component of force will be the buoyant force acting on the dome. It will be the weight of water represented by the cross-hatched region shown in the Fig. (below).

3.127 (continued)

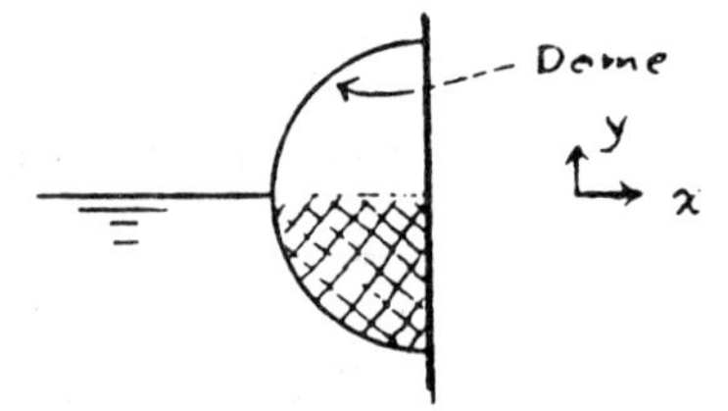

Thus, $F_V = \gamma \forall$

$= (62.4 \text{ lb/ft}^3)((1/6)\pi D^3/4)\text{ft}^3$

$F_V = 8{,}168 \text{ lb}$

The resultant force is then given by:

$F_{result} = 5{,}199\ \mathbf{i} + 8{,}168\ \mathbf{j}$ lb and it acts through the center of curvature of the dome.

3.128 $\Sigma F_z = 0$

$F_{buoy.} - W = 0$

$W = F_{buoy.}$

$= \gamma_w(1 \text{ in}^3 + (2 \text{ in})(0.01 \text{ in}^2))(1/12)^3 \text{ ft}^3/\text{in}^3$

$= (62.4 \text{ lbf/ft}^3)(1.02/1728)\text{ft}^3$

$= \underline{0.0368 \text{ lbf}}$

3.129 $F_B = W$

$0.95\gamma_w \forall = 0.0368 \text{ lbf}$

$0.95\gamma_w(1.00 + .01h)(1/1728) \text{ ft}^3/\text{in}^3 = 0.0368 \text{ lbf}$

$1.00 + 0.01 h = 1.0727$

$\underline{h = 7.27 \text{ in.}}$

3.130 $F_{buoy.} = \text{Wgt.}$

$\forall \gamma_w = \text{Wgt.}$

$(1 \text{ cm}^3 + (5.3 \text{ cm})(0.1 \text{ cm}^2))(.01^3) \text{ m}^3/\text{cm}^3 (\gamma_w) = \text{Wgt.}$

$(1.53 \text{ cm}^3)(10^{-6} \text{ m}^3/\text{cm}^3)(9800 \text{ N/m}^3) = \text{Wgt.}$

$\underline{\text{Wgt.} = 1.499 \times 10^{-2} \text{ N}}$

3.131 $F_{buoy.} = W$

$(1\ cm^3 + (6.3\ cm)(0.1\ cm^2))(0.01^3)\ m^3/cm^3 \gamma_{oil} = 0.015\ N$

$(1 + 0.63) \times 10^{-6}\ m^3 \gamma_{oil} = 0.015\ N$

$\gamma_{oil} = 9202\ N/m^3$

$\underline{\underline{S = \gamma_{oil}/\gamma_w = 9202/9810 = 0.938}}$

3.132 $(1 + .33) \times 10^{-6} m^3 \gamma_b = 0.01499\ N$

$\gamma_b = 11,271\ N/m^3$

$\underline{\underline{S = \gamma_b/\gamma_w = 11,271/9,810 = 1.149}}$

3.133 When only the bulb is submerged;

$F_B = Wt.$

$(\pi/4)[(0.02)^2 \times (0.08)] \times 9,810 \times (sp.gr.) = 0.035 \times 9.81$

sp.gr. = $\underline{\underline{1.39}}$

When the full stem is submerged;

$(\pi/4)[(0.02)^2 \times (0.08) + (0.01)^2 \times (0.08)] 9,810 \times (sp.gr.)$

$= 0.035 \times 9.81$

sp.gr. = $\underline{\underline{1.114}}$; Range $\underline{\underline{1.114 \text{ to } 1.39}}$

3.135 Draft = $400,000/(50 \times 20 \times 62.4) = 6.41\ ft < 8\ ft$

$GM = I_{00}/\forall - CG$

$= \left[(50 \times 20^3/12)/(6.41 \times 50 \times 20)\right] - (8 - 3.205)$

$= 0.40\ ft$

<u>Will float stable</u>

3.136 draft = $1 \times 8{,}000/9{,}810 = 0.8155$ m

$C_{\text{from bottom}} = 0.8155/2 = 0.4077$ m

$G = 0.500$ m; $CG = 0.500 - 0.4077 = 0.0922$ m

$GM = (I/\forall) - CG$

$= ((\pi R^4/4)/(0.8155 \times \pi R^2)) - 0.0922$

$= 0.077$ m $- 0.0922$ m, (negative)

Thus, block is unstable with axis vertical.

3.137 Draft $= 5{,}000/9{,}810 = 0.5097$ m

$GM = I_{00}/\forall - CG$

$= [(\pi \times 0.5^4/4)/(0.5097 \times \pi \times 0.5^2)] - (0.5 - 0.5097/2)$

$= -0.122$ m, negative

So will not float stable with its ends horizontal.

3.138 $GM = I_{00}/\forall - CG$; Let k = block density/water density

$GM = (LB^3/12)/(kB^2L) - ((B/2) - (kB/2))$

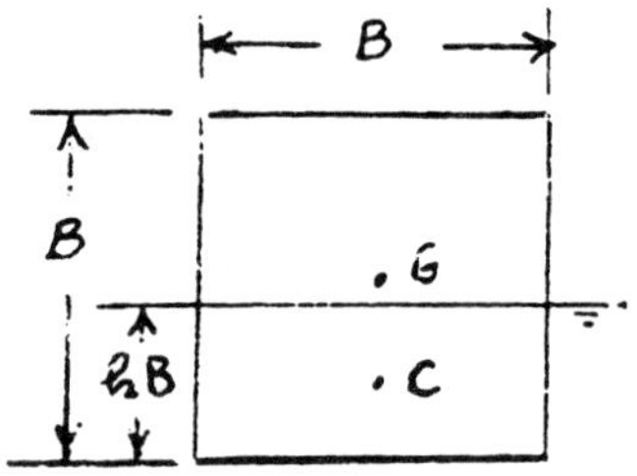

Condition for impending instability is when $GM = 0$

Then solve for k with $GM = 0$.

$0 = (1/12k) - (1/2)(1 - k)$

$k^2 - k + 1/6 = 0$ Solve by quadratic equation

$k = 0.211$ and 0.789 ; Analysis of Eq. (1) reveals that the block will be stable for $K < 0.211$ and for $K > 0.789$.

3.139 $GM = I_{00}/\forall - CG$

$= (3H(2H)^3/(12 \times H \times 2H \times 3H)) - H/2$

$= -H/6$

Not stable about longitudinal axis

$GM = (2H \times (3H)^3/(12 \times H \times 2H \times 3H)) - H/2$

$= +H/4$, positive

Stable about tranverse axis

3.140 With the given assumptions, this is an undamped system where the applied force varies linearly with displacement; therefore, harmonic vibration will occur which has the following solution:

$f = \sqrt{k/m}/(2\pi)$ where k is the proportionality constant between force and displacement $(k = F/x)$.

Here $m = 500\ \text{kg/m}^3 \times V = 500\ LA$

and $F = 9{,}810\ Ax$; $k = 9{,}810\ A\ \text{N/m}$

therefore $f = \sqrt{(9{,}810\ \cancel{A}/500\ L\cancel{A})}/2\pi$

$f = \sqrt{(19.62/L)}/2\pi = \sqrt{19.62/0.2}/2\pi = \underline{\underline{1.58\ \text{Hz}}}$

CHAPTER FOUR

4.1 Non-uniform, unsteady; unsteady, uniform.

4.2 (a) Unsteady, non-uniform.

(b) Local and convective acceleration.

4.3 Non-uniform; steady or unsteady.

4.4

A	B
steady flow	$V_s \partial V_s/\partial s = 0$
unsteady flow	$V_s \partial V_s/\partial s \neq 0$
uniform flow	$\partial V_s/\partial t = 0$
non-uniform flow	$\partial V_s/\partial t \neq 0$

4.5 True statements: (a), (c).

4.6

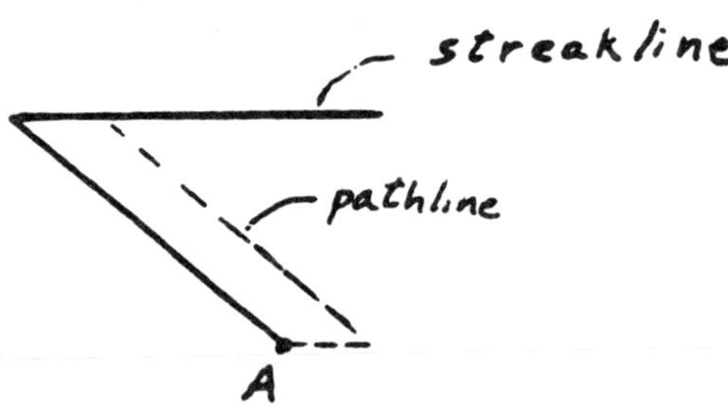

4.7

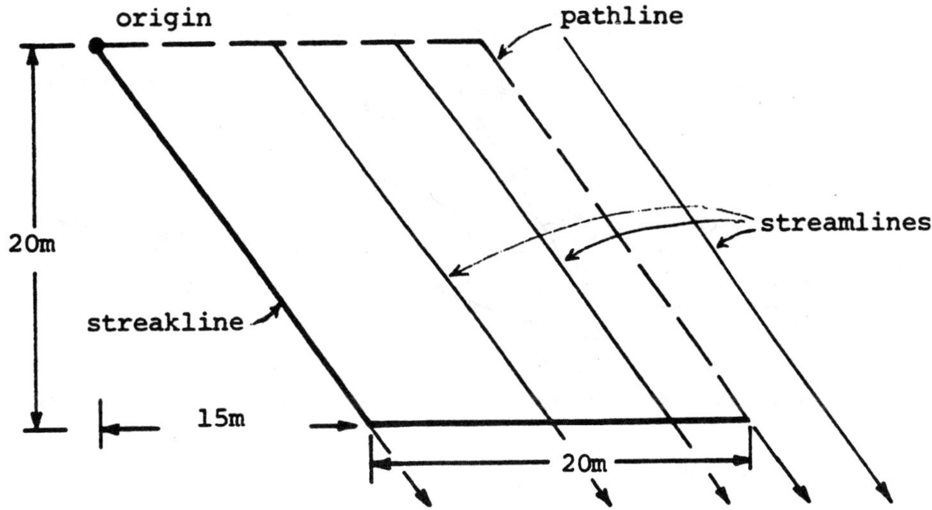

4.8

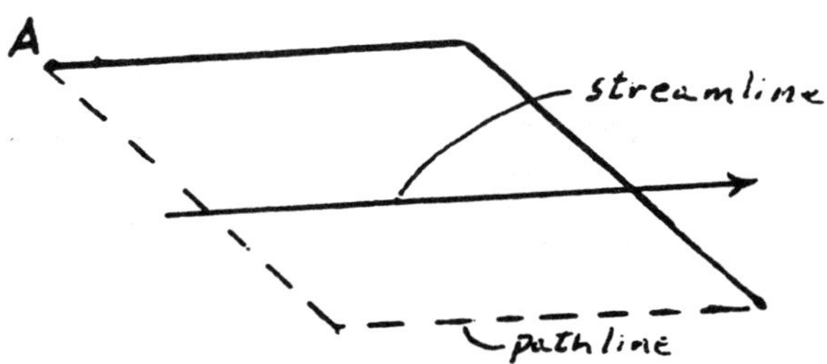

4.9 From time t = 0 to t = 1s dye is emitted from point A and will produce a streak that is 10 meters long (up and to the right of A). See Fig. A below. In the next second the first streak will be transported down and to the right 10 meters and a new streak, 10 ft long, will be generated down and to the right of point A (see Fig. B below). In the next 0.5s streaks in Fig. B will move up and to the right a distance of 5 meters and a new streak 5 meters in length will be generated as shown in Fig. C.

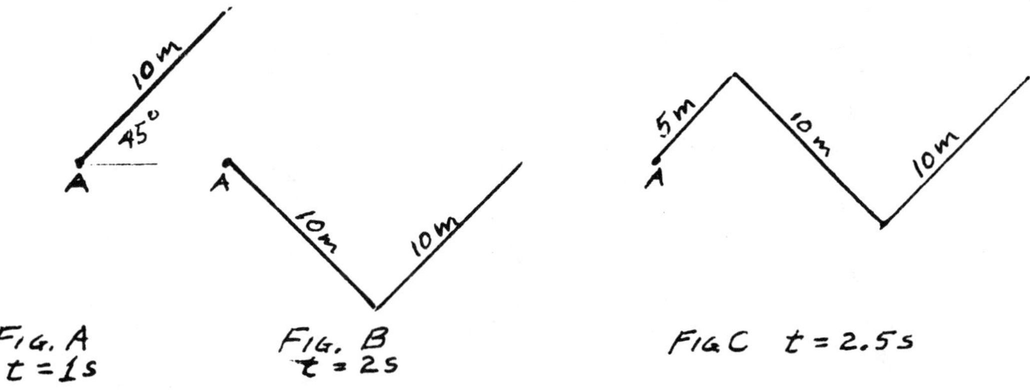

FIG. A $t = 1s$ FIG. B $t = 2s$ FIG C $t = 2.5s$

4.10

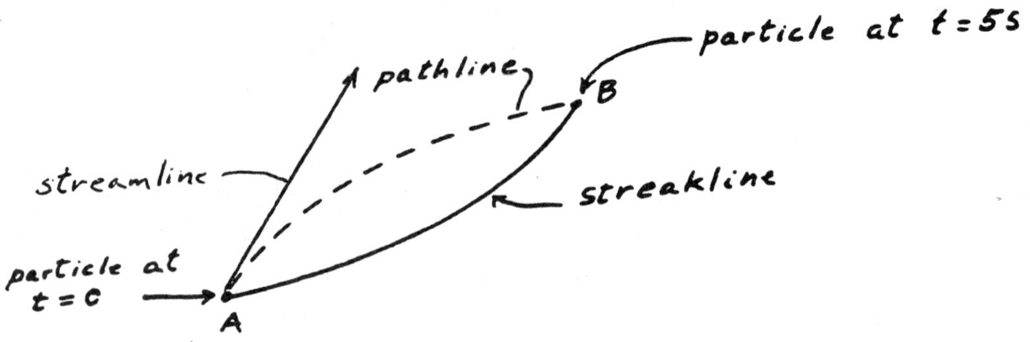

The flow is unsteady.

4.11

			pathline coor's	
t	u	v	x	y
1s	5m/s	- 2m/s	5m	- 1m
2	5	- 4	10	- 4
3	5	- 6	15	- 9
4	5	- 8	20	-16
5	5	-10	25	-25

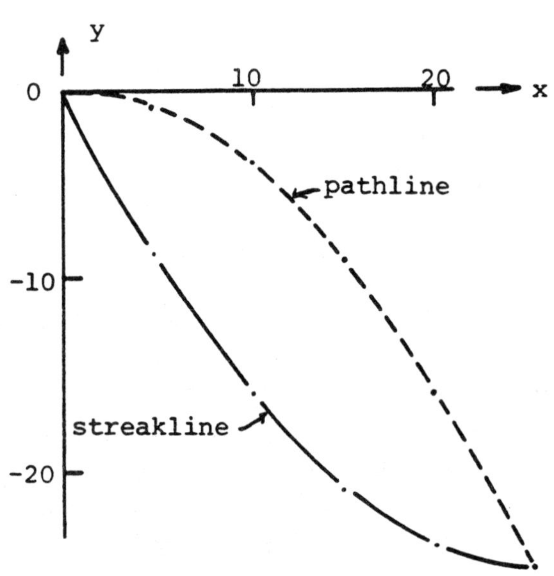

4.12 Pathline: For the first second the particle will follow the circular streamline (clockwise) through an angle of π radians ($\frac{1}{2}$ circle). then for the 2nd second the particle reverses its original path and finally ends up at the starting point. Thus, the pathline will be as shown:

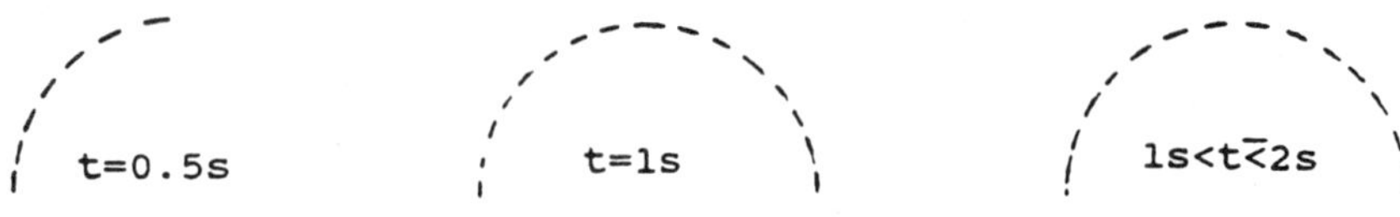

4.12 (Continued)

Streakline: For the first second a stream of dye will be emitted from starting point and the streak from this dye will be generated clockwise along the streamline until the entire top half circle will have a streak of dye at the end of 1 second. When the flow reverses a new dye streak will be generated on the bottom half of the circle and it will be superposed on top of the streak that was generated in the first second. The streakline is shown below for t = ½ sec., 1 sec. & 2 sec.

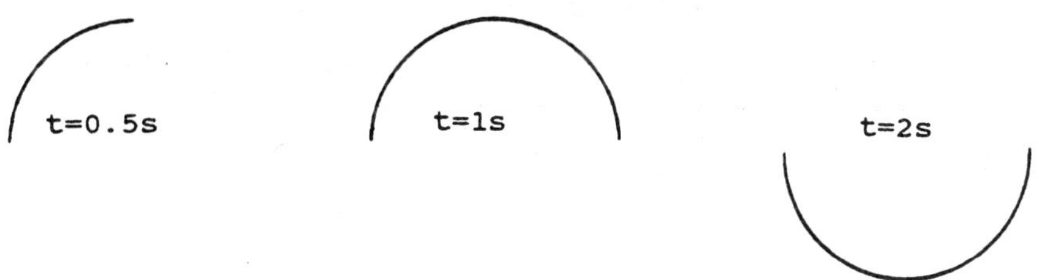

4.13

a. Two dimensional
b. One dimensional
c. One dimensional
d. Two dimensional
e. Three dimensional
f. Three dimensional
g. Two dimensional

4.14 $V = Q/A = 0.4/(\pi \times 0.3 \times 0.3/4) = \underline{5.66 m/s}$

4.15 $Q = VA = ((13 ft/s)(\pi/4)(2\ ft)^2) = 40.84\ ft^3/s$

$Q = (40.84\ ft^3/s)(449\ gpm/ft^3/s = \underline{18,337\ gpm}$

4.16 $Q = VA = (3 m/s)(\pi/4)(2\ m)^2 = 9.425 m^3/s$

$Q = (9.425 m^3/s)\ (1/0.02832)(ft^3/s)/(m^3/s) = \underline{332.8\ ft^3/s}$

4.17 $\rho = p/RT = 200,000/(287 \times 293) = \underline{2.378\ kg/m^3}$

Mass flow rate $= \rho VA = 2.378 \times 20 \times (\pi \times 0.08 \times 0.08/4)$

$= \underline{0.239\ kg/s}$

4.18 Assume $p_{atm} = 101$ kPa

$\rho = p/RT = (101 + 150)10^3/((518) \times (273 + 15)) = 1.682\ kg/m^3$

Mass flow rate $= \rho VA = 1.682 \times 11 \times \pi \times 0.5 \times 0.5 = \underline{14.53\ kg/s}$

4.19 $\rho = p/RT = (50 \times 10^3)/((287)(273 - 18)) = 0.683\ kg/m^3$

$\dot{m} = \rho AV$

or $A = \dot{m}/(\rho V) = (200)/((0.683)(240)) = 1.22\ m^2$

$A = (\pi/4) D^2 = 1.22$; $D = ((4)(1.22)/\pi)^{1/2} = \underline{\underline{1.25\ m}}$

4.20 $V = Q/A = (1{,}100/(60 \times 60))/(1 \times 0.20) = \underline{\underline{1.528 m/sec}}$

4.21 $\bar{V} = Q/A$; $Q = \int v dA$

$v = V_0 (1 - (r/R))$

$dA = 2\pi r dr$

Then $Q = \int_0^R V_0 (1 - (r/R))\, 2\pi r dr$

$= V_0 (2\pi) ((r^2/2) - (r^3/(3R)))_0^R$

$= 2\pi V_0 ((R^2/2) - (R^2/3)) = (2/6) \pi V_0 R^2$

$\therefore \bar{V}/V_0 = (Q/A)/V_0 = [((2/6) \pi V_0 R^2)/(\pi R^2)]/V_0 = (1/3)$

4.22 $Q = \int \underline{V} \cdot d\underline{A} = \int\int V(x,y)\, dxdy$

$= \int_{x=-w/2}^{w/2} \int_{y=0}^{D} V_s (1 - 4x^2/w^2)(1 - y^2/D^2)\, dxdy$

$\underline{\underline{Q = (4/9) V_s WD}}$

4.23 $Q = \int_A VdA = \int_0^{r_0} V 2\pi r dr$, $V = V_{max} - 3r/r_0$

$Q = \int_0^{r_0} (V_{max} - (3r/r_0)) 2\pi r\, dr = 2\pi r_0^2 ((V_{max}/2) - (3/3))$

$= 2\pi \times 4.00 ((15/2) - (3/3)) = \underline{\underline{163.4\ cfs}}$

$= 163.4 \times 449 = \underline{\underline{73{,}350\ gpm}}$

4.24 $Q = 2\pi r_0^2 ((V_{max}/2) - (2/3))$ (see problem 4.23 for derivation)

$= 2\pi \times 1((8/2) - (2/3)) = 20.94\ m^3/s$

$V = Q/A = 20.94/(\pi \times 1) = \underline{\underline{6.67\ m/s}}$

4.25 $dQ = VdA$

$dQ = (20y)\ dy$

$Q = 2\int_0^{0.5} VdA$

$= 2\int_0^{0.5} 20y\ dy$

$= 40\ y^2/2\Big|_0^{0.5} = 20 \times 0.25 = 5\ m^3/s$

$V = Q/A = (5\ m^3/s)/(1\ m^2) = 5\ m/s$

$\dot{m} = \rho Q = (1.3\ kg/m^3)\ (5\ m^3/s) = \underline{\underline{6.5\ kg/s}}$

4.26 $Q = V \times A = 18 \times 4 \cos 30° \times 25 = \underline{\underline{1,559\ cfs}}$

4.27 $Q = \int_0^{0.866} y^{1/3}\ (2\ dy)$

$= 2\int_0^{0.866} y^{1/3}\ dy$

$= (2/(4/3))\ y^{4/3}\Big|_0^{0.866m}$

$Q = \underline{\underline{1.238\ m^3/s}}$

4.28 $Q = \int_0^{0.866} Vdy$

$Q = \int_0^{0.866} (10)(e^y - 1)2dy$

$= [(2)(10)(e^y - y)_0^{0.866}$

$= \underline{\underline{10.23\ m^3/s}}$

$\overline{V} = Q/A = (10.23 m^3/s)/(2 \times 0.866\ m^2) = \underline{\underline{5.906\ m/s}}$

4.29 $Q = \forall/t = (20,000/9,790)/(15 \times 60) = \underline{\underline{2.270 \times 10^{-3} m^3/s}}$

4.30 $\Sigma V_p A_p = V_{rise} \times A_{rise}$

$200 \times V_p \times (2 \times 2) = (5/60) \times (900 \times 85)$

$V_{port} = \underline{\underline{7.97\ ft/s}}$

4.31 $q = \int_0^d u_{max}(y/d)^n dy = u_{max} d/(n+1)$

$= 3 \times 1.2/((1/6) + 1) = 3.09\ m^3/s$

$V = 3.09/1.2 = \underline{\underline{2.57\ m/s}}$

4.32 $Q = \int VdA$ where $V = 3y$ ft/s, $dA = xdy = 0.5\ ydy\ ft^2$

$q = \int_0^1 (3y) \times (0.5\ ydy)$

$= 1.5\ y^3/3|_0^1 = \underline{\underline{0.500\ cfs}}$

4.33 $V/V_c = ((r_0^2 - r^2)/r_0^2)^n$

$V = V_c(1 - (r/r_0^2))^n$

Then $Q = \int VdA$

$= \int V_c(1 - (r/r_0)^2)^n 2\pi r dr$

$= -\pi r_0^2 V_c \int (1 - (r/r_0)^2)^n (-2r/r_0^2) dr$

This is in the form of $K\int u^n du = (Ku^{n+1})/(n+1)$

Thus $Q = -\pi r_0^2 V_c (1 - (r/r_0)^2)^{n+1}/(n+1) \Big|_0^{r_0}$

$Q = \underline{\underline{(1/(n+1))V_c \pi r_0^2}}$ $\quad V = Q/A = \underline{\underline{(1/(n+1))V_c}}$

4.34

r/r_0	$1-(r/r_0)^2$	V(m/s)
0	1.00	12.0
0.2	0.96	11.5
0.4	0.84	10.1
0.6	0.64	7.68
0.8	0.36	4.32
1.0	0.00	0.0

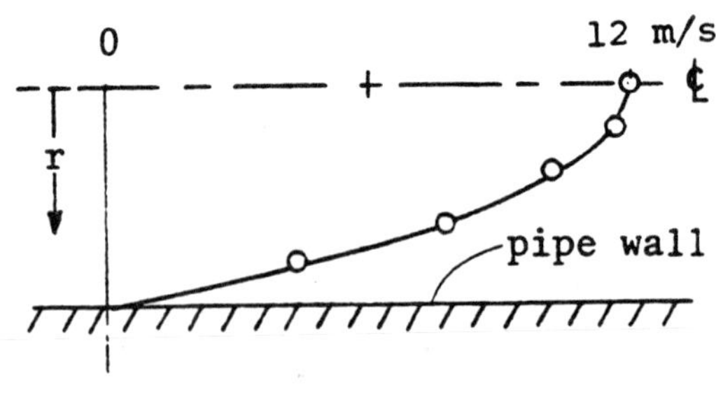

From solution to Prob. 4.33 $V = (1/(n+1))V_c$

$V = (1/2)V_c = \underline{6\ m/s}$; $Q = 4.712\ m^3/s$

4.35 $V = Q/A = (100/(62.37 \times 60))/(\pi \times (1/12)^2) = \underline{\underline{1.22\ ft/s}}$

4.36 $V = Q/A = (1{,}000/(998 \times 60))/(\pi \times 0.20 \times 0.20/4) = \underline{\underline{0.532 \text{ m/s}}}$

4.37 $Q = 4{,}765/(62.37 \times 8 \times 60) = \underline{\underline{0.159 \text{ cfs}}}$

$= 0.159 \times 449 = \underline{\underline{71.5 \text{ gpm}}}$

4.38 $Q = VA = 8(\pi/4)(4/12)^2 = \underline{\underline{0.698 \text{ cfs}}}$

$= 0.698 \times 449 = \underline{\underline{313 \text{ gpm}}}$

$= 0.698 \times 1.94 = \underline{\underline{1.35 \text{ slugs/s}}}$

4.39 (a) Steady. (b) Two-dimensional. (c) No. (d) Yes--see representative vectors below:

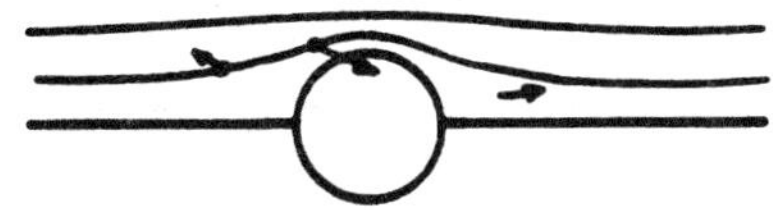

(e) $V_A n_A = V_C n_C$ where n = spacing between streamlines

Then $V_C = V_A(n_A/n_C) = 10(2/1) = \underline{\underline{20 \text{ ft/s}}}$

4.40

$u = xt + 2y$ $\qquad v = xt^2 - yt$

$a_x = u\partial u/\partial x + v\partial u/\partial y + w\partial u/\partial z + \partial u/\partial t$

$= (xt + 2y)(t) + (xt^2 - yt)(2) + 0 + x$

$= (2 + 2)(2) + (4 - 2)(2) + 1$ for x = 1m, y = 1m and t = 2s

$= 8 + 4 + 1 = 13 \text{ m/s}^2$

$a_y = u\partial v/\partial x + v\partial v/\partial y + w\partial v/\partial z + \partial v/\partial t$

$= (xt + 2y)(t^2) + (xt^2 - yt)(-t) + 0 + (2xt - y)$

$= (2 + 2)(4) + (4 - 2)(-2) + (4 - 1)$ for x = 1m, y = 1m and t = 2s

$= 16 - 4 + 3 = 15 \text{ m/s}^2$

$a_{total} = 13\,\mathbf{i} + 15\,\mathbf{j} \text{ m/s}^2$

4.41 $a_x = u\partial u/\partial x + \partial u/\partial t$

$= -U_0(1 - r_0^3/x^3)\partial/\partial x(-U_0(1 - r_0^3/x^3)) + \partial/\partial t(-U_0(1 - r_0^3/x^3))$

$= U_0^2(1 - r_0^3/x^3)(-3r_0^3/x^4) + 0$

$\underline{\underline{= -(3U_0^2 r_0^3/x^4)(1 - r_0^3/x^3)}}$

4.42 $\Sigma\rho \underline{V}\cdot\underline{A} = -d/dt\int_{c.v.} \rho d\forall$

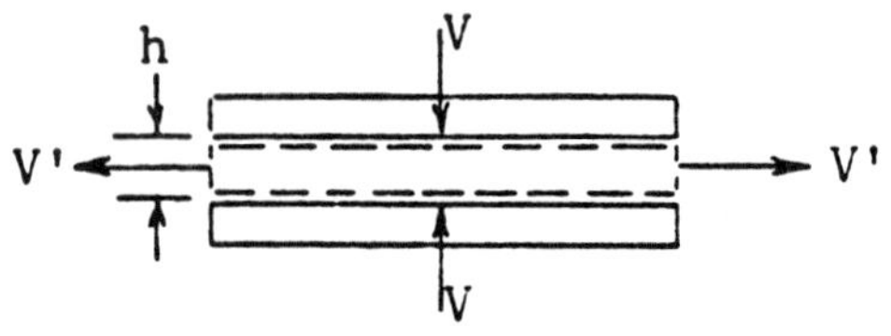

$-2\rho VA + \rho' V' A' = 0 \quad$ but $\rho = \rho'$

$2VA = V'A'$

$V' = 2VA/A' = 2V(\pi D^2/4)/(\pi Dh) = VD/2h = Vr/h$

$a_c = V'\partial/\partial r(V')$

$= Vr/h\ \partial/\partial r(Vr/h) = V^2 r/h^2 = \underline{\underline{V^2 D/2h^2}}$

4.43 $V_{sect} = \pi R^2(2V)/(2\pi Rh) = RV/h$ but $h = h_0 - 2Vt$

so $V_{sect} = RV/(h_0 - 2Vt)$

$\partial V/\partial t = \partial/\partial t[RV(h_0 - 2Vt)^{-1}] = RV(-1)(h_0 - 2Vt)^{-2}(-2V)$

$\partial V/\partial t = 2RV^2/(h_0 - 2Vt)^2$ but $h_0 - 2Vt = h$

so $\partial V/\partial t = 2RV^2/h^2 = \underline{\underline{DV^2/h^2}}$

4.44 $Q_{exit} = Q_A + Q_B$

$V_{exit} = (1/A_{exit})(Q_A + Q_B)$

$= (1/.01\ m^2)(.01t\,m^3/s + 0.005t^2 m^3/s)$

$= t\ m/s + 0.5t^2\ m/s$

Then at t=1s $\underline{\underline{V_{exit} = 1.5\ m/s}}$

$a_{exit} = dV/dt = 1 + t\ m/s^2$

(note that the convective acceleration is zero)

Then at t=1s $\underline{\underline{a_{exit} = 2\ m/s^2}}$

4.45 $V_\theta = 10t$

$a_{tang.} = V\,\partial V_\theta/\partial s + \partial V_\theta/\partial t$ (the term $\partial V_\theta/\partial s$ is crossed out by hand, with an arrow pointing to it)

$a_{tang.} = 0 + 10\ m/s^2$

$a_{normal} = V_\theta^2/r$

$= (10t)^2/r = 100t^2/10 = 10t^2$

at t=1s $a_{normal} = 10\ m/s^2$

$a_{total} = \sqrt{a^2_{tang.} + a^2_{normal}} = \sqrt{200} = \underline{\underline{14.14\ m/s^2}}$

4.46 $Q = Q_0 - Q_1 t/t_0 = 0.985 - 0.5t$

$\partial V/\partial s = +2 m/s/m$ (given)

$V = Q/A = (0.985 - 0.5 \times (1/2))/((\pi/4) \times 0.5^2) = \underline{\underline{3.743\ m/s}}$

$a_\ell = \partial V/\partial t = \partial/\partial t(Q/A) = \partial/\partial t((0.985 - 0.5t)/((\pi/4) \times 0.5^2))$

$= -0.5/((\pi/4) \times 0.5^2) = \underline{\underline{-2.546\ m/s^2}}$

$a_c = V\partial V/\partial s = 3.743 \times 2 = \underline{\underline{+7.487\ m/s^2}}$

4.47 $a_c = VdV/ds$

where $dV/ds = (V_{tip} - V_{base})/L$

$V_{tip} = Q/A_{tip} = 0.40/((\pi/4)(1/12)^2) = 73.34$ ft/s

$V_{base} = Q/A_{base} = 8.149$ ft/s

$dV/ds = (73.34 - 8.149)/1.5 = 43.46\ s^{-1}$

$V_{midway} = 8.149$ ft/s $+ 43.46\ s^{-1} \times (9/12)$ ft $= 40.74$ ft/s

then $a_c = VdV/ds = 40.74$ ft/s $\times 43.46\ s^{-1} = 1{,}770\ ft/s^2$

$a_\ell = \underline{\underline{0}}$

4.48 $a_\ell = \partial V/\partial t$; $V = Q/A$ and $Q = 2t$

then $a_\ell = \partial/\partial t(2t/A)$

$= (2/A)\partial/\partial t(t) = 2/A\ ft/s^2$

4.48 (continued)

From solution to Prob. 4.47: $V_{mid.} = 40.74$ ft/s

Therefore $A_{mid} = Q/V = 0.40\ ft^3/s/40.74\ ft/s = 0.00982\ ft^2$

Finally $a_\ell = 2/A = 2/0.00982 = \underline{\underline{204\ ft/s^2}}$

4.49 $V = q/b = (q_0/t_0)2t/b$ but $b = B - (1/2)B(x/4B)$

$V = (q_0/t_0)(2t)/(B - (1/2)B(x/4B))$

$a_{local} = \partial V/\partial t = (q_0/t_0)(2)/(B - (1/2)B(x/4B))$

then when $x = 2B$

$a_{local} = 2(q_0/t_0)/(B - (1/2)B(1/2)) = 2(q_0/t_0)/(3/4\ B)$

$a_{local} = \underline{\underline{(8/3)(q_0/t_0)/B}}$

4.50 $V = (q_0/t_0)2t/(B - (1/8)x)$

$a_{conv} = V\partial V/\partial x = V(q_0/t_0)2t(-1)(-1/8)/(B - (1/8)x)^2$

$a_{conv} = (1/8)(q_0/t_0)^2 4t^2/(B - (1/8)x)^3$

when $x = 2B$

$a_{conv.} = (1/2)(q_0/t_0)^2t^2/((3/4)B)^3 = \underline{\underline{32/27\ (q_0/t_0)^2t^2/B^3}}$

4.51 $a_\ell = \partial V/\partial t = \partial/\partial t[2t/(1-0.5x/L)^2] = 2/(1-0.5x/L)^2$

$= 2/(1-0.5 \times 0.5L/L)^2 = \underline{\underline{3.56\ ft/s^2}}$

$a_c = V(\partial V/\partial x) = [2t/(1-0.5x/L)^2]\ \partial/\partial x\ [2t/(1-0.5x/L)^2]$

$= 4t^2/((1-0.5x/L)^5 L) = 4(3)^2/((1-0.5 \times 0.5L/L)^5 4)$

$= \underline{\underline{37.93\ ft/s^2}}$

4.52 $V_r = Q/A = Q/(2\pi rh)$

$a_c = V_r\partial V_r/\partial r$

$= (Q/(2\pi rh))(-1)(Q)/(2\pi r^2h) = -Q^2/(r(2\pi rh)^2)$

4.52 (continued)

When D = 0.1m, r = 0.20m, h = 0.01m, and Q = 0.380m^3/s

$$V_{pipe} = Q/A_{pipe} = 0.380/((\pi/4) \times 0.1^2) = \underline{\underline{48.38 \text{ m/s}}}$$

$$\text{Then } a_c = -(0.38)^2/((0.2)(2\pi \times 0.2 \times 0.01)^2) = \underline{\underline{-4{,}572 \text{ m/s}^2}}$$

4.53

$$a_\ell = \partial V/\partial t = \partial/\partial t(Q/(2\pi rh))$$

$$a_\ell = \partial/\partial t(Q_0(t/t_0)/(2\pi rh))$$

$$a_\ell = (Q_0/t_0)/2\pi r h$$

$$a_{\ell;2,3} = (0.1/1)/(2\pi \times 0.20 \times 0.01) = \underline{\underline{7.957 \text{ m/s}^2}}$$

$$a_c = -Q^2/(r(2\pi rh)^2) \quad \text{(from solution to P. 4.52)}$$

at $t = 2\text{s}$, $Q = 0.2 \text{ m}^3/\text{s}$

$$a_{c,2s} = -1{,}266 \text{ m/s}^2$$

$$a_{2s} = a_\ell + a_c = 7.957 - 1{,}266 = \underline{\underline{-1{,}258 \text{ m/s}^2}}$$

$$a_{c,3s} = -2{,}850 \text{ m/s}^2$$

$$a_{3s} = -2{,}850 + 7.957 = \underline{\underline{-2{,}842 \text{ m/s}^2}}$$

4.54 Let the control surface surround the liquid in the tank and let it follow the liquid surface at the top.

Continuity Equation (Eq. 4.25):

$$\sum_{C.S.} \rho \underline{V} \cdot \underline{A} = -\frac{d}{dt}\int_{CV} \rho d\forall$$

$$-\rho V_{in}A_{in} + \rho V_{out}A_{out} = -\frac{d}{dt}(\rho A_{tank}h)$$

$$-V_{in}A_{in} + V_{out}A_{out} = -A_{tank}(dh/dt)$$

$$-V_{in}(.0025) + \sqrt{2g(1)}(.0025) = -0.1(0.1)\times 10^2$$

$$V_{in} = \sqrt{19.62} + (10^{-4}/(.0025))$$

$$\underline{\underline{V_{in} = 4.47 \text{m/s}}}$$

4.55 $Q = 1\ ft^3/s$ $\rho = 0.075\ lbm/ft^2$

$\forall = 0.04\ ft^3$

$\rho_{in\ tire} = 0.40\ lbm/ft^3$

Mass $= (0.40\ lbm/ft^3)(0.04\ ft^3) = (1\ ft^3/min.) \times (0.075\ lbm/ft^3)(t)$

$t = 0.213\ min. = \underline{\underline{12.8\ s}}$

4.56

Case (a)

1) $\beta = \underline{\underline{1}}$

2) $dB_{syst}/dt = \underline{\underline{0}}$

3) $\Sigma\beta\rho\underline{V}\cdot\underline{A} = \Sigma\rho\underline{V}\cdot\underline{A}$

$= -2 \times 10 \times 1.5$

$= \underline{\underline{-30\ slugs/s}}$

4) $d/dt\int_{cv} \beta\rho d\forall = \underline{\underline{+30\ slugs/s}}$

Case (b)

1) $\beta = \underline{\underline{1}}$

2) $dB_{syst}/dt = \underline{\underline{0}}$

3) $\Sigma\beta\rho\underline{V}\cdot\underline{A} = \Sigma\rho\underline{V}\cdot\underline{A}$

$= 2 \times 1 \times 2$

$-1 \times 2 \times 2 = \underline{\underline{0}}$

4) $d/dt\int_{cv} \beta\rho d\forall = \underline{\underline{0}}$

4.57 Valid statement: (b),

4.58

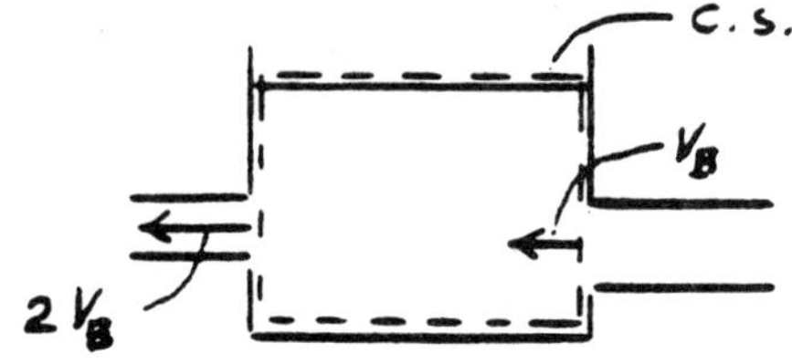

Let the c.s. be as shown and assume it is coincident with and moves with the water surface. Now apply the continuity equation to the problem:

$$\Sigma\rho\mathbf{V}\cdot\mathbf{A} = -d/dt\int_{cv} \rho d\forall$$

$$\rho 2V_B A_A - \rho V_B A_B = -\rho d/dt\int_{cv} d\forall$$

where $A_A = (\pi/4)3^2$; $A_B = (\pi/4)6^2$

$A_A = 9(\pi/4)$ $A_B = 36(\pi/4)$

$A_A = (1/4)A_B$

4.58 (Cont'd)

Then $$2V_B(1/4)A_B - V_BA_B = -d/dt\int_{cv} dV$$

$$V_BA_B((1/2) - 1) = -d/dt\int_{cv} d\forall$$

$$d/dt\int_{cv} d\forall = (1/2)V_BA_B$$

$$d/dt(Ah) = (1/2)V_BA_B$$

$$Adh/dt = (1/2)V_BA_B$$

Because $(1/2)V_BA_B$ is positive dh/dt is positive; therefore, one concludes that the water surface is rising.

4.59 a) True b) True c) True d) True e) True

4.60 1) $\beta = \underline{\underline{1.0}}$

2) $dB_{syst}/dt = \underline{\underline{0}}$

3) $\Sigma\beta\rho\underline{V}\cdot\underline{A} = \Sigma\rho\underline{V}\cdot\underline{A}$

$$\Sigma\rho\underline{V}\cdot\underline{A} = (1.5\ kg/m^3)(-10m/s)(\pi/4) \times (0.04)^2m^2$$
$$+ (1.5\ kg/m^3)(-6m/s)(\pi/4) \times (0.04)^2m^2$$
$$+ (1.2\ kg/m^3)(6\ m/s)(\pi/4) \times (0.06)^2m^2$$
$$= \underline{\underline{-0.00980\ kg/s}}$$

4) Because $\Sigma\beta\rho\underline{V}\cdot\underline{A} + d/dt\int\beta\rho d\forall = 0$

Then $d/dt\int\beta\rho d\forall = -\Sigma\beta\rho\underline{V}\cdot\underline{A}$

or $\underline{\underline{d/dt\int\beta\rho d\forall = +0.00980\ kg/s}}$ (mass is increasing in tank)

4.61 Rate at which liquid is displaced = $V_{up}(D^2-d^2)(\pi/4)$

$$V_{down} \times \pi d^2/4 = V_{up}(D^2 - d^2)\pi/4$$

$$2d^2 = D^2 \quad ; \quad D = \sqrt{2}d \qquad (1)$$

but $y/D = 24d/2d$; $D = y/12$ (2)

Eliminate D between Eqs. (1) & (2); $\underline{\underline{y = 12\sqrt{2}d}}$

4.62

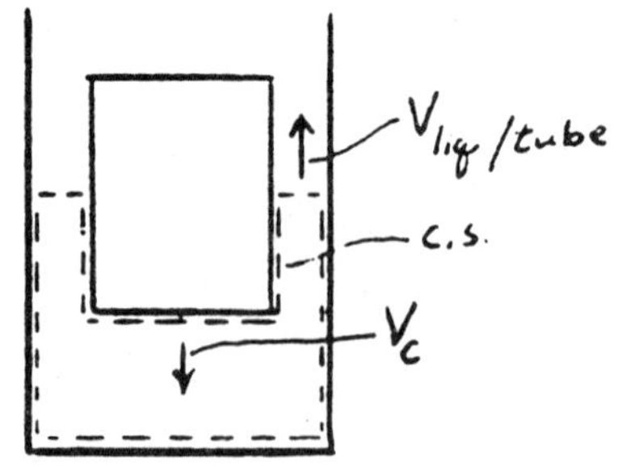

Apply continuity equation and let the c.s. be fixed except at bottom of cylinder where the c.s. follows the cylinder as it moves down.

$$0 = d/dt\int \rho d\forall + \Sigma\rho \mathbf{V}\cdot\mathbf{A}$$

$$0 = d/dt\ (\forall) + V_T A_A$$

$$0 = V_C A_C + V_T\ (\pi/4)\ (8^2 - 6^2)$$

$$0 = -2 \times (\pi/4)\ 6^2 + V_T\ (\pi/4)\ (8^2 - 6^2)$$

$$V_T = 72/(64 - 36) = \underline{\underline{2.57 \text{ ft/s}}} \text{ (upward)}$$

4.63 Apply the continuity equation and let the c.s. move up with the water surface in the tank.

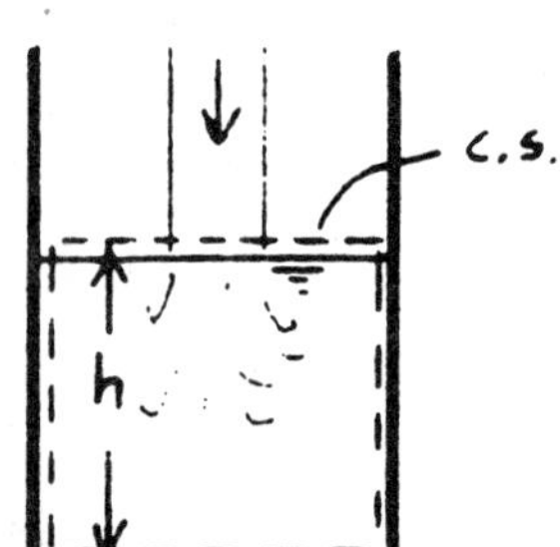

$$0 = d/dt\int_{cv} \rho d\forall + \Sigma\rho \mathbf{V}\cdot\mathbf{A}$$

$$0 = d/dt\ (hA_T) - ((10 + V_R)\ A_p)$$

where A_T = tank area, V_R = rise velocity and A_p = pipe area

$$0 = A_T\ dh/dt - 10A_p - V_R A_p \qquad \text{but } dh/dt = V_R$$

so $$0 = A_T V_R - 10A_p - V_R A_p$$

$$V_R = (10A_p)/(A_T - A_p) = 10\ (\pi/4)\ (1^2)/((\pi/4)4^2 - (\pi/4)1^2)$$

$$\underline{\underline{V_R = (2/3) \text{ ft/s}}}$$

4.64 Apply continuity equation

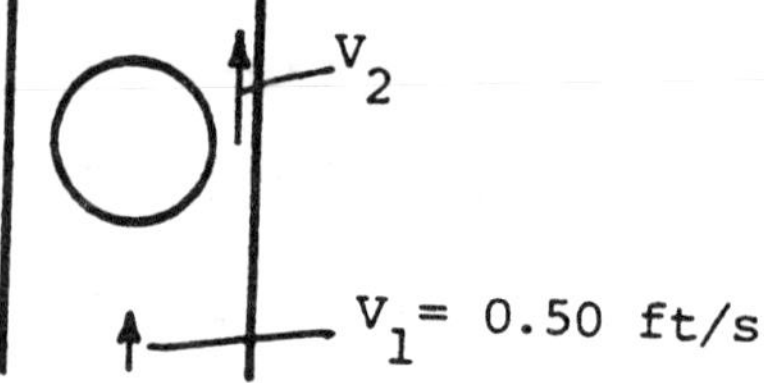

$$-\partial/\partial t\int_{cv} \rho dV = 0 = \Sigma\rho \underline{V}\cdot\underline{A}$$

$A_1V_1 = A_2V_2$ (velocities relative to sphere)

$(\pi \times 1.15^2/4) \times 0.5 = V_2\pi(1.15^2 - 1^2)/4;\quad V_2 = 2.05$ fps

True velocity $V = 2.05 - 0.5 = \underline{\underline{1.55 \text{ ft/s}}}$

4.65 $V_1 = Q/A_1 = 1.5/(0.3 \times 0.5) = \underline{\underline{10.0 \text{ m/s}}}$

$V_2 = 1.5/(0.15 \times 0.4) = \underline{\underline{25 \text{ m/s}}}$

4.66 $V = 0.3/(\pi/4)(0.2^2 + 0.15^2) = 6.11 \text{ m/s}$

$Q_{20\text{ cm}} = VA_{20} = 6.11 \times (\pi \times 0.1 \times 0.1) = \underline{\underline{0.192 \text{ m}^3/\text{s}}}$

$Q_{15\text{ cm}} = VA_{15} = 6.11 \times (\pi \times 0.075 \times 0.075) = \underline{\underline{0.108 \text{ m}^3/\text{s}}}$

4.67 $Q_{tot.} = 0.30 \text{ m}^3/\text{s} = Q_{20} + Q_{15}$

But $Q_{20} = 2Q_{15} \therefore 0.30 = 2\ Q_{15} + Q_{15}$

$Q_{15} = 0.10 \text{ m}^3/\text{s}$; $V_{15} = Q_{15}/A_{15} = \underline{\underline{5.66 \text{ m/s}}}$

$Q_{20} = 0.20 \text{ m}^3/\text{s}$; $V_{20} = 0.20/A_{20} = \underline{\underline{6.37 \text{ m/s}}}$

4.68 $Q = 898 \text{ gpm} = 2 \text{ cfs}$

$V_8 = Q/A_8 = 2/(\pi \times 0.667 \times 0.667/4) = \underline{\underline{5.724 \text{ fps}}}$

$V_6 = Q/A_6 = 2/(\pi \times 0.5 \times 0.5/4) = \underline{\underline{10.186 \text{ fps}}}$

4.69 $V_B = (V_A A_A - V_C A_C)/A_B$

$= [(6 \times \pi \times 4 \times 4/4) - (4 \times \pi \times 2 \times 2/4)]/(\pi \times 4 \times 4/4)$

$= \underline{\underline{5.00 \text{ m/s}}}$

4.70 $V_2 = (\rho_1 A_1 V_1)/(\rho_2 A_2) = (\rho_1 D_1^2 V_1)/(\rho_2 D_2^2)$

$= (2.0 \times 1.0 \times 1.0 \times 20)/(1.6 \times 0.6 \times 0.6) = \underline{\underline{69.4 \text{ m/s}}}$

4.71 Apply the continuity equation and let the c.s. move upward with the water surface as shown.

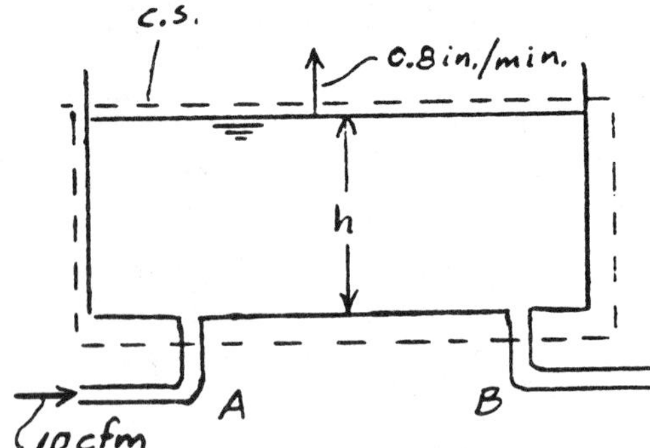

$$0 = d/dt\int_{cv}\rho d\forall + \Sigma\rho\mathbf{A}\cdot\mathbf{V}$$

$$0 = A\, dh/dt + Q_B - Q_A$$

$$Q_B = Q_A - A\, dh/dt$$

$$= 10 - (100)(0.8/12)$$

$$\underline{\underline{Q_B = +3.33 \text{ cfm}}}$$

Because Q_B is positive flow is leaving the tank through pipe B.

4.72 Inflow = 10 x π x 2 x 2/144 = 0.8727 cfs

Outflow = (7 x π x 3 x 3/144) + (4 x π x 1.5 x 1.5/144)= 1.571 cfs

Tank is emptying

$V_{fall} = Q/A = (1.571 - 0.8727)/(\pi \times 3 \times 3) = \underline{0.0247 \text{ fps}}$

4.73 Draw the c.s. so that part of it is coincident with the water surface and moves with it. Apply the continuity equation to see if the water surface is rising or falling:

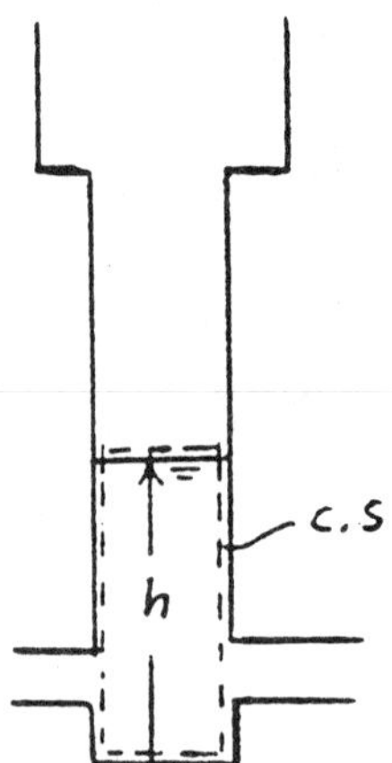

$$0 = d/dt\int_{cv}\rho d\forall + \sum_{cs}\rho\mathbf{V}\cdot\mathbf{A}$$

$$0 = d/dt\,(\rho\forall) + \sum_{cs}\rho\mathbf{V}\cdot\mathbf{A}$$

4.73 (Cont'd)

$$0 = d/dt(\rho Ah) - \rho(\pi/4)(1)(1^2) + \rho(\pi/4)(1/2)^2 \times 2$$

$$0 = Adh/dt - (\pi/4) + (\pi/4)(1/2)$$

$$dh/dt = (\pi/4)(1 - (1/2))/((\pi/4)(1^2)) = 1/2 \text{ ft/s (rising)}$$

Determine the time it takes the water surface to reach the 2 ft. section:

$$10 = (dh/dt)\,t\ ;$$

$$t = (10)/(1/2) = 20 \text{ sec.}$$

Therefore, at the end of 20 sec. the water surface will be in the 2 ft. section. Then the rise velocity will be:

$$dh/dt = (\pi/4)(1/2)/((\pi/4)(2^2)) = 1/8 \text{ ft/sec.}$$

4.74 At equilibrium $Q_{Evap.} = Q_{in.}$

$15\ ft^3/s/mi^2 (4.5 + 5.5\ h) mi^2 = 1{,}000\ ft^3/s$

Equilibrium h = 11.30 ft

Lake will dry up when h=0 and $Q_{Evap.} = Q_{in.}$ at this level

$15(4.5 + 5.5 \times 0) = Q_{in.}$

Lake will dry up when $Q_{in.} = 67.5\ ft^3/s$

4.75 Refer velocities to moving plate, then $V_R = 3V_0/2$

$Q_{plate} = AV_R = AV_0(3/2) = (3/2)Q_{jet} = (3/2) \times (6) = 9.0$ cfs

---

4.76 $Q_{in.} = Q_{out}$ at equilibrium

$20\ ft^3/s = V_{out}A_{out}$

$20 = (\sqrt{2gh})(\pi/4)(d^2_{out})$ where d = 1 ft

Solving for h yields: h = 10.07 ft

4.77 Assuming steady flow,

$\Sigma \rho \underline{V} \cdot \underline{A} = 0$

$-\rho_A V_A A_A - \rho_B V_B A_B + \rho_C V_C A_C = 0$

$\rho_C V_C A_C = 0.95 \times 1.94 \times 3 + 0.85 \times 1.94 \times 1 = \underline{\underline{7.18 \text{ slugs/s}}}$

Assuming incompressible flow,

$V_C A_C = V_A A_A + V_B A_B = 3 + 1 = 4 \text{ cfs} \qquad V_C = Q/A = 4/(\pi/4(1/2)^2) = \underline{\underline{20.4 \text{ ft/s}}}$

$\rho_C = 7.18/4 = 1.80 \text{ slugs/ft}^3 \qquad \underline{S} = 1.80/1.94 = \underline{\underline{0.925}}$

4.78 $\rho_{O_2} = p/RT = 200{,}000/(260 \times 373) = 2.06 \text{ kg/m}^3$

$\rho_{CH_4} = 200{,}000/(518 \times 373) = 1.03 \text{ kg/m}^3$

$V_{exit} = (2.06 \times 5 \times 3 + 1.03 \times 5 \times 1)/(2.2 \times 3) = \underline{\underline{5.46 \text{ m/s}}}$

4.79 Apply Eq. (4.25):

$$\sum \rho \underline{V} \cdot \underline{A} = -\frac{d}{dt} \int_{cv} \rho d\forall$$

$$-d/dt(\rho\forall) = -\dot{m}$$

$$\forall(d\rho/dt) = 0.5\rho_o/\rho$$

$$\rho d\rho = 0.5\rho_o dt/\forall$$

$$\rho^2/2\Big|_o^f = 0.5\rho_o \Delta t/\forall$$

$$(\rho_f^2 - \rho_o^2)/2 = 0.5\rho_o \Delta t/\forall$$

$$\therefore \Delta t = \forall\rho_o\left((\rho_f^2/\rho_o^2) - 1\right)$$

$$= 10(2)(2^2 - 1)$$

$$= \underline{\underline{60\text{s}}}$$

4.80 $\dot{m} = -d/dt(\rho\forall) = (\forall/RT)(dp/dt)$

$(1/p)(dp/dt) = -(0.68A\sqrt{RT})/\forall$

$\ell n(p_0/p) = (0.68A\sqrt{RT}t)/\forall$

$A = (\forall/0.68t\sqrt{RT})\ell n(p_0/p)$

$= (0.5/(0.68 \times 3 \times 3{,}600))(1{,}716 \times 520)^{-0.5}\ell n(44/39)$

$= \underline{\underline{8.69 \times 10^{-9}\ ft^2}} = \underline{\underline{1.25 \times 10^{-6} in.^2}}$

4.81 From problem 4.80:

$t = (\forall/0.68A\sqrt{RT})\ell n(p_0/p)$

$= 0.1\ \ell n(10/5)/(0.68(\pi/4)(1.5 \times 10^{-4})^2\sqrt{260 \times 291}) = 21{,}000\ s$

$= \underline{\underline{5\ hr. 50\ min.}}$

4.82 From example 4-9:

$t = (2A_T/\sqrt{2g}\ A_2)(h_1^{1/2} - h^{1/2})$

$= (2 \times \pi \times 0.6 \times 0.6/4)(\sqrt{3} - \sqrt{0.3})/(\sqrt{2 \times 9.81}\,\pi \times 0.03 \times 0.03/4)$

$= \underline{\underline{214\ s}}$

4.83 Given:

$$D = 2ft, R = 1ft, V = \sqrt{2gh}; L = 4ft$$
$$d = 2in. = 0.167ft, h_o = 1ft$$

Find: time to empty tank

Apply the continuity equation (Eq. 4.25):

$$\sum_{cs} \rho \underline{V} \cdot \underline{A} = -d/dt \int_{cv} \rho d\forall$$

Let the control surface surround the water in the tank as shown below. Thus, the control volume will decrease as the tank is being emptied.

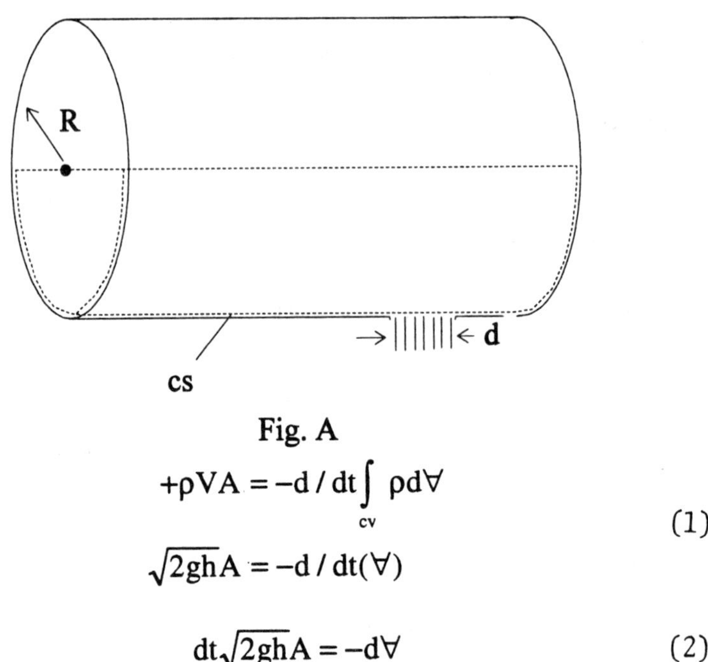

Fig. A

$$+\rho VA = -d/dt\int_{cv} \rho d\forall \qquad (1)$$

$$\sqrt{2gh}A = -d/dt(\forall)$$

$$dt\sqrt{2gh}A = -d\forall \qquad (2)$$

The control volume decreases with time because the volume of water in the tank becomes smaller; therefore, $d\forall$ for time dt is always negative. Physically one can visualize $d\forall$ as shown in the Fig. below.

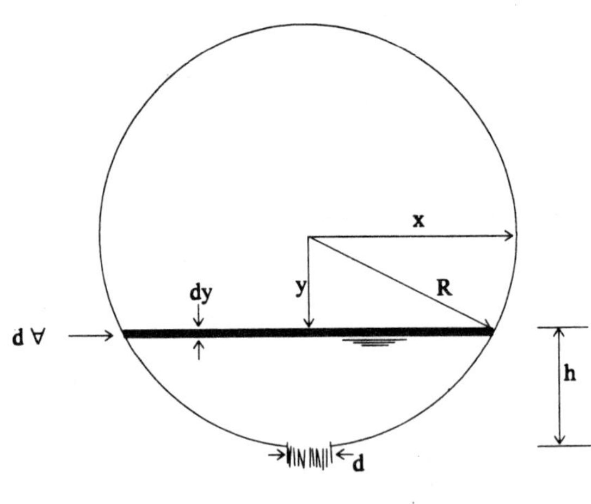

Fig. B

Thus $d\forall = -L(2x)dy$. Then when this is substituted into Eq. (2) we have

$$dt\sqrt{2gh}A = 2Lxdy \qquad (3)$$

4.83 (Cont'd)

But h can be expressed as a function of y:

$$h = R - y$$

or $$dt\sqrt{2g(R-y)}A = 2Lxdy$$

$$\text{Also } R^2 = x^2 + y^2; x = \sqrt{y^2 - R^2} = \sqrt{(y-R)(y+R)}$$

$$dt\sqrt{2g(R-y)}A = 2L\sqrt{(y-R)(y+R)}dy$$

$$dt = (2L/(\sqrt{2g}A))\sqrt{(y+R)}dy \qquad (4)$$

Integrate Eq. (4) $$t\Big|_0^t = (2L/(\sqrt{2g}A))\int_0^R \sqrt{R+y}dy$$

$$= (2L/(\sqrt{2g}A))[(2/3)(R+y)^{3/2}]_0^R$$

$$t = (2L/(\sqrt{2g}A))(2/3)((2R)^{3/2} - R^{3/2})$$

For $R = 1$

$$t = (2L/(\sqrt{2g}A))(2/3)(2^{2/3} - 1) \qquad (5)$$

In Eq. (5) $$A = (\pi/4)d^2 = 0.0219 \text{ ft}^2$$

$$\therefore t = (2 \times 4/(\sqrt{64.4} \times 0.0219))(2/3)(1.828)$$

$$= \underline{\underline{55.5 \text{ s}}}$$

Note: The above solution assumes that the velocity of water is uniform across the jet just as it leaves the tank. This is not exactly so but the solution should yield a reasonable approximation.

4.84

$$Q = -A_T(dh/dt); \quad dt = -A_T dh/Q$$

$$\text{where } Q = \sqrt{2gh}\, A_j = \sqrt{2gh}\, (\pi/4)d_j^2$$

$$A_T = (\pi/4)(d + C_1h)^2 = (\pi/4)(d^2 + 2\,dC_1h + C_1^2h^2)$$

$$dt = -(d^2 + 2dC_1h + C_1h^2)dh/(\sqrt{2g}\, h^{1/2}d_j^2)$$

$$t = -\int_{h_0}^{h} (d^2 + 2dC_1h + C_1^2h^2)dh/(\sqrt{2g}\, h^{1/2}d_j^2)$$

$$t = (1/(d_j^2\sqrt{2g}))\int_h^{h_0} (d^2h^{-1/2} + 2dC_1h^{1/2} + C_1^2h^{3/2})dh$$

$$t = (2/(d_j^2\sqrt{2g}))\left[d^2h^{1/2} + (2/3)dC_1h^{3/2} + (1/5)C_1^2h^{5/2}\right]_h^{h_0}$$

4.84 (Cont'd)

$$t = (2/(d_j^2\sqrt{2g}))\left[(d^2(h_0^{1/2}-h^{1/2}) + (2/3)dC_1(h_0^{3/2}-h^{3/2}) + (1/5)C_1^2(h_0^{5/2}-h^{5/2}))\right]$$

Then for $h_0 = 1\text{ m}$, $h = 0.20\text{ m}$, $d = 0.20\text{ m}$, $C_1 = 0.4$, and $d_j = 0.05\text{ m}$

$\underline{\underline{t = 18.4\text{ s}}}$

4.85 Continuity equation:

$$\sum_{c.s.} \rho\underline{V}\cdot\underline{A} = -d/dt\int_{c.v.}\rho d\forall \qquad (4.25)$$

$$\rho\sqrt{2gh}\, A_e = -d/dt\int_{c.v.}\rho d\forall$$

Mass of water in c.v. = $\rho B \times$ Face area

$$M = \rho B(W_0 h + h^2\tan\alpha)$$

Then $\rho\sqrt{2gh}\, A_e = -d/dt\, \rho B(W_0 h + h^2\tan\alpha)$

$$\sqrt{2gh}\, A_e = -BW_0(dh/dt) - 2Bh\tan\alpha(dh/dt)$$

$$dt = (1/(\sqrt{2g}\, A_e))(-BW_0 h^{-1/2}dh - 2\, B\tan\alpha h^{1/2}dh)$$

Integrate: $t = (1/\sqrt{2g}\, A_e)\int_{h_0}^{h} -BW_0 h^{-1/2}dh - 2B\tan\alpha h^{1/2}dh$

$$t = (1/(\sqrt{2g}\, A_e))(-2BW_0 h^{1/2} - (4/3)B\tan\alpha h^{3/2})_{h_0}^{h}$$

$$t = (\sqrt{2}Bh_0^{3/2}/(\sqrt{g}A_e))((W_0/h_0)(1-(h/h_0)^{0.5}) + (2/3)\tan\alpha(1-(h/h_0)^{1.5}))$$

for $W_0/h_0 = 0.2$, $\alpha = 30°$, $A_e g^{0.5}/(h_0^{1.5}B) = 0.01\text{ sec.}^{-1}$ and $h/h_0 = 0.5$ we get $\underline{t = 43.4\text{ seconds}}$

4.86

$$u = -\omega y \qquad v = \omega x$$

Check continuity:

$$\partial u/\partial x = 0 \text{ and } \partial v/\partial y = 0$$

$$\therefore \underline{\underline{\partial u/\partial x + \partial v/\partial y = 0 \text{ continuity is satisfied}}}$$

$$\text{Rate of rotation} = \omega_z = (1/2)(\partial v/\partial x - \partial u/\partial y)$$
$$= (1/2)(\omega - (-\omega))$$
$$= (1/2)(2\omega)$$
$$= \underline{\underline{\omega}}$$

Vorticity is twice the average rate of rotation; therefore, the $\underline{\underline{\text{vorticity} = 2\omega}}$

4.87 $\partial u/\partial x + \partial v/\partial y = (-2Cx/(y^2+x^2)^2) - (2C(y^2-x^2)(2x)/(y^2+x^2)^3)$

$-(2Cx/(y^2+x^2)^2) + (4Cxy(2y)/(y^2+x^2)^3)$

$= 0$ Continuity is satisfied

$\partial u/\partial y - \partial v/\partial x = (2Cy/(y^2+x^2)^2) - (2C(y^2-x^2)2y/(y^2+x^2)^3)$

$+ (2Cy/(y^2+x^2)^2) - (4Cxy(2x)/(y^2+x^2)^3)$

$= 0$ The flow is irrotational

4.88

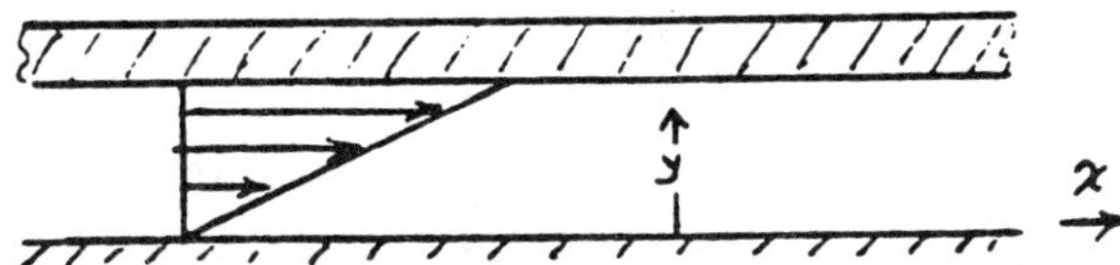

Equation for velocity: $u = ky$ where k is a constant.

Continuity Eq.: $\partial u/\partial x + \partial v/\partial y = 0$

$0 + 0 = 0$ continuity is satisfied

Check to see if flow is irrotational:

$\partial v/\partial x = \partial u/\partial y$

$0 \neq k$ flow is rotational

4.89 $(\partial u/\partial x) + (\partial v/\partial y) + (\partial w/\partial z) = U(3x^2+y^2) + U(3y^2+x^2) + 0$

$\neq 0$ Continuity is not satisfied

4.90 $u = xt + 2y \qquad V = xt^2 - yt$

Check for irrotationality:

$\partial u/\partial y = 2 \;;\; \partial v/\partial x = t^2 \qquad \partial u/\partial y \neq \partial v/\partial x$ Therefore,

the flow is rotational.

Determine acceleration:

$a_x = u\partial u/\partial x + v\partial u/\partial y + \partial u/\partial t$

$a_x = (xt + 2y)t + 2(xt^2 - yt) + x$

$a_y = u\partial v/\partial x + v\partial v/\partial y + \partial v/\partial t$

$= (xt + 2y)t^2 + (xt^2 - yt)(-t) + (2xt - y)$

4.90 (Cont'd)

$\underline{a} = ((xt+2y)t + 2t\ (xt-y) + x)\underline{i} + (t^2(xt+2y) - t^2(xt-y) + (2xt-y))j$

Then for x = 1m, y = 1m, and t = 1s the acceleration is:

$\underline{a} = ((1+2) + 0 + 1)\underline{i} + ((1+2) + 0 + (2-1))j$ m/s

$\underline{a} = \underline{\underline{4\underline{i} + 4\underline{j}\ m/s^2}}$

4.91 $\partial u/\partial x + \partial v/\partial y = -3xy/(x^2+y^2)^{5/2} + 3xy/(x^2+y^2)^{5/2}$

= 0 Continuity is satisfied

$\partial u/\partial y - \partial v/\partial x = -3y^2/(x^2+y^2)^{5/2} + 1/(x^2+y^2)^{3/2}$

$-3x^2/(x^2+y^2)^{5/2} + 1/(x^2+y^2)^{3/2}$

$\neq 0$ Flow is not irrotational

4.92 u = Axy

$\partial u/\partial x + \partial v/\partial y = 0$

$Ay + \partial v/\partial y = 0$

$\partial v/\partial y = -Ay$

$\underline{\underline{V = (-1/2)Ay^2 + C}}$

for irrotationality

$\partial u/\partial y - \partial v/\partial x = 0$

$Ax - \partial v/\partial x = 0$

$\partial v/\partial x = Ax$

or $V = 1/2\ Ax^2 + C(y)$

If we let $C(y) = -1/2\ Ay^2$ then the equation will also satisfy continuity

$\underline{\underline{V = 1/2\ A(x^2-y^2)}}$

4.93 $\rho_e = 10{,}000/((415 \times (2{,}000 + 273)) = 0.0106\ kg/m^3$

$V_e = V_m\rho_m A_m/(\rho_e A_e) = 0.01 \times 1{,}800 \times (\pi/4)(0.1)^2/(0.0106 \times (\pi/4)(0.08)^2)$

$= \underline{\underline{2{,}653\ m/s}}$

4.94 $A_g = \pi DL + 2\ (\pi/4)(D_0^2 - D^2)$

$= \pi \times 0.12 \times 0.4 + (\pi/2)(0.2^2 - 0.12^2) = 0.191\ m^2$

$\rho_e = V_g \rho_g A_g/(V_e A_e) = 0.012 \times 2{,}000 \times 0.191/(2{,}000 \times (\pi/4) \times (0.20)^2)$

$= \underline{\underline{0.073\ kg/m^3}}$

4.95 $\rho_p \dot{r} A_g = \dot{m}$

$\rho_p a p_c^n A_g = 0.65\ p_c A_t/\sqrt{RT_c}$

$p_c^{1-n} = (a\rho_p/0.65)(A_g/A_t)(RT_c)^{1/2}$

$p_c = \underline{\underline{(a\rho_p/0.65)^{1/(1-n)}(A_g/A_t)^{1/(1-n)}(RT_c)^{1/(2(1-n))}}}$

$p_c = 3.5\ (1+0.20)^{1/(1-0.3)} = \underline{\underline{4.541\ MPa}}$

4.96 $d/dt(\rho V) + 0.65\ p_c A_v/\sqrt{RT_c} = 0$

$\forall\ d\rho/dt + \rho d\forall/dt + 0.65\ p_c A_v/\sqrt{RT_c} = 0$

$d\rho/dt = -\rho/\forall\ d\forall/dt - 0.65\ p_c A_v/\forall\sqrt{RT_c}$

$\forall = (\pi/4)(0.1)^2(0.1) = 7.854 \times 10^{-4} m^3$

$(d\forall/dt) = -(\pi/4)(0.1)^2(30) = -0.2356\ m^3/s$

$\rho = p/RT = 300{,}000/(350 \times 873) = 0.982\ kg/m^3$

$d\rho/dt = (-0.982/7.854 \times 10^{-4}) \times (-0.2356) - 0.65 \times 300{,}000$

$\times 1 \times 10^{-4}/(7.854 \times 10^{-4} \times \sqrt{350 \times 873})$

$= \underline{\underline{249.7\ kg/m^3 \cdot s}}$

4.97 $Vr = \text{Const.}$

(15 mph) (200 mi.) = Const.

V_{100} = Const./100 mi.

V_{100} = (15 mph) (200 mi./100 mi.) = $\underline{\underline{30\ mph}}$

V_{50} = (15 mph) (200/50) = $\underline{\underline{60\ mph}}$

5.1 $\partial/\partial \ell(p + \gamma z) = -\rho a_\ell$

$\partial p/\partial \ell + \gamma \partial z/\partial \ell = -\rho a_\ell$

$\partial p/\partial \ell = -\rho a_\ell - \gamma \partial z/\partial \ell$

$= -(\gamma/g) \times (-0.30g) - \gamma \sin 30°$

$= \gamma(0.30 - 0.50)$

$= \underline{\underline{-0.20\gamma}}$

30°

ℓ

5.2 $\partial(p + \gamma z)/\partial z = -\rho a_z = -(\gamma/g) \times 0.20g$

$\partial p/\partial z + \gamma = -0.20\gamma$

$\partial p/\partial z = \gamma(-1 -0.20) = 0.81 \times 62.4(-1.20) = \underline{\underline{-60.7 \text{ lbs/ft}^3}}$

5.3 $\rho a_\ell = -\partial/\partial \ell\ (p + \gamma z)$ (5-3)

$a_\ell = (1/\rho)\ (-\partial p/\partial \ell - \gamma \partial z/\partial \ell)$

Let ℓ be positive upward. Then $\partial z/\partial \ell = +1$ and $\partial p/\partial \ell = (p_A - p_B)/1 =$ $-9{,}000$ Pa/m

Thus $a_\ell = (g/\gamma)\ (9{,}000 - \gamma)$

$a_\ell = g((9{,}000/\gamma) - 1)$

$a_\ell = g(0.900 - 1.0)\ \text{m/s}^2$

a_ℓ has a negative value;

therefore, acceleration is <u>downward</u>.

5.4 $\rho a_\ell = -\partial/\partial \ell\ (p + \gamma z)$; Let ℓ be positive upward.

$\rho(0.5g) = -\partial p/\partial \ell - \gamma \partial z/\partial \ell$

$(\gamma/g)\ (0.5g) = -\partial p/\partial \ell - \gamma(1)$

$\partial p/\partial \ell = -\gamma(0.5 + 1) = -1.5\gamma$

Thus the pressure decreases upward at a rate of 1.5γ.

Then at a depth of 2 ft: $p_2 = (1.5\gamma)\ (2) = 3\gamma$

$p_2 = 3 \text{ ft} \times 62.4 \text{ lb/ft}^3 = \underline{\underline{187.2 \text{ lb/ft}^2}}$

5.5 $\partial/\partial s\ (p+\gamma z) = -\rho a_s$; $-\Delta(p+\gamma z) = 1.94 \times 10 \times a_s$

$-(p_2-p_1) - \gamma(z_2-z_1) = 19.4a_s$

$a_s = (9 \times 144 - 62.4 \times 10)/19.4 = \underline{\underline{34.6\ \text{ft/s}^2}}$

5.6 $(\partial p/\partial s) = -\rho a_s = -1{,}000 \times 5 = \underline{\underline{-5{,}000\ \text{N/m}^3}}$

5.7 $(\partial p/\partial s) = -\rho a_s = -1{,}000 \times 6 = -6{,}000\ \text{N/m}^3$

$p_{upstream} = 90{,}000 + 6{,}000 \times 100 = 690{,}000\ \text{Pa} = \underline{\underline{690\ \text{kPa}}}$

5.8 $\partial/\partial z\,(p+\gamma z) = -\rho a_z$

$\Delta(p+\gamma z) = -\rho a_z \Delta z$

$(p+\gamma z)_{\text{at water surface}} - (p+\gamma z)_{\text{at piston}} = -\rho a_z(z_{surface} - z_{piston})$

$p_{atm} - p_v + \gamma(z_{surface} - z_{piston}) = -12\rho a_z$

$14.7 \times 144 - 0 + 62.4\ (12) = -12 \times 1.94\ a_z$

$a_z = \underline{\underline{-123.1\ \text{ft/s}^2}}$

5.9 $-\partial/\partial s\ (p + \gamma z) = \rho a_s$

$-\partial p/\partial s - \gamma \partial z/\partial s = \rho a_s$

where $\partial p/\partial s = (100 - 170)/2 = -35\ \text{lb/ft}^3$

$\partial z/\partial s = \sin 30° = 0.5$

then $a_s = (1/\rho)\ (35\ \text{lb/ft}^3 - (100)\ (0.5))$

$= (1/\rho)\ (-15\ \text{lb/ft}^3)$

Because a_s has a negative value we conclude that the <u>acceleration is in the negative s direction.</u> <u>The flow direction cannot be discerned.</u>

5.10 $d/dx(p + \gamma z) = -\rho a_x$

but z = const.; therefore,

$dp/dx = -\rho a_x$

$a_x = a_{convective} = vdv/dx$

$= (50\ ft/s)(40\ ft/s/ft) = 2{,}000\ ft/s^2$

Finally

$dp/dx = (-1.94\ slug/ft^3) \times (2{,}000\ ft/s^2) = \underline{\underline{-3{,}880\ psf/ft}}$

5.11 The valid statement is (b).

5.12 Let y = vertical dimension in the duct

Then $V_x = q/y$ where $y = b - 0.1x$; $V_x = 0.2t/(b - 0.1x)$

$a_{local} = \partial V_x/\partial t = 0.2/(b - 0.1x)$; At point A x = -1

So $a_{local} = 0.2/0.3 = 0.667\ m/s^2$

$a_{conv} = V_x \partial V_x/\partial x$; $\partial V_x/\partial x = \partial/\partial x[q(b - 0.1x)^{-1}]$

$\partial V_x/\partial x = 0.1q/(b - 0.1x)^2 = 0.1 \times 0.2t/(b - 0.1x)^2$

and $V_x \partial V_x/\partial x = [0.2t/(b - 0.1x)][0.1 \times 0.2t/(b - 0.1x)^2]$

$a_{conv} = 0.004t^2/(b - 0.1x)^3$

$= 0.5926\ m/s^2$ for t = 2s and x = -1m

Then $a_{tot} = 0.5926\ m/s^2 + 0.667\ m/s^2 = 1.260\ m/s^2$

$\partial p/\partial x = -\rho a_x = -1{,}000 \times 1.260 = \underline{\underline{-1{,}260\ Pa/m}}$

5.13 From solutions to Probs. 4.49 and 4.50.

$a = (8q_0/3t_0 B) + (32t^2 q_0^2/(27B^3 t_0^2))$

Then for $q_0 = 0.10\ m^3/s$, $t_0 = 0.1s$, t = 0.5s , and B = 0.40 m

$a = 11.29\ m/s^2$

Then $\partial p/\partial x = -\rho a_\ell$

$= -(1{,}000\ kg/m^3)(11.29\ m/s^2)$

$= \underline{\underline{-11.29 kPa/m}}$

5.14 From the solution to Prob. 4.46 the acceleration is

$$a_\ell + a_c = (-2.546 + 7.487)\text{m/s}^2$$

Therefore $dp/dx = -\rho a_x$

$$= -\rho(-2.546 + 7.487) \text{ Pa/m}$$

$$= \underline{-4.94\rho \text{ Pa/m}}$$

5.15 $V = Q/A = 0.03\, t/\pi r^2; \quad r = D_0/2 - x/20 = 0.05(1-x)$

$$V = 0.03\, t/(\pi \times 0.0025(1-x)^2) = 3.820\, t(1-x)^{-2}$$

$$a_{conv} = V\partial V/\partial x = (3.820\, t)^2(1-x)^{-2} \times (-2) \times (1-x)^{-3}(-1)$$

$$= 29.18\, t^2(1-x)^{-5}$$

at t = 2s and x = 0.30m $a_{conv} = \underline{694.5 \text{ m/s}^2}$

$$a_{local} = \partial V/\partial t = 3.820(1-0.3)^{-2} = \underline{7.796 \text{ m/s}^2}$$

$$(dp/dx)_{accn} = -\rho a_x = -1.6 \times 1{,}000(694.5 + 7.796) = -1{,}124 \text{ kN/m}^3$$

$$(dp/dz) = 1.124 - 9.810 \times 1.6 = \underline{1.108 \text{ kN/m}^3}$$

$$(dp/dx)_{accn} = -\rho[116.7\,(1-x)^{-5} + 3.820\,(1-x)^{-2}]$$

$$p_{accn} = +\rho[29.18\,(1-x)^{-4} + 3.820\,(1-x)^{-1}] + c$$

at x = 0.6 m, p = 0

$$c = 1{,}000 \times 1.6\,[29.18\,(1-0.6)^{-4} + 3.820\,(1-0.6)^{-1}] = 1.839 \text{ MPa}$$

$$p_A = -1{,}000 \times 1.6\,[29.18\,(1-0.3)^{-4} + 3.820\,(1-0.3)^{-1}]$$

$$+ 1{,}839{,}000 - 9{,}810 \times 1.6 \times 0.3$$

$$= 1{,}631{,}000 \text{ Pa} = \underline{1.631 \text{ MPa}}$$

5.16 Possible method of design:

1. Arbitrarily choose L_1
2. Arbitrarily choose $V_n^2/2g$; solve for V_n
3. Solve for $V_0^2/2g$ with Bernoulli's equation
4. Solve for V_0 and then A_0 and D_0 for given Q
5. Solve for d_n for given Q

5.17 $(\partial P/\partial r) = -\rho a_r = -1.4 \times (-4{,}572) = \underline{\underline{6{,}401\ N/m^3}}$

$a = -Q^2/(4\pi^2 h^2 r^3)$, $(\partial p/\partial r) = (\rho Q^2/4\pi^2 h^2) r^{-3}$

Integrating $p = -(\rho Q^2/8\pi^2 h^2) r^{-2} + c$

At $r = r_0$, $p = p_{atm}$, So $c = p_{atm} + (\rho Q^2/8\pi^2 h^2) r_0^{-2}$

$p = p_{atm} + (\rho Q^2/8\pi^2 h^2)(r_0^{-2} - r^{-2})$

$p = 100{,}000 + (1.4 \times 0.380 \times 0.380/(8\pi^2 \times 0.01 \times 0.01))(0.5^{-2} - 0.2^{-2})$

$= 99{,}462$ Pa $= 99.46$ kPa absolute

or $p_A = \underline{\underline{-538\ \text{Pa gage}}}$

5.18 $\tan\alpha = a_x/g$, $a_x = g\tan\alpha = 9.81 \times 3/5 = \underline{\underline{5.886\ m/s^2}}$

5.19 $(dp/dz) = -\rho(g + a_z) = -1.1 \times 1.94\ (32.2 - 1.4g) = 27.5$ psf/ft

$p_B - p_A = -27.5 \times 4 = \underline{\underline{-110.0\ \text{psf}}}$

$p_C - p_B = \rho a_x L = 1.1 \times 1.94 \times 0.9g \times 3 = 185.5$ psf

$p_C - p_A = 185.5 - 110.0 = \underline{\underline{75.5\ lbf/ft^2}}$

5.20 $(dp/dz) = -1.3 \times 1{,}000\ (9.81 - 6.54) = -4{,}251\ N/m^3$

$p_B - p_A = 4{,}251 \times 3 = 12{,}753$ Pa $= \underline{\underline{12.75\ \text{kPa}}}$

$p_C - p_B = \rho a_x L = 1.3 \times 1{,}000 \times 9.81 \times 2 = 22{,}506$ Pa

$p_C - p_A = 22{,}506 + 12{,}753 = 38{,}259$ Pa $= \underline{\underline{38.26\ \text{kPa}}}$

5.21 $\tan\alpha = a_x/g = 8.02/32.2 = 0.2491$

$\tan\alpha = h/9$, $h = 9\tan\alpha = 9 \times 0.2491 = 2.242$ ft

Maximum depth $= 7 - 2.242 = \underline{\underline{4.758\ \text{ft}}}$

5.22 (1st part) $\tan\alpha = a_x/g$

$= (1/3)\ g/g$

$= 1/3$

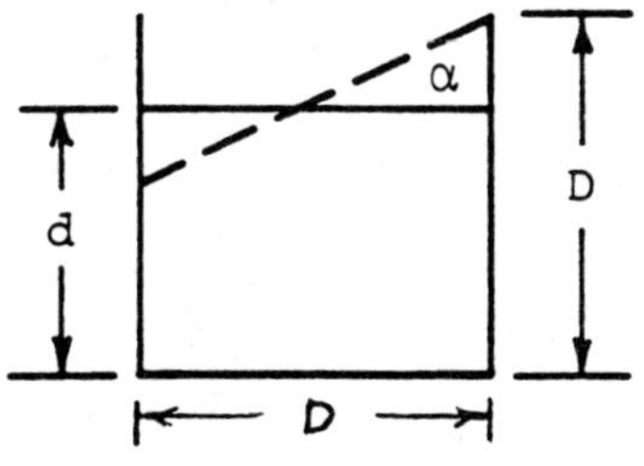

$\tan\alpha = 1/3 = (D-d)/(0.5D)$

thus $d = D - (1/6)D = (5/6)D$

Tank can be 5/6 full without spilling

(2nd part) $\tan\alpha = 1/3$

Then $1/3 = a_n/g$

$a_n = (1/3)g$

$V^2/r = (1/3)g$

or $V = \sqrt{(1/3)gr}$

$= 12.8\ m/s$

5.23 The valid statement is (b).

5.24 $\tan\theta = a_s/g = 1$

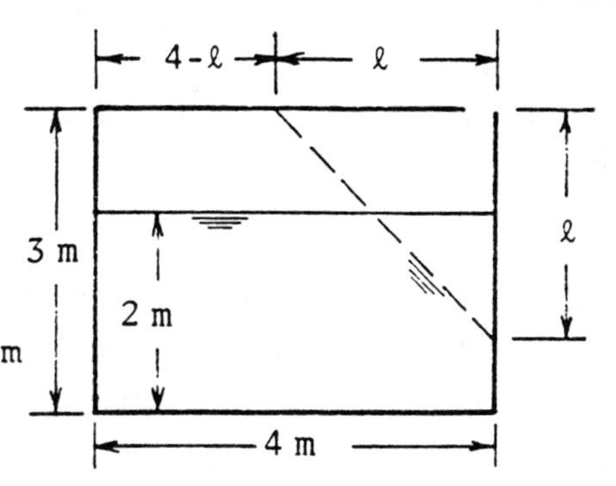

area of air space $= 1/2\ell^2 = 4 \times 1$

$\ell^2 = 8\ ;\ \ell = \sqrt{8}\ m$

$p_{max}/\gamma = 4 - \sqrt{8} + 3 = 4.17\ m$

$p_{max} = 33.95\ kPa\ gage$

5.25 Write Eq. 5.8 from the tank (center bottom) to the water surface in the piezometer.

$$p + \gamma z - \rho r^2\omega^2/2 = p_p + \gamma z_p - \rho r_p^2\omega^2/2$$

where $p_p = 0\,\text{gage},\ r_p = 3\text{ft and } r = 0$

Then

$$p = -(\rho/2)(9 \times 225) + \gamma(z_p - z)$$

$$= -(1.94/2)(2025) + 62.4 \times 2.5$$

$$= -1808\text{psf} = -12.56\text{psi}$$

5.26 Write Eq. (5.8) from point A to point B

$$p_A + \gamma z_A - \rho r_A^2 \omega^2 / 2 = p_B + \gamma z_B - \rho r_B^2 \omega^2 / 2$$

$$p_B = p_A + (\rho / 2)(\omega^2)(r_B^2 - r_A^2) + \gamma(z_A - z_B)$$

where $\omega = V_A / r_A = 20 / 1.5 = 13.333 \text{rad} / \text{s}; \rho = 0.8 \times 1.94 \text{slugs/ft}^3$

Then

$$p_B = 30 + (1.94 \times 0.80 / 2)(13.33^2)(2.5^2 - 1.5^2) + 62.4 \times 0.8(-1)$$

$$p_B = 30 + 551.5 - 49.9 = \underline{\underline{531.6 \text{psf}}}$$

5.27 Let point C be at the center bottom of the tank. Write Eq. (5.8) between points B & C.

$$p_B - \rho r_B^2 \omega^2 / 2 = p_c - \rho r_c^2 \omega^2 / 2$$

where $r_B = 0.5\text{m}$, $r_c = 0$ and $\omega = 10\text{rad/s}$

Then

$$p_B - p_C = (\rho / 2)(\omega^2)(0.5^2)$$

$$\doteq (1200 / 2)(100)(0.25)$$

$$= 15{,}000 \text{Pa}$$

$$p_C - p_A = 2\gamma + \rho a_z \ell$$

$$= 2 \times 11{,}772 + 1{,}200 \times 4 \times 2$$

$$= 33{,}144 \text{Pa}$$

Then $p_B - p_A = \underline{\underline{48{,}144 \text{Pa}}} = \underline{\underline{48.14 \text{kPa}}}$

5.28 At the condition of imminent spilling the outside leg will have liquid to the top of the outside leg and the leg on the axis of rotation will have the liquid surface at the bottom of its leg. Writing the equation for a rotating fluid we have

$$p_1 + \gamma z_1 - \rho r_1^2 \omega^2 / 2 = p_2 + \gamma z_2 - \rho r_2^2 \omega^2 / 2$$

Let point 1 be at top of outside leg and point 2 be at surface of liquid of inside leg. $\therefore p_1 = p_2$; $z_1 = .5\text{m}$ and $z_2 = 0$

$$\gamma \times 0.5 - (\gamma / g) \times .5^2 \omega^2 / 2 = 0$$

$$\omega^2 = 4g$$

$$\omega = 2\sqrt{g} = \underline{\underline{6.26 \text{rad} / \text{s}}}$$

$0.5 = \frac{0.5^2 \omega^2}{2g}$

5.29 Write Eq. (5.8) between points 1 and 2

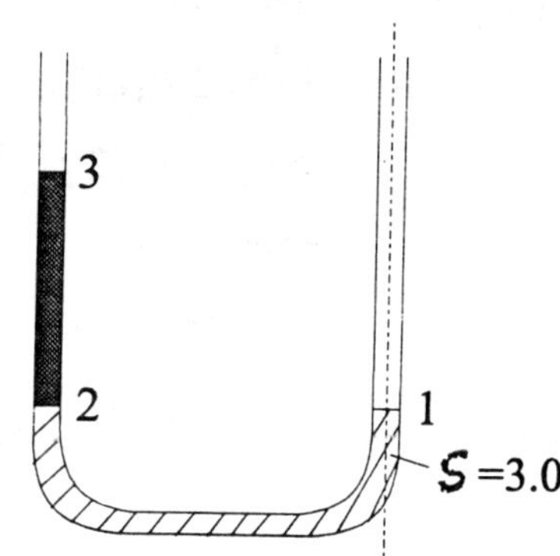

$$p_2 + \gamma z_2 - \rho r_2^2\omega^2/2 = p_1 + \gamma z_1 - \rho r_1^2\omega^2/2$$

where $z_2 = z_1$; $r_1 = 0$, $r_2 = 1\text{ft}$ and $\omega = (60/60)\times 2\pi = 2\pi\text{rad/s}$

Then $$p_2 = (1.94\times 3)(1^2)(2\pi)^2/2 = 114.9\text{psf} \quad (1)$$

Also by hydrostatics because there is no acceleration in the vertical we have

$$p_2 = 0 + \tfrac{1}{2}\times\gamma_f \quad (2)$$

where γ_f is the specific weight of the other fluid. Solve for γ_f between Eqs. (1) and (2)

$$\gamma_f = 229.8\text{lbf/ft}^3$$

$$S = \gamma_f/\gamma_{H_2O} = 229.8/62.4 = \underline{\underline{3.68}}$$

5.30 A preliminary check shows that the water will evacuate the axis leg. Thus this Fig. applies.

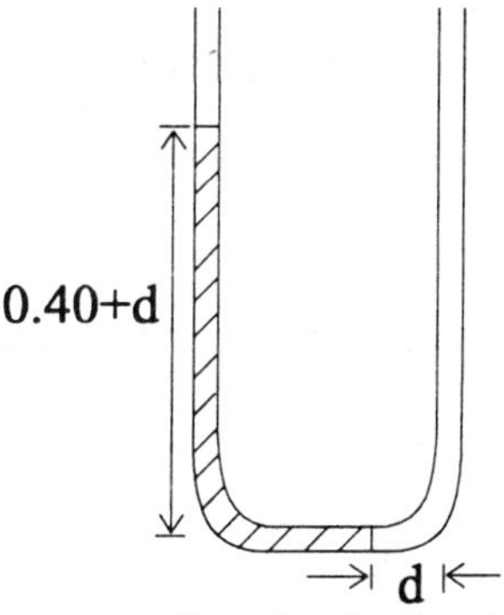

Write Eq. (5.8) between the water surface in the horizontal part of the tube and the water surface in the vertical part of the tube.

$$p_1 + \gamma z_1 - \rho r_1^2\omega^2/2 = p_2 + \gamma z_2 - \rho r_2^2\omega^2/2$$

where $r_1 = d$, $r_2 = 0.30\text{m}$ and $(z_2 - z_1) = 0.40 + d$

Then

$$(\rho\omega^2/2)(r_2^2 - r_1^2) = \gamma(0.40 + d)$$

$$(1000\times 16.06^2/2)(0.3^3 - d^2) = (0.40 + d)9{,}810$$

Solving for d yields $\underline{d = 0.209\text{m}}$

Then $z_2 = 0.40 + 0.209 = \underline{0.609\text{m}}$

5.31 Write Eq. (5.8) between the liquid surface in the large tube and the liquid surface in the small tube for conditions after rotation occurs.

$$\gamma z_1 - (\rho/2)r_1^2\omega^2 = \gamma z_2 - (\rho/2)r_2^2\omega^2$$

$$z_1 - z_2 = (\rho/2\gamma)(\omega^2)(r_1^2 - r_2^2)$$

$$= ((\gamma/g)/(2\gamma))\omega^2(r_1^2 - r_2^2)$$

$$= (\omega^2/(2g))(.4^2 - .2^2)$$

$$= (4^2/(2g))(0.12)$$

$$= 0.0978\text{m} = 9.79\text{cm}$$

1 2

Because of the different tube sizes a given increase in elevation in tube (1) will be accompanied by a fourfold decrease in elevation in tube (2). Then $z_1 - z_2 = 5\Delta z$ where Δz = increase in elevation in (1)

$$\Delta z = 9.79\text{cm}/5 = 1.96\text{cm or } \underline{\underline{z_1 = 21.96\text{cm}}}$$

Decrease in elevation of liquid in small tube = $4\Delta z$ = 7.83cm. Final elevation in small tube = 20cm - 7.83cm = $\underline{12.17\text{cm.}}$

5.32

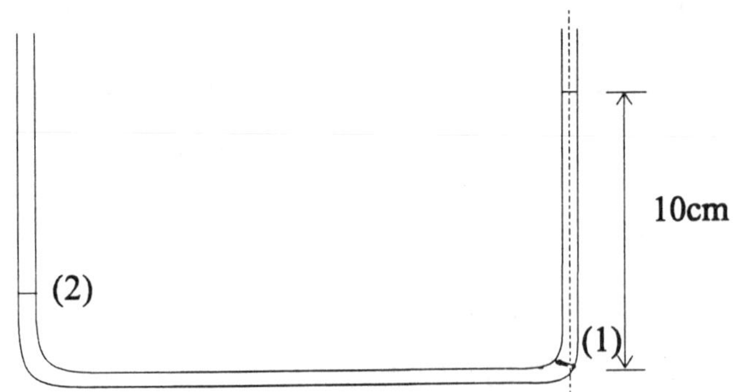

p_1 = (0.10m) (γ_{H_2O}) because of hydrostatic pressure distribution in the vertical direction (no acceleration).

5.32 (Cont'd)

Write Eq. 5.8 between pts. (1) and (2);

$$p_1 + \gamma z_1 - \rho r_1^2 \omega^2 / 2 = p_2 + \gamma z_2 - \rho r_2^2 \omega^2 / 2$$

$$\text{where } p_2 = 0, z_2 - z_1 = 0.01\text{m}, r_1 = 0 \text{ and } r_2 = 1\text{m}$$

$$\text{Then } 0.1\gamma_{H_2O} + 0 + 0 = 0 + \gamma_{Hg} \times 0.01 - (\gamma_{Hg} / g) \times 1^2 \omega^2 / 2$$

$$\omega^2 = ((2g)(.01\gamma_{Hg} - 0.1\gamma_{H_2O})) / \gamma_{Hg}$$

$$\omega = (2 \times 9.81)(.01 - (0.1 / 13.6))$$

$$= \underline{\underline{0.228 \text{rad} / \text{s}}}$$

5.33 Let leg 1 be the leg on the axis of rotation. Let leg 2 be the other leg of the manometer. Using Eq. 5.8 written between the liquid surfaces of 1 & 2 we have

$$p_1 + \gamma z_1 - \rho r_1^2 \omega^2 / 2 = p_2 + \gamma z_2 - \rho r_2^2 \omega^2 / 2$$

$$0 + \gamma z_1 - 0 = \gamma z_2 - (\gamma / g) r_2^2 \omega^2 / 2$$

$$\omega^2 r_2^2 / (2g) = z_2 - z_1$$

$$a_n = r\omega^2$$

$$= (z_2 - z_1)(2g) / r$$

$$= (0.2)(2g) / r_2$$

$$= (0.2)(2g) / 0.1$$

$$\underline{\underline{a_n}} = \underline{\underline{4g}}$$

5.34 Write Eq. 5.8 from the liquid surface to point A. Call the liquid surface point 1.

$$p_1 + \gamma z_1 - \rho r_1^2 \omega / 2 = p_A + \gamma z_A - \rho r_A^2 \omega^2 / 2$$

$$p_A = p_1 + (\rho \omega^2 / 2)(r_A^2 - r_1^2) + \gamma(z_1 - z_A)$$

However $\gamma(z_1 + z_A) = 0$ in zero -g environment. Thus

$$p_A = p_1 + ((800\text{kg}/\text{m}^3)/2)(6\pi/60\text{rad}/\text{s})^2(1.5^2 - 1^2)$$

$$= 100\text{Pa} + 49.3\text{Pa}$$

$$= \underline{\underline{149.3\text{Pa}}}$$

5.35 Start with Eq. (5.8)

$$p_1 + \gamma z_1 - \rho r_1^2 \omega^2 / 2 = p_2 + \gamma z_2 - \rho r_2^2 \omega / 2$$

Let point 1 be at the liquid surface in the large tube and point 2 be at the liquid surface in the small tube. Then $p_1 = p_2 = 0$ gage, $r_1 = 0$, $r_2 = \ell$, and $z_2 - z_1 = 3.75\ \ell$.

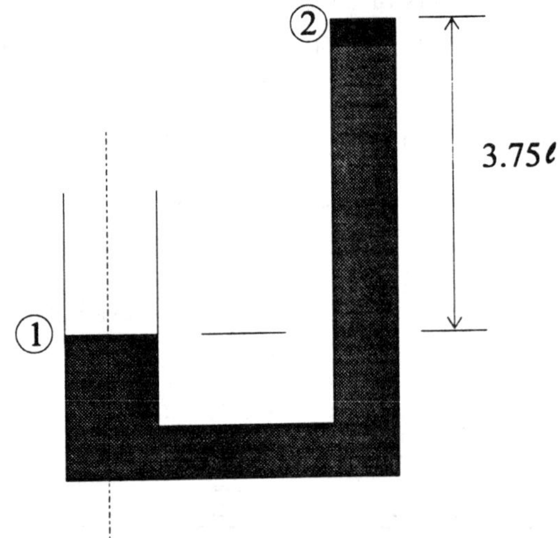

$$\rho r_2^2 \omega^2 / 2 = \gamma(3.75\ell)$$

$$(\gamma / (2g))(\ell^2)\omega^2 = 3.75\gamma\ell$$

$$\omega^2 = \frac{7.5g}{\ell}$$

$$\underline{\underline{\omega = \sqrt{7.5g/\ell}}}$$

5.36 Write Eq. (5.8) from the liquid surface in the small tube to the liquid surface in the large tube。

$$p_s + \gamma z_s - \rho r_s^2 \omega^2 / 2 = p_L + \gamma z_L - \rho r_L^2 \omega^2 / 2$$

But $p_s = p_L$ and $r_s = 0.5\ell$, $r_L = 1.5\ell$

Then

$$(\rho / 2)\omega^2 (r_L^2 - r_s^2) = \gamma(z_L - z_s)$$

$$(\gamma / 2g)\omega^2 (1.5^2 \ell^2 - 0.5^2 \ell^2) = \gamma(2\ell)$$

$$\omega^2 = 2g / \ell$$

$$\omega = \sqrt{2g / (2/3)}$$

$$= \underline{\underline{9.829 \text{rad/s}}}$$

Change in volume of Hg in small tube is same as in large tube.

Or

$$\forall_s = \forall_L$$

$$\Delta z_s \pi d^2 / 4 = \Delta z_L \pi (2d)^2 / 4$$

$$\Delta z_s = 4 \Delta z_L$$

Also

$$\Delta z_s + \Delta z_L = 2\ell$$

$$4\Delta z_L + \Delta z_L = 2 \times 0.667 \text{ft}$$

$$\Delta z_L = 1.33 \text{ft} / 5 = 0.267 \text{ft}$$

Mercury level in large tube will drop 0.267 ft from its original level.

5.37 Write Eq. (5.8) from the open end of the tube to the closed end.

$$p_1 = \gamma z_1 - \rho r_1^2 \omega^2 / 2 = p_2 + \gamma z_2 - \rho r_2^2 \omega^2 / 2$$

where $z_1 = z_2$. Also let point 2 be at the closed end; therefore, r_1=0 and r_2=0.40m

$$p_2 = (\rho / 2)(0.40^2)(60.8)^2$$

$$= 500 \times 0.16 \times 3697$$

$$= 295.73 \text{kPa}$$

$$\text{Then } F = p_2 A = 295{,}730 \times (\pi / 4)(.01)^2$$

$$= \underline{\underline{23.2 \text{N}}}$$

5.38 Write Eq. (5.8) from the mercury surface in the left tube to the mercury surface in the right one. The $p_\ell = p_r$ and

$$\gamma z_\ell - \rho r_\ell^2 \omega^2 / 2 = \gamma z_r - \rho r_r^2 \omega^2 / 2$$

$$\omega^2 (\gamma / 2g)(r_r^2 - r_\ell^2) = \gamma(z_r - z_\ell)$$

$$\omega^2 = 2g(z_r - z_\ell) / (r_r^2 - r_\ell^2)$$

$$= 2g(\ell) / (g\ell^2 - \ell^2)$$

$$\omega = \sqrt{g / (4\ell)}$$

5.39 (a) rotation at 5rad/s

Write Eq. (5.8) between the water surface in the left leg and the water surface in the right leg. At these surfaces $p_\ell = p_r = 0$ gage; therefore, Eq. (5.8) is

$$\gamma z_\ell - \rho r_\ell^2 \omega^2 / 2 = \gamma z_r - \rho r_r^2 \omega^2 / 2 \quad (1)$$

$$z_\ell - z_r = -r_r^2 \omega^2 / 2g = -25\ell^2 / 2g$$

Also
$$z_\ell + z_r = 1.4\ell \quad (2)$$

Solving Eqs. (1) and (2) for $\ell = 0.30$m yields

$$\underline{\underline{z_\ell = 15.3\text{cm}}} \text{ and } \underline{\underline{z_r = 26.7\text{cm}}}$$

(b) rotation at 15rad/s

$$\gamma z_1 - \rho r_1^2 \omega^2 / 2 = \gamma z_r - \rho r_r^2 \omega^2 / 2$$

where
$$r_1 = d \text{ and } r_r = \ell$$

Then
$$z_1 - z_r = (\omega^2 / 2g)(d^2 - \ell^2) \quad (3)$$

Also
$$z_r - z_1 = 1.4\ell + d \quad (4)$$

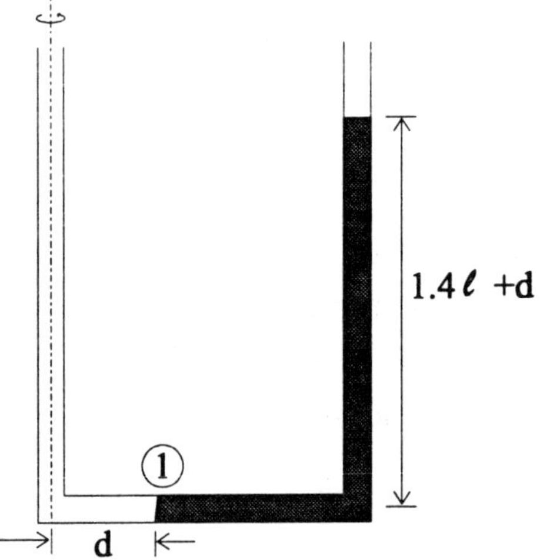

Solving Eqs. (3) and (4) for $\ell = 0.30$ yields d = 0.1915m

Thus $z_r - z_\ell = 1.4\ \ell + d = 1.4 \times 0.3 + 0.1915 = \underline{0.612\text{m}}$

5.40 $$p_A + \gamma z_A - \rho r_A^2 \omega^2 / 2 = p_B + \gamma z_B - \rho r_B^2 \omega^2 / 2$$

where $p_B = 0\text{gage}, r_A = 0,\ r_B = 0.64\text{m},\ z_B - z_A = 0.32\text{m}$

Then

$$p_A = 0.32\gamma - (\gamma / g) \times 0.64^2 \times 8^2 / 2$$

$$p_A = \gamma(0.32 - 0.64^2 \times 8^2 / (2g))$$

$$= 2 \times 9{,}810(.32 - 1.336)$$

$$= -19{,}936\text{Pa}$$

$$= \underline{\underline{-19.94\text{kPa}}}$$

$$p_B = 0.32 \times 2 \times 9{,}810 = \underline{\underline{6.278\text{kPa}}}$$

Now for $\omega = 20\text{rad/s}$ solve for p_A as above

$$p_A = \gamma(0.32 - 0.64^2 \times 20^2 / (2g))$$

$$= 2 \times 9{,}810(-8.031) = -157{,}560\text{Pa};$$

not possible. The liquid will vaporize. Therefore, $p_A = p_V = -101\text{kPa}$ assuming $p_V = 0$ abs.
To get p_B visualize the liquid as shown in Fig.

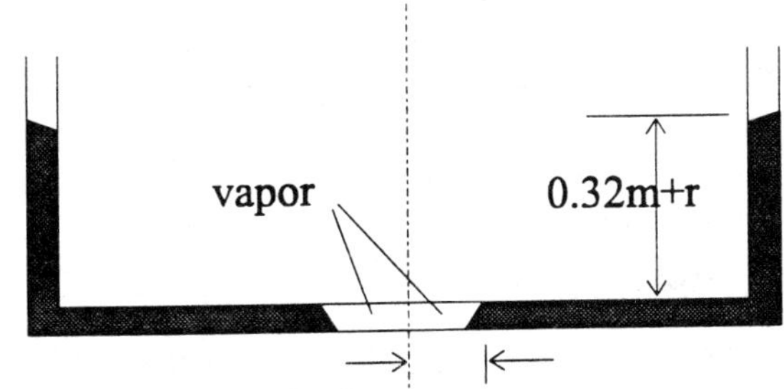

Now $p_r - \rho r^2 \omega^2 / 2 = p_B - \rho \times 0.64^2 \omega^2 / 2$ where $p_r = p_V = -101\text{kPa}$

Then

$$-101{,}000 - \rho r^2 \omega^2 / 2 = \gamma(0.32 + r) - \rho \times 0.64^2 \omega^2 / 2$$

$$-101{,}000 - 1000r^2 \times 20^2 = 2 \times 9{,}810(.32 + r) - 1000 \times 0.64^2 \times 20^2 \quad (1)$$

Solving Eq. (1) for r yields r = 0.352m

Therefore, $p_B = (0.32 + 0.352) \times 2 \times 9{,}81 = 13{,}184\ \text{Pa} = \underline{13.18\text{kPa}}$

5.41 When the water is one the verge of spilling from the open tube, the air volume in the closed part of the tube will have doubled. Therefore, get the pressure in the air volume with this condition.

$$p_i \forall_i = p_f \forall_f$$

where i and f refer to initial and final conditions

$$p_f = p_i \forall_i / \forall_f = 101\text{kPa} \times \tfrac{1}{2}$$

$$p_f = 50.5\text{kPa, abs} = -50.5\text{kPa, gage}$$

Now write Eq. 5.8 between water surface in leg A-A to water surface in open leg after rotation.

$$p_A + \gamma z_A - \rho r_A^2 \omega^2 / 2 = p_{open} + \gamma z_{open} - \rho r_{open}^2 \omega^2 / 2$$

$$-50.5 \times 10^3 + 0 - 0 = 0 + \gamma \times 6\ell - (\gamma / g)(6\ell)^2 \omega^2 / 2$$

$$+50.5 \times 10^3 / 9{,}810 = -6\ell + (36\ell^2 / (2 \times 9.81))\omega^2$$

$$\omega^2 = (+101 + 11.772)/.36 = 313.3$$

$$\underline{\underline{\omega = 17.7\text{rad / s}}}$$

5.42 Write equation for rotational flow from point 1 in vertical pipe at level of water to point 2 at the outer edge of the rotating disk.

$$p_1 + \gamma z_1 - \rho r_1^2 \omega^2 / 2 = p_2 + \gamma z_2 - \rho r_2^2 \omega^2 / 2$$

$$0 + 0 - 0 = 0 + \gamma z_2 - (\gamma / g)(.05^2)\omega^2 / 2$$

$$0 = z_2 - .05^2 \omega^2 / 2g$$

But

$$\omega = (2000\,\text{rev / min})(1\,\text{min./60s})(2\pi\text{rad / rev}) = 209.44\text{rad / s}$$

$$z_2 = ((0.05)(209.44))^2 / (2 \times 9.81)$$

$$\underline{\underline{z_2 = z = 5.59\text{m}}}$$

5.43 $\partial p/\partial r + \gamma(\partial z/\partial r) = -\rho r\omega^2$

$\partial p/\partial z = -\gamma - \rho r\omega^2$

when $z = -1$ m

$\partial p/\partial z = -\gamma - \rho\omega^2 = -\gamma - 25\rho = -\gamma(1 + 25/g)$

$= -9{,}810\ (1 + 2.548) = \underline{\underline{-34.8\text{ kPa/m}}}$

when $z = +1$ m

$\partial p/\partial z = -\gamma + 25\rho = -9{,}810\ (1 - 2.548) = \underline{\underline{15.186\text{ kPa/m}}}$

$(\partial p/\partial z)_0 = \underline{\underline{-9.810\text{ kPa/m}}}$

5.44 Below the axis both gravity and acceleration cause pressure to increase with decrease in elevation; therefore, the maximum pressure will occur at the bottom of the cylinder. Above the axis the pressure initially decreases with elevation (due to gravity); however, this is counteracted by acceleration due to rotation. Where these two effects completely counter-balance each other is where the minimum pressure will occur ($\partial p/\partial z = 0$). Thus, above the axis:

$$\partial p/\partial z = 0 = -\gamma + r\omega^2\rho \quad \text{minimum pressure condition}$$

Solving: $\underline{\underline{r = \gamma/\rho\omega^2}}$; $\underline{\underline{p_{min} \text{ occurs at } z = +\gamma/\rho\omega^2}}$

Then $p_{max} - p_{min} = \Delta p_{max} = (\rho\omega^2/2)(r_0^2 - r^2) + \gamma(r_0 + r)$

where r_0 = radius of cylinder

$r = (\gamma/\rho\omega^2)$ = radius to point of minimum pressure

$$\Delta p_{max} = (\rho\omega^2/2)[r_0^2 - (\gamma/\rho\omega^2)^2] + \gamma[r_0 + (\gamma/\rho\omega^2)]$$

$$\underline{\underline{\Delta p_{max} = \rho\omega^2 r_0^2/2 + \gamma r_0 + \gamma^2/(2\rho\omega^2)}}$$

5.45 From solution to Prob. 5.44

p_{min} occurs at $z = \gamma/\rho\omega^2 = g/\omega^2$ where $\omega = (20 \text{ ft/s})/2.5 \text{ ft} = 8 \text{ rad/s}$

Then $z_{min} = 32.2/(8)^2 = \underline{0.503 \text{ ft}}$ above axis

$$\Delta p_{max} = 1.94 \times 8^2 \times 2.5^2/2 + 62.4 \times 2.5 + 62.4^2/(2 \times 1.94 \times 8^2)$$

$$\Delta p_{max} = \underline{\underline{560 \text{ lbs/ft}^2}}$$

5.46 $(p_1/\gamma) + (V_1^2/2g) + z_1 = (p_2/\gamma) + (V_2^2/2g) + z_2$

$p_1 = p_2 + \gamma[(V_2^2 - V_1^2)/2g + z_2 - z_1] = 0$

$p_2 = ((-256/(2 \times 32.2)) + 15) \times 62.4 = 688 \text{ psf} = \underline{\underline{4.78 \text{ psi}}}$

5.47 $p_2 = \gamma(z_1 - V_2^2/2g) = 9{,}810(15 - 36/2 \times 9.81) = 129{,}150 \text{ Pa}$

$= \underline{129.15 \text{ kPa}}$

5.48 First write the continuity equation between the duct sections:

$$V_cA_c = V_sA_s$$
$$100(\pi D^2/4) = V_sD^2$$
$$V_s = 100(\pi/4)$$
$$= 78.54 \text{ ft/s}$$

Now write the Bernoulli equation between the two sections

$$p_c + \rho V_c^2/2 + z_c = p_s + \rho V_s^2/2 + z_s$$
$$\rho = \gamma/g = 0.075/32.2$$
$$= 0.00233 \text{ slugs/ft}^3$$

Then $$p_c - p_s = (\rho/2)(V_s^2 - V_c^2)$$
$$= (0.00233/2)(78.54^2 - 100^2)$$
$$p_c - p_s = -4.46 \text{ lbf/ft}^2$$

5.49 Let section 1 be in the large duct where the manometer pipe is connected and section 2 in the smaller duct at the level where the upper manometer pipe is connected. Assume uniform air density.

Continuity equation:

$$V_1A_1 = V_2A_2$$
$$V_2 = V_1(A_1/A_2)$$
$$= 100(2)$$
$$= 200 \text{ ft/s}$$

Write Bernoulli's equation from 1 to 2

$$p_1 + \rho V_1^2/2 = p_2 + \rho V_2^2/2$$
$$p_1 - p_2 = (1/2)\rho(V_2^2 - V_1^2)$$
$$= (1/2)(0.0644/32.2)(40{,}000 - 10{,}000)$$
$$= 30 \text{ psf}$$

Manometer deflection:

$$p_1 - p_2 = \Delta h(\gamma_{\text{liquid}} - \gamma_{\text{air}})$$
$$30 = \Delta h(100 - .0644)$$
$$\underline{\underline{\Delta h = 0.300 \text{ ft}}}$$

5.50 $p_A - p_B = \gamma[(V_B^2 - V_A^2)/2g - z_A] = 62.4\ [(400 - 64)/2 \times 32.2 - 1] = \underline{\underline{263.2\ \text{psf}}}$

5.51 Write the Bernoulli equation from the nozzle to the top of the jet. Let point 1 be in the jet at the nozzle and point 2 at the top.

$$p_1/\gamma + V_1^2/2g + z_1 = p_2/\gamma + V_2^2/2g + z_2$$

$$\text{where} \quad p_1 = p_2 = 0 \text{ gage}$$

$$V_1 = 20\text{ft}/\text{s}$$

$$V_2 = 0$$

$$0 + (20)^2/2g + z_1 = 0 + 0 + z_2$$

$$z_2 - z_1 = h = 400/64.4$$

$$\underline{\underline{= 6.21\text{ft}}}$$

5.52 $V_1A_1 = V_2A_2 \quad ; \quad V_2 = V_1(D/d)^2 = 5 \times (4/2)^2 = 20$ ft/s

$p_1/\gamma + V_1^2/2g = V_2^2/2g$

$p_1 = \gamma(V_2^2/2g - V_1^2/2g) = 363$ psf

Then $F_{piston} = p_1A_1 = 363\ (\pi/4) \times (4/12)^2 = \underline{\underline{31.7\ \text{lb}}}$

5.53 $p_1/\gamma + V_1^2/2g + z_1 = p_j/\gamma + V_j^2/2g + z_j$

where 1 and j refer to conditions in pipe and jet, respectively

$V_1 = Q/A_1$

$= 20/((\pi/4) \times 1.0^2) = 25.5$ ft/s

$V_jA_j = V_1A_1 \quad ; \quad V_j = V_1A_1/A_j$

$V_j = 25.5 \times 4 = 101.9$ ft/s

Also $z_1 = z_j$ and $p_j = 0$

Then $p_1/\gamma = (V_j^2 - V_1^2)2g$

$p_1 = \gamma(V_j^2 - V_1^2)/2g$

$= 62.4\ (101.9^2 - 25.5^2)/64.4$

$= \underline{9,423\ \text{psf}}$

$= \underline{65.4\ \text{psi}}$

5.54 $p_0 + \rho V_0^2/2 = p_x + \rho V_x^2/2$

$p_0 = 0$ gage

$p_x = (\rho/2)(V_0^2 - V_x^2)$

$V_x = u = U_0(1 - r_0^3/x^3)$

Then $V_{x=r_0} = U_0(1-1) = 0$

$V_{x=1.1r_0} = U_0(1-1/1.1^3) = 7.46$ m/s

$V_{x=2r_0} = U_0(1-1/2^3) = 26.25$ m/s

Finally $p_{x=r_0} = (1.2/2)(30^2-0) = $ 540 Pa, gage

$p_{x=1.1r_0} = (1.2/2)(30^2-7.46^2) = $ 507 Pa,gage

$p_{x=2r_0} = (1.2/2)(30^2-26.25^2) = $ 127 Pa,gage

5.55 $V = K/r$

$Q = \int VdA = \int VLdr = L\int(K/r)dr = KL\ell n\ (r_2/r_1)$ (1)

$\Delta p = (1/2)\rho(V_1^2 - V_2^2)$

$\Delta p = (1/2)\rho\ ((K^2/r_1^2) - (K^2/r_2^2)) = (K^2\rho/2)\ ((r_2^2) - (r_1^2))/(r_1^2 r_2^2)$ (2)

Eliminate K between Eqs. (1) and (2) yielding:

$(2\Delta p/\rho) = ((Q^2)/(L^2\ (\ell n\ (r_2/r_1))^2))\ (r_2^2 - r_1^2)/(r_1^2 r_2^2)$

$A_c = L\ (r_2 - r_1)$

$\therefore\ 2\Delta p/\rho = (Q^2/A_c^2)\ (r_2 - r_1)^2\ (r_2^2 - r_1^2)/(r_1^2 r_2^2(\ell n\ (r_2/r_1))^2)$

$Q = A_c\sqrt{2\Delta p/\rho}\ \ (r_1 r_2\ \ell n\ (r_2/r_1))/((r_2 - r_1)\ (r_2^2 - r_1^2)^{0.5})$

$Q = A_c\sqrt{2\Delta p/\rho}\ \ ((r_2/r_1)\ell n\ (r_2/r_1))/((r_2/r_1 - 1)\ ((r_2^2/r_1^2) - 1)^{0.5})$

For $r_2/r_1 = 1.5$ the $f(r_2/r_1)$ is evaluated:

$f(r_2/r_1) = 1.5\ \ell n\ 1.5/(0.5\ (1.25^{0.5}) = 1.088$

5.56 Write Bernoulli's equation between the outside of the bend at the surface (point 2) and the inside of the bend at the surface (point 1):

$$(p_2/\gamma) + V_2^2/2g + z_2 = (p_1/\gamma) + V_1^2/2g + z_1$$

$$0 + V_2^2/2g + z_2 = 0 + V_1^2/2g + z_1$$

$$z_2 - z_1 = V_1^2/2g - V_2^2/2g$$

where $V_2 = (1/2)\text{m/s}$; $V_1 = (1/1)\text{m/s}$

Then $z_2 - z_1 = (1/2g)(1^2 - 0.5^2) = \underline{0.038 \text{ m}}$

5.57 $V_A = Q/A_A = 70/((\pi/4) \times 6^2) = 2.476$ ft/s

$V_B = Q/A_B = 70/((\pi/4) \times 2^2) = 22.28$ ft/s

$$p_A/\gamma + V_A^2/2g + z_A = p_B/\gamma + V_B^2/2g + z_B$$

$$p_B/\gamma = 3{,}500/62.4 + 2.48^2/64.4 - 22.28^2/64.4 - 4$$

$p_B = 2{,}775 \text{ lb/ft}^2 = \underline{19.3 \text{ lb/in}^2}$

5.58 $p_C/\gamma + V_C^2/(2g) + z_C = p_E/\gamma + V_E^2/(2g) + z_E$

$$(10 \times 144)/\gamma + 10^2/(2g) + z_c = p_E/\gamma + 50^2/(2g) + z_E$$

$$p_E/\gamma = ((10 \times 144)/\gamma) + (1/2g)(10^2 - 50^2) + z_C - z_E$$

$$p_E = 10 \times 144 + ((62.4)/(64.4)(-2400)) + 62.4(3-1)$$

$= 1440$ psf $- 2325$ psf $+ 124.8$ psf

$p_E = \underline{\underline{-760 \text{ psf}}} = -5.28$ psi

5.59 Write Bernoulli's equation from a point at radius r to the outlet:

$$p_r + \rho V_r^2/2 = 0 + \rho V_{outlet}^2/2$$

$$V = Q/A$$

$$= Q/(2\pi rh)$$

$$= Q/(2\pi \times 0.01r)$$

$$= (0.380\ m^3/s)/(0.02\pi rm^2)$$

$$= (6.048/r)\ m/s$$

$$V_{outlet} = 6.048/0.25 = 24.19\ m/s$$

Then

$$p_r = (\rho/2)(24.19^2 - V_r^2)$$

$$p_r = (\rho/2)(24.19^2 - (6.048/r)^2)$$

$$p_r = (1.2/2)(585.2 - (36.58)/r^2)$$

p_r vs. r given in table below:

r (m)	0.06	0.10	0.15	0.20	0.25
p_r(Pa)	-5,745	-1,844	-624	-198	0

Because the center of the disk is a stagnation point, the pressure there will be $\rho V_{outlet}^2/2$ or +351 Pa. The pressure variation on the disk is plotted below:

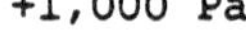

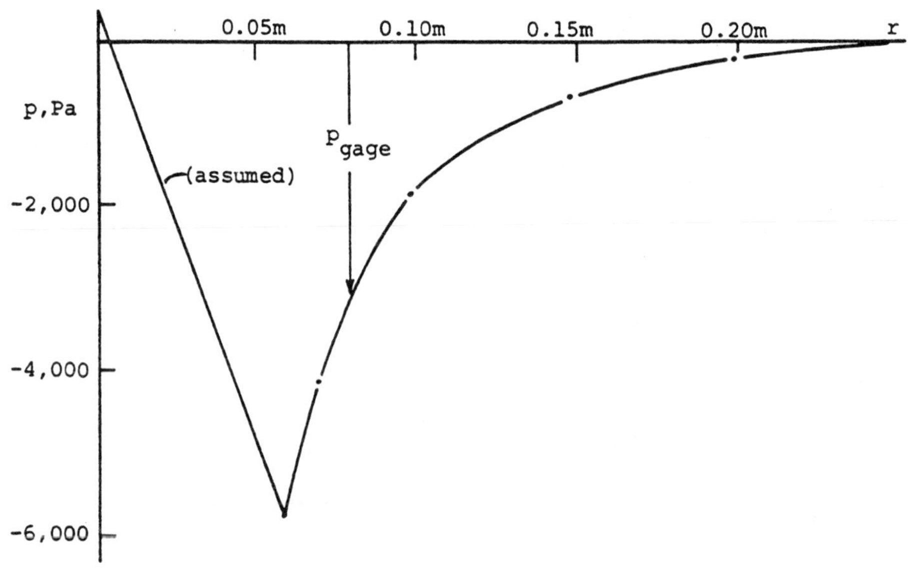

5.59 (continued)

The force on the disk will be $\int p dA = \int p 2\pi r dr$. So plot $p \times (2\pi r)$ vs. r. The area under the curve will be desired force. This assumes that zero gage pressure prevails on the other side of the disk.

r,m	0	0.04m	0.06	0.08	0.10	0.14	0.16	0.20	0.25
p,Pa	351	-3,500	-5,745	-2,800	-1,844	-810	-560	-198	0
$2\pi rp$	0	- 880	-2,166	-1,407	-1,159	-712	-563	-249	0

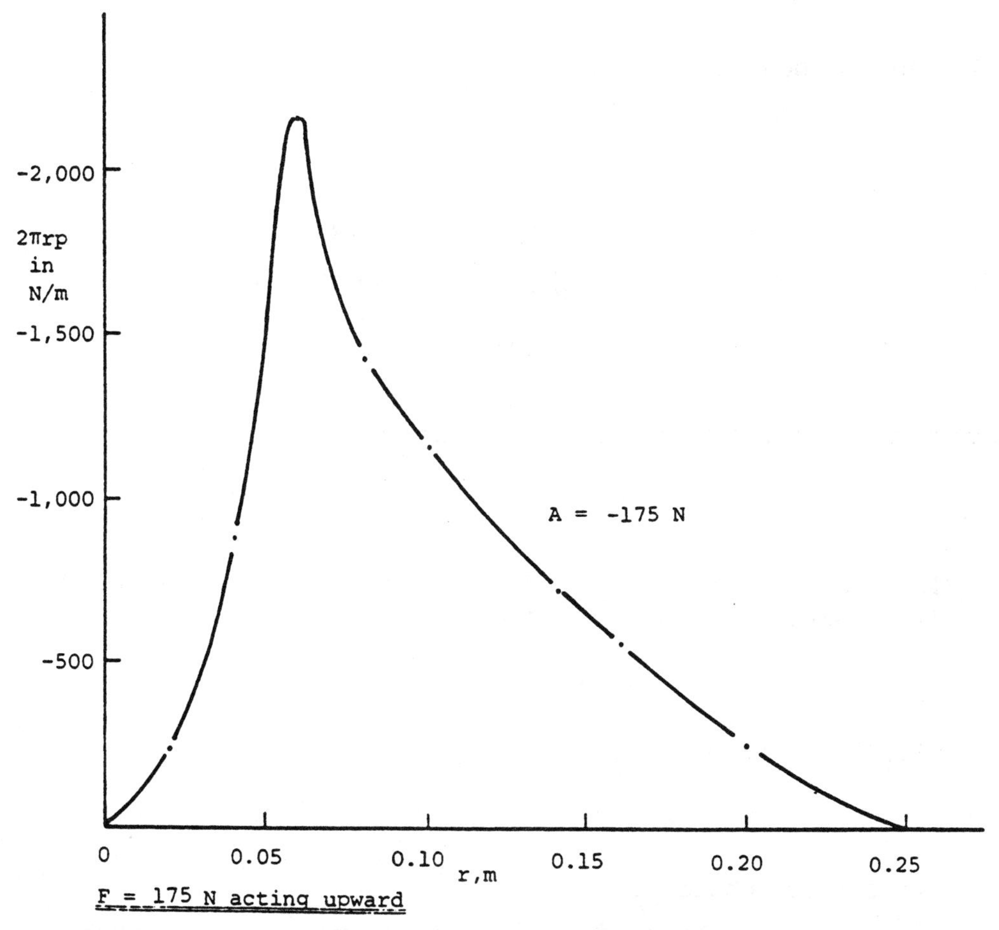

F = 175 N acting upward

5.60

$$C_p = 1 - 4 \sin^2 \theta$$

$$C_{p,50} = 1 - 4 \sin^2 50^\circ$$

$$= 1 - 4\ (0.766)^2 = -1.347$$

$$C_{p,10} = 1 - 4\ (0.174)^2 = +0.879$$

$$C_{p,10} - C_{p,50} = 0.879 - (-1.347) = 2.226$$

$$\Delta p = \gamma_{H_2O}\ \Delta h = \Delta C_p\ \rho_{air}\ V_0^2/2$$

$$\Delta h = 2.226\ (1.2\ kg/m^3)\ (50\ m/s)^2/(2 \times 9{,}810\ N/m^3)$$

$$\Delta h = 0.340\ m = \underline{\underline{34.0\ cm}}$$

5.61 The maximum pressure will occur at the point of least velocity which will be at the stagnation point where V = 0. The point of lowest pressure will be where the velocity is highest (V = 60m/s).

Write the Bernoulli equation between points of highest and lowest pressure and assume hydrostatic effects are negligible.

$$p_h + \rho V_h^2/2 = p_\ell + \rho V_\ell^2/2$$

$$p_h + 0 = p_\ell + (\rho/2)(60^2)$$

$$p_h - p_\ell = (1.2/2)(3600)$$

$$= 2{,}160 Pa$$

$$\underline{\underline{= 2.160 kPa}}$$

5.62 The flow is non-uniform. Check to see if it is irrotational by seeing if it satisfies Bernoulli's equation.

$$p_1/\gamma + V_1^2/2g + z_1 = p_2/\gamma + V_2^2/2g + z_2$$

$$(10{,}000/9{,}810) + (1/(2 \times 9.81)) + 0 = (7{,}000/9{,}810) + 2^2/(2 \times 9.81)$$

$$1.070 \neq 0.917$$; Flow is rotational

The correct choice is c.

5.63 $\frac{1}{2}\rho V^2 = \Delta p = \gamma_{H2O}$ (9/12)

$\rho = p/(RT) = (10)(144)/((1716)(483)) = 0.00173$ slugs/ft^3

$V^2 = 2(62.4$ lbf/ft$^3)$ (9/12) ft/(0.00173 slugs/ft^3)

$= 54{,}104$ ft^2/s^2

$\underline{\underline{V = 233 \text{ ft/sec}}}$

5.64

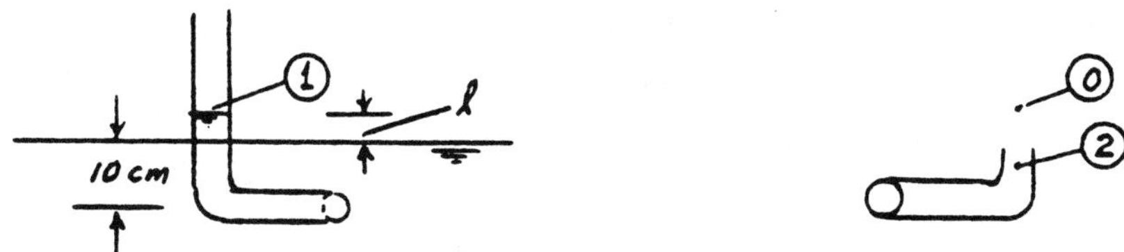

Write rotational flow equation from pt. 1 to pt. 2 where pt. 1 is at liquid surface in vertical part of tube and pt. 2 is just inside the open end of the pitot tube.

$$p_1/\gamma - V_1^2/2g + z_1 = p_2/\gamma - V_2^2/2g + z_2$$

$$0 - 0 + (0.10 + \ell) = p_2/\gamma - r^2\omega^2/2g - 0 \qquad (1)$$

If we reference the velocity of the liquid to the tip of the pitot tube then we have steady flow and Bernoulli's equation will apply from pt. 0 (point ahead of the pitot tube) to point 2 (point at tip of pitot tube).

$$p_0/\gamma - V_0^2/2g + \not{z}_0 = p_2/\gamma - V_2^2/2g + \not{z}_2$$

$$0.1\gamma/\gamma + r^2\omega^2/2g = p_2/\gamma + 0 \qquad (2)$$

solve Eqs. (1) & (2) for ℓ: $\underline{\underline{\ell = 0}}$ Liquid surface in the tube is the same as elevation as outside liquid surface.

5.65 $V = (2 \times 32.2 \times 8/12)^{1/2} = \underline{\underline{6.55 \text{ fps}}}$

5.66 $V = \sqrt{2gh}$

$\underline{\underline{h = V^2/2g = 3^2/(2 \times 9.81) = 0.459 \text{ m} = 45.9 \text{ cm}}}$

5.67 Because it is a Bourdon tube gage, the difference in pressure that is sensed will be between the stagnation point and the separation zone downstream of the plate.

Therefore $\Delta C_p = 1 - (C_{p,\text{back of plate}})$

$\Delta C_p = 1 - (\text{neg. number})$

$\therefore \quad \Delta p/(\rho V_0^2/2) = 1 + \text{positive number}$

$\Delta p = (\rho V_0^2/2)\ (1 + \text{positive number})$

case (c) is the correct choice.

5.68 $\rho = p/RT = 15 \times 144/(1{,}715)(60 + 460) = 0.00242$ slugs/ft

$V = (2\Delta p/\rho)^{1/2} = (2 \times 62.4 \times (1.8/12)/0.00242)^{1/2} = \underline{87.95 \text{ fps}}$

5.69 $$p_1/\gamma + V_1^2/2g = p_2/\gamma + V_2^2/2g = p_t/\gamma \qquad \text{(Bernoulli Eq.)}$$

neglect

$p_1 + (0.10)\ (9{,}810) - (0.10)\ (1.2)\ (9.81) = p_T$

$p_T - p_1 = 981\ \text{N/m}^2 = V_1^2\rho/2$

$V_1^2 = 2 \times 981/1.2 \quad ; \quad V_1 = 40.4$ m/s

$V_2A_2 = V_1A_1 \quad ; \quad V_2 = V_1\ (A_2/A_1) = 2V_1 = \underline{80.8 \text{ m/s}}$

5.70 Let point 1 be the stagnation point and point 2 at 90° around the sphere. Write Bernoulli's equation between the two points.

$p_1 + \rho V_1^2/2 = p_2 + \rho V_2^2/2$

$p_1 + 0 = p_2 + \rho(1.5\ V_0)^2/2 \quad ; \quad p_1 - p_2 = 1.125\ \rho V_0^2$

$V_0^2 = 2{,}000/(1.125 \times 1{,}000) = 1.78$

$\underline{V_0 = 1.33 \text{ m/s}}$

5.71 Write Bernoulli's equation between the stagnation point and the second pressure tap. Assume $z_1 = z_2$:

$$p_1 + \rho V_1^2/2 = p_2 + \rho V_2^2/2$$

$$p_1 - p_2 = \rho V_2^2/2$$

$$= (1.5\ V_0 \sin\theta)^2(\rho/2)$$

$$p_1 - p_2 = \Delta p = 2.25\ V_0^2 \sin^2\theta\ \ (\rho/2) \qquad (1)$$

Given: $V_0 = \sqrt{\Delta p/\rho}\ ;\ V_0^2 = \Delta p/\rho$ (2)

Solve Eqs. (1) & (2) for θ

$\underline{\theta = 70.53°}$

5.72 Check minimum pressure value. Minimum pressure will occur where streamlines have smallest spacing (inside of bend).

Thus: $p_{min} + \rho V_m^2/2 = p_1 + \rho V_1^2/2;\ \ V_{min} \times n_{min} = V_2 n_2$

where n is streamline spacing

$V_{min} = V_1 \times n_1/n_{min} = V_1 \times 2.6/1.3$, scaled from Fig.; $V_{min} = 2V_1$

$p_{min} = p_1 + (\rho/2)(-3V_1^2) = 110{,}000 + 500\ (-3 \times 13^2) = \underline{-143\text{ kPa abs}}$

p_{min} is less than p_{vapor}; therefore, cavitation will occur.

5.73 Cavitation will occur when the pressure reaches the vapor pressure of the liquid (p_V = 1,230 Pa abs).

$$p_A + \rho V_A^2/2 = p_{throat} + \rho V_{throat}^2/2$$

where $V_A = Q/A_A = Q/((\pi/4) \times 0.40^2)$

$V_{throat} = Q/A_{throat} = Q/((\pi/4) \times 0.10^2)$

$$\rho/2\,(V_{throat}^2 - V_A^2) = p_A - p_{throat}$$

$$(\rho Q^2/2)\,[1/((\pi/4) \times 0.10^2)^2 - 1/((\pi/4) \times 0.40^2)^2]$$

$$= 200{,}000 - 1{,}230$$

$$500\ Q^2(16{,}211 - 63) = 198{,}770$$

$\underline{Q = 0.157\ m^3/s}$

5.74 $V = \sqrt{2g\Delta h} = \sqrt{2\Delta p/\rho}$

The Δp is the same for both; however,

$\rho_w >> \rho_a$. Therefore $V_A > V_W$ (case b).

5.75 $\Delta p = \Delta h(\gamma_{HG} - \gamma_{ker}) = (7/12)(847 - 51) = 464$ psf

$V = (2\Delta p/\rho)^{1/2} = (2 \times 464/1.58)^{1/2} = \underline{24.3 \text{ fps}}$

5.76 $V = (2\Delta p/\rho)^{1/2} = (2 \times 3{,}000/1.2)^{1/2} = \underline{70.7 \text{ m/s}}$

5.77 $V = (2 \times 11/0.00237)^{1/2} = \underline{96.3 \text{ fps}}$

5.78 $p_{stagn} - p_{static} = \rho V_0^2/2$

$0.9 \times 144 \text{ psf} = (0.1/32.2)\, V_0^2/2$

$\underline{V_0 = 289 \text{ ft/sec}}$

5.79

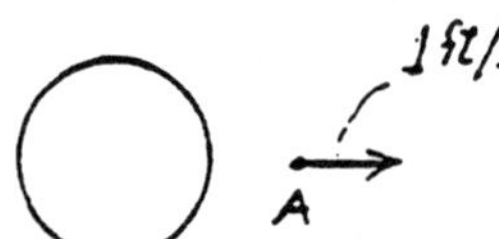

By referencing velocities to the sphere a steady flow case will be developed. Thus, for the steady flow case $V_0 = 11$ ft/s and $V_A = 10$ ft/s. Then when Bernoulli's equation is applied between points 0 and A it will be found that $\underline{p_A/p_0 > 1 \text{ (case c)}}$.

5.80 $p_B - p_C = (1{,}000/2)[(13-3)^2 - (13-5)^2] = 18{,}000 \text{ Pa} = \underline{18 \text{ kPa}}$

5.81 Bernoulli's equation can be applied if all velocities are relative to ship. Thus,

$V_{A,rel} = \sqrt{0.2^2 + 7^2} = 7.003$ m/s

$V_{B,rel} = 7.70$ m/s

$p_A + \rho V_A^2/2 = p_B + \rho V_B^2/2$

$p_A - p_B = (\rho/2)(V_B^2 - V_A^2)$

$= (1{,}050/2)(7.7^2 - 7.003^2)$

$= 5{,}380 \text{ Pa} = \underline{5.380 \text{ kPa}}$

5.82 Both gage A and B will read the same, due to hydrostatic pressure distribution in the vertical in both cases.

5.83 The side tube samples the pressure for the undisturbed flow and the central tube senses the stagnation pressure. Thus, we have

$$p_0 + \rho V_0^2/2 = p_{stagn.} + 0$$

or $$V_0 = \sqrt{(2/\rho)(p_{stagn.} - p_0)}$$

But $p_{stagn.} - p_0 = (0.067-0.023)\sin 30° \times 0.8 \times 9{,}810 = 172.7\text{ Pa}$

$$\rho = p/RT = 150{,}000/(287 \times (273 + 20)) = 1.784\text{ kg/m}^3$$

Then $$V_0 = \sqrt{(2/1.784)(172.7)} = \underline{\underline{13.92\text{ m/s}}}$$

$$Q = VA = 13.92 \times (\pi/4) \times 0.10^2 = \underline{\underline{0.1093\text{ m}^3/\text{s}}}$$

5.84 $$\Delta C_p = 1.4 = (p_A - p_B)/(\rho V_0^2/2)$$

$$V_0^2 = 2(4{,}000)/((1.5) \times 1.4);\quad V_0 = \underline{\underline{61.7\text{ m/s}}}$$

5.85 $$\rho = p/RT = 101{,}000/(200 \times (250 + 273)) = 0.966\text{ kg/m}^2$$

$$\Delta p = \gamma_{water}\Delta h = 9{,}790 \times 0.005 = 48.95\text{ Pa}$$

$$(p_A - p_B) = \Delta p = 48.95\text{ Pa} \qquad (1)$$

$$(p_A - p_0)/(\rho V_0^2/2) = 1.0 \leftarrow C_{p_A}$$

$$(p_B - p_0)/(\rho V_0^2/2) = -0.3 \leftarrow C_{p_B}$$

Then $(p_A - p_B)/(\rho V_0^2/2) = 1.3$ (2)

Solving Eq's (1) and (2) yields

$$\rho V_0^2/2 = 48.95/1.3;\quad V_0 = \underline{\underline{8.83\text{ m/s}}}$$

5.86 Let point 1 be at the stagnation point and point 2 be at the 90° position. At the 90° position U = 1.5 U sin Θ = 1.5 U. Write Bernoulli's equation between points 1 and 2.

$$p_1 + \rho V_1^2/2 = p_2 + \rho V_2^2/2 \quad (V_1 = 0)$$

$$p_1 - p_2 = \rho V_2^2/2$$

$$(\gamma_{Hg} - \gamma_{H_2O})\Delta h = (\rho/2)\,(1.5U)^2$$

$$((\gamma_{Hg}/\gamma_{H_2O}) - 1)\Delta h = (1/2g)\,(1.5U)^2$$

$$(13.6 - 1)\Delta h = (1/2g)\,(2.25)U^2$$

$$\underline{\underline{U = 2.34 \text{ m/s}}}$$

5.87

$$C_p = (p - p_o)/(\rho V^2/2)$$

Write the Bernoulli equation from the free stream to the point of separation:

$$p_0 = \rho U^2/2 = p + \rho u^2/2$$

$$p - p_o = (\rho/2)(U^2 - u^2)$$

or

$$(p - p_0)/(\rho U^2/2) = (1 - (u/U)^2)$$

but

$$u = 1.5U \sin\theta$$

or at separation point

$$u = 1.5U \sin 120^\circ$$

$$u = 1.5U \times 0.866$$

$$(u/U) = 1.299$$

$$(u/U)^2 = 1.687$$

$$C_p = 1 - 1.687$$

$$= \underline{\underline{-0.687}}$$

$$p_{gage} = C_p(\rho/2)U^2$$

$$= (-0.687)(1.2/2)(100^2)$$

$$= -4{,}122 \text{Pa}$$

$$= \underline{\underline{-4.122 \text{kPa}}}$$

5.88

$$u = 1.5U\sin\theta$$
$$u_{\theta=90^\circ} = 1.5U(1)$$
$$= 1.5U$$

Write Bernoulli equation between the stagnation point (forward tap) and the side tap where u = 1.5U. Neglect elevation difference.

$$p_1 + \rho V_1^2/2 = p_2 + \rho V_2^2/2$$
$$p_1 - p_2 = (\rho/2)(V_2^2 - V_1^2)$$
$$p_1 - p_2 = (1.2/2)((1.5U)^2 - 0)$$
$$120 = 1.35U^2$$
$$\underline{\underline{U = 9.43 m/s}}$$

5.89

$$V = K\sqrt{2\Delta p/\rho}$$

then $V_{calibr.} = (K/\sqrt{\rho_{calibr.}})\sqrt{2\Delta p}$

$$V_{true} = (K/\sqrt{\rho_{true}})\sqrt{2\Delta p} \qquad (1)$$

$$V_{indic.} = (K/\sqrt{\rho_{calibr.}})\sqrt{2\Delta p} \qquad (2)$$

divide Eq. (1) by Eq. (2):

$$V_{true}/V_{indic.} = \sqrt{\rho_{calib.}/\rho_{true}} = [(101/70) \times (273-5)/(273+17)]^{1/2} = 1.15$$

$$V_{true} = 60 \times 1.15 = 69.3 m/s$$

5.90 Assume the bottom of the tube through which water will be drawn is 5 in. below the neck of the atomizer. Therefore if the atomizer is to operate at all, the pressure in the necked down portion must be low enough to draw water 5 in. up the tube. In other words p_{neck} must be $-(5/12)\gamma_{water}$ = -26psfg. Let the outlet diameter of the atomizer be 0.5 in. and the neck diameter be 0.25 in. Assume that the change in area from neck to outlet is gradual enough to prevent separation so that the Bernoulli equation will be

5.90 (Cont'd)

valid between these sections. Thus

$$p_n + \rho V_n^2/2 = p_0 + \rho V_0^2/2$$

where n and 0 refer to the neck and outlet sections respectively. But

$$p_n = -26\text{psfg and } p_0 = 0$$

or

$$-26 + \rho V_n^2/2 = \rho V_0^2/2 \qquad (1)$$

$$\begin{aligned} V_n A_n &= V_0 A_0 \\ V_n &= V_0 A_0 / A_n \\ &= V_0 (.5/.25)^2 \\ V_n &= 4V_0 \end{aligned} \qquad (2)$$

Eliminate V_n between Eqs. (1) and (2)

$$\begin{aligned} -26 + \rho(4V_0)^2/2 &= \rho V_0^2/2 \\ -26 + 16\rho V_0^2/2 &= \rho V_0^2/2 \\ 15\rho V_0^2/2 &= 26 \\ V_0 &= ((52/15)/\rho)^{1/2} \end{aligned}$$

Assume $\rho = 0.0024$ slugs/ft^2

$$\begin{aligned} V_0 &= ((52/15)/0.0024)^{1/2} \\ &= 38\text{ft/s} \\ Q = VA &= 38 \times (\pi/4)(.5/12)^2 \\ &= .052\text{cfs} \\ &= 3.11\text{cfm} \end{aligned}$$

One could use a vacuum cleaner (one that you can hook the hose to the discharge end) to provide the air source for such an atomizer.

5.91 Venturi exit area, A_e, = $10^{-3} m^2$
Venturi throat area, $A_t = (1/4)\ A_e$
Suction cup area, A_s, = $0.1 m^2$

$$p_{atm} = 100 kPa$$
$$T_{water} = 15^\circ C$$

Find $V_{e\ max}$; Q_{max}, $Lift_{max}$

Write the Bernoulli equation for the Venturi from the throat to exit with the pressure at the throat equal to the vapor pressure of the water. This will establish the maximum lift condition. Cavitation would prevent any lower pressure from developing at the throat.

$$p_v / \gamma + V_t^2 / 2g + z_t = p_e / \gamma + V_{e\max}^2 / 2g + z_e$$
$$\text{at } T = 15^\circ C \ p_v = 1700 Pa \ \text{(from Table A.5)} \qquad (1)$$
$$\rho = 999 kg/m^3 \text{(from Table A.5)}$$

Continuity Equation:

$$V_t A_t = V_e A_e$$
$$V_t = V_e (A_e / A_t) \qquad (2)$$
$$V_t = 4V_e$$

Then Eq. (1) can be written as

$$1700/\gamma + (4V_{e\max})^2 / 2g = 100{,}000/\gamma + V_{e\max}^2 / 2g$$
$$V_{e\max} = ((1/15)(2g/\gamma)(98{,}300))^{1/2}$$
$$= ((1/15)(2/\rho)(98{,}300))^{1/2}$$
$$= \underline{\underline{3.62 m/s}}$$

$$Q_{max} = V_e A_e$$
$$= (3.62 m/s)(10^{-3} m^2)$$
$$= \underline{\underline{0.00362 m^3/s}}$$

Find pressure in the suction cup at the level of the suction cup.

$$p_t + \gamma \Delta h = p_{suction}$$
$$p_{suction} = 1{,}700 Pa + 9800 \times 2$$
$$= 21{,}300 Pa$$

But the pressure in the water surrounding the suction cup will be $p_{atm} + \gamma \times 1 = (100 + 9.80) kPa$

5.91 (Cont'd)

or

$$p_{water} - p_{suction} = (109{,}800 - 21{,}300)\text{Pa}$$
$$= 88{,}500\text{Pa}$$

Thus the maximum lift will be:

$$\text{Lift}_{max} = \Delta p A_s = (p_{water} - p_{suction})A_s$$
$$= (88{,}500\text{N}/\text{m}^2)(0.1\text{m}^2)$$
$$= \underline{\underline{8{,}850\text{N}}}$$

Note: Buoyancy on object being lifted was neglected.

5.92 $p_{max} = \rho V_0^2/2$

$= 1.1 \times 50^2/2 = \underline{\underline{1{,}375 \text{ Pa, gage}}}$

$p_{min} = -0.45\ \rho V_0^2/2 = -\ 0.45 \times (1{,}375) = \underline{\underline{-619 \text{ Pa, gage}}}$

5.93 $p_2 - p_1 = (1.2/2)(90 \times 90 - 70 \times 70) = 1{,}920 \text{ Pa} = \underline{\underline{1.920 \text{ kPa}}}$

5.94 Assume $p_{atm} = 101$ kPa abs; $p_{vapor} = 1{,}230$ Pa abs

Considering a point ahead of the foil (at same depth as foil) and the point of minimum pressure on the foil, and applying the definition of C_p between these two points yields:

$C_p = (p_{min} - p_0)/(\rho V_0^2/2)$

where $p_0 = p_{atm} + 1.8\gamma = 101{,}000 + 1.8 \times 9{,}810 = 118{,}658$ Pa, abs.

$p_{min} = 70{,}000$ Pa abs; $V_0 = 7$ m/s

Then $C_p = (70{,}000 - 118{,}658)/(500 \times 7^2) = -1.986$

Now use $C_p = -1.986$ (constant) for evaluating V for cavitation where p_{min} is now p_{vapor}:

$-1.986 = (1{,}230 - 118{,}658)/((1{,}000/2)V_0^2)$; $V_0 = \underline{\underline{10.87 \text{ m/s}}}$

5.95 See solution for Prob. 5.94 for preliminaries. We have the same C_p, but $p_0 = 101,000 + 3\gamma = 130,430$. Then:

$$-1.986 = (1,230 - 130,430)/((1,000/2)V_0^2)$$

$$V_0 = \underline{\underline{11.41 \text{ m/s}}}$$

5.96 Solution similar to solution for Prob. 5.94.

$$p_{min} = -2.5 \times 144 = -360 \text{ psf gage}$$

$$p_0 = 4\gamma = 4 \times 62.4 = 249.6 \text{ psf}$$

Then $C_p = (p_{min} - p_0)/(\rho V_0^2/2) = (-360 - 249.6)/((1.94/2) \times 20^2)$

$$C_p = -1.571$$

Now let $p_{min} = p_{vapor} = 0.178$ psia $= -14.52$ psi gage $= -2,091$ psf gage

Then $-1.571 = -(249.6 + 2,091)/((1.94/2)V_0^2)$

$$V_0 = \underline{\underline{39.2 \text{ ft/s}}}$$

5.97 From solution of Prob. 5.96 we have $C_p = -1.571$

but now $p_0 = 10\gamma = 624$ psf

Then: $-1.571 = -(624 + 2,091)/((1.94/2)V_0^2)$

$$V_0 = \underline{\underline{42.2 \text{ ft/s}}}$$

5.98 **Write Bernoulli's equation between a point in the free stream to the 90° position where $V = 1.5\ V_0$. The free stream velocity is the same as the sphere velocity (reference velocities to sphere).**

$$\rho V_0^2/2 + p_0 = p + \rho(1.5\ V_0)^2/2$$

where $p_0 = 18$ psia and $p = 0.18$ psia (Table A.5)

$$\rho V_0^2/2\ (2.25 - 1) = (18 - 2)(144)$$

$$V_0^2 = 2\ (17.8)(144)/((1.25)\ (1.94))\ \text{ft}^2/\text{s}^2$$

$$V_0 = \underline{\underline{46.0 \text{ ft/sec}}}$$

5.99 $C_p = (p - p_0)/(\rho V_0^2/2)$

$p_0 = 100{,}000 + 1 \times 9{,}810 \text{ Pa} = 109{,}810 \text{ Pa}$

$p = 90{,}000 \text{ Pa}$

Thus $C_p = -1.585$

For cavitation to occur $p = 1{,}230$ Pa (assumed)

$-1.585 = (1{,}230 - 109{,}810)/(1{,}000\, V_0^2/2)$; $V_0 = \underline{\underline{11.7 \text{ m/s}}}$

5.100 $C_p = (p-p_0)/(\rho V_0^2/2)$

From Fig. 5-13 $C_{p_{min}} = -0.45$; therefore,

$$p_{min} = p_0 + C_{p_{min}} \rho V_0^2/2$$

$$= 100 - 0.45 \times 1.94 \times 32^2/2$$

$$= \underline{\underline{-347 \text{ psf, gage}}}$$

Cavitation will not occur because the minimum pressure is much greater than the vapor pressure of the water.

5.101

If one connected a manometer between wall reference points P_1 and P_{13} then from Fig. 5.8 we see that

$$(h-h_0)/(V_0^2/2g = -3$$

or $h_0-h = 3V_0^2/2g$ (1)

For $V_0 = 5\text{m/s}$; $h_0-h = 3.823$ meters of water

One could use a mercury-water manometer for sensing the difference in head. Thus, the deflection on such a manometer would be $\Delta = (h_0-h)/12.55$ (2)

Therefore, for $V_0 = 5\text{m/s}$, Δ would be $3.823/12.55 = 0.305\text{m}$

Solving Eqs. (1) and (2) for V_0 yields

$$V_0 = 9.06\sqrt{\Delta}$$

5.101 (Cont'd)

The discharge would be given as

$$Q = V_0 A$$

$$= V_0 (4n_0 B)\sqrt{\Delta}$$

or $$Q = 9.06 \sqrt{\Delta}(4n B)$$

$$= 36.24 \; n_0 B \sqrt{\Delta}$$

where Δ is the deflection on a mercury-water manometer in meters, n_0 and B are lineal dimensions in meters and Q is the discharge in m^3/s.

The maximum deflection of 0.305m is not large; therefore, one may want to increase the accuracy of reading by utilizing an inclined tube manometer such as given in Problem 3.32. Also, one could connect the manometer between other points on the wall (such as between points points $P_{5.3}$ and P_9) to achieve a greater difference in head for a given discharge.

5.102 One possible design is shown. The manometer is an air-water manometer. Then for this setup

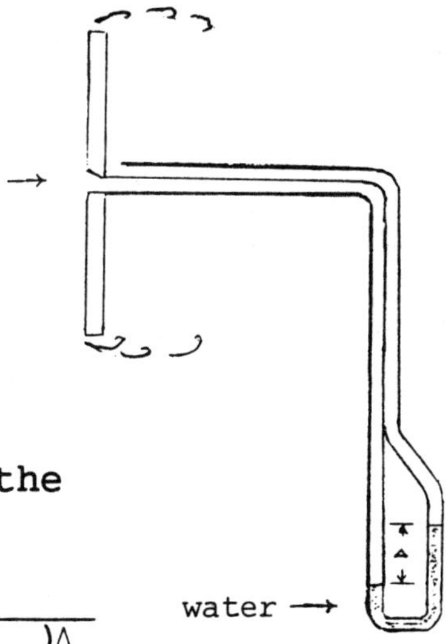

$$\Delta p = 1.45 \, \rho V_0^2/2$$

or $$V_0 = \sqrt{(2/1.45)\Delta p/\rho_{air}}$$

but $\Delta p = \gamma_{H_2O}\Delta$ where Δ is the manometer deflection.

Then $$V_0 = \sqrt{(2/1.45)(\gamma_{H_2O}/\rho_{air})\Delta}$$

One could also use an inclined tube manometer such as given for Problem 3.32 to increase the accuracy of reading.

5.103 The main point to this question is that while inhaling, the air is drawn into your mouth without any separation occurring in the flow that is approaching your mouth. Thus there is no concentrated flow; all air velocities in the vicinity of your face are relatively low. However, when exhaling as the air passes by your lips separation occurs thereby concentrating the flow of air which allows you to easily blow out a candle.

5.104 If a building has a flat roof as air flows over the top of the building separation will occur at the sharp edge between the wall and roof. Therefore, most if not all of the roof will be in the separation zone. Because the zone of separation will have a pressure much lower than the normal atmospheric pressure a net upward force will be exerted on the roof thus tending to lift the roof.

Even if the building has a peaked roof much of the roof will be in zones of separation. These zones of separation will occur downwind of the peak. Therefore, peaked roof buildings will also tend to have their roofs uplifted in high winds.

CHAPTER SIX

6.1 $50{,}000 = 1{,}000\ V_1^2/2;\quad V_1 = 10\ \text{m/s}$

$F_H = \rho Q(V_{2x} - V_{1x})$

$= 1{,}000 \times 0.4\ (0 - 10) = -4{,}000\ \text{N} = \underline{\underline{-4\ \text{kN}}}$

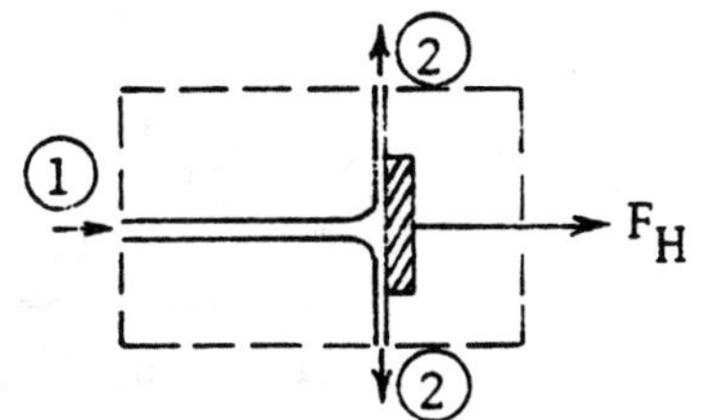

6.2 $\Sigma F_x = \rho Q(V_2 - V_1)\quad ;\quad V_1 = 160/(1.94 \times 2) = \underline{\underline{41.24\ \text{ft/s}}}$

6.3 $V_1^2 = 10 \times 144 \times 2/1.94 = 1{,}485;\quad V_1 = 38.54\ \text{fps}$

$\Sigma F_x = \rho Q(V_2 - V_1)$

$-400 = 1.94 \times 38.54 \times A_1 \times (0\text{-}38.54);\ A_1 = 0.1388\ \text{ft}^2$

$d_1 = \underline{\underline{0.420\ \text{ft}}}$

6.4 $V_1^2 = 80{,}000 \times 2/1{,}000 = 160\ \text{m}^2/\text{s}^2;\ V_1 = 12.65\ \text{m/s}$

$\Sigma F_x = \rho Q(V_2 - V_1)$

$-2{,}000 = 1{,}000 \times 12.65 \times A_1 \times (0\text{-}12.65);\ A_1 = 0.0125\ \ \text{m}^2$

$d_1 = 0.1262\ \text{m} = \underline{\underline{12.62\ \text{cm}}}$

6.5 $\Sigma F_x = \Sigma v_x(\rho \underline{V}\cdot\underline{A}) + \cancel{d/dt\int_{cv} v_x \rho dV} \rightarrow 0$

$F_{\text{of wall on jet}} = V_{x,\text{approaching plate}}(-\rho VA) + 0\ (+\rho VA)$

$-2\cancel{,000} = -V^2\ (1\cancel{,000})(.01)$

$\underline{\underline{V = 14.14\ \text{m/s}}}$

Find pressure in tank with Bernoulli's equation

$p_j + \rho V_j^2/2 = p_A + \rho V_A^2/2$ (assume $z_A = z_j$)

Assume V_A is negligible

Then $p_A = \rho V_j^2/2 = (1{,}000\ \text{kg/m}^3)\ (14.14^2\text{m}^2/\text{s}^2)/2$

$= \underline{\underline{100\ \text{kPa}}}$

6.6 $V_1 = V_{2x} = Q/A_1 = 1.2/((\pi/4) \times (1/3)^2) = 13.75$ ft/s

$\Sigma F_x = \rho Q(V_{3x} - V_{1x})$

$= 1.94 \times 1.2\ (0 - 13.75)$

$F_B = -32.01$ lb

$\Sigma F_y = \rho Q(V_{3y} - V_{2y})$; $V_{2y} = -\sqrt{2g \times 9} = -24.07$ ft/s

$F_A - Wgt = 1.94 \times 1.2\ (0 - (-24.07))$

$F_A - 300 - 4 \times 1 \times 62.4 = 1.94 \times 1.2 \times 24.07$

$F_A = \underline{605.6 \text{ lbf}}$

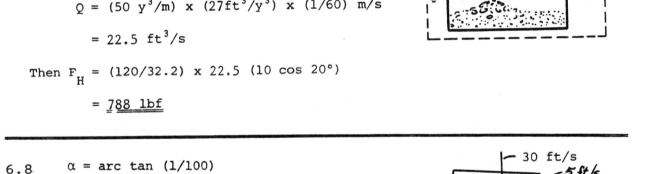

6.7 $\Sigma F_x = \rho Q(V_{2x} - V_{1x})$

$-F_H = \rho Q(0 - V_1 \cos 20°)$

$\rho = \gamma/g = (120/32.2)$ slugs/ft^3

$Q = (50\ y^3/m) \times (27 ft^3/y^3) \times (1/60)$ m/s

$= 22.5$ ft^3/s

Then $F_H = (120/32.2) \times 22.5\ (10 \cos 20°)$

$= \underline{788 \text{ lbf}}$

6.8 $\alpha = \arctan (1/100)$

$= 0.573°$

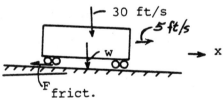

Assume rolling friction just balances the component of weight in the direction of motion.

Let the control volume move with the box car. Then reference all velocities to the box car. Thus the velocity of the grain with respect to the box car will be as shown below.

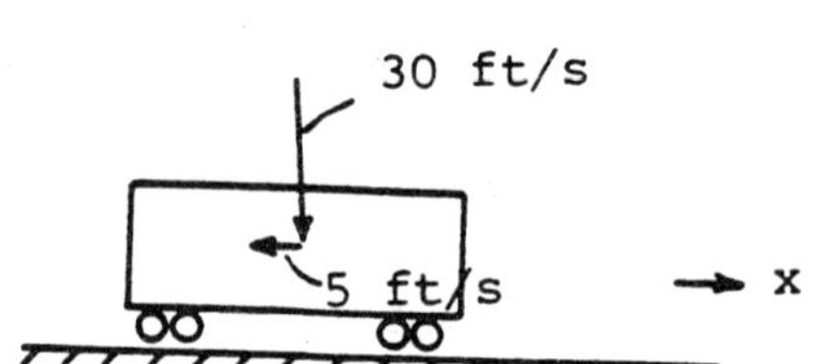

Apply the momentum equation in the x direction and include a fictitious force F_x needed to maintain the 5 ft/s velocity.

$\Sigma F_x = \rho Q(V_{2x} - V_{1x})$

6.8 (continued)

$$-F_{frict} + W \sin\alpha + F_x = (500/32.2)(0 - (-5+30 \sin\alpha))$$

$$F_x = 77.5 \text{ lbf}$$

The 77.5 lbf is the force that would be required to maintain a speed of 5 ft/s. However, such a force is not present; therefore, the box car will decelerate.

6.9 $F_x = 0.9 \times 1{,}000 \times 0.10\,(-27 \cos 30° - 20) = \underline{\underline{-4{,}624\text{ N}}}$

$F_y = 0.9 \times 1{,}000 \times 0.10\,(-27 \sin 30° - 0) = \underline{\underline{-1{,}215\text{ N}}}$

6.10 $F_x = 0.9 \times 1.94 \times 2 \times (-85 \cos 30° - 90) = \underline{-572\text{ lbf}}$

$F_y = 0.9 \times 1.94 \times 2 \times (-85 \sin 30° - 0) = \underline{-148\text{ lbf}}$

6.11 $\Sigma F_x = \Sigma V_x\ \rho \underline{V}\cdot\underline{A}$

$$F_x = 50 \times \rho(-50 \times 0.3) + 50 \times \cos 60° \times \rho \times (50 \times 0.2) + (-50 \cos 30°) \times \rho \times (50 \times 0.1)$$

$$= \rho \times 50^2(-0.3 + 0.2 \cos 60° - 0.1 \cos 30°)$$

$$F_x = 1.94 \times 2{,}500(-0.287) = \underline{\underline{-1{,}391\text{ lbf}}}$$

$$\Sigma F_y = \Sigma V_y \rho \underline{V}\cdot\underline{A}$$

$$= 40 \sin 60° \times \rho \times (40 \times 0.2) + (-40 \sin 30°) \times \rho \times (40 \times 0.1)$$

$$= 1.94 \times 2{,}500\ (0.2 \sin 60° - 0.1 \sin 30°) = \underline{\underline{+\ 597\text{ lbf}}}$$

6.12 $\Sigma \underline{F} = \int \underline{V}(\rho \underline{V}\cdot\underline{A})$; Assume $\underline{V}$ does not vary across a section

$$F_{V_x} = \Sigma V_x(\rho \underline{V}\cdot\underline{A})$$

where F_{V_x} = force of the vane on the fluid in the x direction

$$V_{initial,x} = (V-V_V)_{1x}$$

$$V_{final,x} = (V-V_V)_{2x}$$

Finally, $F_{Vx} = (V-V_V)_{1x}(\rho(V-V_V) \times A) + (V-V_V)_{2x}(\rho \times (-(V-V_V) \times A))$

$$= \underline{\underline{\rho(V-V_V)A[(V-V_V)_{2x} - (V-V_V)_{1x}]}}$$

6.13 $Q_{rel} = (20 - 7) \times \pi \times (0.03)^2 = 0.0368 \text{ m}^3/\text{s}$

$F_x = \rho Q_{rel}(V_{2x} - V_{1x}) = 1{,}000 \times 0.0368 \times (-13 \cos 45° - 13) = -817 \text{ N}$

$F_y = 1{,}000 \times 0.0368 \times (13 \cos 45° - 0) = 338 \text{ N}$

Force on vane $\underline{\underline{F_x = 817 \text{ N}}}$ and $\underline{\underline{F_y = -338 \text{ N}}}$

6.14 Make the flow steady by referencing all velocities to the moving vane and let the c.v. move with the vane as shown.

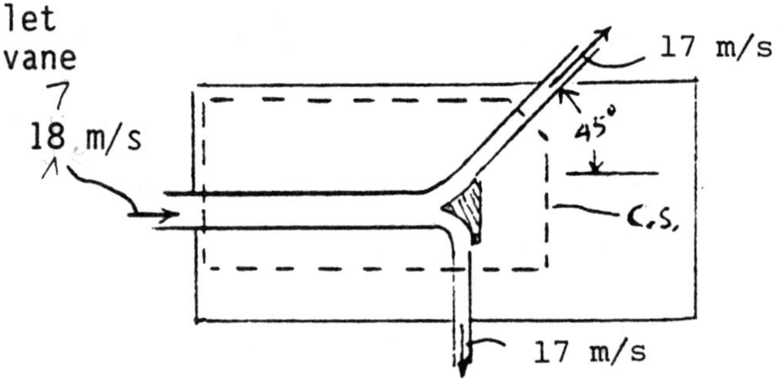

18 m/s

17 m/s

45°

C.S.

17 m/s

$\Sigma \mathbf{F} = \Sigma \mathbf{V} \rho \mathbf{V} \bullet \mathbf{A}$

$\Sigma F_x = \Sigma V_x \rho \mathbf{V} \bullet \mathbf{A}$

ΣF_x, vane on water $= (17)\ (1{,}000)\ (-17)\ (\pi/4)\ (0.1^2)$

$+ (17^2 \cos 45°)\ (1{,}000)\ (\pi/4)\ (0.1^2)/2$

$= -2{,}270 + 802 = -1{,}645 \text{ N}$

F_x, water on vane $= +1{,}468$ N

$\Sigma F_y = \Sigma V_y \rho \mathbf{V} \bullet \mathbf{A}$

F_y, vane on water $= (17)\ (1{,}000)\ (\sin 45°)\ (17)\ (\pi/4)\ (0.1^2)/2$

$-(17)^2 (1{,}000)\ (\pi/4)\ (0.1^2)/2$

$= -333$ N

F_y, water on vane $= +333$ N

$\underline{\underline{\mathbf{F}_{\text{water on vane}} = 1{,}468\ \mathbf{i} + 333\ \mathbf{j} \text{ N}}}$

6.15 Let the control surface surround the cart and let it move with the cart at 5ft/s. Then we have a steady flow situation and the relative jet velocities are shown below.

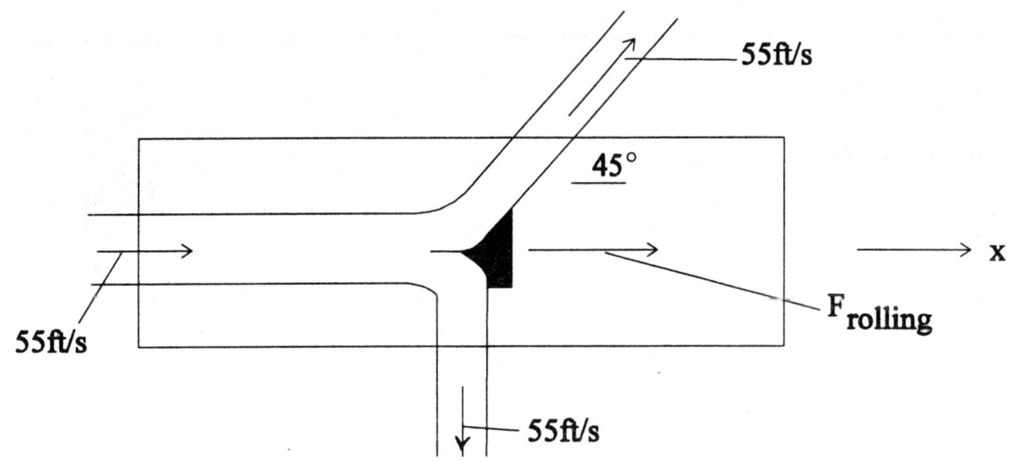

Apply the momentum equation in the x-direction:

$$\sum F_x = \sum V_x \rho \underline{V} \cdot \underline{A}$$

$$F_{rolling} = 55(1.94)(-55 \times (\pi/4) \times 0.1^2)$$

$$+0.707 \times 55(1.94)(55 \times 0.5(\pi/4) \times 0.1^2)$$

$$F_{rolling} = 55^2 \times (1.94)(\pi/4) \times 0.1^2(-1 + (0.5 \times 0.707))$$

$$= \underline{\underline{-29.8\text{lbf}}}$$

<u>Rolling resistance is equal to 29.8lbf</u>

6.16 Let section 2 be the outlet section.

$V_1A_1 = V_2A_2$; $V_2 = V_1 (A_1/A_2) = 30 \times 2 = 60$ ft/s

Let the c.s. surround the fluid adjacent to the blade and assume this flow has its velocity changed by only this blade.

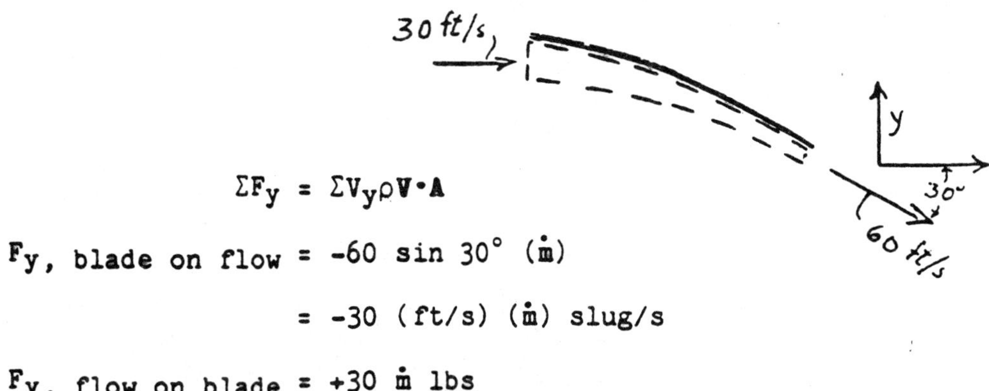

$$\Sigma F_y = \Sigma V_y \rho \mathbf{V} \cdot \mathbf{A}$$

F_y, blade on flow $= -60 \sin 30° \; (\dot{m})$

$= -30$ (ft/s) $(\dot{m})$ slug/s

or F_y, flow on blade $= +30\,\dot{m}$ lbs

where $\dot{m}$ is the mass rate of flow in slugs/s.

6.17 Relative to cone $V_{fx} = 43 \cos 50° = 27.64$ m/s

$F_x = 1{,}000 \times \pi \times (0.05)^2 \times 43 \times (27.64 - 43) = 5{,}187$ N = 5,187 kN

6.18 $\Sigma F_x = \rho Q(V_{fx} - V_{0x})$

100 ft/s
0.30 ft
50°
x
60 ft/s

Let velocities be relative to vane.

$F_x = 1.94 \times 40 \times 0.3(40 \cos 50° - 40)$

$F_x = -332.6$ lbf

The force acting on the vane will be +332.6 lbf

Power = FV

= 332.6 x 60

= 19,958 ft-lbf/s = 36.3 horsepower

6.19

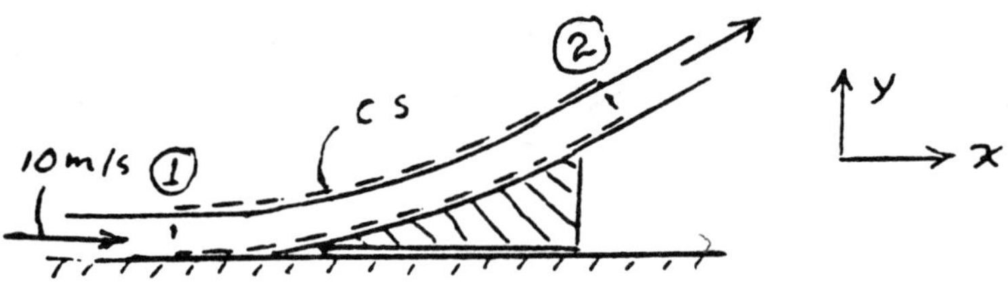

Force of block on jet:

$$\Sigma F_x = \Sigma V_x \rho \mathbf{V} \cdot \mathbf{A}$$

$$F_x = V_{x,1}(-\dot{m}) + V_{x,2}(\dot{m})$$

$$= \dot{m}(V_{x,2} - V_{x,1})$$

$$= (1 \text{ kg/s})(10 \cos 30° - 10) \text{ m/s}$$

$$F_x = -1.34 \text{ N} \quad ;$$

Force on block in x direction is $F_x = +1.34$ N

Force on jet in y direction:

$$\Sigma F_y = \dot{m}(V_{y,2} - V_{y,1})$$

$$F_y - Wgt = \dot{m}(10 \sin 30°)$$

$$= (1 \text{ kg/s})(5 \text{ m/s}) = +5 \text{ N}$$

$$F_y = 9.81 \text{ N} + 5 \text{ N} = 14.81 \text{ N}$$

6.19 (continued)

Force on block in y direction is $F_y = -14.81$ N

Friction force = (0.1) (14.81 N) = 1.481 N > F_x

∴ block will not move.

6.20 Although there will be some forward thrust per engine pod itself, this computation will be the thrust produced by the deflecting vanes only.

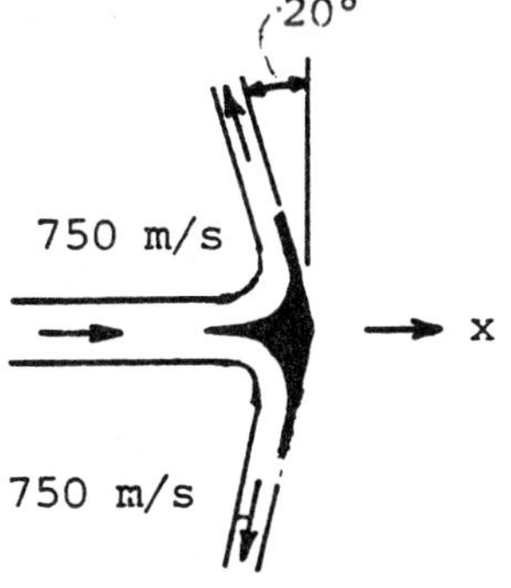

$\Sigma F_x = \Sigma V_x \rho \underline{V} \cdot \underline{A}$

$F_x = \dot{m}(V_{fx} - V_{ox})$; $\dot{m} = 150\ (1 + .025) = 154$ kg/s

$F_x = 154\ (-750 \sin 20° - 750)$

$F_x = -155{,}000$ N

Reverse thrust per engine = +155,000 N = 155 kN

6.21 Reference jet velocity to disk for steady flow

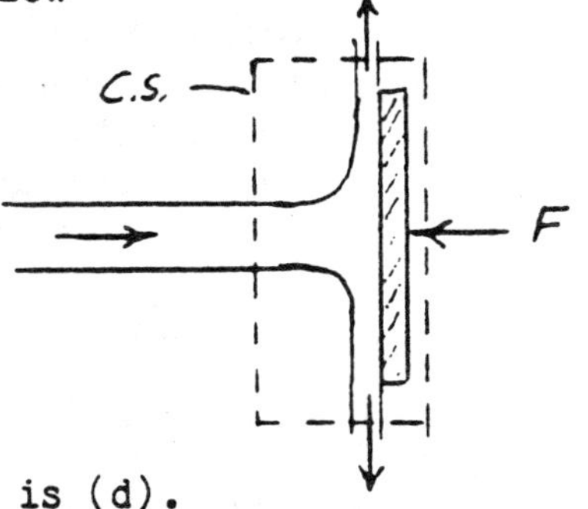

Then $V_{j,x,rel.} = 2V_j$

$\Sigma F_x = \Sigma V_x \rho \mathbf{V} \cdot \mathbf{A}$

$-F = 2V_j\rho(-2V_jA) = -4\rho V_j^2 A$

$F = 4\rho V_j^2 A$ Correct choice is (d).

6.22 $\Sigma F_x = \Sigma V_x \rho \underline{V} \cdot \underline{A}$

$F_x = -30 \times 1{,}000 \times 30 \times \pi \times 0.05 \times 0.05 + 30 \times 1{,}000 \times 30 \times \pi$

$\times\ 0.025 \times 0.025 = -5{,}300$ N $=$ -5.30 kN

6.23 $\Sigma F_x = \Sigma V \rho \underline{V} \cdot \underline{A}$

$0 = -V_1^2\ b \cos\theta - V_2^2 d_2 + V_3^2 d_3$

$d_2 + b_1 \cos\theta = d_3$ but $d_2 + d_3 = b_1$ by continuity equation

$b_1 - d_3 + b \cos\theta = d_3$

$d_3 = \underline{\underline{(b_1/2)(1 + \cos\theta)}}$

$d_2 = b_1 - d_3 = \underline{\underline{(b_1/2)(1 - \cos\theta)}}$

6.24

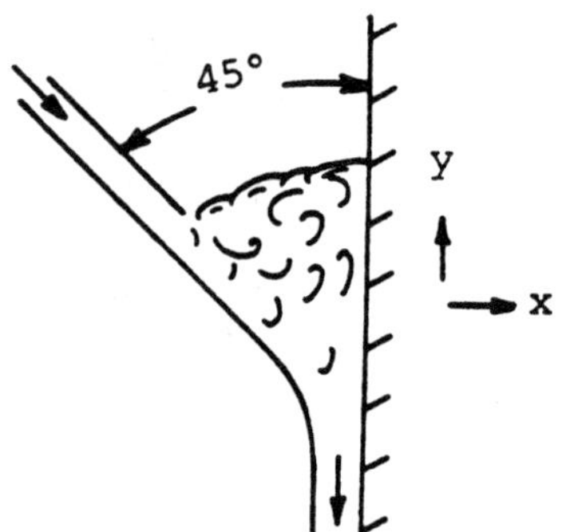

$\Sigma F_x = \rho Q(V_{2x} - V_{1x})$

$F_x = \rho V_1 t(0 - V_1 \cos 45°)$

$F_x = -\rho V_1^2 t \cos 45°$

Then force on wall is $\underline{\underline{F_{x,wall} = +\rho V_1^2 t \cos 45°}}$

In the vertical direction there will be a slight viscous shear force acting on the wall in the downward direction--it would be very small. However, there will have to be a downward vertical force acting on the liquid to produce the momentum change that occurs when the jet turns through the 45° angle. This force is effected by the weight of liquid that "piles up" between the jet at an angle of 45° and the vertical wall as depicted above. There will be a counter-clockwise circulation of flow in this mass of liquid because it will be driven by the jet on its underside. Therefore, the liquid surface will be higher at the wall than where the jet at 45° first hits it--the liquid will flow from the wall downhill until the jet impinges upon it.

6.25 $\Sigma F_x = \rho Q(V_{2x} - V_{1x})$

$\Sigma F = 1{,}000 \times (0.20) \times (0.08) \times 100 \times (100 \cos 60° - 100)$

$= -80{,}000$ N; $a = F/M = 80{,}000/1{,}000 = \underline{\underline{80.0 \text{ m/s}^2}}$

6.26 $\Sigma F_x = \rho Q(V_{2x} - V_{1x})$

Rel. speed = 40 ft/s

$V_{1x} = +40$ ft/s

$V_{2x} = -40 \cos 60° \cos 30°$

$= -17.32$ ft/s

6.26 (Cont'd) $\Sigma F_x = 1.94 \times 0.2 \times 40 \times 2 \times (1/4)(-17.32 - 40)$

$= -444.8$ lbf

$P = FV = 444.8 \times 40 = \underline{17{,}792 \text{ ft-lbf/s}}$

$HP = \underline{32.3 \text{ h.p.}}$

6.27 Maximum force occurs at the beginning; hence, the tank will accelerate immediately after opening the cap. However, as water leaves the tank the force will decrease, but acceleration may decrease or increase because mass will also be decreasing. In any event, the tank will go faster and faster until the last drop leaves, assuming no wind friction.

6.28 $\dot{m} = 60$ kg/s $\quad V_0 = 200$ m/s

$V_e = \dot{m}/\rho_e A = 60/(0.25 \times 0.5) = 480$ m/s

$T = \dot{m}(V_e - V_0)$

$= 60\,(480 - 200) = 16{,}800 \text{ N} = \underline{16.8 \text{ kN}}$

6.29 Assume the resistive force is caused primarily by rolling resistance (bearing friction, etc); therefore, the resistive force will act on the wheels at the ground surface. Reference the jet velocity to the cart to produce a steady flow situation. The velocities and resistive force are shown below

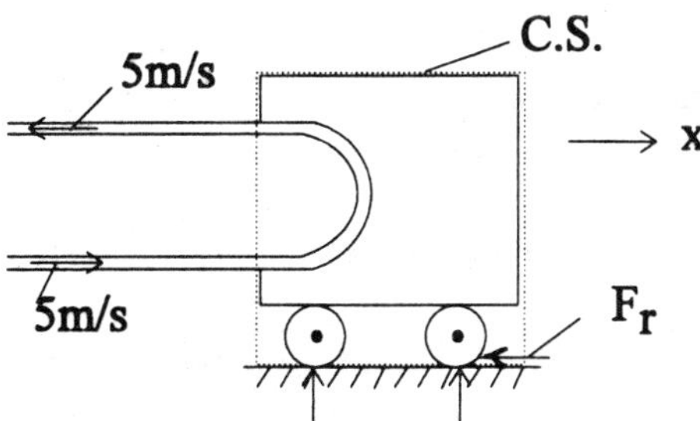

Apply the momentum equation:

$$\sum F_x = \sum \rho \underline{V} \cdot \underline{A}$$

$$-F_r = 5(1000)(-5 \times 0.001) + (-5)(1000)(5 \times .001)$$

$$-F_r = 5^2(1000)(0.001)(-1-1)$$

$$\underline{F_r = 50\text{N acting to the left}}$$

6.30

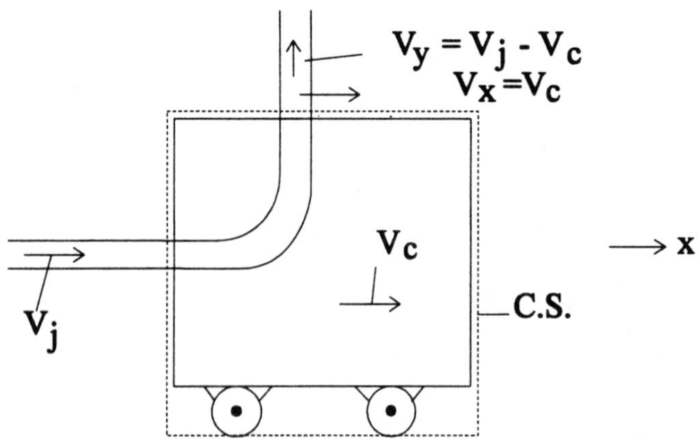

Apply the momentum equation and let the control surface surround the cart as shown above and let the C.S. move with the velocity of the cart. Let V_j = velocity of the jet with respect to the earth; V_c = velocity of cart with respect to earth.

$$\sum F_x = \sum_{cs} V_x(\rho \underline{V} \cdot \underline{A}) + d/dt \int_{cv} V_x \rho d\forall \qquad (6.4)$$

But $\Sigma F_x = 0$ (given in problem statement). The V_x in the first term on the right hand side of the equation is the velocity of the jet with respect to the earth (inertial reference). The $\underline{V}$ in $\underline{V} \cdot \underline{A}$ is the velocity with respect to the C.S.; therefore; it is given by V_j-V_c. The V_x in the second term on the right hand side of the equation is the velocity of the cart with respect to the earth.

Thus

$$0 = V_j[\rho(-(V_j - V_c)A] + [\rho(V_j - V_c)(V_j - V_c)A]$$
$$+ d/dt \int_{cv} V_x \rho d\forall \qquad (1)$$

In the last term of Eq. (1) V_x and ρ inside the C.V. are assumed to be constant (we're neglecting the motion of the water in the C.V.); therefore, the integral becomes $V_x \rho \forall = V_x M$, where M is the mass of the cart and V_x is the velocity of the cart with respect to the earth (inertial reference). Eq. (1) is now written as

$$0 = -\rho V_j A(V_j - V_c) + \rho V_c A(V_j - V_c) + d/dt[MV_x]$$
$$0 = -\rho V_j A(V_j - V_c) + \rho(V_j V_c A / V_j)(V_j - V_c) + Ma$$
$$a = (\rho V_j A / M)[-(V_c / V_j)(V_j - V_c) + (V_j - V_c)]$$
$$\text{But } V_j A = Q = 0.1 m^3/s; \; M = 10 kg; \; \rho = 1000 kg/m^3$$
$$\text{so } \underline{\underline{a = (\rho Q / M)[(V_j - V_c)(1 - (V_c / V_j))]}} \qquad (2)$$
$$a = (1000 \times 0.1 / 10)[(V_j - V_c)(1 - (V_c / V_j))]$$
$$\text{When the cart speed, } V_c \text{, is } 5 m/s$$
$$\underline{\underline{a = 25 m/s^2}}$$

6.31 Apply momentum equation in the x direction for the c.v. shown.

$\Sigma F_x = \rho Q(V_{2x} - V_{1x})$, neglect friction

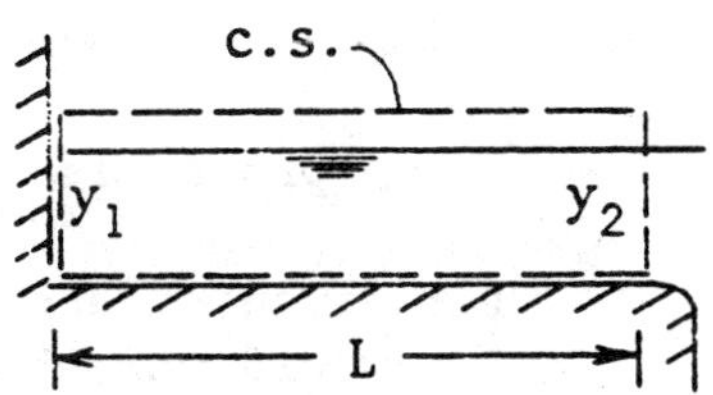

$(By_1^2\gamma/2) - (By_2^2\gamma/2) = \rho Q(Q/(y_2B) - 0)$

$$\underline{\underline{y_1 = \sqrt{y_2^2 + (2/(gy_2)) \times (Q/B)^2}}}$$

6.32 Obtain the pressure variation along the pipe by applying the momentum equation in steps along the pipe (numerical scheme). The first step would be for the end segment of the pipe. Then move up the pipe solving for the pressure change (Δp) for each segment. Then $p_{end} + \Sigma\Delta p$ would give the pressure at a particular section. The momentum equation for a general section is developed below.

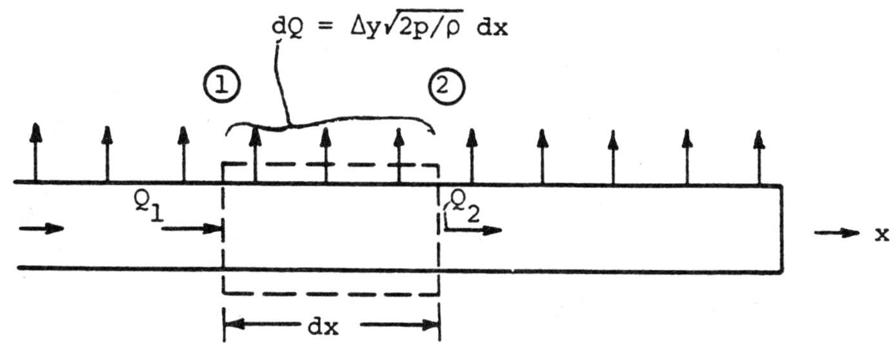

$\Sigma F_x = \Sigma v_x\,\rho \underline{V}\cdot\underline{A}$

$p_1A_1 - p_2A_2 = (Q_1/A_1)\rho(-Q_1) + (Q_2/A_2)\rho Q_2$

but $A_1 = A_2 = A$ so we get

$p_1 - p_2 = (\rho/A^2)(Q_2^2 - Q_1^2)$

$$dp = (\rho/A^2)(Q_2^2 - Q_1^2) \qquad (1)$$

However $Q_1 = Q_2 + dQ$

$Q_1 = Q_2 + \Delta y\sqrt{2p/\rho}\,dx$

Assume that the average pressure over the section is given by

$p_2 + (dp/2)$.

Then $Q_1 = Q_2 + \Delta y\sqrt{2(p_2 + (dp/2))/\rho}\,dx$

Eq. (1) is then $dp = (\rho/A^2)[Q_2^2 - (Q_2 + \Delta y\sqrt{2(p_2 + (dp/2))/\rho}\,dx)^2]$ (2)

In finite difference form use finite values for $dx, \Delta y$ to solve for Δp.

6.32 **(continued)**

Solution procedure:

1. Given p_2 and Q_2 solve for Δp. An iteration procedure would undoubtedly be used.
2. Add Δp to p_2 to get p_1
3. Move upstream along pipe to get a new solution for Δp, etc.

Conclusions:

1. Inspection of Eq. (1) reveals the Δp will be negative--the pressure will increase in the direction of flow.
2. If friction effects were included, this pressure distribution trend might be reversed.

6.33

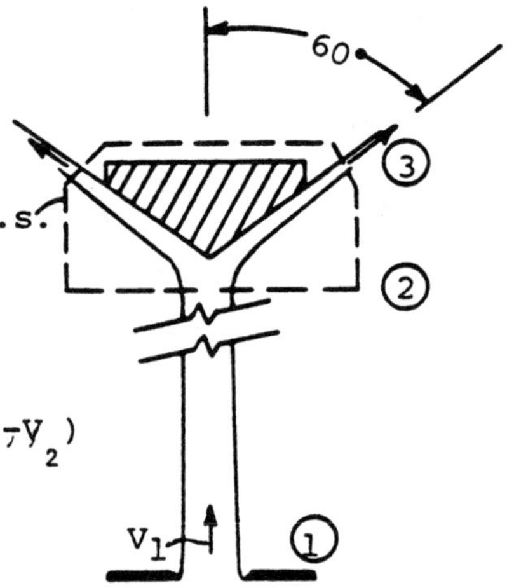

$$(V_1^2/2g) + 0 = (V_2^2/2g) + h$$

$$V_2^2 = (15)^2 - 2gh$$

$$\Sigma F_y = \rho Q(V_{3y} - V_{2y})$$

$$V_2^2 = (15)^2 - 2gh = 225 - 2 \times 9.81h$$

$$= 225 - 19.62h$$

$$-40 = 1{,}000 \times 15 \times \pi \times (0.015)^2 (V_2 \sin 30° - V_2)$$

$$V_2 = 7.55 \text{ m/s}$$

so $(7.55)^2 = 225 - 19.62\ h$; $\underline{h = 8.56 \text{ m}}$

6.34

$$V = Q/A = 6/(\pi \times 0.25 \times 0.25) = 30.56 \text{ fps}$$

$$\Sigma F_x = \rho Q(V_{2x} - V_{1x})$$

$$p_1A_1 + p_2A_2 + F_x = 1.94 \times 6 \times (-30.56 - 30.56)$$

$15 \times 144 \times \pi \times (0.25)^2 \times 2 + F_x = -711.4$; $\underline{F_x = -1560 \text{ lbf}}$

6.35

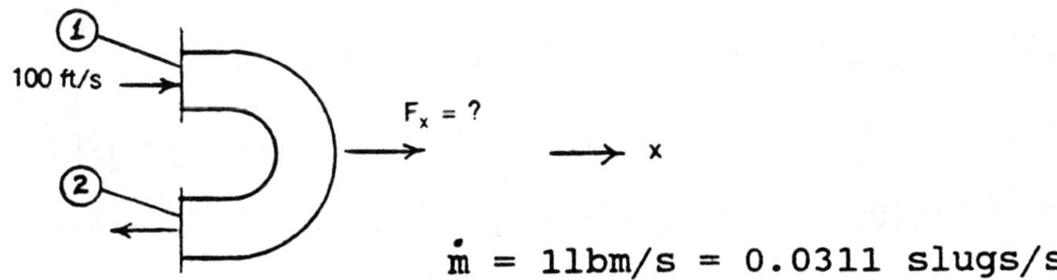

$\dot{m}$ = 1lbm/s = 0.0311 slugs/s

At section (1):

$V_x = +100$ ft/s , $\rho_1 = 0.02$ lbm/ft^3 = 0.000621 slugs/ft^3

At section (2):

$V_x = -?$ ft/s , $\rho_2 = 0.06$ lbm/ft^3 = 0.000186 slugs/ft^3

Continuity Eq.:

$\rho_1 V_1 A_1 = \rho_2 V_2 A_2$; $V_2 = (\rho_1/\rho_2)(A_1/A_2)V_1$

$= (1/3)V_1 = 33.33$ ft/s

$\Sigma F_x = \Sigma v_x \rho \underline{V}\cdot\underline{A}$

$F_x = (+100)(-0.0311) + (-33.33)(-0.0311)$

$\underline{F_x = -4.147 \text{ lbf}}$

6.36 Correct choice is (d).

6.37 $V = Q/A = 15/(\pi \times 0.5 \times 0.5) = 19.10$ fps

$\Sigma F_x = \rho Q(V_{2x} - V_{1x})$

$p_1A_1 + p_2A_2 + F_x = 1.94 \times 15(V_{2x} - V_{1x})$

$(\pi \times 0.5 \times 0.5)(10 \times 144 + (10 \times 144 + 2 \times 62.4)) + F_x$

$= 1.94 \times 15 \times (-19.10 - 19.10)$

$F_x = -3,472$ lbf

$\Sigma F_y = \rho Q(V_{2y} - V_{1y})$

$-Wt_{bend} - Wt_{H_2O} + F_y = 0$

$F_y = 200 + 3 \times 62.4 = 387.2$

Force required $\underline{F} = -3,472\ \underline{i} + 387.2\ \underline{j}$ lbf

---

6.38 $\Sigma F_x = \Sigma V_x \rho \mathbf{V} \cdot \mathbf{A}$; $V = Q/A = 8.49$ m /s

$$2pA + F_x = (8.49)(1{,}000)(-0.60) + (-8.49)(1{,}000)(0.60)$$

$$(2)(100{,}000)(\pi/4)(0.3^2) + F_x = (1{,}000)(0.60)(-8.49 - 8.49)$$

$$F_x = -24{,}325 \text{ N}$$

$$F_y = 0$$

$$F_z = 500 + (0.1)(9{,}810) = 1.481 \text{ kN}$$

$$\underline{F = (-24.33\mathbf{i} + 1.48\mathbf{k}) \text{ kN}}$$

6.39 Assume zero head loss so the pressure in the pipe will be zero gage.

$$\Sigma F_x = \rho Q(V_{fx} - V_{0x})$$

$$V = Q/A = 10/((\pi/4) \times 1.0^2) = 12.73 \text{ ft/s}$$

$F_x = 1.94 \times 10(0-12.73) = -247$ lbf ; Also $F_y = -247$ lbf

$\Sigma F_z = 0$; $-100 - 4 \times 62.4 + F_z = 0$; $F_z = +350$ lbf

Force required to hold bend = $\underline{(-247\ \mathbf{i} - 247\ \mathbf{j} + 350\ \mathbf{k}) \text{ lbf}}$

6.40 $V = Q/A = 10/(\pi \times 0.5 \times 0.5) = 12.73$ m/s

$$\Sigma F_x = \rho Q(V_{2x} - V_{1x})$$

$$300{,}000 \times \pi \times 0.5 \times 0.5 + F_x = 1{,}000 \times 10 \times (0 - 12.73)$$

$F_x = -362{,}919$ N = $\underline{\underline{-362.9 \text{ kN}}}$

6.41 $V = Q/A = 31.4/(\pi \times 1 \times 1) = 9.995$ ft/sec

$$\Sigma Fy = \rho Q(V_{2y} - V_{1y})$$

$$F_{anch} - Wt_{water} - Wt_{bend} - p_2A_2 \sin 30° = \rho Q(V \sin 30° - V \sin 0°)$$

$$F_{anch} = \pi \times 1 \times 1 \times 4 \times 62.4 + 300 + 8.5 \times 144 \times \pi \times 1 \times 1 \times 0.5$$

$$+ 1.94 \times 31.4 \times (9.995 \times 0.5 - 0)$$

$$= \underline{3{,}311 \text{ lbs}}$$

6.42 $(V_1^2/2g) + (p_1/\gamma) + z_1 = (V_2^2/2g) + (p_2/\gamma) + z_2$

$(V_1^2/2g) + (25 \times 144/62.4) + 0 = 16(V_1^2/2g) + 0 + 0$

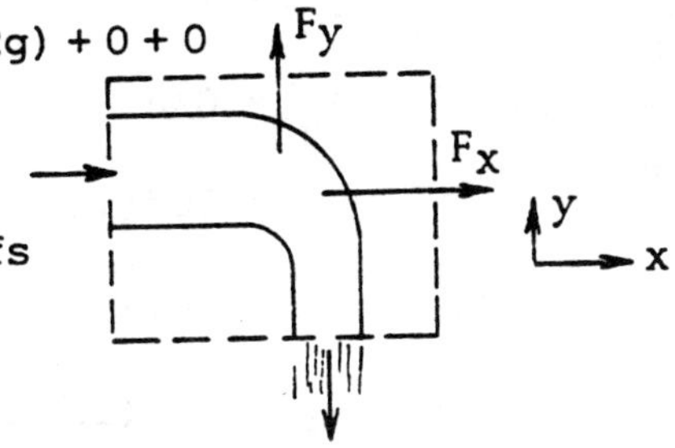

$V_1 = 15.74$ fps; $V_2 = 62.95$ ft/sec

$Q = A_1V_1 = \pi \times 1 \times 1 \times 15.74 = 49.44$ cfs

$\Sigma F_x = \rho Q(V_{2x} - V_{1x})$

$p_1A_1 + F_x = \rho Q(V_{2x} - V_{1x})$

$F_x = 1.94 \times 49.44(0 - 15.74) - 25 \times 144 \times \pi \times 1 \times 1 = \underline{\underline{-12,820 \text{ lbf}}}$

6.43

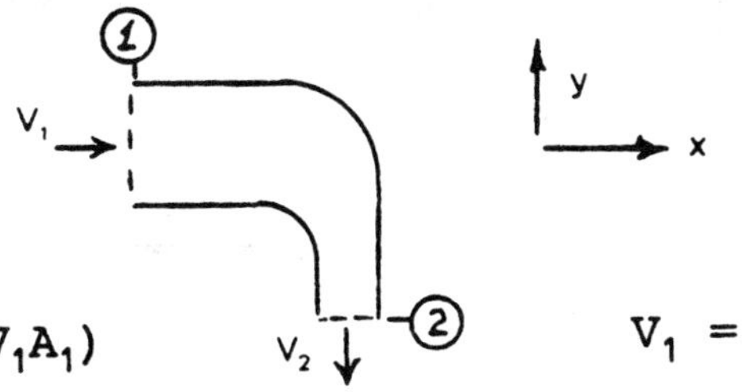

$\Sigma F_x = \Sigma v_x \rho \underline{V} \cdot \underline{A}$

$p_1A_1 + F_x = V_1(-\rho V_1 A_1)$

$F_x = -p_1A_1 - V_1\rho Q$

$= (-150,000)A_1 - (3.183)(1,000)(0.1)$

$F_x = -5,031\text{N}$

$V_1 = Q/A_1 = 3.183$ m/s

$V_2 = Q/A_2 = 12.73$ m/s

$\Sigma F_y = \Sigma V_y \rho \underline{V} \cdot \underline{A}$

$F_y + p_2A_2 = -V_y\rho Q$

$F_y = (-73,000)A_2 + (-12.73)(1,000)(0.1)$

$F_y = -573.3 - 1273 = -1846$ N

$\Sigma F_z = 0$

$F_z - \text{Wgt} = 0$

$F_z = 200 + 0.02\gamma = 396.2$ N

$\underline{\underline{\mathbf{F} = -5031\mathbf{i} - 1846\mathbf{j} + 396\mathbf{k}\ \mathbf{N}}}$

6.44 $p_1/\gamma + V_1^2/2g + z_1 = p_2/\gamma + V_2^2/2g + z_2$

$V_1 = Q/A$

$= 20/((\pi/4) \times (14/12)^2)$

$V_1 = 18.71$ ft/s $\qquad V_2 = 18.71 \times (14/10)^2 = 36.67$ ft/s

Then $p_1/\gamma + (18.71)^2/64.4 + 0 = 0 + (36.67)^2/64.4 + 0$

$p_1 = 963.7$ psf

$\Sigma F_x = \rho Q(V_{2x}-V_{1x})$

$p_1 A_1 + F_x = \rho Q(V_{2x}-V_{1x})$

$F_x = 1.94 \times 20(-36.67-18.71) - 963.7 \times (\pi/4) \times (14/12)^2$

$F_x = \underline{\underline{-3,179 \text{ lbf}}}$

6.45 $V_1 = 10/4 = 2.5$ m/s; $Q = A_1 V_1 = \pi \times 0.3 \times 0.3 \times 2.5 = 0.707$ m³/s

$p_1 = p_2 + (\rho/2)(V_2^2 - V_1^2) = 0 + (1,000/2)(10 \times 10 - 2.5 \times 2.5) = 46,875$ Pa

$F_x = -46,875 \times \pi \times 0.3 \times 0.3 + 1,000 \times 0.707 \times (-10 \cos 60° - 2.5)$

$= -18,560$ N

$F_y = 1,000 \times 0.707(-10 \sin 60° - 0) = -6,123$ N

$F_z = 0.25 \times 9,810 = 2,452$ N; Total force $\underline{\underline{\mathbf{F} = (-18.56\,\mathbf{i} - 6.123\,\mathbf{j} + 2.452\,\mathbf{k}) \text{ kN}}}$

6.46

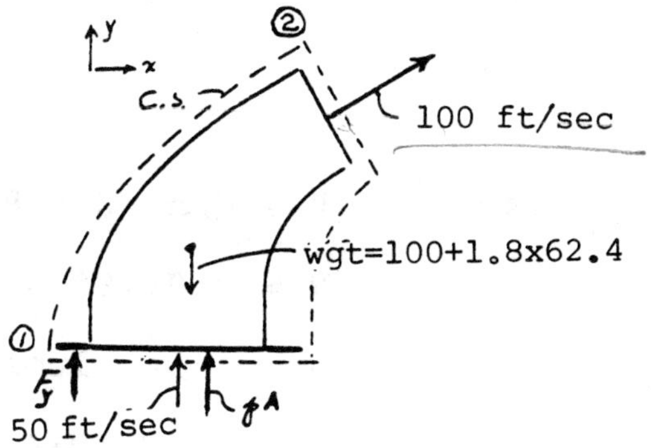

$Q = VA = (100 \text{ ft/s})(0.5 \text{ ft}^2)$

$= 50 \text{ ft}^3/\text{s}$

$V_1A_1 = V_2A_2 \;;\; V_1 = V_2A_2/A_1 = 50 \text{ ft/s}$

$p_1/\gamma + V_1^2/2g + z_1 = p_2/\gamma + V_2^2/2g + z_2$

$p_1/\gamma = 0 + (100^2/2g) + 2 - (50^2/2g)$

$p_1 = 62.4\,(155.28 + 2 - 38.82) = 7{,}391.9 \text{ lbf/ft}^2$

$\Sigma F_y = \Sigma V_y \rho \mathbf{V}\cdot\mathbf{A}$

$-Wgt + p_1A_1 + F_y = (50)(1.94)(-50) + (100 \sin 30°)(1.94)(50)$

$-100 - (1.8)(62.4) + (7391.9)(1.0) + F_y = (50)(1.94)(-50 + 50)$

$F_y = -7{,}379.4 + 224.8 = \underline{\underline{-7{,}155 \text{ lbf}}}$

6.47

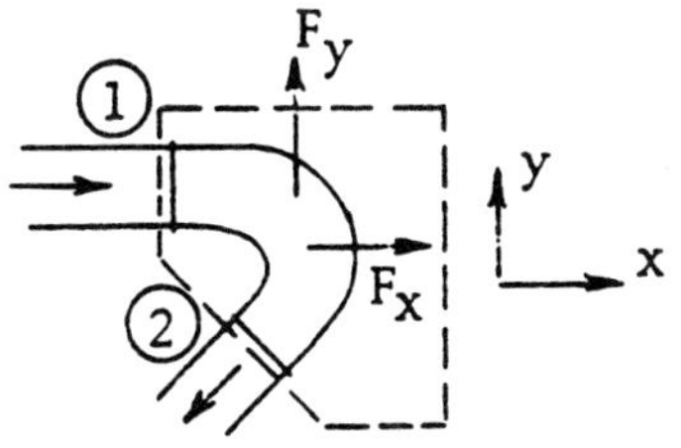

$Q = VA = 15 \times \pi \times 0.5 \times 0.5 = 11.78 \text{ cfs}$

$\Sigma F_x = \rho Q(V_{2x} - V_{1x})$

$p_1A_1 + p_2A_2 \cos 45° + F_x = \rho Q(V_{2x} - V_{1x})$

$F_x = -10 \times 144 \times \pi \times 0.5 \times 0.5\,(1 + \cos 45°) + 1.552 \times 11.78 \times 15\,(-\cos 45° - 1) = -2{,}399 \text{ lbf}$

$\Sigma F_y = \rho Q(V_{2y} - V_{1y})$

$p_2A_2 \cos 45° + F_y = \rho Q(V_{2y} - V_{1y})$

$F_y = -10 \times 144 \times \pi \times 0.5 \times 0.5 \cos 45° + 1.552 \times 11.78 \times 15\,(-\cos 45° - 0)$

$= -994$ lbf; Total force $\mathbf{F}$ = $\underline{\underline{(-2{,}399\, i - 994\, j) \text{ lbf}}}$

6.48

$Q = 8 \times \pi \times 0.15 \times 0.15 = 0.565 \text{ m}^3/\text{s}$

$F_x = -100{,}000 \times \pi \times 0.15 \times 0.15(1 + \cos 45°) + 800 \times 0.565 \times 8(-\cos 45° - 1) = -18{,}239 \text{ N}$

$F_y = -100{,}000 \times \pi \times 0.15 \times 0.15 \cos 45° + 800 \times 0.565 \times 8(-\cos 45° - 0)$

$= -7{,}555$ N; Total force $\underline{F} = \underline{\underline{(-18.239\, \underline{i} - 7.555\, \underline{j})\,\text{kN}}}$

6.49 $V = Q/A$; $Q = 8$ cfs, $A = \pi D^2/4 = \pi/4\ ft^2 = 0.785\ ft^2$

$V = 8/0.785 = 10.19$ ft/s

$\rho = 1.94\ slugs/ft^3$

$\Sigma F_x = \Sigma V_x \rho \underline{V} \cdot \underline{A}$

$F_{ext,x} + p_1 A_1 + p_2 A_2 \cos 45° = \Sigma V_x \rho \underline{V} \cdot \underline{A}$

but $p_1 = p_2$ and $A_1 = A_2$

so $F_{ext,x} + pA\,(1 + \cos 45°) = V_1 \times 1.94\ (-V_1 A)$

$+ (V_{2,x} \times 1.94\ (V_2 A))$

$F_{ext,x} + pA\,(1 + \cos 45°) = -1.94\ V^2 A$

$+ ((-V\cos 45°) \times 1.94 \times VA)$

$F_{ext,x} = -pA(1 + \cos 45°) - 1.94\ V^2 A(1 + \cos 45°)$

$= -10 \times 144 \times 0.785 \times 1.707 - 1.94 \times 10.19^2$

$\times 0.785 \times 1.707$

$= -1930 - 270$

$= \underline{\underline{-2200\ lbf}}$

$\Sigma F_y = \Sigma V_y \rho \underline{V} \cdot \underline{A}$

$F_{ext,y} + p_2 A_2 \sin 45° = -10.19 \sin 45 \times 1.94 \times 10.19$

$\times 0.785$

$F_{ext,y} = -10 \times 144 \times 0.785 \times 0.707 - 10.19^2 \times 0.707$

$\times 1.94 \times .785$

$= -799 - 112$

$= \underline{\underline{-911\ lbf}}$

6.50 Use Bernoulli's equation to get p_{inlet} and let $z_i = z_0$

$p_i + \rho V_i^2/2 = p_0 + \rho V_0^2/2$ where i = inlet and 0 = outlet

$p_0 = 0$ gage $\qquad V_0 = 50$ m/s (given)

$V_i = V_0 A_0 / A_i = 50(1/10) = 5$ m/s

$p_i = -(1000/2)(5^2) + 0 \text{ gage} + (1000/2)(50^2)$

$p_i = 1237.5$ kPa

$\Sigma F_x = \Sigma V_x \rho \underline{V} \cdot \underline{A}$

$p_i A_i + F_{exter.} = (5)(1000)(-5 \times 0.001) + (50\cos 60°)(1000)(50 \times .0001)$

$1238 + F_{exter.} = -25 + 125$

$F_{ext.} = 100 - 1238 = \underline{-1138 \text{ N}}$

6.51

$V_1 = Q/A_1 = 0.1/(\pi \times 0.1 \times 0.1) = 3.183$ m/s

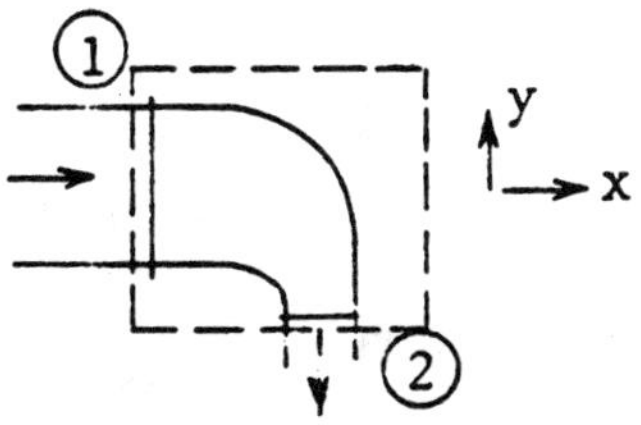

$V_2 = 3.183 \times 4 = 12.73$ m/s

$p_2 = p_1 + (\rho/2)(V_1^2 - V_2^2) = 150{,}000$

$+ (1{,}000/2)(3.813 \times 3.183 - 12.73 \times 12.73) = 74{,}040$ Pa

$\Sigma F_x = \rho Q(V_{2x} - V_{1x})$

$F_x = -150{,}000 \times \pi \times 0.1 \times 0.1 + 1{,}000 \times 0.1(0 - 3.183) = -5{,}031$ N

$\Sigma F_y = \rho Q(V_{2y} - V_{1y})$; $F_y = -74{,}040 \times \pi \times 0.05 \times 0.05 + 1{,}000 \times 0.1(-12.73 - 0$

$= -1{,}855$ N; Total force $\underline{F} = (-5.03\,\underline{i} - 1.855\,\underline{j})$ kN

6.52

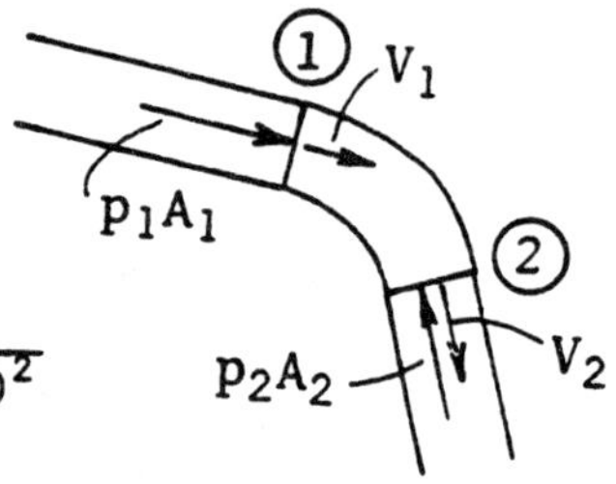

Preliminaries:

Velocities:

$\underline{V}_1 = (Q/A)[(13/\ell_1)\underline{j} - (10/\ell_1)\underline{k}]$ where $\ell_1 = \sqrt{13^2+10^2}$

Thus $\underline{V}_1 = (Q/A)[0.793\,\underline{j} - 0.6097\,\underline{k}]$

$\underline{V}_2 = (Q/A)[(13/\ell_2)\underline{i} + (19/\ell_2)\underline{j} - (20/\ell_2)\underline{k}]$ where $\ell_2 = \sqrt{13^2+19^2+20^2}$

Then $\underline{V}_2 = (Q/A)[0.426\underline{i} + 0.623\underline{j} - 0.656\underline{k}]$

Pressure forces:

$\underline{F}_{p_1} = p_1A_1(0.793\underline{j} - 0.6097\underline{k})$; $\underline{F}_{p_2} = p_2A_2(-0.426\underline{i} - 0.623\underline{j} + 0.656\underline{k})$

Weight:

$\underline{W} = -3 \times 9{,}810\underline{k}$

Solution:

$\Sigma\underline{F} = \rho Q(\underline{V}_2 - \underline{V}_1)$

$F_{block,x} - 0.426\ p_2A_2 = \rho Q[0.426\ Q/A - 0]$; where $p_2A_2 = 25{,}000 \times (\pi/4) \times (1.3)^2 = 33{,}183$ N; $Q/A = 15/((\pi/4) \times (1.3)^2) = 11.30$ m/s

Then $F_{block,x} = 1{,}000 \times 15 \times 0.426 \times 11.3 + 0.426 \times 33{,}183 = \underline{\underline{86{,}343\text{N}}}$

$F_{block,y} + 0.793\ p_1A_1 - 0.623\ p_2A_2 = \rho Q[0.623(Q/A) - 0.793\ Q/A]$

where $p_1A_1 = 20{,}000 \times (\pi/4)(1.3)^2 = 26{,}546$ N

Then $F_{block,y} = 1{,}000 \times 15(11.3)(-0.170) - 0.793 \times 26{,}546 + 0.623 \times 33{,}183 = -28{,}815 - 21{,}051 + 20{,}673 = \underline{\underline{-29{,}193\text{N}}}$

$F_{block,z} - 0.6097\ p_1A_1 + 0.656\ p_2A_2 - \text{Wgt} = 1{,}000 \times 15[-0.656(Q/A) - (-0.6097\ Q/A)]$

$F_{block,z} = -7{,}848 + 3 \times 9{,}810 + 10{,}000 - 0.656 \times 33{,}183 + 0.6097 \times 26{,}546 = \underline{\underline{25{,}999\text{N}}}$

Then the total force which the thrust block exerts on the bend to hold it in place is

$\underline{\underline{\underline{F} = (86.35\underline{i} - 29.19\underline{j} + 26.00\underline{k})\text{kN}}}$

6.53

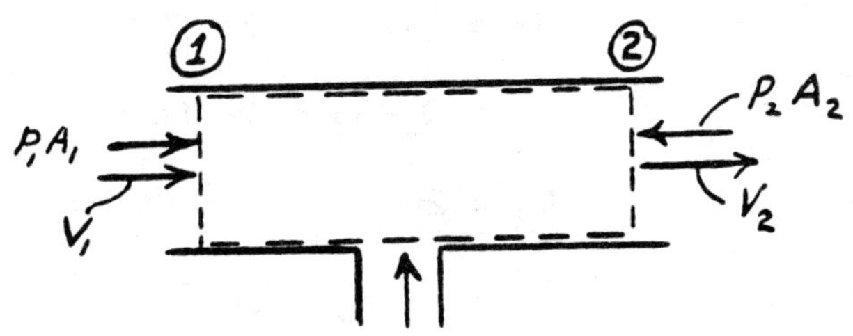

Continuity equation:

$\dot{m}_1$ + 500 kg/s = $\dot{m}_2$; $\dot{m}_1$ = (10 m/s) (0.10 m^2) (1,000 kg/m^3)

= 1,000 kg/s

(1,000 + 500) kg/s = $\dot{m}_2$

$V_2 = (\dot{m}_2)/(\rho A_2)$ = (1,500)/((1,000) (0.1)) = 15 m/s

Momentum equation in x-direction:

$\Sigma F_x = V_x \rho \mathbf{V} \bullet \mathbf{A}$

$p_1A_1 + p_2A_2 = -\dot{m}_1V_1 + \dot{m}_2V_2$

$A(p_1 - p_2)$ = (-1,000) (10) + (1,500) (15)

$p_1 - p_2$ = (22,500 - 10,000)/0.10

$p_1 - p_2$ = 125,000 P_a = 125 kPa

6.54 $Q_2 = Q_1 + Q_3 = 3.40 \text{ m}^3/\text{s}$

$V_2 = Q_2/A_2 = (3.40 \text{ m}^3/\text{s})/A_2 = 13.6$ m/s

Momentum Eq:

$\Sigma F_y = \Sigma V_y \rho \underline{V} \cdot \underline{A}$

$-(1 \times A_1 \times \gamma) + p_1 A_1 = 10 \times 1,000 \times (-10 \times 0.25) + 13.6 \times 1,000\ (13.6 \times 0.25)$

p_1 = (9,810 - 100,000 + 185,000) N/m^2 = 94.8 kPa

6.55

$$V_1 = Q_1/A_1 = 20 \text{ ft/s}$$

$$V_2 = Q_2/A_2 = 12 \text{ ft/s}$$

$$Q_3 = 20 - 12 = 8 \text{ ft}^3\text{/s}$$

$$V_3 = Q_3/A_3 = 32 \text{ ft/s}$$

$$\Sigma F_x = \Sigma V_x \rho \mathbf{V} \cdot \mathbf{A}$$

$$F_x + p_1A_1 - p_2A_2 = (20\rho)\ (-20) + (12\rho)\ (+12) + (32 \cos 30°)\ (\rho)\ (8)$$

$$F_x + (1{,}000)\ (1) - (900)\ (1) = -400\rho + 144\rho + \rho(8)\ (32)\ (0.866)$$

$$F_x + 100 = 1.94\ (-34.3)$$

$$\underline{\underline{F_x = -166.5 \text{ lbf}}}$$

6.56

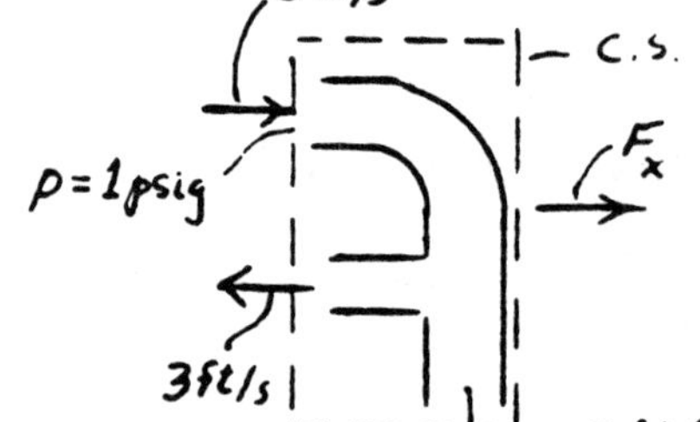

Momentum equation:

$$p_aA_a + p_bA_b + F = \Sigma V_x \rho \mathbf{V} \cdot \mathbf{A}$$

$$(1)\ (5) + 0 + F = (6)\ (1.94)\ ((-6)\ (5/144)) + (-3)\ (1.94)\ (3)\ (5/144)$$

$$5 + F = -2.425 - 0.606$$

$$\underline{\underline{F = -8.03 \text{ lbf}}}$$

6.57 $\Sigma F_x = \Sigma V_x \rho \underline{V} \cdot \underline{A} + \cancelto{0}{d/dt \int \rho V d\forall}$

$$F_x + p_A A_A + p_B A_B = V_A(-\rho V_A A_A) + (-V_B)(\rho V_B A_B)$$

$$F_x = -A(p_A + \rho V_A^2 + \rho V_B^2)$$

$$= -0.2\ ((48 + (1000)(6^2) + (1000)(3^2))$$

$$\underline{F_x = -9010 \text{ N}}$$

$$\Sigma F_y = \Sigma V_y(\rho \underline{V} \cdot \underline{A})$$

$$F_y = (-3)(1000)(3 \times 0.2)$$

$$\underline{F_y = -1800 \text{ N}}$$

6.58 $V_1 = 0.25/(\pi \times 0.075 \times 0.075) = 14.15$ m/s

$V_2 = 0.15/(\pi \times 0.05 \times 0.05) = 19.10$ m/s

$V_3 = (0.25 - 0.15)/(\pi \times 0.075 \times 0.075) = 5.66$ m/s

$F_x = -100{,}000 \times \pi \times 0.075 \times 0.075 + 80{,}000 \times \pi \times 0.075 \times 0.075$

$-1{,}000 \times 14.15 \times 0.25 + 1{,}000 \times 5.66 \times 0.10 = -3{,}325\text{N} = -3.325\text{kN}$

$F_y = -1{,}000 \times 19.10 \times 0.15 - 70{,}000 \times \pi \times 0.05 \times 0.05$

$= -3{,}415\text{N} = -3.415\text{kN};$ Total force $\underline{\underline{\underline{F} = (-3.325\underline{i} - 3.415\underline{j})\text{kN}}}$

6.59

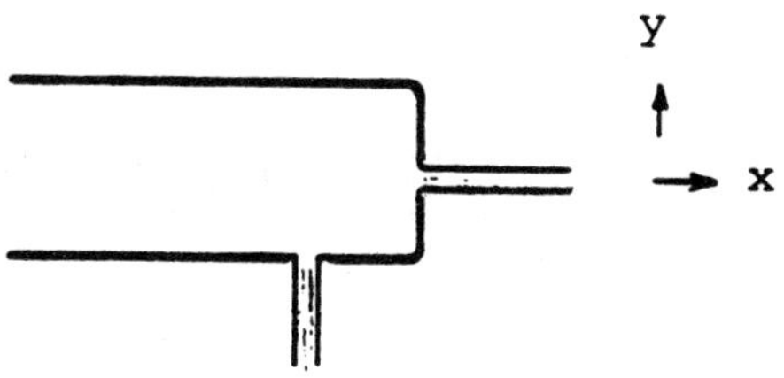

$p_{pipe}/\gamma + V_p^2/2g = p_{jet}/\gamma + V_j^2/2g$

$V_pA_p = \Sigma V_jA_j$

$V_p = 2 \times 30 \times 0.01/0.10 = 6.00$ m/s

Then $p_p = (\gamma/2g)(V_j^2 - V_p^2)$

$= 500(900-36) = 432{,}000$ Pa

$\Sigma F_x = \Sigma V_x \rho \underline{V} \cdot \underline{A}$

$p_pA_p + F_x = -V_p\rho V_pA_p + V_j\rho V_jA_j$

$F_x = -1{,}000 \times 6^2 \times 0.10 + 1{,}000 \times 30^2 \times 0.01 - 432{,}000 \times 0.1$

$F_x = -37{,}800$ N

$\Sigma F_y = \Sigma V_y \rho \underline{V} \cdot \underline{A}$

$F_y = V_y \rho V_y A_y$

$= -30 \times 1{,}000 \times 30 \times 0.01 = -9{,}000\text{N}$

$\Sigma F_z = 0$

$-200 - \gamma V\!\!\!\!- + F_z = 0$

$F_z = 200 + 9{,}810 \times 0.1 \times 0.4 = \underline{\underline{592 \text{ N}}}$

Force required = $\underline{\underline{-37.8\underline{i} - 9.0\underline{j} + 0.59\,\underline{k}\text{ kN}}}$

6.60

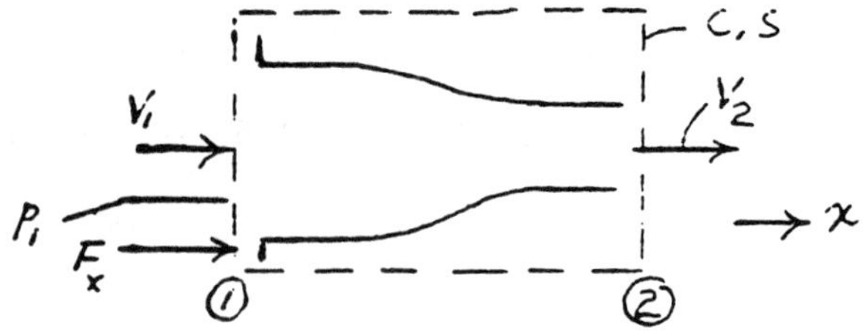

$$V_1 = Q/A_1 = 15/((\pi/4) \times 1^2) = 19.098 \text{ ft/s}$$

$$V_2 = Q/A_2 = 15/((\pi/4)(8/12)^2) = 42.97 \text{ ft/s}$$

$$p_1 + \rho V_1^2/2 = p_2 + \rho V_2^2/2$$

$$p_1 = 0 + (\rho/2)(V_2^2 - V_1^2)$$

$$= 1{,}437 \text{ lbf/ft}^2$$

$$\Sigma F_x = \Sigma V_x \rho \mathbf{V} \cdot \mathbf{A}$$

$$p_1 A_1 + F = (19.098)(1.94)(-15) + 42.97(1.94)(15)$$

$$(1{,}437)(\pi/4)(1^2) + F = (1.94)(15)(42.97 - 19.098)$$

$$\underline{\underline{F = -434 \text{ lbf } \mathbf{i}}}$$

6.61

$V_1 = 0.3/(\pi \times 0.15 \times 0.15) = 4.244$ m/s; $V_2 = 4.244 \times 4 = 16.976$ m/s

$p_1 = 0 + (1{,}000/2)(16.976 \times 16.976 - 4.244 \times 4.244) = 135{,}086$ Pa

$F_x = -135{,}086 \times \pi \times 0.15 \times 0.15 + 1{,}000 \times 0.3(16.976 - 4.244)$

$= -5{,}729\text{N} = \underline{\underline{-5.729 \text{ kN}}}$

6.62

$V_A = V_B = 16 \times 144/[(\pi/4)(4 \times 4 + 4.5 \times 4.5)] = 80.93$ fps

$V_1 = 16/(\pi \times 0.5 \times 0.5) = 20.37$ fps

$p_1 = 0 + (1.94/2)(80.93 \times 80.93 - 20.37 \times 20.37) = 5{,}951$ psf

$\Sigma F_x = \Sigma V_x \rho \underline{V} \cdot \underline{A}$

$F_x = -5{,}951 \times \pi \times 0.5 \times 0.5 \times \sin 30° - 80.93 \times 1.94 \times 80.93 \times \pi \times 2$

$\times 2/144 - 20.37 \times 1.94 \times 16.0 \sin 30° = \underline{\underline{-3{,}762 \text{ lbf}}}$

6.63 $V_A = V_B = 0.5/(\pi \times 0.05 \times 0.05 + \pi \times 0.06 \times 0.06) = 26.1$ m/s

$V_1 = 0.5/(\pi \times 0.15 \times 0.15) = 7.07$ m/s

$p_1 = (1{,}000/2)(26.1 \times 26.1 - 7.07 \times 7.07) = 315{,}612$ Pa

$F_x = -315{,}612 \times \pi \times 0.15 \times 0.15 \times \sin 30° - 26.1 \times 1{,}000 \times 26.1$

$\times \pi \times 0.05 \times 0.05 - 7.07 \times 1{,}000 \times 0.5 \sin 30° = -18{,}270\text{N} = \underline{\underline{-18.27\text{kN}}}$

6.64 $V_2 = 4\ V_1$

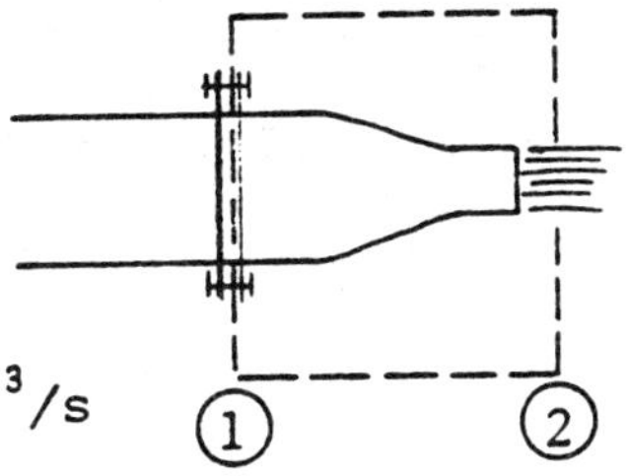

$(V_1^2/2g) + (p_1/\gamma) = (V_2^2/2g) + (p_2/\gamma)$

$15(V_1^2/2g) = (200{,}000/9{,}810)$

$V_1 = 5.16$ m/s; $V_2 = 20.66$ m/s; $Q = 0.365\ \text{m}^3/\text{s}$

$\Sigma F_x = \rho Q(V_{2x} - V_{1x})$

$F_{bolts} = -200{,}000 \times \pi \times 0.15 \times 0.15 + 1{,}000 \times 0.365(20.66 - 5.16)$

$= -7{,}441$ N; Force per bolt = $\underline{\underline{1{,}240\text{ N}}}$

6.65 $Q = 300/449 = 0.668$ cfs; $V_p = 13.61$ ft/sec; $V_n = 54.44$ ft/sec

$p_p = (\rho/2)(V_n^2 - V_p^2) = (1.94/2)(54.44^2 - 13.61^2) = 2{,}695$ psf

$\Sigma F_x = \rho Q(V_{nx} - V_{px})$

$F_x = -(2{,}695 \times \pi \times 1.5 \times 1.5/144) + 1.94 \times 0.668(54.44 - 13.61) = \underline{\underline{-79.38\text{ lbf}}}$

6.66 $V_h = 7.073$ m/s; $V_n = 28.29$

$p_h = (1{,}000/2)(28.29 \times 28.29 - 7.073 \times 7.073) = 375{,}148$ Pa

$F_x = -(375{,}148 \times \pi \times 0.015 \times 0.015) + 1{,}000 \times 0.005(28.29 - 7.073)$

$= \underline{\underline{-159.1\text{ N}}}$

6.67 $V_b = 5/(1/4) = 20$ ft/sec; $V_B = 20 \times (3/8) = 7.5$ ft/sec

$p_B = (\rho/2)(V_b^2 - V_B^2) = (1.94/2)(20 \times 20 - 7.5 \times 7.5) = 333.4$ psf

$p_{gage} = 333.4 - 62.4 \times 4/12 = \underline{\underline{312.6\text{ psf} = 2.171\text{ psi}}}$

$\Sigma F_x = \rho Q(V_b - V_B)$

$F_x = -333.4 \times 8/12 + 1.94 \times 5 \times (20 - 7.5) = \underline{\underline{-101\text{ lbf}}}$

6.68 $V_b = 0.4/0.07 = 5.71$ m/s; $V_B = 0.40/0.20 = 2.00$ m/s

$p_B = (1{,}000/2)(5.71 \times 5.71 - 2.00 \times 2.00) = 14.326$ Pa

$p_{gage} = 14{,}326 - 9.810 \times 0.1 + 13{,}345\ P_a = 13.345$ kPa

$F_x = -14{,}326 \times 0.2 + 1{,}000 \times 0.4(5.71 - 2.00) = -1{,}381$ N $= \underline{\underline{-1.381 \text{ kN}}}$

6.69 $V_2 = 4/(\pi \times 1.25 \times 1.25/144) = 117.34$ fps

$V_1 = 117.34 \times (2.5 \times 2.5/6 \times 6)\ 20.37$ fps

$p_1 = (\rho/2)(V_2^2 - V_1^2) = (1.94/2)(117.34 \times 117.34 - 20.37 \times 20.37) = 12{,}953$ psf

$\Sigma F_x = \rho Q(V_{2x} - V_{1x})$

$F_x = -12{,}953 \times \pi \times 0.25 \times 0.25 + 1.94 \times 4(117.34 - 20.37) = \underline{\underline{-1791 \text{ lbf}}}$

6.70 $V_2 = 0.05/(\pi \times 0.0175 \times 0.0175) = 51.97$ m/s

$V_1 = 51.97 \times (3.5/10)^2 = 6.37$ m/s

$p_1 = (1{,}000/2)(51.97 \times 51.97 - 6.37 \times 6.37) = 1{,}330{,}200$ Pa

$F_x = -1{,}330{,}200 \times \pi \times 0.05 \times 0.05 + 1{,}000 \times 0.05(51.97 - 6.37)$

$= \underline{\underline{-8{,}167 \text{ N}}}$

6.71 $\Sigma F_x = \Sigma V_x \rho \underline{V} \cdot \underline{A}$

$p_1A_1 + F_x = V_1(-\rho V_1 A_1) + V_2(\rho V_2 A_2)$

$F_x = -p_1A_1 - \rho V_1^2 A_1 + \rho V_2^2 A_2$

Get p_1 by Bernoulli's equation. Assume $z_1 = z_2$

$p_1 + \rho V_1^2/2 = p_2 + \rho V_2^2/2$

$p_1 = \rho V_2^2/2 - \rho V_1^2/2$

Then $F_x = -((\rho/2)(V_2^2 - V_1^2))A_1 - \rho V_1^2 A_1 + \rho V_2^2 A_2$

$= -(500)(30^2-3^2)(0.001) - (1000)(3^2)(.001) + 1000(30^2)(.0001)$

$\underline{F_x = -364.5 \text{ N}}$

6.72

$Q_1 = 40 \times \pi \times 1 \times 1/144 = 0.872$ cfs

$Q_2 = 40 \times \pi \times 0.5 \times 0.5/144 = 0.218$ cfs

$Q_3 = 40 \times \pi \times 1.5 \times 1.5/144 = \underline{1.963 \text{ cfs}}$

$Q_t = 3.053$ cfs

$V = 3.053/(\pi \times 0.5 \times 0.5) = 3.887$ fps

$p = 0 + (1.94/2)(40 \times 40 - 3.887 \times 3.887) = 1{,}537$ psf

$\Sigma F_x = \Sigma V_x \rho \underline{V} \cdot \underline{A}$

$F_x = -1{,}537 \times \pi \times 0.5 \times 0.5 - 3.887 \times 1.94 \times 3.053 + 40 \times 1.94$

$\times\, 0.872 + 40 \times 1.94 \times 0.218 \cos 30° = -1{,}148$ lbf

$F_y = 40 \times 1.94 \times 0.218 \sin 30° + 40 \times 1.94 \times 1.963 = 160.8$ lbf

$\underline{F} = \underline{\underline{-1{,}148i + 160.8j \text{ lb}}}$

6.73

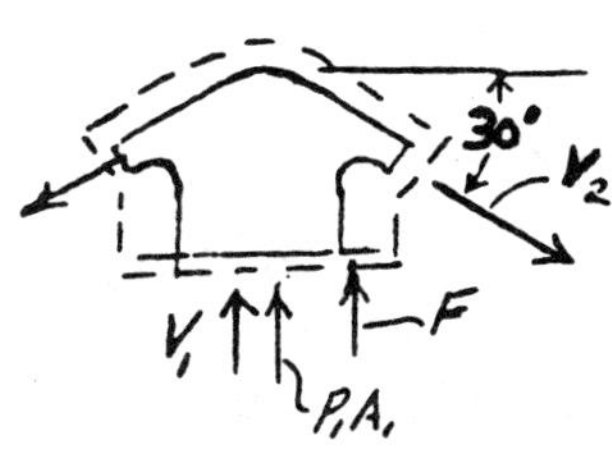

$V_2 = 65$ ft/s

$V_1 = Q/A_1 = 3/((\pi/4)(0.5^2)) = 15.28$ ft/s

Bernoulli Eq:

$p_1 + \rho V_1^2/2 = p_2 + \rho V_2^2/2$

$p_1 = (\rho/2)(V_2^2 - V_1^2) = (1.94/2)(65^2 - 15.28^2) = 3872.8$ lbf/ft^2

Momentum Eq.:

$\Sigma F_y = \Sigma V_y \rho \mathbf{V} \cdot \mathbf{A}$

$p_1 A_1 + F_y = (15.28)(1.94)(-3.0) + (-65 \sin 30°)(1.94)(3)$

$F_y = -(3872.8)(\pi/4)(.5^2) - 278.1 = \underline{\underline{-1039 \text{ lbf}}}$

6.74 $V_p = Q/A = (2 \times 80.2 \times \pi \times 0.5 \times 0.5)/(\pi \times 2 \times 2) = 10.025$ fps

$\Sigma F_x = \Sigma V_x \rho \underline{V} \cdot \underline{A}$

$F_x = -43 \times \pi \times 2 \times 2 + (80.2 \times 1.94 \times 80.2 \times \pi \times 0.5 \times 0.5/144)$

$-(80.2 \times 1.94 \times 80.2 \times \pi \times 0.5 \times 0.5/144)\sin 30°$

$-(10.025 \times 1.94 \times 10.025 \times \pi \times 0.1667 \times 0.1667) = -523.3$ lbf

$F_y = -1.94 \times 80.2 \times \cos 30° \times 80.2 \times \pi \times 0.5 \times 0.5/144 = -58.94$ lbf

Total force $\underline{F} = -523.3\underline{i} - 58.94\underline{j}$ lbf

(Weight of water and nozzle are neglected)

6.75 $V_3 = (40 \times 2 \times 2 - 100 \times 1 \times 1)/(1 \times 1) = 60$ ft/s

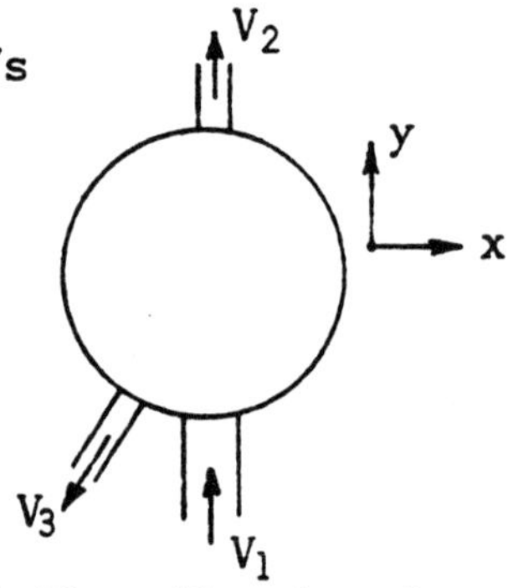

$\Sigma F_x = \Sigma V_x \rho \underline{V} \cdot \underline{A}$

$F_x = -60 \sin 30° \times 1.5 \times 1.94 \times 60 \times \pi \times 0.5$

$\times 0.5/144 = -28.57$ lbf

$\Sigma F_y = \Sigma V_y \rho \underline{V} \cdot \underline{A}$

$F_y = -60 \times \pi \times 1 \times 1 + 200 + (1.94 \times 1.5\pi/144)(-40 \times 40 \times 1 \times 1$

$+ 100 \times 100 \times 0.5 \times 0.5 - 60 \times 60 \times 0.5 \times 0.5 \times \cos 30)$

$= +19.16$ lbf; Total force $\underline{F} = -28.57\underline{i} + 19.16\underline{j}$ lbf

---

6.76 $V_3 = (10 \times 5 \times 5 - 30 \times 2.5 \times 2.5)/(2.5 \times 2.5) = 10$ m/s

$F_x = -10 \sin 30° \times 1{,}500 \times 10 \times \pi \times 0.0125 \times 0.0125 = -36.8$N

$F_y = -400{,}000 \times \pi \times 0.025 \times 0.025 + 600 + (1{,}500\pi)$

$\times (-10 \times 10 \times 0.025 \times 0.025 + 30 \times 30 \times 0.0125 \times 0.0125$

$- 10 \times 10 \times 0.0125 \times 0.0125 \cos 30°) = 119$ N

Total force $\underline{F} = (-36.8i + 119\ j)$ N

6.77 $\Sigma F_x = \Sigma V_x \underline{V} \cdot \underline{A}$

$F_x = 0$

$\Sigma F_y = \Sigma V_y \underline{V} \cdot \underline{A} = 2{,}000 \times 20 \times 0.1$

$-2{,}000 \times 15 \times 0.5 = 4{,}000 - 15{,}000$

$= -11{,}000$ N; $\underline{F} = -11.0\underline{j}$ kN

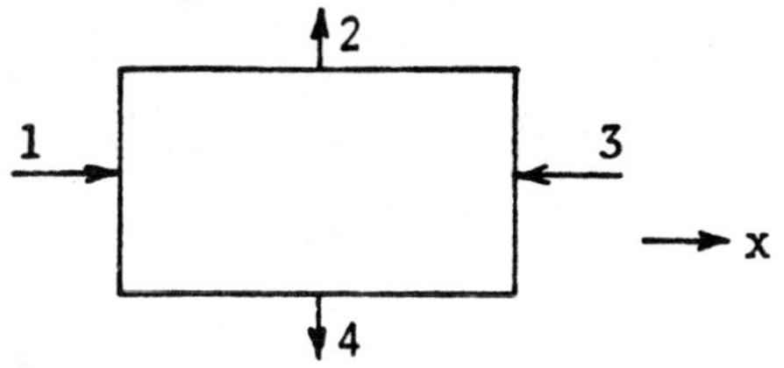

$Q_4 = 0.6 - 0.10$
$= 0.50 \text{ m}^3/\text{s}$

6.78 To verify Eq. (6.16) the quantities Q, V_1, V_2, b, y_1, y_2 and F_G will have to be measured. Since a laboratory is available for your experiment it is assumed that the laboratory has discharge measuring equipment to obtain Q. The width b can be measured by a suitable scale. The depths y_1 and y_2 can be measured by means of piezometer tubes attached to openings in the bottom of the channel or by means of point gages by which the actual level of the surface of the water can be determined. Then V_1 and V_2 can be calculated from $V = Q/A = Q/(by)$.

The force on the gate can be indirectly evaluated by measuring the pressure distribution on the face of the gate. This pressure may be sensed by piezometers or pressure transducer attached to small openings (holes) in the gate. The pressure taps on the face of the gate could all be connected to a manifold and by appropriate valving the pressure at any particular tap could be sensed by a piezometer or pressure transducer. The pressures at all the taps should be measured for a given run. Then by integrating the pressure distribution over the surface of the gate one can obtain F_G. Then compare the measured F_G with the value obtained from the right hand side of Eq. 6.16. The design should be such that air bubbles can be purged from tubes leading to piezometer or transducer so that valid pressure readings are obtained.

6.79

$$V_1^2/2g + z_1 = V_2^2/2g + z_2$$

$$(0.6/3)^2\ V_2^2/2g + 3 = V_2^2/2g + 0.6$$

$$V_2 = 12.69 \text{ fps};\quad V_1 = 2.54;\ Q = 7.614 \text{ cfs/ft}$$

$$\Sigma F_x = \rho Q(V_{2x} - V_{1x})$$

$$F_x = -62.4 \times 3.0 \times 3.0/2 + 62.4 \times 0.6 \times 0.6/2 + 1.94 \times 7.614$$

$$\times\ (12.69 - 2.54) = \underline{\underline{-119.6 \text{ lbf/ft}}}$$

6.80

$$\Sigma F_x = \Sigma V_x \rho \underline{V} \cdot \underline{A}$$

$$p_1 A - p_2 A - F_\tau = -\rho U^2 A + \int_{A_2} \rho u_2^2 dA \qquad (1)$$

$$u_2 = u_{max}\ (1 - (r/r_0)^2)$$

$$u_2^2 = u_{max}^2\ (1 - (r/r_0)^2)^2$$

$$\int_{A_2} \rho u_2^2 dA = \int_0^{r_0} \rho u_{max}^2\ (1 - (r/r_0)^2)^2 2\pi r dr)$$

$$= -\rho u_{max}^2 \pi r_0^2 \int_0^{r_0} (1 - (r/r_0)^2)^2 (-2r/r_0^2) dr$$

The above integral is in the form $\int u^n du = u^{n+1}/(n+1)$

Thus $$\int_{A_2} \rho u_2^2 dA = -\rho u_{max}^2 \pi r_0^2 (1 - (r/r_0)^2)^3/3 \Big|_0^{r_0}$$

$$= -\ \rho u_{max}^2 \pi r_0^2 (0 - 1/3)$$

$$= +\ \rho u_{max}^2 \pi r_0^2/3 \qquad (2)$$

Now determine U in terms of u_{max}.

$$UA = \int u dA$$

$$= \int_0^{r_0} u_{max}(1 - (r/r_0)^2)\ 2\pi r dr$$

$$= -u_{max} \pi r_0^2 \int_0^{r_0} (1 - (r/r_0)^2)(-2r/r_0^2) dr$$

$$= -u_{max} \pi r_0^2 (1 - (r/r_0)^2)^2/2 \Big|_0^{r_0}$$

6.80 (Cont'd) $UA = U_{max}\pi r_0^2/2$

or $U_{max} = 2U$ (3)

Substituting $U_{max} = 2U$ into Eq. (2) yields:

$$\int \rho u_2^2 dA = (4/3)\rho U^2 A \quad (4)$$

Then when Eq. (4) is substituted into Eq. (1) we get

$$p_1 A - p_2 A - F_T = -\rho U^2 A + (4/3)\ \rho U^2 A$$

$$\underline{\underline{F_T = A(p_1 - p_2 - (1/3)\rho U^2)}}$$

6.81 $Q = 80 \times \pi \times 1.5 \times 1.5 = 565.5$ cfs; $V_2 = 80$ fps; $V_1 = 0$

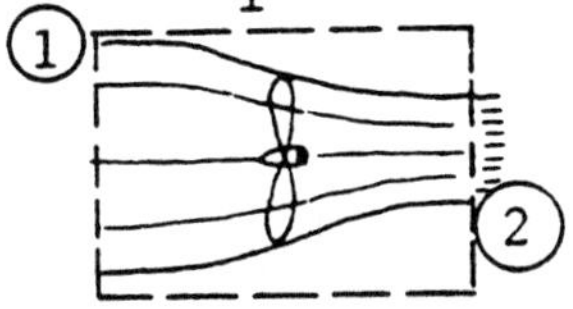

$\Sigma F_x = \rho Q(V_{2x} - V_{1x})$

$F_x = 0.00228 \times 565.5(80 - 0) = 103.1$ lbf

When $V_1 = 20$ fps, $F_x = 0.00228 \times 565.6\ (80\text{–}30) = \underline{\underline{64.5\ \text{lbf}}}$

6.82 $V_2 = 10 \times (3/4.5)^2 = 4.44$ m/s

$\Sigma F_x = \Sigma V_x \rho \underline{V} \cdot \underline{A}$

$T = 1.2 \times 10 \times \pi \times 1.5 \times 1.5\ (10 - 4.44) = \underline{\underline{471\ \text{N}}}$

6.83 $V_1 = V_0 D_o^2/(D_c^2 - D_j^2)$; $V_2 = (V_0 D_c^2 + V_j D_j^2)/D_c^2$

$\Sigma F_x = \Sigma \rho V_x \mathbf{V} \cdot \mathbf{A}$

Neglecting friction between sections (1) & (2)

$(p_1 - p_2)\pi D_0^2/4 = -\rho V_1^2 \pi (D_0^2 - D_j^2)/4 - \rho V_j^2 \pi D_j^2/4 + \rho V_2^2 \pi D_0^2/4$

$(p_2 - p_1) = \underline{\underline{\rho V_1^2 (D_0^2 - D_j^2)/D_0^2 + \rho V_j^2 \cdot D_j^2/D_0^2 - \rho V_2^2}}$

$V_j = 15$m/s ; $V_1 = 3$m/s

$V_2 A_2 = V_j A_j + V_1 \times (2/3) A_0$

$= V_j \times (1/3)(A_0) + 3\ (2/3) A_0$

$= 1/3 \times 15\ A_0 + 2A_0$

$= 7A_0$

$V_2 = 7$ m/s

Solving yields: $p_2 - p_1 = \underline{\underline{32\ \text{kPa}}}$

6.84

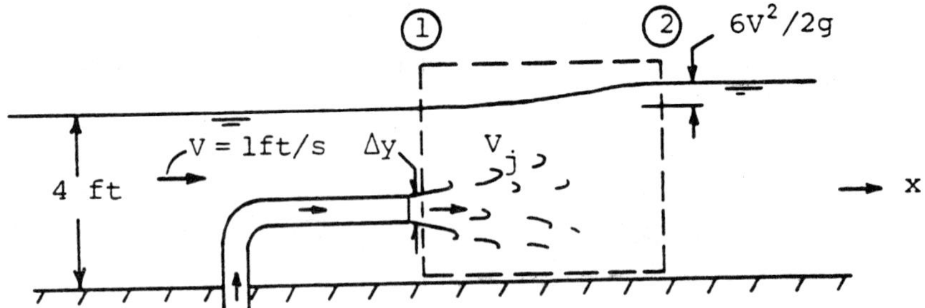

First carry out the analysis for a section 1 ft wide (unit width) and neglect bottom friction.

$$\Sigma F_x = \Sigma V_x \rho \underline{V} \cdot \underline{A}$$

$$\gamma y_1^2/2 - \gamma y_2^2/2 = -1\,\rho\,(1 \times (4-\Delta y)) - V_j \rho (V_j \Delta y) + V_2 \rho (V_2 y_2) \qquad (1)$$

but $\quad y_2 = 4 \text{ ft} + 6\ V^2/2g$

$$= 4 + 6/2g = 4.0932 \text{ ft}$$

Also $\quad V_2 y_2 = V_1 (4-\Delta y) + V_j \Delta y$

$$V_2 = V_1 (4-\Delta y)/\bar{y}_2 + V_j \Delta y / y_2$$

Assume $\quad \Delta y = 0.10$ ft

Then $\quad V_2 = 1(3.9)/(4.0932) + V_j \times 0.1/4.0392 = 0.9528 + 0.02476\ V_j$

Eq. (1) then reduces to:

$$V_j^2 - (0.9528 + 0.02476\ V_j)^2 \times 40.932 = 5g(y_2^2 - y_1^2) - 39.0$$

$$= 82.44 \text{ ft}^2/\text{s}^2$$

Solving: $V_j = 12.1$ ft/s $\quad A_j = 0.10 \text{ ft}^2$

If circular nozzles were used, then $A_j = (\pi/4) d_j^2$; $d_j = 4.28$ in. Therefore, one could use 8 nozzles of about 4.3 in. in diameter discharging water at 12.1 ft/s.

Other combinations of d_j, V_j and number of jets are possible to achieve the desired result.

6.85

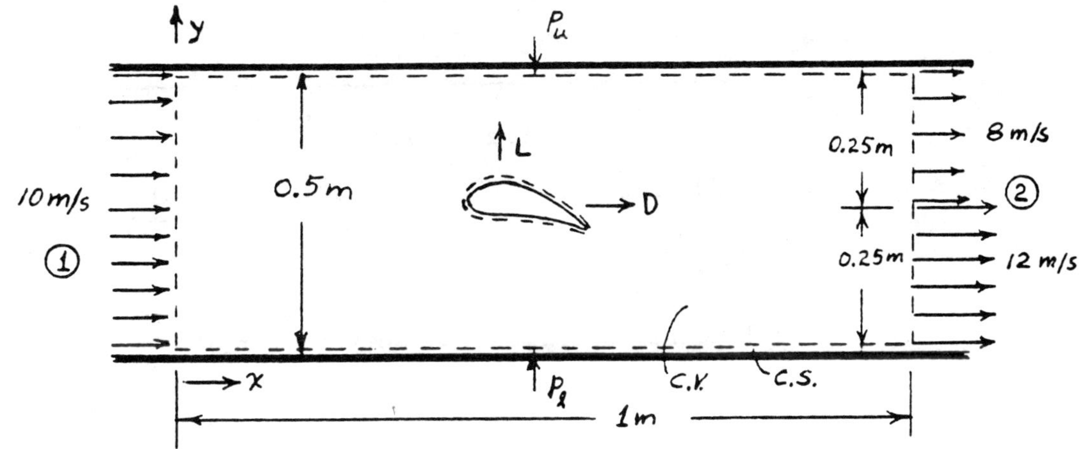

Momentum equation in *x* direction (along tunnel)

$\Sigma F_x = \Sigma V_x \rho \underline{V} \cdot \underline{A}$

$-D + p_1A_1 - p_2A_2 = V_1(-\rho V_1A_1) + V_a(\rho V_aA/2) + V_b(\rho V_bA/2)$

$-D/A = p_2 - p_1 - \rho V_1^2 + \rho V_a^2/2 + \rho V_b^2/2$

where $p_1 = p_u(x=0) = p_\ell(x=0) = 100$ Pa, gage

$p_2 = p_u(x=1) = p_\ell(x=1) = 90$ Pa, gage

then $-D/A = 90 - 100 + 1.2\ (-100 + 32 + 72)$

$-D/A = -5.2$

$\underline{D = (5.2)(0.5 \times .5) = 1.3\text{ N}}$

y direction:

$\Sigma F_y = 0$

$-L + \int_0^1 p_\ell B dx - \int_0^1 p_u B dx = 0$ where B is depth of tunnel

$-L + \int_0^1 (100-10x+20x(1-x))0.5dx - \int_0^1 (100-10x-20x(1-x))0.5dx = 0$

$-L + 0.5\ (100x - 5x^2 + 10x^2 - (20/3)x^3)|_0^1 - 0.5(100x - 5x^2 - 10x^2 + (20/3)x^3|_0^1 = 0$

$-L + 49.167 - 45.833 = 0$

$\underline{L = 3.334\text{ N}}$

6.86

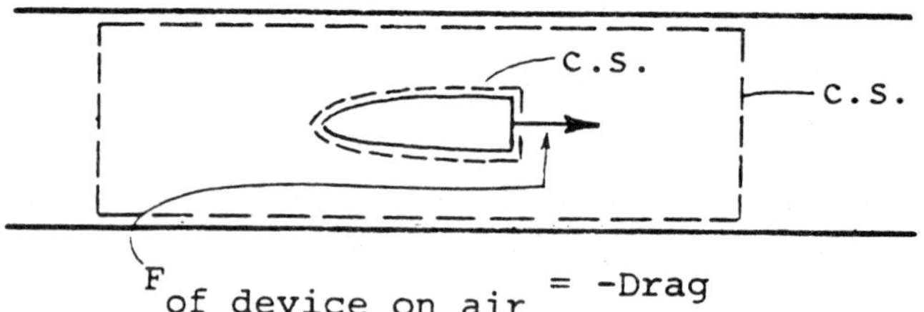

$\dot{m} = \rho VA = 0.0026 \times 100 \times (\pi/4)(3.0)^2 = \underline{\underline{1.838 \text{ slugs/s}}}$

$$\int_0^{r_0} VdA = Q$$

But V is linearly distributed, so $V = (r/r_0)V_{max}$

Thus $$\int_0^{r_0} ((r/r_0)V_{max})2\pi r dr = \bar{V}A$$

$$2V_{max}r_0^2/3 = \bar{V}r_0^2$$

$$V_{max} = 1.5\bar{V} = \underline{\underline{150 \text{ ft/s}}}$$

Write momentum equation to determine drag:

$p_1A_1 - p_2A_2 - \text{Drag} = -100 \times .0026 \times (\pi/4) \times 3^2 \times 100 + \int_0^{r_c} \rho V^2 dA$

$144 \times (0.24 \times \pi \times 1.5^2 - 0.10 \times \pi \times 1.5^2) - \text{Drag} = -184 + \int_0^{r_0} \rho$

$1 \times ((r/r_0)\, 1.5\bar{V})^2\, 2\pi r dr;\ 142 - \text{Drag} = -184 + (4.5\pi/4)\bar{V}^2 r_0^2 \rho$

Then for $\bar{V} = 100$ ft/s and $r_0 = 1.5$ ft

$\text{Drag} = 142 + 184 - 207 = \underline{\underline{119 \text{ lbf}}}$

6.87 This type of problem is directly analogous to the rocket problem except that the weight does not directly enter as a force term and $p_e = p_0$. Therefore, the appropriate equation is

$$M\, dv_s/dt = \rho V_e^2 A_e - \text{Friction}$$

$$a = (1/M)(\rho V_e^2 (\pi/4) d_e^2 - Wf)$$

where f = coeffic. of sliding friction

$$W = \text{wgt} = 350 + 981 = 1{,}331 \text{ N}$$

$$a = (g/W)(1{,}000 \times 25^2 (\pi/4) \times 0.015^2 - (1{,}331 \times 0.05))$$

$$a = (9.81/1{,}331)(43.90) \text{ m/s}^2 = \underline{0.324 \text{ m/s}^2}$$

6.88 $F_y = \int_{c.s.} V_y \rho \underline{V}\cdot d\underline{A}$

$$F_y = \int_0^{\pi} V \sin\theta \rho V t r d\theta = \rho V^2 t r \int_0^{\pi} \sin\theta d\theta = \underline{\underline{2\rho V^2 t r}}$$

6.89

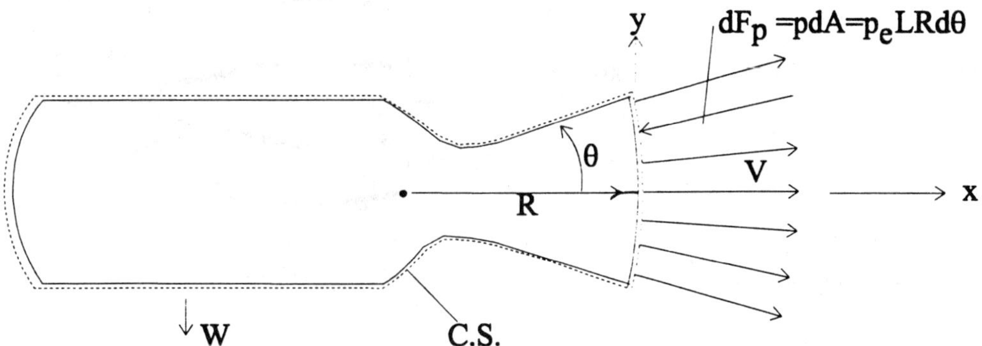

Define A_e as the projection of the exit area on the y plane. Use the momentum equation to solve this problem and let the control surface surround the nozzle and fuel chamber as shown above. The forces acting on the system are the pressure forces and thrust, T. The pressure forces in the x-direction are from p_0 and p_e. Writing the momentum equation in the x-direction we have:

$$T + p_o A_e - p_e A_e = \int_A V_x \rho \underline{V} \cdot d\underline{A}$$

$$T + p_0 A_e - p_e A_e = \int 2(V\cos\theta)\rho(-VLRd\theta)$$

$$T + p_0 A_e - p_e A_e = -2V^2 \rho LR \int_0^\theta \cos\theta d\theta$$

$$T + p_0 A_e - p_e A_e = -2V^2 \rho LR \sin\theta$$

$$\text{But } \dot{m} = 2\int_0^\theta \rho V dA = 2\int_0^\theta \rho VLRd\theta = 2\rho VLR\theta$$

$$T + p_0 A_e - p_e A_e = -2\rho V^2 LR\theta(\sin\theta/\theta)$$

$$T + p_0 A_e - p_e A_e = -V\dot{m}\sin\theta/\theta$$

$$T = \dot{m}V(-\sin\theta/\theta) + p_e A_e - p_0 A_e$$

$$T = \dot{m}Vf(\theta) + A_e(p_e - p_0)\lambda(\theta)$$

$$\text{where } f(\theta) = -\sin\theta/\theta$$

$$\text{and } \lambda(\theta) = 1$$

6.90

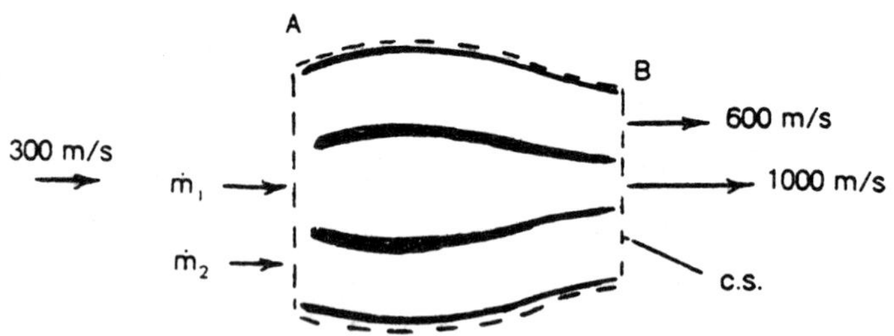

$\dot{m} = \rho VA = \rho Q$

$\dot{m}_A = \dot{m}_B = 300 \text{ kg/s}$

$\dot{m}_A = \dot{m}_1 + \dot{m}_2$ but $\dot{m}_2 = 2\dot{m}_1$

then $300 \text{ kg/s} = \dot{m}_1 + 2\dot{m}_1$

$\dot{m}_1 = 100 \text{ kg/s}$ and $\dot{m}_2 = 200 \text{ kg/s}$

Momentum equation:

$\Sigma F_x = \Sigma V_x \rho \underline{V} \cdot \underline{A}$

$T = \Sigma V_B \rho (V_B A_B) + V_A \rho (V_A A_A)$

$= 600(200) + 1000(100) - (300)(300)$

$\underline{T = 130{,}000 \text{ N}}$

6.91 $M_0 = M_f \exp (V_{bo}\lambda/T) = 50 \exp(7{,}200/3{,}000) = \underline{551.2 \text{ kg}}$

6.92 $T - \text{wt.} = ma$ where T = thrust

$T = \dot{m}V_e$

$\dot{m}V_e - mg = m dV_R/dt$

$dV_R/dt = (T/m) - g$

$= (T/(m_j - \dot{m}t)) - g$

$dV_R = ((Tdt)/(m_i - \dot{m}t)) - gdt$

$V_R = (-T/\dot{m}) \ln (m_i - \dot{m}t) - gt + \text{const.}$

where $V_R = 0$ when $t = 0$ $\therefore$ const. $= (T/\dot{m}) \ln (m_i)$

$\therefore$ $V_R = (T/\dot{m}) \ln ((m_i)/(m_i - \dot{m}t)) - gt$

$V_{Rmax} = (T/\dot{m}) \ln (m_i/m_f) - gt_f$

$T/\dot{m} = \dot{m}V_e/\dot{m} = V_e$

6.92 (Cont'd)

Apply Bernoulli's equation (neglecting hydrostatic pressure)

$$p_i + (\rho_f/2)V_i^2 = p_e + (\rho_f/2)V_e^2$$

(neglect: $(\rho_f/2)V_i^2$; 0: p_e)

$$V_e^2 = 2p_i/\rho_f = (2)\ (100)\ (10^3)/998 = 200\ m^2/s^2$$

$$V_e = 14.14\ m/s$$

$$\dot{m} = \rho_e V_e A_e = (1{,}000)(14.14)(0.10)(0.05)^2(\pi/4)$$

$$= 2.77\ kg/s$$

Time for water to exhaust: $t = m_w/\dot{m} = 0.10/2.77 = 0.036\ s$

$\therefore\ V_{max} = 14.14\ \ell n((100 + 50)/50) - (9.81)(0.036) = \underline{15.18\ m/s}$

6.93

$$\Sigma F_z = \Sigma V_z \rho (\underline{V} \cdot \underline{A})$$

$$T - p_a A_e \cos 30° + p_e A_e \cos 30° = -V_e \cos 30° \rho V_e A_e$$

$$T = -1 \times 0.866 \times (50{,}000 - 10{,}000 + 0.3 \times 2{,}000 \times 2{,}000)$$

$$= -1.074 \times 10^6 N$$

Thrust of four engines $= 4 \times 1.074 \times 10^6 = 4.3 \times 10^6 N = \underline{\underline{4.3\ MN}}$

6.94 $\Sigma \underline{F} = \Sigma V \rho \underline{V} \cdot \underline{A}$

$$\underline{F} = -220(-100 + 2{,}000) + 1.5 \times 10^6 \times 1 + (2 - 1) \times 10^5 - 8 \times 10^4 \times 2$$

$$= 1.022 \times 10^6 N = \underline{\underline{1.022\ MN}}$$

6.95

$$\Sigma \underline{F} = \int \underline{V} \rho \underline{V} \cdot d\underline{A}$$

$$T = \int_0^{\alpha} V_e \cos\theta \rho V_e \int_0^{2\pi} \sin\theta r d\phi r d\theta$$

$$T = 2\pi r^2 \rho V_e^2 \int_0^{\alpha} \cos\theta \sin\theta\ d\theta$$

$$= 2\pi r^2 \rho V_e^2 \sin^2\alpha/2$$

$$= \rho V_e^2 2\pi r^2 (1 - \cos\alpha)(1 + \cos\alpha)/2$$

Exit Area $A_e = \int_0^{\alpha}\int_0^{2\pi} \sin\theta\ r d\phi\ r d\theta = 2\pi r^2 (1 - \cos\alpha)$

$$T = \rho V_e^2 A_e (1 + \cos\alpha)/2 = \underline{\underline{\dot{m}\ V_e (1 + \cos\alpha)/2}}$$

6.96 $\Delta p = \rho V c$

$c = \sqrt{E_v/\rho} = ((715)(10^6)/(680))^{0.5} = 1{,}025$ m/s

$\Delta p = (680)(10)(1{,}025) = \underline{\underline{6.97 \text{ MPa}}}$

6.97 $c = (2.2 \times 10^9/1{,}000)^{1/2} = 1{,}483$ m/s

$t_{crit} = 2L/c = 2 \times 10{,}000/1{,}483 = 13.5 \text{ s} > 10 \text{ s}$

Then $\Delta p = \rho V c = 1{,}000 \times 3 \times 1{,}483 = 4{,}449{,}000 \text{ Pa} = \underline{\underline{4.45 \text{ MPa}}}$

6.98 $c = 1{,}483$ m/s (from solution to Prob. 6.94)

$t = 4L/c \therefore 3 = 4L/1{,}483$; $\underline{\underline{L = 1{,}112 \text{ m}}}$

6.99 $c = (320{,}000 \times 144/1.94)^{1/2} = 4{,}874$ ft/s

$t_{crit} = 2L/c = 2 \times 5 \times 5{,}280/4{,}874 = 10.83 \text{ s} > 10 \text{ s}$

Then $\Delta p_{max} = \rho V c = 1.94 \times 8 \times 4{,}874 = \underline{\underline{75{,}644 \text{ psf} = 525 \text{ psi}}}$

6.100 $t_{crit} = 2L/c = 2 \times 4{,}000/1{,}485.4 = 5.385 \text{ s} > 3 \text{ s}$

$F_{valve} = A\Delta p = A\rho(Q/A)c = \rho Q c = 998 \times 0.025 \times 1{,}483 = \underline{\underline{37{,}000 \text{ N} = 37.0 \text{ kN}}}$

6.101 From continuity equation

$(V + c)\rho = c(\rho + \Delta\rho)$

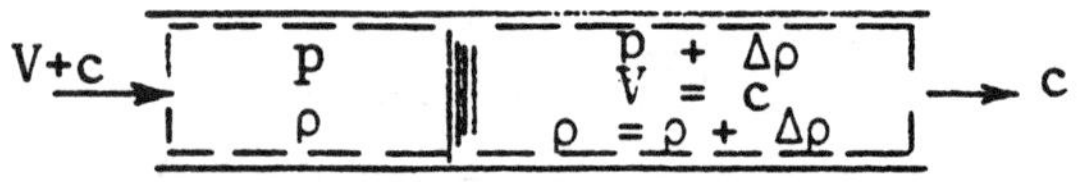

$\therefore \Delta\rho = V\rho/c$

$\Sigma F_x = \Sigma V_x \rho \underline{V} \cdot \underline{A}$

$p A - (p + \Delta p)A = -(V + c)\rho(V + c)A + c^2(\rho + \Delta\rho)A$

$\Delta p = 2\rho V c - c^2\Delta\rho + V^2\rho = 2\rho V c - c^2 V\rho/c + V^2\rho = \rho V c + \rho V^2$

Here ρV^2 is very small compared to $\rho V c$

$\underline{\underline{\therefore \Delta p = \rho V c}}$

6.102 $V = 0.1$ m/s; $c = 1{,}483$ m/s from solution to Prob. 6.94;

$p_{pipe} = 10\gamma - \rho V^2_{pipe}/2 \approx 98{,}000$ Pa

$\Delta p = \rho V c = 1{,}000 \times 0.10 \times 1{,}483$

$\Delta p = 148{,}000$ Pa

6.102 (continued)

Thus $p_{max} = p + \Delta p = 98{,}000 + 148{,}000 = 246$ kPa gage

$p_{min} = p - \Delta p = -50$ kPa gage

The sequence of events are as follows: Σt

Pressure wave reaches pt. B at t = 1,000m/1,483 m/s = 0.674s 0.67s
Time period of high pressure at B = 600/1,483 = 0.405s 1.08s
Time period of static pressure at B = 2,000/1,483 = 1.349s 2.43s
Time period of negative pressure at B = 600/1,483 = 0.405s 2.83s
Time period of static pressure at B = 2.000/1,483 = 1.349s 4.18s
Time period of high pressure at B = 600/1,483 = 0.405s 4.59s
Time period of static pressure at B = 2,000/1,483 = 1.349s 5.94s

Results are plotted below:

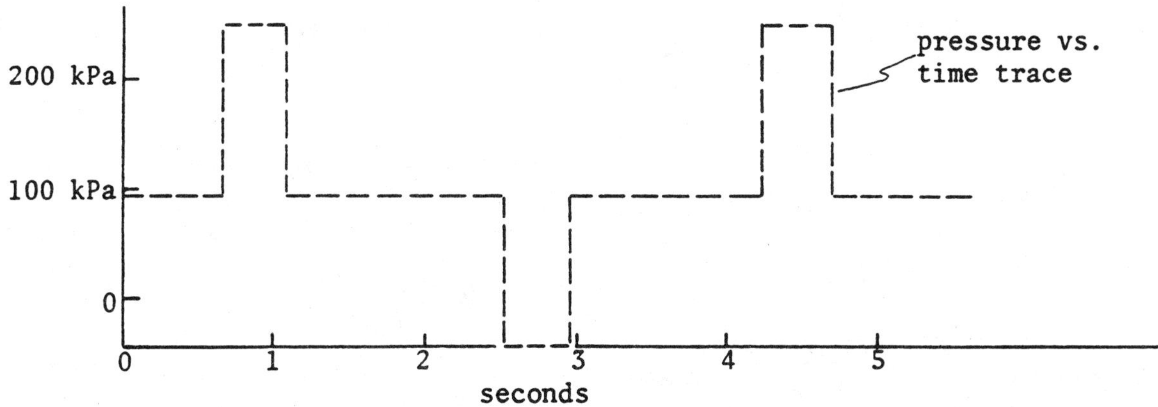

At t = 1.5s high pressure wave will have travelled to reservoir and static wave will be travelling toward valve.

Time period for wave to reach reservoir = 1,300/1,483 = 0.877s.
Then static wave will have travelled for 1.5 - 0.877s = 0.623s.
Distance static wave has travelled = 0.623s x 1,483 m/s = 924 m.
The pressure vs. position plot is shown below:

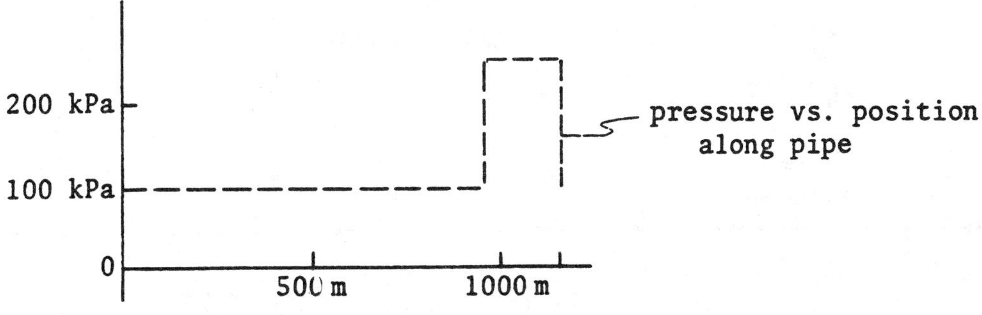

6.103 $c = 1{,}483$ m/s; $\Delta p = \rho \Delta V c$; $t = L/c$

$L = tc = 1.46\text{s} \times 1{,}483 = \underline{\underline{2{,}165 \text{ m}}}$

$\Delta V = \Delta p/\rho c$

$= (2.5 - 0.2) \times 10^6 \text{Pa}/1.483 \times 10^6 \text{kg/m}^2\text{s} = 1.551 \text{ m/s}$

$Q = VA = 1.551 \times \pi/4 = \underline{\underline{1.22 \text{ m}^3/\text{s}}}$

6.104 $V_y = -(3.1 + 3x)$ m/s

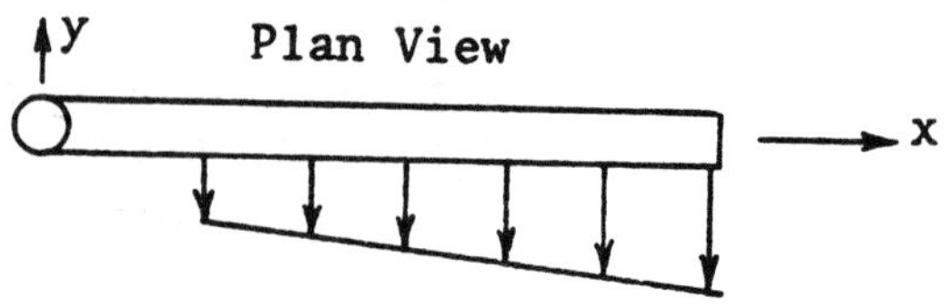

$\Sigma F_y = \int V_y \rho \underline{V} \cdot d\underline{A}$

$R_y = -\int_{0.3}^{1.3} (3.1 + 3x) \times 1{,}000 \times (3.1 + 3x) \times 0.015\, dx = \underline{\underline{-465 \text{ N}}}$

$Q = \int V dA = 0.015 \int_{0.3}^{1.3} (3.1 + 3x) dx = 0.0825 \text{ m}^3/\text{s}$

$\Sigma F_z = \Sigma V_z \rho \underline{V} \cdot \underline{A}$

$R_z = 30{,}000 \times \pi \times 0.04 \times 0.04 + 0.08 \times \pi \times 0.04 \times 0.04 \times 9{,}810$

$+ 1.3 \times \pi \times 0.025 \times 0.025 \times 9{,}810 + 1{,}000 \times 0.0825 \times 0.0825/$

$(\pi \times 0.04 \times 0.04) = \underline{\underline{1{,}534 \text{ N}}}$

$\Sigma F_x = \Sigma V_x \rho \underline{V} \cdot \underline{A}$; $\underline{\underline{R_x = 0}}$

$\Sigma M_z = 0$

$T_z = \int_{c.s.} rV\rho \underline{V} \cdot d\underline{A}$

$= 15 \int_{0.3}^{1.3} (3.1 + 3r)^2 r dr = \underline{\underline{413.2 \text{ N}\cdot\text{m}}}$

$\Sigma M_y = 0$

$T_y = -1.3\, \pi \times 0.025 \times 0.025 \times 9{,}810 \times 0.65 = \underline{\underline{-16.28 \text{ N}\cdot\text{m}}}$

$\Sigma M_x = 0$

$\underline{\underline{T_x = 0}}$

6.105 $V_1 = (0.1 \times 50 + 0.2 \times 50)/0.6 = 25$ fps

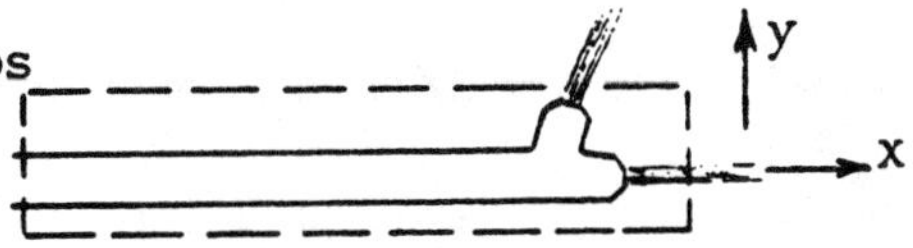

$\Sigma F_x = \Sigma V_x \rho \underline{V} \cdot \underline{A}$

$F_{1x} = -20 \times 144 \times 0.6 - 1.94 \times 25 \times 25 \times 0.6 + 1.94 \times 50 \times 50 \times 0.2$

$+ 1.94 \times 50 \times 50 \times 0.1 \times \sin 30° = \underline{\underline{-1,243 \text{ lbf}}}$

$\Sigma F_y = \Sigma V_y \rho \underline{V} \cdot \underline{A}$

$F_{1y} = 1.94 \times 50 \times 50 \times 0.1 \times \cos 30° = \underline{\underline{420 \text{ lbf}}}$

$\Sigma F_z = 0$; therefore $\underline{\underline{F_{1z} = 0}}$

$\Sigma M_z = 0$

$M_z = \Sigma(\underline{r} \times \underline{V})\rho \underline{V} \cdot \underline{A} = (3\underline{i} \times 50 \cos 30° \underline{j})\ 1.94 \times 50 \times 0.1 = \underline{\underline{1,260 \underline{k} \text{ ft-lbf}}}$

$\Sigma M_y = 0$; therefore $\underline{\underline{M_y = 0}}$

6.106 $V_1 = (0.01 \times 20 + 0.02 \times 20)/0.1 = 6$ m/s

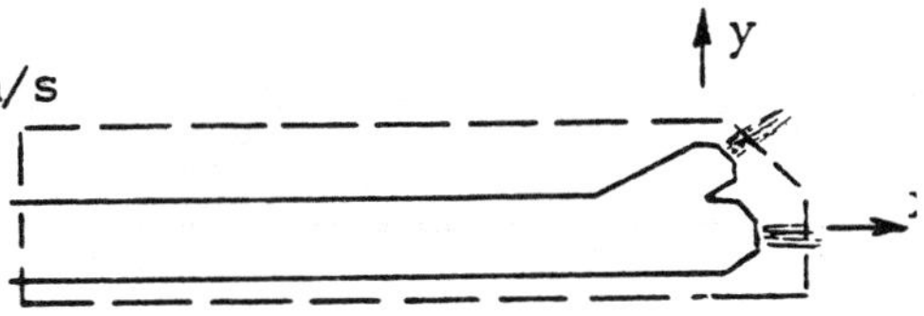

$F_{1x} = -200,000 \times 0.1 - 1,000 \times 6 \times 6$

$\times 0.1 + 1,000 \times 20 \times 20 \times 0.02$

$+ 1,000 \times 20 \times 20 \times 0.01 \times \cos 30°$; Assume $W_{water} = 0.1 \times 1 \times \gamma$

$= 981$ N

$= \underline{-12,135 \text{ N}}$

$F_{1y} = 1,000 \times 20 \times 20 \times 0.01 \times \sin 30° + 1071 = \underline{\underline{3,071 \text{ N}}}$; $\underline{\underline{F_{1z} = 0}}$

$M_z = (1\underline{i} \times 20 \sin 30° \underline{j}) \times 1,000 \times 20 \times 0.01 + 0.5\underline{i} \times 1,071\underline{j}$

$= \underline{\underline{2,535\underline{k} \text{ N·m}}}$; $\underline{\underline{M_y = 0}}$

6.107

$V_1 = 2/(\pi \times 3 \times 3/144) = 10.19$ ft/sec

$V_2 = 2/(\pi \times 2 \times 2/144) = 22.92$ ft/sec

$p_2 = 20 \times 144 + (1.94/2)(10.19 \times 10.19 - 22.92 \times 22.92) = 2{,}471$ psf

z, y, x, ③

$\Sigma M_z = 0$

$M_{z3} = (\underline{r}_1 \times \underline{V}_1)\rho(-Q) + (\underline{r}_2 \times \underline{V}_2)\rho Q$

$= \rho Q(V_1 - V_2) + p_1A_1 - p_2A_2$

$= 1.94 \times 2 \times (10.19 - 22.92) + 20 \times 144 \times \pi \times 3 \times 3/144$

$- 2{,}471 \times \pi \times 2 \times 2/144 = \underline{\underline{-300.5 \text{ ft-lb}}}$

$\Sigma F_y = \rho Q(V_{2y} - V_{1y})$

$F_{y3} = -(20 \times 144 \times \pi \times 3 \times 3/144) - (2{,}471 \times \pi \times 2 \times 2/144)$

$+(1.94 \times 2 \times (-22.92 - 10.19)) = -909.6$ lbf

Shear force at section 3, $F_{y3} = \underline{\underline{909.6 \text{ lbf}}}$

$M_{x3} = r \times F_y = -2 \times 909.6 = \underline{\underline{-1{,}819.2 \text{ ft-lb}}}$

6.108

$V + \frac{\partial V}{\partial s}\Delta s$, $p + \frac{\partial p}{\partial s}\Delta s$, $\rho + \frac{\partial \rho}{\partial s}\Delta s$, $A + \frac{\partial A}{\partial s}\Delta s$, V, ρ, p, A, $W = \bar{\rho}\bar{\forall}g$, Δs

$$\sin\alpha = \frac{\left(z + \frac{\partial z}{\partial s}\Delta s\right) - z}{\Delta s} = \frac{\partial z}{\partial s}$$

average values

$$\bar{V} = \frac{1}{2}\left[V + \left(V + \frac{\partial V}{\partial s}\Delta s\right)\right] = V + \frac{\partial V}{\partial s}\frac{\Delta s}{2}$$

$$\bar{p} = p + \frac{\partial p}{\partial s}\frac{\Delta s}{2}$$

$$\bar{A} = A + \frac{\partial A}{\partial s}\frac{\Delta s}{2}$$

$$\bar{\rho} = \rho + \frac{\partial \rho}{\partial s}\frac{\Delta s}{2}$$

$$\forall = \bar{A}\Delta s = \left(A + \frac{\partial A}{\partial s}\frac{\Delta s}{2}\right)\Delta s$$

let $x' = \frac{\partial x}{\partial s}$

Apply continuity,

$$\frac{\partial}{\partial t}\int \bar{\rho}\, d\forall + \int \rho \underline{V}\cdot \underline{dA} = 0$$

$$\frac{\partial}{\partial t}\left[\left(\rho + \rho'\frac{\Delta s}{2}\right)\left(A + A'\frac{\Delta s}{2}\right)\Delta s\right] - \rho VA + (\rho + \rho'\Delta s)(V + V'\Delta s)(A + A'\Delta s) = 0$$

$$\frac{\partial}{\partial t}\left[\rho A\Delta s + \rho' A\frac{(\Delta s)^2}{2} + \rho A'\frac{(\Delta s)^2}{2} + \rho' A'\frac{(\Delta s)^3}{4}\right] - \rho VA + \rho VA + \rho V'\Delta s A$$

$$+ \rho VA'\Delta s + \rho V'A'(\Delta s)^2 + \rho'\Delta s VA + \rho' VA'(\Delta s)^2 + \rho' V'A(\Delta s)^2 + \rho' V'A'(\Delta s)^3 = 0$$

dropping higher order differentials and dividing by Δs,

$$\frac{\partial}{\partial t}(\rho A) + \frac{\partial}{\partial s}(\rho VA) = 0$$

6.108 (continued)

Applying the momentum equation along the streamtube direction,

$$\Sigma F_s = \frac{\partial}{\partial t}\int V_s \rho d\forall + \int V_s \rho \underline{V}\cdot\underline{A}$$

$$pA - (p+p'\Delta s)(A+A'\Delta s) + (p+p'\tfrac{\Delta s}{2})A'\Delta s$$
$$-g(\rho+\rho'\tfrac{\Delta s}{2})(A+A'\tfrac{\Delta s}{2})(\Delta s)\sin\alpha = \frac{\partial}{\partial t}\left[(V+V'\tfrac{\Delta s}{2})(\rho+\rho'\tfrac{\Delta s}{2})(A+A'\tfrac{\Delta s}{2})\Delta s\right]$$
$$-\rho V^2 A + (\rho+\rho'\Delta s)(V+V'\Delta s)^2(A+A'\Delta s)$$

$$pA - pA - pA'\Delta s - p'A\Delta s - p'A'(\Delta s)^2 + pA'\Delta s + p'A'\frac{(\Delta s)^2}{2}$$
$$-\gamma A\Delta s z' - \gamma' A\frac{(\Delta s)^2}{2}z' - \gamma A'\frac{(\Delta s)^2}{2}z' - \gamma' A'\frac{(\Delta s)^3}{4}z' =$$

$$\frac{\partial}{\partial t}\Big[V\rho A\Delta s + V'\rho A\frac{(\Delta s)^2}{2} + V\rho' A\frac{(\Delta s)^2}{2} + V\rho A'\frac{(\Delta s)^2}{2} + V'\rho' A\frac{(\Delta s)^3}{4}$$
$$+ V'\rho A'\frac{(\Delta s)^3}{4} + V\rho' A'\frac{(\Delta s)^3}{4} + V'\rho' A'\frac{(\Delta s)^4}{8}\Big] - \rho V^2 A + \rho V^2 A$$
$$+\rho V^2 A'\Delta s + 2\rho VAV'\Delta s + \rho' V^2 A\Delta s + \rho' V^2 A'(\Delta s)^2 + 2\rho VV'A'(\Delta s)^2$$
$$+2\rho' VV'A(\Delta s)^2 + \rho(V')^2 A(\Delta s)^2 + \rho'(V')^2 A(\Delta s)^3 + \rho(V')^2 A'(\Delta s)^3$$
$$+2\rho' VV'A'(\Delta s)^3 + \rho'(V')^2 A'(\Delta s)^4$$

dropping higher order differentials and dividing by Δs,

$$-\frac{\partial p}{\partial s}A - \gamma A\frac{\partial z}{\partial s} = \frac{\partial}{\partial t}(V\rho A) + \rho V^2\frac{\partial A}{\partial s} + 2\rho VA\frac{\partial V}{\partial s} + \frac{\partial \rho}{\partial s}V^2 A$$
$$= V\frac{\partial}{\partial t}(\rho A) + V\frac{\partial}{\partial s}(\rho VA) + \rho A\frac{\partial V}{\partial t} + \rho VA\frac{\partial V}{\partial s}$$
$$= V\left[\frac{\partial}{\partial t}(\rho A) + \frac{\partial}{\partial s}(\rho VA)\right] + \rho A\frac{\partial V}{\partial t} + \rho VA\frac{\partial V}{\partial s}$$

dividing by A and noting that

1) $\frac{\partial}{\partial t}(\rho A) + \frac{\partial}{\partial s}(\rho VA) = 0$ by continuity

2) $\frac{DV}{Dt} = \frac{\partial V}{\partial t} + V\frac{\partial V}{\partial s}$

$$-\frac{\partial}{\partial s}(p+\gamma z) = \rho\frac{DV}{Dt} \quad \equiv \quad \text{Euler's equation.}$$

6.109

$V_i = 2\pi \times 4 = 25.13$ m/s

$\Sigma F_x = \Sigma V_x \rho \underline{V}\cdot\underline{A}$

$F = \rho Q(V_{ex} - V_{ix}) = 1.2 \times 25.13 \times 20 \times 10^{-4}(500 - 25.13) = 28.64$ N

Power $P = 2FV = 2 \times 28.64 \times 25.13 = 1{,}440$ W $= \underline{\underline{1.44 \text{ kW}}}$

7.1 $\dot{Q} - \dot{W}_s = \dot{m}(h_2 + V_2^2/2 - h_1 - V_1^2/2)$

$-2,500 - \dot{W}_s = 5,800\ [1,098 + 200^2/(2 \times 778 \times 32.2)] - 1,268 - [50^2/(2 \times 778 \times 32.2)]\text{BTU/hr}$

$\dot{W}_s = 9.79 \times 10^5\ \text{BTU/hr} = \underline{\underline{384\ \text{hp}}}$

7.2 $\dot{Q} - \dot{W}_s = \dot{m}[(h_2 - h_1) + (V_2^2 - V_1^2)/2]$

$-5 - \dot{W}_s = 4,000[(2,621 - 3,062) + (50^2 - 10^2)/(2 \times 1,000)]\text{kJ/hr}$

$\dot{W}_s = \underline{\underline{489\ \text{kW}}}$

7.3 $\dot{Q} - \dot{W}_s = \dot{m}[(h_2 - h_1) + (V_2^2 - V_1^2)/2]$

$= 1[(2,630 - 3,470) + (70^2 - 5^2)/(2 \times 1,000)]$

$= -837.6\ \text{kJ/s}$

Heat loss $= -\dot{Q} = -(\dot{W}_s - 837.6) = -(830 - 837.6) = \underline{\underline{7.6\ \text{kJ/s}}}$

7.4 **Energy equation:**

$\dot{W} = \dot{Q} + \dot{m}\ (V_1^2/2 - V_2^2/2 + h_1 - h_2)$ (V_1^2: negligible)

$\dot{W} = \dot{m}\ (-V_2^2/2 + h_1 - h_2)$

$= 1.5\ (-200^2/2 + 300 \times 10^3 - 450 \times 10^3)$

$\underline{\underline{\dot{W} = -255\ \text{kW}}}$ *Note:* *Work done to system is indicated by the negative sign.*

7.5 $h_1 + V_1^2/2 = h_2 + V_2^2/2$ or $h_1 - h_2 = V_2^2/2 - V_1^2/2$ (1)

$\dot{m} = \rho_1 V_1 A = (p_1/RT_1)V_1 A$ or $T_1 = p_1 V_1 A/(R\dot{m})$

where $A = (\pi/4) \times (0.08)^2 = 0.005024\ \text{m}^2$

$h_1 - h_2 = c_p(T_1 - T_2) = [c_p p_1 V_1 A/(R\dot{m})] - [c_p p_2 V_2 A/(R\dot{m})]$ (2)

$c_p p_1 A/(R\dot{m}) = 1,004 \times 150 \times 10^3 \times 0.005024/(287 \times 0.5)$

$= 5,272$ m/s and $c_p p_2 A/(R\dot{m}) = (100/150) \times (5,272) = 3,515$ m/s

7.5 (continued)

Continuity equation: $V_1 = \dot{m}/\rho_1 A$

where $\rho_1 = 150 \times 10^3/(287 \times 293) = 1.784\ \text{kg/m}^3$

Then $V_1 = 0.50/(1.784 \times 0.005024) = 55.8\ \text{m/s}$ (3)

Utilizing Eqs. (1), (2) and (3), we have

$55.8 \times 5{,}272 - 3{,}515\ V_2 = V_2^2/2) - (55.8^2/2)$ (4)

Solving Eq. (4) yields $V_2 = \underline{\underline{83.15\ \text{m/s}}}$

$c_p(T_1 - T_2) = (83.15^2 - 55.8^2)/2 = 1{,}900\ \text{m}^2/\text{s}^2$

$T_2 = T_1 - (1{,}900/c_p) = 20°\text{C} - 1{,}900/1{,}004 = 18.1°\text{C}$

7.6 $\overline{V} = Q/A = \int_A VdA/A$ where $V = V_{max} - 0.3\ V_{max} r/r_0$

$= V_{max}(1 - 0.3\ r/r_0)$

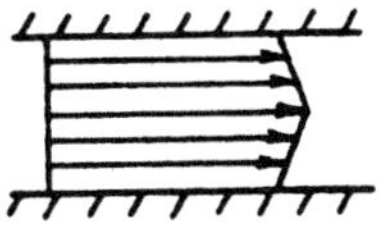

Then $\overline{V} = (V_{max}/\pi r_0^2)\int_0^{r_0} (1 - 0.3\ r/r_0)2\pi r dr$

$= (2\pi V_{max}/\pi r_0^2)\int_0^{r_0} (1 - 0.3r/r_0) r dr$

$= (2V_{max}/r_0^2)\int_0^{r_0} (rdr - (0.3\ r^2/r_0))dr$

$= (2V_{max}/r_0^2)(r_0^2/2 - 0.3r_0^2/3) = 0.800\ V_{max}$

$\alpha = (1/\pi r_0^2)\int_0^{r_0} [(1 - 0.3r/r_0)V_{max}/0.800\ V_{max}]^3 2\pi r dr$

$\alpha = 2\pi/((0.800)^3 \pi r_0^2)\int_0^{r_0} (1 - 0.3\ r/r_0)^3 r dr$

Integrating yields $\underline{\underline{\alpha = 1.022}}$

7.7 $\overline{V} = V_{max}/2$ and $V = V_{max}\ y/d$

Then $\alpha = (1/d)\int_0^d (V_{max}\ y/((V_{max}/2)d))^3 dy = (1/d)\int_0^d (2y/d)^3 dy = \underline{\underline{2}}$

7.8 a) $\underline{\underline{\alpha = 1.0}}$; b) $\underline{\underline{\alpha > 1.0}}$ c) $\underline{\underline{\alpha > 1.0}}$; d) $\underline{\underline{\alpha > 1.0}}$

7.9 $\alpha = (1/A)\int_A (V/\bar{V})^3 dA;\quad V = V_{max} - (r/r_0)V_{max}$

$V = V_{max}(1 - (r/r_0))$

$Q = \int V dA = \int_0^{r_0} V(2\pi r dr) = \int_0^{r_0} V_m(1 - r/r_0)2\pi r dr$

$= 2\pi V_m \int_0^{r_0} r dr - (r^2/r_0)dr$

Integrating yields $Q = 2\pi V_m[(r^2/2) - (r^3/(3r_0))]_0^{r_0}$

$Q = 2\pi V_m[(1/6)r_0^2]$

$Q = (1/3)V_m A$ or $V = Q/A = 1/3\ V_m$

Then $\alpha = (1/A)\int_0^{r_0} [V_m(1 - r/r_0)/((1/3)V_m)]^3 2\pi r dr$

$\alpha = (54\pi/\pi r_0^2)\int_0^{r_0} (1 - (r/r_0))^3 r dr = \underline{\underline{2.7}}$

7.10 $V = kr;\quad Q = \int_0^{r_0} V(2\pi r dr) = \int_0^{r_0} 2\pi k r^2 dr = 2\pi k\ r_0^3/3$

$\bar{V} = Q/A = ((2/3)k\pi r_0^3)/\pi r_0^2 = 2/3\ k\ r_0$

Then $\alpha = (1/A)\int_A (V/\bar{V})^3 dA$

$\alpha = (1/A)\int_0^{r_0} (kr/(2/3\ kr_0))^3 2\pi r dr$

$\alpha = ((3/2)^3 2\pi/(\pi r_0^2))\int_0^{r_0} (r/r_0)^3 r dr$

$\alpha = ((27/4)/r_0^2)(r_0^5/(5r_0^3)) = \underline{\underline{27/20}}$

7.11 b) turbulent

7.12 $u/u_{max} = (y/r_0)^n = ((r_0 - r)/r_0)^n = (1 - r/r_0)^n$

$Q = \int_A u dA = \int_0^{r_0} u_{max}(1 - r/r_0)^n 2\pi r dr = 2\pi u_{max}\int_0^{r_0}(1 - r/r_0)^n r dr$

Upon integration $Q = 2\pi u_{max} r_0^2[(1/(n+1)) - (1/(n + 2))]$

Then $\bar{V} = Q/A = 2u_{max}[(1/(n + 1)) - (1/(n + 2))] = 2u_{max}$

$/[(n + 2)(n + 1)]$

$\alpha = (1/A)\int_0^{r_0} [u_{max}(1 - r/r_0)^n/(2u_{max}/((n + 2)(n + 1)))]^3 2\pi r dr$

Upon integration one gets $\underline{\underline{\alpha = (1/4)[((n+2)(n+1))^3/((3n+2)(3n+1))]}}$

If $n = 1/6$, then $\alpha = (1/4)[((1/6 + 2)(1/6 + 1))^3/((3 \times 1/6 + 2)(3 \times 1/6 + 1))]$

$\underline{\underline{\alpha = 1.078}}$

7.13

$$u/u_{max} = (y/d)^n$$

Solve for q first in terms of u_{max} and d

$$q = \int_0^d u\,dy = \int_0^d u_{max}(y/d)^n dy = u_{max}/d^n \int_0^d y^n dy$$

Integrating: $q = (u_{max}/d^n)[y^{n+1}/(n+1)]_0^d = u_{max} d^{n+1} d^{-n}/(n+1)$

$$q = u_{max} d/(n+1)$$

Then $\bar{u} = q/d = u_{max}/(n+1)$

$$\alpha = (1/A)\int_A (u/\bar{u})^3 dA = 1/d \int_0^d [u_{max}(y/d)^n/(u_{max}/(n+1))]^3 dy$$

$$= ((n+1)^3/d^{3n+1})\int_0^d y^{3n} dy$$

Integrating: $\alpha = ((n+1)^3/d^{3n+1})[d^{3n+1}/(3n+1)] = \underline{\underline{(n+1)^3/(3n+1)}}$

When $n = 1/7$, $\alpha = (1+1/7)^3/(1+3/7) = \underline{\underline{1.045}}$

7.14 Write energy equation from water surface in tank to outlet:

$(p_1/\gamma + (V_1^2/2g) + z_1 = (p_2/\gamma) + (V_2^2/2g) + z_2 + h_L$

$(100{,}000/9{,}810) + 0 + 10 = 0 + (10^2/2g) + 0 + K_L V^2/2g$; $\underline{\underline{K_L = 2.96}}$

7.15 Write the energy equation from the water surface in the tank to the outlet:

$p_1/\gamma + V_1^2/2g + z_1 = p_2/\gamma + V_2^2/2g + z_2 + h_L$

$p_1/\gamma = V_2^2/2g + h_L - z_1 = 6V_2^2/2g - 10$

$V_2 = Q/A_2 = 0.1/((\pi/4)(1/12)^2) = 18.33$ ft/s

$p_1/\gamma = (6(18.33^2)/64.4) - 10 = 2.13$ ft

$p_1 = 62.4 \times 21.3 = \underline{\underline{1{,}329 \text{ lbf/ft}^2 = 9.23 \text{ lb/in}^2}}$

7.16 $p_A = -\gamma y = -62.4 \times 4 = \underline{\underline{-250 \text{ lb/ft}^2}}$

7.17 $p_A = 9{,}810(-3) = \underline{\underline{-29.43 \text{ kPa}}}$

7.18 $p_A - p_B = 1\gamma + (\rho/2)(V_B^2 - V_A^2)$; $V_A = Q/A_1 = 1.910$ m/s

$V_B = 4V_A = 7.64$ m/s

Then $p_A - p_B = 1 \times 9{,}810 \times 0.9 + (900/2)(7.64^2 - 1.91^2) = \underline{\underline{33.45 \text{ kPa}}}$

7.19 Write the energy equation from the water surface in the reservoir (pt. 1) to the outlet end of the pipe (pt. 2). Also assume α=1.

$$\overset{0}{\cancel{p_1/\gamma}} + \overset{\text{negligible}}{\cancel{V_1^2/2g}} + z_1 = \overset{0}{\cancel{p_2/\gamma}} + V_2^2/2g + \overset{0}{\cancel{z_2}} + h_L$$

$$z_1 = V_2^2/2g + h_L$$

$$11 = V_2^2/2g + 10V_2^2/2g$$

$$V_2^2 = (2g/11)\ (11)$$

$$V_2 = 4.429 \text{ m/s}$$

$$Q = V_2A_2 = (4.429 \text{ m/s})(5 \text{ cm}^2)(10^{-4}\text{m}^2/\text{cm}^2)$$

$$\underline{Q = 2.21 \times 10^{-3}\text{m}^3/\text{s}}$$

7.20 The pipe will have to <u>decrease in elevation</u> at a rate greater than the head loss per given length of pipe.

7.21

$$V_1 = Q/A_1 = 5/0.8 = 6.25 \text{ ft/s};\ V_1^2/2g = 0.606 \text{ ft}$$

$$V_2 = Q/A_2 = 5/0.2 = 25 \text{ ft/s};\ V_2^2/2g = 9.70 \text{ ft}$$

$$p_1/\gamma + V_1^2/2g + z_1 = p_2/\gamma + V_2^2/2g + z_2 + 4$$

$$(10 \times 144)/(0.8 \times 62.4) + 0.606 + 12 = p_2/\gamma + 9.70 + 0 + 4$$

$$p_2/\gamma = 27.75 \text{ ft};\quad p_2 = 27.75 \times 0.8 \times 62.4 = \underline{1{,}385 \text{ lb/ft}^2 \text{ gage}}$$

$$= \underline{9.62 \text{ psi gage}}$$

7.22

$$p_{\text{reser.}}/\gamma + V_r^2/2g + z_r = p_{\text{outlet}}/\gamma + V_0^2/2g + z_0$$

$$0 + 0 + 5 = 0 + V_0^2/2g;\quad V_0 = 9.90 \text{ m/s}$$

$$Q = V_0A_0 = 9.90 \times (\pi/4) \times 0.20^2 = \underline{0.311 \text{ m}^3/\text{s}}$$

Write energy equation from reservoir surface to point B:

$$0 + 0 + 5 = p_B/\gamma + V_B^2/2g + 3.5$$

where $V_B = Q/V_B = 0.311/(\pi/4) \times 0.4^2 = 2.48 \text{ m/s};\ V_B^2/2g = 0.312 \text{ m}$

$p_B/\gamma = 5 - 3.5 - 0.312;\quad \underline{p_B = 11.7 \text{ kPa}}$ assuming $\gamma = 9810$ N/m

---

7.23 First solve for h_L from reservoir to C. Let point 1 be reservoir surface.

$p_1/\gamma + V_1^2/2g + z_1 = p_C/\gamma + V_C^2/2g + z_C + h_L$; $V_C = Q/A_2$

$0 + 0 + 3 = 0 + 8.02^2/64.4 + 0 + h_L$

$V_C = 2.8/((\pi/4) \times (8/12)^2) = 8.02$ ft/s

$h_L = \underline{\underline{2.00 \text{ ft}}}$

Now get p_B by writing energy equation from reservoir surface to B.

$0 + 0 + 3 = p_B/\gamma + V_B^2/2g + 6 + (2/3) \times 2$; $V_B = V_C = 8.02$ ft/s

$p_B/\gamma = 3 - 1 - 6 - 1.33 = -5.33$ ft

$p_b = -5.33 \times 62.4 = -333$ psf, gage = $\underline{\underline{-2.31 \text{ psig}}}$

7.24

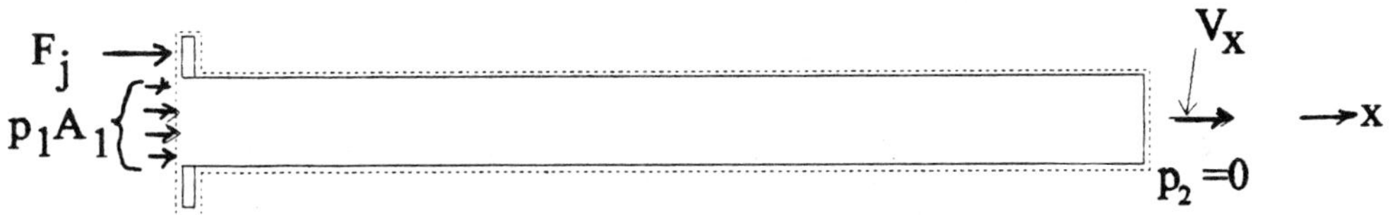

$F_j \rightarrow$, p_1A_1, V_x, $\rightarrow x$, $p_2 = 0$

Momentum Eq.

$$\sum F = \sum \rho V_x \underline{V} \cdot \underline{A}$$

$$F_j + p_1A_1 = -\rho V_x^2 A + \rho V_x^2 A \qquad (1)$$

$$F_j = -p_1A_1$$

Energy Eq.

$$p_1/\gamma + V_1^2/2g + z_1 = p_2/\gamma + V_2^2/2g + z_2 + h_2$$

$$p_1 - p_2 = \gamma h_2$$

$$p_1 = \gamma(1) = 62.4\text{psf} \qquad (2)$$

$$F_j = -62.4 \times (10/144)$$

$$\underline{\underline{F_j = -4.33\text{lbf}}}$$

7.25 $h_{\ell_{pipe}} = V^2/2g$; $h_{total} = h_{\ell_{pipe}} + h_{\ell_{outlet}} = 2V_p^2/2g$

Write the energy equation from A to C:

$0 + 0 + 30 = 0 + 0 + 27 + 2V_p^2/2g$

$V_p = 5.42$ m/s

$Q = V_p A_p = 5.42 \times (\pi/4) \times 0.30^2$

$\underline{\underline{Q = 0.383 \text{ m}^3/\text{s}}}$

Write the energy equation to point B:

$30 = p_B/\gamma + V_p^2/2g + 32 + 0.75\, V_p^2/2g$; $p_B/\gamma = -2 - 1.5 \times 1.75$ m

$p_B = \underline{\underline{-45.3 \text{ kPa gage}}}$

7.26 Write energy equation from point A to point B.

$p_A/\gamma + V_A^2/2g + z_A = p_B/\gamma + V_B^2/2g + z_B + h_L$

p_A = atmospheric pressure = 100 kPa abs.

p_B = vapor pressure of water = 1.23 kPa abs.

$V = Q/A = (7)(10^{-4})\text{m}^3/\text{s}/((10^{-4})\text{m}^2) = 7$ m/s

Then $V^2/2g = 7^2/(2\times 9.81) = 2.497$ m

$h_{L,A\to B} = 2V^2/2g = 4.995$ m

Let z=0 at bottom of reservoir

Then $100{,}000/9810 + 0 + z_A = 1{,}230/9{,}810 + 2.497 + 10 + 4.995$

$\underline{\underline{z_A = \text{depth} = 7.42 \text{ m}}}$

7.27 Assume the flow is from A to B. Then write the energy equation from A to B:

$p_A/\gamma + V_A^2/2g + z_A = p_B/\gamma + V_B^2/g + z_B + h_L$

$(10{,}000/9{,}810) + 10 = (98{,}100/9{,}810) + 0 + h_L$

$h_L = 1.02 + 10 - 10.0 = +1.02$

Because the value for head loss is positive it verifies our assumption of downward flow.

7.28 Writing the energy equation from the reservoir surface to the pipe outlet and assuming that the machine is a pump, one finds that h_p is positive (h_p = 24.3 ft); therefore the machine is a pump.

7.29 Let V_n = velocity of jet from nozzle:

$V_n = Q/A_n = 0.20/((\pi/4) \times 0.15^2) = 11.32$ m/s; $V_n^2/2g = 6.53$ m

$V_2 = Q/A_2 = 0.20/((\pi/4) \times 0.3^2) = 2.83$ m/s; $V_2^2/2g = 0.408$ m

$p_2/\gamma + 0.408 + 2 = 0 + 6.53 + 7$

p_2/γ = 11.1 m

7.30 Write the energy equation from the reservoir water surface to the outlet:

$0 + 0 + 0 + h_p = 0 + h + V_c^2/2g + 2.0\, V_c^2/2g$ (1)

where $V_c^2/2g = 12^2/64.4 = 2.24$ ft

$H.P. = Q\gamma h_p/(550 \times 0.7)$

Then $h_p = (30 \times 550 \times 0.7)/(62.4 \times 12 \times (\pi/4) \times 0.5^2) = 78.56$ ft

Solve Eq. (1) for h:

$h = 78.56 - 3.0 \times 2.24 =$ 71.8 ft

7.31 Follow the same solution process as in P7.30:

$0 + 0 + 0 + h_p = 0 + h + 3.0\, V_c^2/2g$

$V_c^2/2g = 4^2/(2 \times 9.81) = 0.815$ m

$P = Q\gamma h_p/0.7$

$h_p = 35{,}000 \times 0.7/((4 \times \pi/4 \times 0.15^2)(9{,}810)) =$ 35.3 m

$h = 35.3 - 3.0 \times 0.815 =$ 32.9 m

7.32 $V_A = Q/A_A = 3.92/((\pi/4) \times 1^2) = 4.99$ ft/sec; $V_A^2/2g = 0.387$ ft

$V_B = Q/A_B = 3.92/((\pi/4) \times 0.5^2) = 19.96$ ft/s; $V_B^2/2g = 6.19$ ft

Write the energy equation from A to B:

$$p_A/\gamma + V_A^2/2g + z_A + h_p = p_B/\gamma + V_B^2/2g + z_B$$

$$10 \times 144/62.4 + 0.387 + 0 + h_p = 40 \times 144/62.4 + 6.19 + 0$$

$h_p = 75.04$ ft; H.P. $= Q\gamma h_p/550 = 3.92 \times 62.4 \times 75.04/550$

Power = 33.4 H.P.

7.33 Write energy equation from reservoir surface to end of pipe:

$$p_1/\gamma + V_1^2/2g + z_1 + h_p = p_2/\gamma + V_2^2/2g + z_2 + h_L$$

$$0 + 0 + 40 + h_p = 0 + V^2/2g + 20 + 7V^2/2g$$

$$V = Q/A = 7.85/((\pi/4) \times 1^2) = 10.0 \text{ m/s}$$

$$V^2/2g = 10^2/(2 \times 9.81) = 5.09 \text{ m}$$

Then $h_p = 8 \times 5.1 + 20 - 40 = 20.8$ m

$$P = Q\gamma h_p = 7.85 \times 9{,}810 \times 20.8$$

$$= 1.60 \text{ MW}$$

7.34 $V = Q/A = 0.25/((\pi/4) \times 0.3^2) = 3.54$ m/s; $V^2/2g = 0.638$ m

Write energy equation from reservoir surface to 10 m elevation:

$$0 + 0 + 6 + h_p = 100{,}000/9{,}810 + V^2/2g + 10 + 2.0V^2/2g$$

$$h_p = 10.19 + 10 - 6 + 3.0 \times 0.638$$

$$h_p = 16.1 \text{ m}$$

$$P = Q\gamma h_p = 0.25 \times 9.810 \times 16.1 = 39.5 \text{ kW}$$

7.35 $V_{12} = Q/A_{12} = 5/((\pi/4) \times 1^2) = 6.366$ ft/sec; $V_{12}^2/2g = 0.629$ ft

$V_6 = 4V_{12} = 25.46$ ft/sec; $V_6^2/2g = 10.07$ ft

$(p_6/\gamma + z_6) - (p_{12}/\gamma + z_{12}) = (13.55 - 0.88)(46/12)/0.88$

$(p_{12}/\gamma + z_{12}) + V_{12}^2/2g + h_p = (p_6/\gamma + z_6) + V_6^2/2g$

$h_p = (13.55/0.88 - 1) \times 3.833 + 10.07 - 0.629$

$h_p = 64.6$ ft

Power $= Q\gamma h_p/550$

Power $= 5 \times 0.88 \times 62.4 \times 64.6/550 =$ 32.3 hp

---

7.36 Write the energy equation from the upstream water surface to the downstream water surface:

$p_1/\gamma + V_1^2/2g + z_1 = p_2/\gamma + V_2^2/2g + z_2 + h_L + h_T$

$0 + 0 + 35 = 0 + 0 + 0 + 1.5V^2/2g + h_T$

here $V = Q/A_6 = 250/((\pi/4) \times 6^2) = 8.84$ ft/sec; $V^2/2g$ 1.21 ft

$h_t = 35 - 1.82 = 33.19$ ft

H.P. $= Q\gamma h_t \times 0.8/550$

Power = 753 h.p.

3.37 **Write the energy equation from the upstream water surface to the downstream water surface. Assume all head loss is expansion loss. Also assume 100% efficiency.**

$p_1/\gamma + V_1^2/2g + z_1 = p_2/\gamma + V_2^2/2g + z_2 + h_t + h_L$

$0 + 0 + 11 \text{ m} = 0 + 0 + 0 + h_t + V^2/2g$

$h_t = 11 \text{ m} - (5^2/2g) = 9.725 \text{ m}$

$P = Q\gamma h_t = (1 \text{ m}^3/\text{s})(9{,}810 \text{ N/m}^3)(9.725\text{m}) =$ 95.4 kW

7.38 Write the energy equation from the upper water surface to the lower water surface:

$$p_1/\gamma + V_1^2/2g + z_1 = p_2/\gamma + V_2^2/2g + z_2 + \Sigma h_L + h_t$$

$$0 + 0 + 100 \text{ ft} = 0 + 0 + 4 \text{ ft} + h_t$$

$$h_t = 96 \text{ ft}$$

$$P = (Q\gamma h_t)\ (\text{efficiency})$$

$$\text{Horsepower} = Q\gamma h_t\ (\text{eff.})/550 = 1{,}000 \times 62.4 \times 96 \times 0.85/550$$

$$P = \underline{\underline{9{,}258 \text{ hp}}}$$

7.39 Write the energy equation from the reservoir water surface to point B:

$$p/\gamma + V^2/2g + z + h_p = p_B/\gamma + V_B^2/2g + z_B$$

$$0 + 0 + 40 + h_p = 0 + 0 + 65; \quad h_p = 25 \text{ m}$$

$$P = Q\gamma h_p; \quad Q = V_j A_j = 30 \times 10^{-4} \text{m}^2 \times V_j$$

where $V_j = \sqrt{2g \times (65 - 35)} = 24.3 \text{ m/s}$; $Q = 30 \times 10^{-4} \times 24.3 = 0.0728 \text{ m}^3/\text{s}$

Then Power $= 0.0728 \times 9{,}810 \times 25\text{W} = \underline{\underline{17.85 \text{ kW}}}$

7.40 Solution procedure is the same as for P7.39:

$$0 + 0 + 110 + h_p = 0 + 0 + 200; \quad h_p = 90 \text{ ft}$$

$P = Q\gamma h_p/550$ where $Q = V_j A_j = 0.10\ V_j$

and $V_j = \sqrt{2g \times (200 - 110)} = 76.13 \text{ ft/sec}$; $Q = 7.613 \text{ ft}^3/\text{sec}$

Then $P = 7.613 \times 62.4 \times 90/550 = \underline{\underline{77.7 \text{ h.p.}}}$

7.41 Write the energy equation from water surface in A to water surface in B.

$$p_A/\gamma + V_A^2/2g + z_A = p_B/\gamma + V_B^2/2g + z_B + \sum H_L$$

$$p_A = p_B = p_{atm} \text{ and } V_A = V_B = 0$$

Let the pipe from A be called pipe 1

Let the pipe from B be called pipe 2

Then $\sum h_L = (V_1 - V_2)^2/2g + V_2^2/2g$

7.41 (Cont'd)

But $V_1A_1 = V_2A_2$ or $V_1 = V_2(A_2/A_1)$

However $A_2 = 2A_1 \therefore V_1 = 2V_2$

Then Eq. (1) becomes

$$z_A - z_B = (2V_2 - V_2)^2/2g + V_2^2/2g$$
$$= 2V_2^2/2g$$
$$V_2 = \sqrt{g(z_A - z_B)}$$
$$= \sqrt{10g}\ \text{m/s}$$

Then

$$Q = V_2A_2$$
$$= (\sqrt{10g}\ \text{m/s})(20\text{cm}^2)(10^{-4}\text{m}^2/\text{cm}^2)$$
$$= \underline{\underline{0.0198\text{m}^3/\text{s}}}$$

7.42

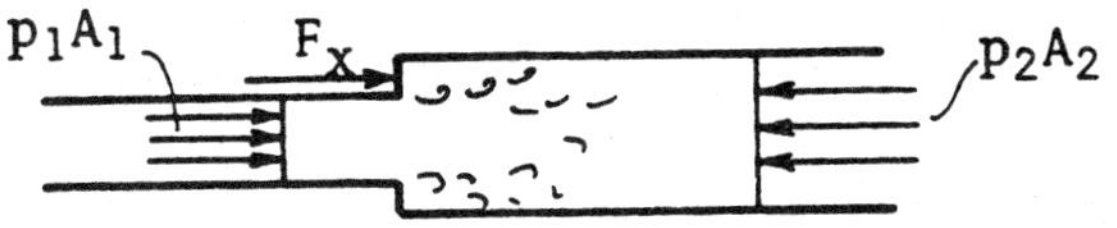

p_1A_1 F_x p_2A_2

$V_{40} = Q/A_{40} = 0.80/((\pi/4) \times 0.40^2) = 6.366$ m/s; $V_{40}^2/2g = 2.066$m

$V_{60} = V_{40} \times (4/6)^2 = 2.829$ m/s; $V_{60}^2/2g = 0.408$ m

$h_L = (V_{40} - V_{60})^2/2g = 0.638$ m

$p_{40}/\gamma + V_{40}^2/2g = p_{60}/\gamma + V_{60}^2/2g + h_L$

$p_{60} = 70{,}000 + 9{,}810(2.066 - 0.408 - 0.638) = 80{,}006$ Pa

Momentum equation:

$70{,}000 \times \pi/4 \times 0.4^2 - 80{,}006 \times \pi/4 \times (0.6^2) + F_x = 1{,}000 \times 0.80$

$\times (2.829 - 6.366)$

$F_x = -8{,}796 + 22{,}619 - 2{,}830 = 10{,}993$ N = $\underline{\underline{10.99\ \text{kN}}}$

7.43 $V_{10}A_{10} = V_{15}A_{15}$; $V_{15} = V_{10}A_{10}/A_{15} = 7 \times (10/15)^2 = 3.11$ m/s

$h_L = (7 - 3.11)^2/(2 \times 9.81) = \underline{\underline{0.771\ \text{m}}}$

7.44 $V_6 = Q/A_6 = 5/((\pi/4) \times (1/2)^2) = 25.46$ ft/s;

$V_{12} = (1/4)V_6 = 6.37$ ft/s

$h_L = (25.46 - 6.37)^2/(2 \times 32.2) = \underline{5.66 \text{ ft}}$

7.45 $h_L = (V_1 - V_2)^2/(2g)$; $V_2 = V_1 (A_1/A_2) = 25 (1/4) = 6.25$ ft/s

$h_L = (25 - 6.25)^2/64.4 = 5.46$ ft

a) H.P. = $Q\gamma h/550$; $Q = VA = 25 (\pi/4) (5^2) = 490.9 \text{ ft}^3/\text{s}$

H.P. = (490.9) (62.4) (5.46)/550 = $\underline{\underline{\text{304 horsepower}}}$

b) $$p_1/\gamma + V_1^2/2g + z_1 = p_2/\gamma + V_2^2/2g + z_2 + h_L$$

$$(5 \times 144)/62.4 + 25^2/64.4 = p_2/\gamma + 6.25^2/64.4 + 5.46$$

$$p_2/\gamma = 15.17 \text{ ft.}$$

$$p_2 = 15.17 \times 62.4 = 946.6 \text{ psf} = \underline{\underline{6.57 \text{ psig}}}$$

c) $$\Sigma F_x = V_x \rho \mathbf{V} \cdot \mathbf{A}$$

$$p_1A_1 - p_2A_2 + F_x = (25)(1.94)((-25)(\pi/4)(5^2) + (6.25)^2(1.94)(\pi/4)(10^2))$$

$$(5)(144)(\pi/4)(5^2) - (6.57)(144)(\pi/4)(10^2) + Fx = -23,807 + 5952$$

$$F_x = 74,305 - 14,137 - 23,807 + 5952$$

$$F_x = \underline{\underline{42,313 \text{ lbf}}}$$

7.46 $p_1/\gamma + V_1^2/2g + z_1 = p_2/\gamma + V_2^2/2g + z_2 + h_L$

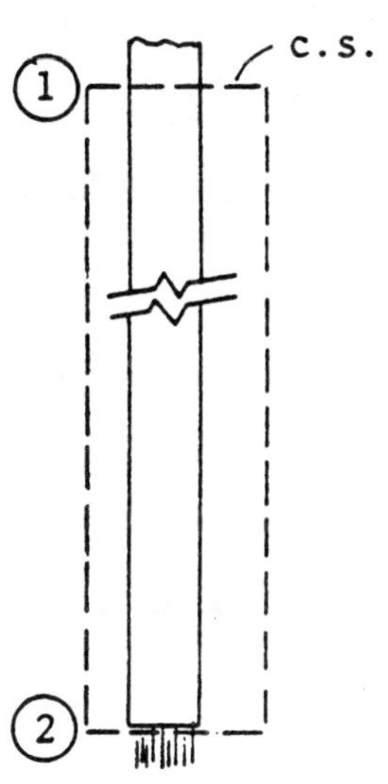

but $V_1 = V_2$ and $p_2 = 0$

Therefore, $p_1/\gamma = -50 + 10$; $p_1 = -2{,}496\ \text{lb/ft}^2$

$\Sigma F_y = \rho Q(V_{2_y} - V_{1_y})$

$-p_1A_1 - \gamma AL - 2L + F_{wall} = 0$

$F_{wall} = 1.5L + \gamma A_1 L - p_1A_1$

$= 75 + (\pi/4) \times 0.5^2(62.4 \times 50 - 2{,}496)$

$= 75 + 122.5$

$\underline{\underline{F_{wall} = 197.5\ \text{lbf}}}$

7.47 $p_{50}/\gamma + V_{50}^2/2g + z_{50} = p_{80}/\gamma + V_{80}^2/2g + z_{80} + h_L$

where $p_{50} = 650{,}000$ Pa; $z_{50} = z_{80}$

$V_{80} = Q/A_{80} = 5/((\pi/4) \times 0.8^2) = 9.947$ m/s; $V_{80}^2/2g = 5.04$ m

$V_{50} = V_{80} \times (8/5)^2 = 25.46$ m/s; $V_{50}^2/2g = 33.05$ m; $h_L = 10$ m

Then $p_{80}/\gamma = 650{,}000/\gamma + 33.05 - 5.04 - 10$

$p_{80} = 650{,}000 + 9{,}810(33.05 - 5.04 - 10) = 826{,}700$ Pa

$= \underline{\underline{826.7\ \text{kPa}}}$

$\Sigma F_x = \rho Q(V_{80_x} - V_{50_x})$

$p_{80}A_{80} + p_{50}A_{50} \times \cos 60° + F_x = 1{,}000 \times 5(-9.947 - 0.5 \times 25.46)$

$F_x = -415{,}540 - 63{,}814 - 113{,}385 = -592{,}700\ \text{N} = \underline{\underline{-592.7\ \text{kN}}}$

7.48 $V = Q/A = 10/((\pi/4) \times 1^2) = 12.73$ ft/sec

$h_L = V^2/2g = \underline{\underline{2.52\ \text{ft}}}$

7.49 $V = Q/A = 0.60/((\pi/4) \times 0.5^2) = 3.056$ m/s

$h_L = V^2/2g = (3.056)^2/(2 \times 9.81) = \underline{\underline{0.476\ \text{m}}}$

7.50 Take section 1 at reservoir surface and section 2 at section of d diameter.

$$p_1/\gamma + V_1^2/2g + z_1 = p_2/\gamma + V_2^2/2g + z_2$$

$0 + 0 + 5 = p_{2,vapor}/\gamma + V_2^2/2g + 0$ where $p_{2,vapor} = 2,340$ Pa abs.

$p_{2,vapor} = -97,660$ Pa gage

Then $V_2^2/2g = 5 + 97,660/9,790 = 14.97$ m; $V_2 = 17.1$ m/s

$Q = V_2A_2 = 17.1 \times \pi/4 \times 0.15^2 = \underline{\underline{0.303 \text{ m}^3/\text{s}}}$

7.51 First write the energy equation from the Venturi section to the end of the pipe:

$$p_1/\gamma + V_1^2/2g + z_1 = p_2/\gamma + V_2^2/2g + z_2 + h_L$$

$$p_{vapor}/\gamma + V_1^2/2g = 0 + V_2^2/2g + 0.9V_2^2/2g$$

where $p_{vapor} = 2,340$ Pa abs. $= -97,660$ Pa gage

$$V_1A_1 = V_2A_2; \quad V_1 = V_2A_2/A_1 = 4V_2; \quad V_1^2/2g = 16V_2^2/2g$$

Then $-97,660/9,790 + 16V_2^2/2g = 1.9V_2^2/2g$; $V_2 = 3.73$ m/s

$Q = V_2A_2 = 3.73 \times \pi/4 \times 0.4^2 = \underline{\underline{0.468 \text{ m}^3/\text{s}}}$

Now write the energy equation from reservoir water surface to outlet:

$$z_1 = V_2^2/2g + h_L$$

$H = 1.9\ V_2^2/2g = \underline{\underline{1.34 \text{ m}}}$

7.52

a) Flow is from right to left.

b) Machine is a pump.

c) Pipe CA is smaller because of steeper H.G.L.

d)

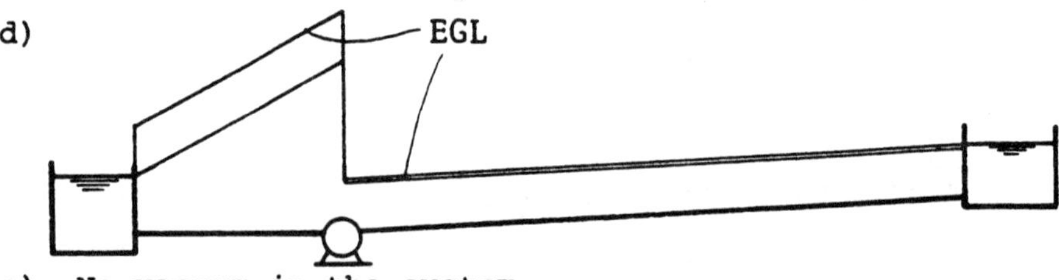

e) No vacuum in the system.

7.53

Write the energy equation from the reservoir water surface to the jet from the nozzle.

$$p_1/\gamma + V_1^2/2g + z_1 = p_2/\gamma + V_2^2/2g + z_2 + h_L$$

$$0 + 0 + 100 = 0 + V_6^2/2g + 60 + 0.02\ (1{,}000/1)\ V_{12}^2/2g$$

Continuity equation: $V_6A_6 = V_{12}A_{12}$; $V_6 = V_{12}\ (A_{12}/A_6)$

$V_6 = 4V_{12}$; $V_6^2/2g = 16\ V_{12}^2/2g$

$$40 = (V_{12}^2/2g)(16 + 20)$$

$V_{12}^2 = (40/36)\ 2g$; $V_{12} = 8.46$ ft/s

$Q = V_{12}A_{12} = (8.46)\ (\pi/4)\ (1^2) = \underline{6.64\ ft^3/s}$

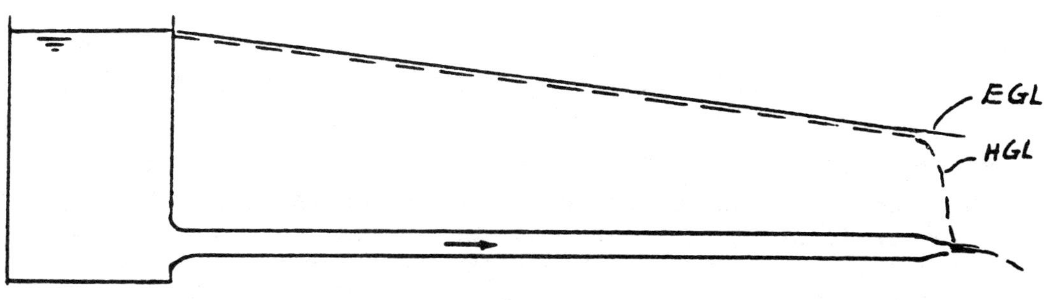

7.54

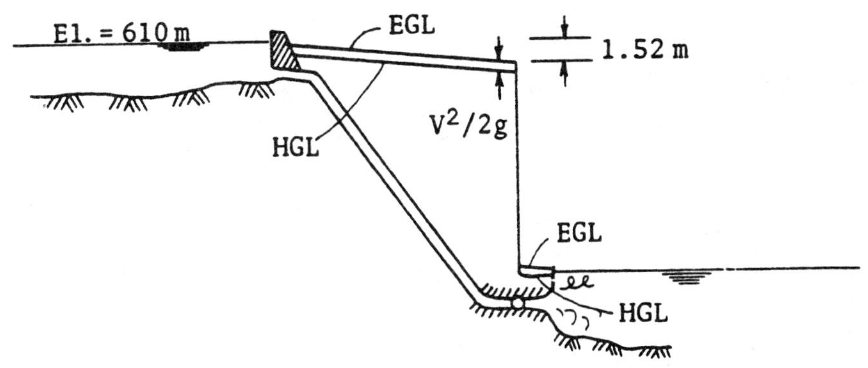

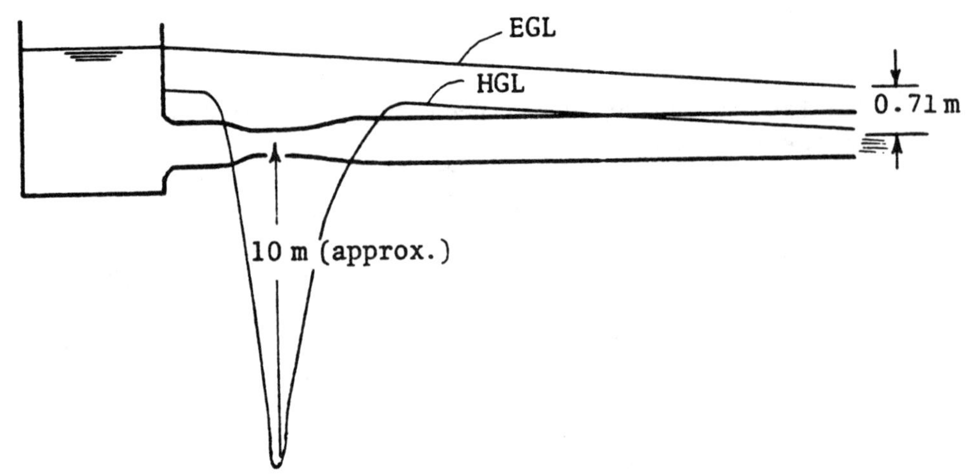

7.56

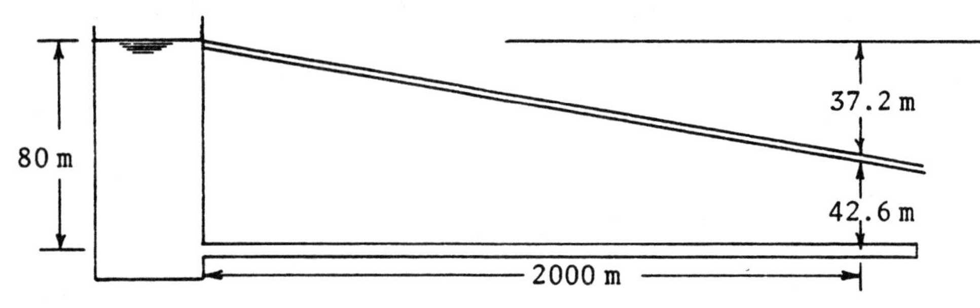

7.57 Because the E.G.L. slopes downward to the left, it is obvious that the flow is from right to left and that in the "black box" there could either be a turbine, an abrupt expansion or a partially closed valve. Circle b, c, d.

7.58 This is possible if the fluid is being accelerated to the left.

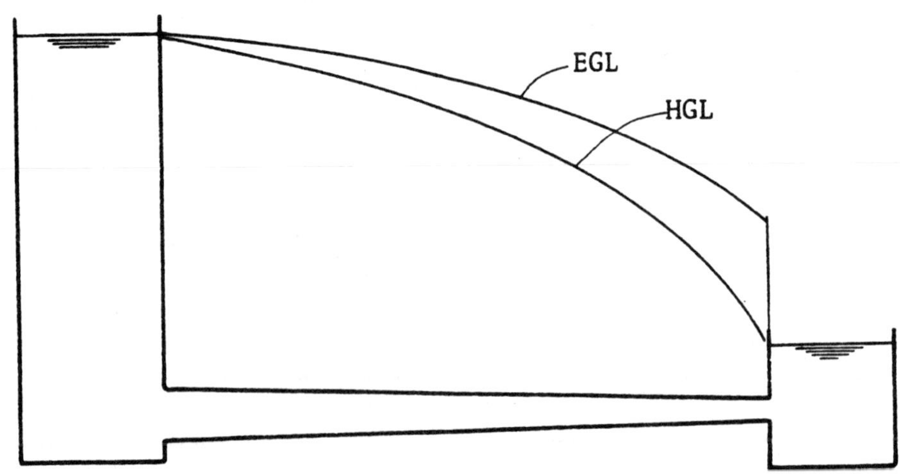

7.60 a) Solid line is EGL, dashed line is HGL

b) No; AB is smallest.

c) from B to C

d) p_{max} is at the bottom of the tank

e) p_{min} is at the bend C.

f) A nozzle.

g) above atmospheric pressure

h) abrupt expansion

7.61 Write energy equation from upper to lower reservoir:

$$p_1/\gamma + V_1^2/2g + z_1 = p_2/\gamma + V_2^2/2g + z_2 + \Sigma h_L$$

$$0 + 0 + 100 = 0 + 0 + 70 + \Sigma h_L$$

$$\Sigma h_L = 30 \text{ m}$$

$$0.02 \times (200/0.3)(V_u^2/2g) + (0.02(100/0.20) + 1.0)\, V_d^2/2g = 30 \quad (1)$$

but $V_u = Q/A_u = Q/((\pi/4) \times 0.3^2)$ (2)

$V_d = Q/A_d = Q/((\pi/4) \times 0.2^2)$ (3)

Substituting Eq.(2) and Eq.(3) into (1) and solving for Q yields:

$\underline{Q = 0.206 \text{ m}^3/\text{s}}$

7.62 $V = Q/A = 2.5/((\pi/4) \times (2/3)^2) = 7.16$ ft/sec

$$h_L = 0.015 \times 3{,}000 \times (7.16)^2/((2/3)\times(2 \times 32.2)) + (7.16)^2/(2 \times 32.2)$$

$= 54.6$ ft

Write the energy equation from water surface to water surface:

$$p_1/\gamma + V_1^2/2g + z_1 + h_p = p_2/\gamma + V_2^2/2g + z_2 + h_L$$

$0 + 0 + 90 + h_p = 0 + 0 + 140 + 54.6$; $h_p = 104.6$ ft

Power supplied $= Q\gamma h_p = 2.5 \times 62.4 \times 104.6 = \underline{16{,}318 \text{ ft-lb/s}}$

$P = \underline{29.7 \text{ horsepower}}$

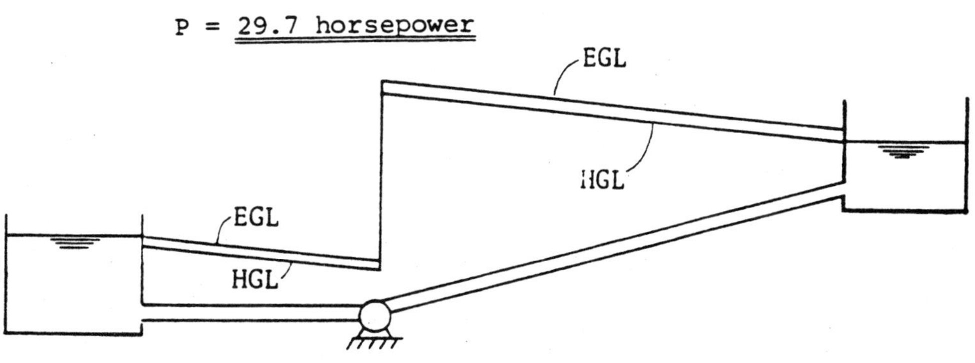

7.63 Write the energy equation from the water surface in A to the water surface in B:

$$p_A/\gamma + V_A^2/2g + z_A = p_B/\gamma + V_B^2/2g + z_B + h_L$$

$$0 + 0 + H = 0 + 0 + 0 + 0.01 \times (300/1)\ V_P^2/2g + V_P^2/2g$$

$$16 = 4\ V_p^2/2g;\quad V_p = \sqrt{4 \times 2 \times 9.81} = 8.86 \text{ m/s}$$

$$Q = VA = 8.86 \times (\pi/4) \times 1^2 = \underline{\underline{6.96 \text{ m}^3/\text{s}}}$$

To determine p_P write the energy equation between the water surface in A and point P:

$$0 + 0 + H = p_P/\gamma + V_P^2/2g - h + 0.01 \times (150/1) V_P^2/2g$$

$$16 = p_P/\gamma - 2 + 2.5\ V_p^2/2g \text{ where } V_p^2/2g = 4 \text{ m}$$

Then $p_P = 9.810\ (16 + 2 - 10) = \underline{\underline{78.5 \text{ kPa}}}$

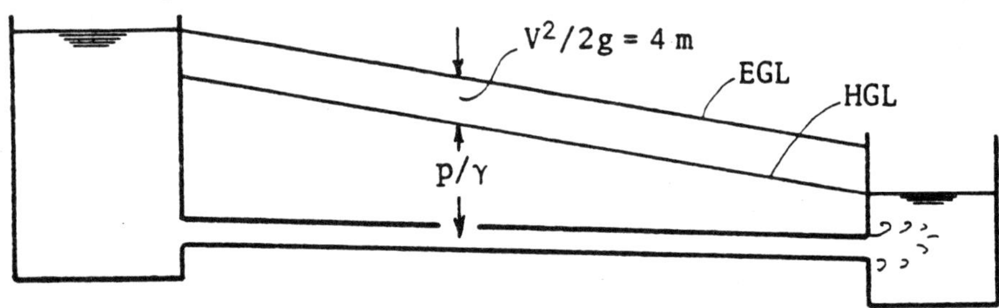

7.64 Write the energy equation from the left reservoir to the right reservoir:

$$p_L/\gamma + V_L^2/2g + z_L = p_R/\gamma + V_R^2/2g + z_R + h_L$$

$$0 + 0 + z_L = 0 + 0 + 110 + 0.02\ (200/1.128)\ (V_1^2/2g) +$$
$$0.02\ (300/1.596)\ (V_2^2/2g) + (V_1 - V_2)^2/2g + V_2^2/2g$$

where $V_1 = Q/A_1 = 16/1 = 16$ ft/s ; $V_2 = 8$ ft/s

$$z_L = 110 + (0.02/2g)\ ((200/1.128)\ (16^2) + (300/1.596)\ (8^2)) +$$
$$((16 - 8)^2/64.4) + 8^2/64.4$$
$$= 110 + 17.83 + 0.99 + 0.99$$

$$\underline{\underline{z_L = 129.8 \text{ ft}}}$$

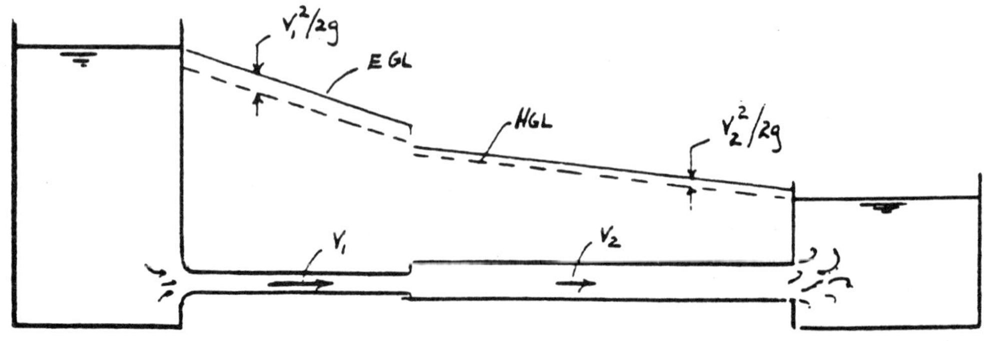

7.65 Write the energy equation from the lower reservoir surface to the upper reservoir surface:

$$p_\ell/\gamma + V_\ell^2/2g + z_\ell + h_p = p_u/\gamma + V_u^2/2g + z_u + h_L$$

$$0 + 0 + 150 + h_p = 0 + 0 + 250 + \Sigma 0.018(L/D)(V^2/2g) + V^2/2g$$

where $V_1 = Q/A_1 = 2/((\pi/4) \times 1^2) = 2.55$ m/s; $V_1^2/2g = 0.330$ m

$V_2 = Q/A_2 = 4V_1 = 10.19$ m/s; $V_2^2/2g = 5.29$ m

Then $h_p = 250 - 150 + 0.018\,[(100/1) \times 0.330 + (1{,}000/0.5)$

$\times\ 5.29] + 5.29 = 296.3$ m

Power required: $P = Q\gamma h_p/\text{eff.} = 2 \times 9{,}810 \times 296.3/0.74 = \underline{\underline{7.86\text{ MW}}}$

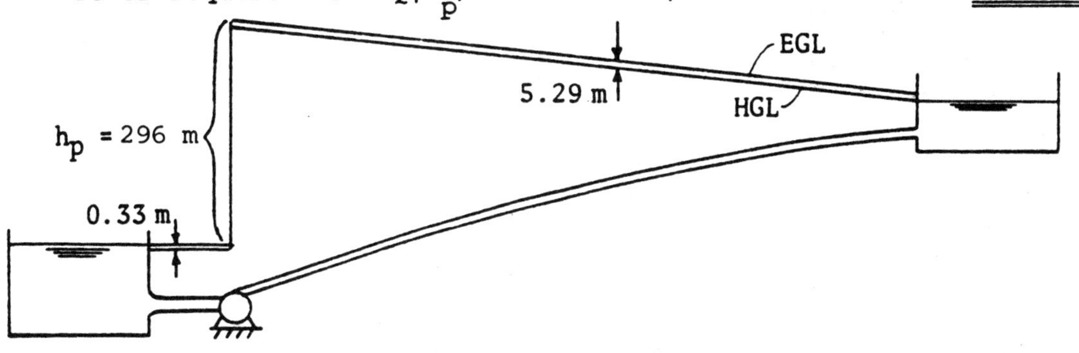

7.66 First write energy equation from reservoir water surface to end of pipe:

$$p_1/\gamma + V_1^2/2g + z_1 = p_2/\gamma + V_2^2/2g + z_2 + h_L$$

$$0 + 0 + 200 = 0 + V^2/2g + 185 + 0.02(200/0.30)\,V^2/2g$$

$$14.33\,V^2/2g = 15 \quad ; V^2/2g = 1.047\text{m} \quad \text{and } V = 4.53 \text{ m/s}$$

$Q = VA = 4.53 \times (\pi/4) \times 0.30^2 = \underline{\underline{0.320\text{ m}^3/\text{s}}}$

To solve for the pressure midway along pipe, write the energy equation to the midpoint:

$$p_1/\gamma + V_1^2/2g + z_1 = p_m/\gamma + V_m^2/2g + z_m + h_L$$

$$0 + 0 + 200 = p_m/\gamma + V_m^2/2g + 200 + 0.02(100/0.30)V^2/2g$$

$$p_m/\gamma = -(V^2/2g)(1 + 6.667)$$

$$= -(1.047)(7.667) = -8.027\text{m}$$

$p_m = -8.027\gamma = -78{,}745 \text{ Pa} = \underline{\underline{-78.7\text{ kPa}}}$

7.67 Write the energy equation between the water surface in the river and water surface in the tank. Let point 1 be the river water surface.

$$p_1/\gamma + \cancelto{0}{V_1^2/2g} + z_1 + h_p = p_2/\gamma + \cancelto{\text{neglect}}{V_2^2/2g} + z_2 + h_L$$

$$0 + 0 + 0 + h_p = 0 + 0 + (2\text{ m} + h) + h_L$$

where h = depth of water in the tank

$$20 - (5)(10^4)Q^2 = h + 2 + V^2/2g + 10\, V^2/2g$$

(abrupt expansion)

$$\therefore \quad 18 = (5)(10^4)Q^2 + 11\,(V^2/2g) + h \quad ; \quad V = Q/A$$

$$(11\, V^2)/2g = (11/2g)(Q^2/A^2) = (1.46)(10^5)Q^2$$

$$18 = 1.96 \times 10^5 Q^2 + h$$

$$Q^2 = (18 - h)/((1.96)(10^5)) \quad ; \quad Q = (18 - h)^{0.5}/443$$

But $Q = A_T\, dh/dt$ where A_T = tank area

$$\therefore \quad dh/dt = (18 - h)^{0.5}/((443)(\pi/4)(5)^2) = (18 - h)^{0.5}/8{,}693$$

$$dh/(18 - h)^{0.5} = dt/8{,}693$$

Integrate:

$$-2\,(18 - h)^{0.5} = (t/8{,}693) + \text{const.}$$

$$t = 0 \text{ when } h = 0 \quad \cdot \quad \text{const.} = -2(18)^{0.5}$$

$$\therefore \quad t = (18^{0.5} - (18 - h)^{0.5})(17{,}386)$$

For $h = 10$ m

$$t = (18^{0.5} - 8^{0.5})(17{,}386) = \underline{\underline{24{,}588 \text{ s}}} = \underline{\underline{6.83 \text{ hrs}}}$$

7.68 (a) Flow is from A to E because EGL slopes downward in that direction.

(b) Yes, at D, because EGL and HGL are coincident there.

(c) Uniform diameter because $V^2/2g$ is constant (EGL and HGL uniformly spaced).

(d) No, because EGL is always dropping (no energy added).

(e)

(f) Nothing else.

7.69 **Write energy equation from reservoir water surface to jet:**

$p_1/\gamma + V_1^2/2g + z_1 = p_2/\gamma + V_2^2/2g + z_2 + h_L$

$0 + 0 + 100 = 0 + V_2^2/2g + 30 + 0.014(L/D)(V_p^2/2g)$

$100 = 0 + V_2^2/2g + 30 + 0.014(500/0.60)V_p^2/2g$

$V_2A_2 = V_pA_p; \quad V_2 = V_pA_p/A_L = 4V_p$

Then $V_p^2/2g(16 + 11.67) = 70$; $V_p = 7.046$ m/s; $V_p^2/2g = 2.53$ m

$Q = V_pA_p = 7.046 \times (\pi/4) \times 0.60^2 = \underline{\underline{1.992 \text{ m}^3/\text{s}}}$

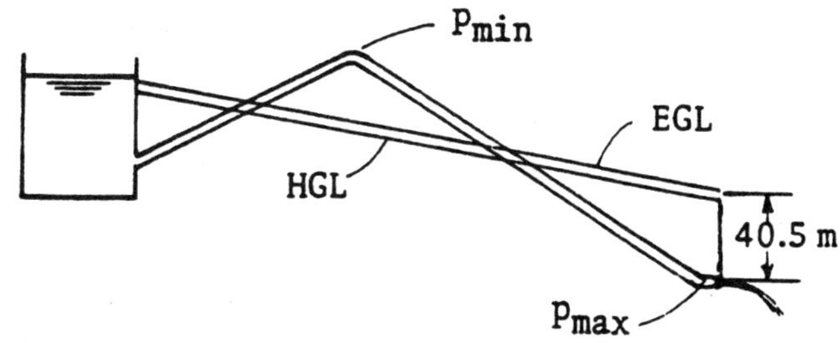

p_{min}: $100 = p_{min}/\gamma + V_p^2/2g + 100 + 0.014(100/0.60)V_p^2/2g$

$100 = p_{min}/\gamma + 100 + 3.33 \times 2.53; \quad p_{min} = \underline{\underline{-82.6 \text{ kPa, gage}}}$

$p_{max}/\gamma = 40.5 - 2.53$ m; $p_{max} = \underline{\underline{372.5 \text{ kPa}}}$

7.70 **Assume negligible head loss:**

$p_1/\gamma + V_1^2/2g = p_2/\gamma + V_2^2/2g + h_t; \quad h_t = V_1^2/2g - V_2^2/2g$

where $V_2 = V_1A_1/A_2 = V_1(3/4.5)^2 = 0.444\ V_1$; $V_2^2/2g = 0.197\ V_1^2/2g$

Then $h_t = 10^2/(2 \times 9.81)[1 - 0.197] = 4.09$ m

Then Power $= Q\gamma h_t = 10(\pi/4) \times 3^2 \times 1.2 \times 9.81 \times 4.09 = \underline{\underline{3.40 \text{ kW}}}$

7.71 **Write energy equation from upstream end to downstream end:**

$p_1/\gamma + V_1^2/2g + z_1 + h_p = p_2/\gamma + V_2^2/2g + z_2 + h_L$

$0 + 0 + 0 + h_p = 0 + V_2^2/2g + 0 + 0.02\ V_T^2/2g$

$V_TA_T = V_2A_2; \quad V_2 = V_TA_T/A_2 = V_T \times 0.4; \quad V_2^2/2g = 0.16\ V_T^2/2g$

$h_p = V_T^2/2g(0.18) = (50^2/(2 \times 9.81))(0.18); \quad h_p = 22.94$ m

$P = Q\gamma h_p = 50 \times 4 \times 1.2 \times 9.81 \times 22.94 = \underline{\underline{54.0 \text{ kW}}}$

7.72 **Write energy equation from section (1) to section (2):**

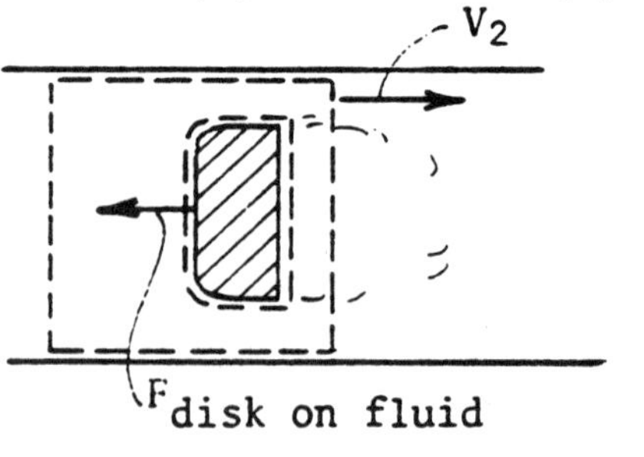

$$p_1 + \rho U_1^2/2 = p_2 + \rho U_2^2/2$$

$$p_1 - p_2 = \rho U_2^2/2 - \rho U_1^2/2$$

but $U_1 A_1 = U_2 (\pi/4)(D^2 - d^2)$

$$U_1 (\pi/4) D^2 = U_2 (\pi/4)(D^2 - d^2)$$

$$U_2 = U_1 D^2/(D^2 - d^2) \qquad (1)$$

Then $p_1 - p_2 = (\rho/2) U_1^2 [(D^4/(D^2 - d^2)^2 - 1]$ (2)

Now write the momentum equation for the C.V.

$$\Sigma F_x = \rho Q (U_{2_x} - U_{1_x})$$

$$p_1 A - p_2 A + F_{\text{disk on fluid}} = \rho Q (U_2 - U_1)$$

$$F_{\text{fluid on disk}} = F_d = \rho Q (U_1 - U_2) + (p_1 - p_2) A$$

Eliminate $p_1 - p_2$ by Eq. (2), and U_2 by Eq. (1):

$$F_d = \rho U A (U_1 - U_1 D^2/(D^2 - d^2)) + (\rho U^2/2)[(D^4/(D^2 - d^2)^2 - 1] A$$

$$\underline{\underline{F_d = \rho U^2 \pi D^2/8 [1/(D^2/d^2 - 1)^2]}}$$

When $U = 10$ m/s, $D = 5$ cm, $d = 4$ cm and $\rho = 1.2$ kg/m^3

$$F_d = (1.2 \times 10^2 \pi \times (0.05)^2/8)[1/((0.05/0.04)^2 - 1)^2] = \underline{\underline{0.372\text{N}}}$$

7.73

Let the control volume include the sphere and fluid in a given length of tube. Also let the control volume move with the sphere; thus, steady flow conditions will prevail.

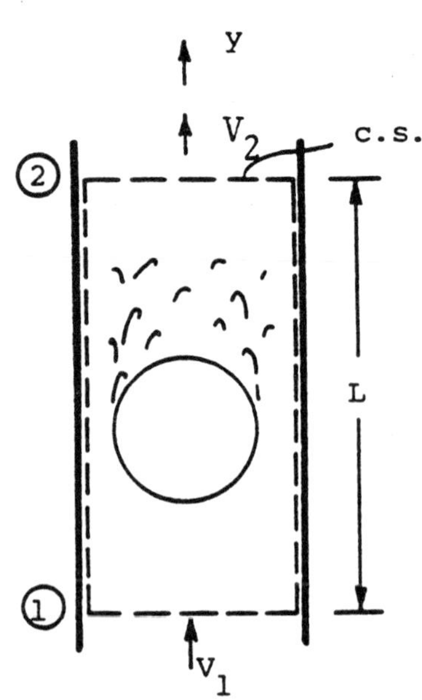

First write the momentum equation for this control volume. Neglect viscous forces in the solution:

$$\Sigma F_y = \Sigma V_y \rho \underline{V} \cdot \underline{A}$$

Because $V_1 = V_2$ this equation reduces to

$$p_1 A_1 - p_2 A_2 - W_{\text{water}} - W_{\text{sphere}} = 0$$

$$(p_1 - p_2) A = \gamma_f (L A_C - \forall_s) - \gamma_s \forall_s = 0$$

7.73 (continued)

where A_c = cross-sectional area of cylinder

$\forall$ = sphere volume

γ_f = specific wgt. of fluid

γ_s = specific wgt. of sphere

$$p_1 - p_2 = \gamma_f(LA_c - \forall_s)/A_c + \gamma_s\forall_s/A_c \quad (1)$$

Now write the energy equation from section (1) to section (2)

$$p_1/\gamma + V_1^2/2g + z_1 = p_2/\gamma + V_2^2/2g + z_2 + h_L$$

but $V_1^2/2g = V_2^2/2g$ and $z_2 - z_1 = L$

Also $h_L = (V_a - V_c)^2/2g$

where V_a = velocity in annulus between the sphere and cylinder

V_c = velocity in unobstructed cylinder

Then $p_1 - p_2 = \gamma_f(L + ((V_a - V_c)^2/2g))$ (2)

Now eliminate $p_1 - p_2$ between Eqs. (1) and (2) yielding:

$$\gamma_s\forall_s/A_c + (\gamma_f(LA_c - \forall_s)/A_c) = \gamma_f(L + ((V_a - V_c)^2/2g))$$

$$\gamma_s/\gamma_f = ((A_c/\forall_s)(V_a - V_c)^2/2g) + 1$$

$$V_c = 0.50 \text{ ft/s}$$

and $V_aA_a = V_cA_c$; $V_a = V_c(A_c/A_a)$

$$= 0.5(1.05^2/(1.05^2-1))$$

$$V_a = 5.38 \text{ ft/s}$$

$$A_c/\forall_s = (\pi D_c^2/4)/(\pi D_s^3/6)$$

$$= (3/2)(D_c^2/D_s^3)$$

$$= (3/2)(1.05^2/1^3)$$

$$= 1.654 \text{ ft}^{-1}$$

Then $\gamma_s/\gamma_f = ((1.654 \text{ ft}^{-1})(5.38 - 0.50)^2\text{ft}^2/\text{s}^2/(64.4 \text{ ft/s}^2)) + 1$

$\gamma_s/\gamma_f = S = \underline{1.612}$

CHAPTER EIGHT

8.1 a) $Q = (2/3)\ CL\ \sqrt{2g}\ H^{3/2}$

$[Q] = L^3/T = L(L/T^2)^{1/2}L^{3/2}$

$L^3/T = L^3/T$ homogeneous

b) $V = (1.49/n)R^{2/3}S^{1/2}$

$[V] = L/T = L^{-1/6}L^{2/3}$ not homogeneous

c) $h_f = f(L/D)V^2/2g$

$[h_f] = L = (L/L)(L/T)^2/(L/T^2)$ homogeneous

d) $D = 0.074\ R^{-0.2}Bx\rho V^2/2$

$[D] = F = L \times L \times (FT^2/L^4)(L/T)^2$ homogeneous

8.2 a) $[T] = FL;\quad [T] = (ML/T^2) \times L = ML^2/T^2$

b) $[\rho V^2/2] = (FT^2/L^4)(L/T)^2 = F/L^2; [\rho V^2/2] = (M/L^3)(L^2/T^2) = M/LT^2$

c) $[\sqrt{T/\rho}\ ] = \sqrt{(F/L^2)/(FT^2/L^4)} = L/T$

d) $[Q/ND^3] = (L^3/T)/(T^{-1}L^3) = 1 \rightarrow$ Dimensionless

---

8.3 $\Delta h = f(t, \rho, D, d, \gamma, h)$

F: $\Delta h = f(t, \gamma/\rho, D, d, h)$ where $[\gamma/\rho] = L/T^2$

T: $\Delta h = f(\gamma t^2/\rho, D, d, h)$

L: $\Delta h/d = f(\gamma t^2/\rho d, d/D, h/d)$

or $\Delta h/d = f(gt^2/d, d/D, h/d)$

8.4 $V = f(h, \sigma, \gamma, g)$

where $[V] = L/T, [h] = L, [\sigma] = M/T^2, [\gamma] = M/(L^2T^2), [g] = L/T^2$

$[V] = [h^a\sigma^b\gamma^c g^d]$

$L/T = (L^a)(M^b/T^{2b})(M^c/(L^{2c}T^{2c})(L^d/T^{2d})$

8.4 (continued)

L: $1 = a - 2c + d$

M: $0 = b + c$

T: $1 = 2b + 2c + 2d$

Determine the exponents b, c & d in terms of a

$$0 - 2c + d = 1 - a$$

$$b + c + 0 = 0$$

$$2b + 2c + 2d = 1$$

Solution yields: $b = -c$, $d = 1/2$

$$-2c + 1/2 = 1 - a \Rightarrow -2c = 1/2 - a \Rightarrow c = -1/4 + a/2$$

$$b = 1/4 - a/2$$

Thus $V = h^a \sigma^{(1/4 - a/2)} \gamma^{(-1/4 + a/2)} g^{1/2}$

$$= (g^{1/2}\sigma^{1/4}/\gamma^{1/4}) (h\gamma^{1/2}/\sigma^{1/2})^a$$

Which can also be written as

$$V^4\gamma/(g^2\sigma) = f(h^2\gamma/\sigma)$$

Alternate forms: $(V^4\gamma/(g^2\sigma)) (\sigma/h^2\gamma) = f(h^2\gamma/\sigma)$

$$V^4/(gh)^2 = f(h^2\gamma/\sigma)$$

or $$V/\sqrt{gh} = f(h^2\gamma/\sigma)$$

8.5 $h = f(d, \sigma, \gamma)$

F: $h = f(d, \sigma/\gamma)$ where $[\sigma/\gamma] = (F/L)/(F/L^3) = L^2$

L: $h/d = \underline{\underline{f(\sigma/\gamma d^2)}}$

8.6 $F_D = f(V, \mu, d)$

F: $F_D/\mu = f(V, d)$

where $[F_D/\mu] = F/(FT/L^2) = L^2/T$

T: $F_D/(\mu V) = f(d)$

where $[F_D/(\mu V)] = (L^2/T)/(L/T) = L$

L: $\underline{\underline{F_D/(\mu V d) = C}}$

8.7 $F_D = f(D,\rho,\mu,V,k)$

F: $F_D/\rho = f(D,\rho/\mu,V,k)$

where $[F_D/\rho] = F/(FT^2/L^4) = L^4/T^2$

$[\rho/\mu] = (FT^2/L^4)/(FT/L^2) = T/L^2$

T: $F_D/(\rho V^2) = f(D,\rho V/\mu,k)$

where $[F_D/(\rho V^2)] = (L^4/T^2)/(L^2/T^2) = L^2$

$[\rho V/\mu] = (T/L^2)(L/T) = L^{-1}$

L: $\underline{\underline{F_D/(\rho V^2 D^2) = f(k/D,\ VD\rho/\mu)}}$ ←One possible solution

---

8.8

$$F = f(D,V_o,\rho,\mu,k_s,\omega)$$

Eliminate force dimension by combining ρ with other variables

$$[\rho] = FT^2/L^4$$
$$[\mu] = FT/L^2$$

Then $F/\rho = f(D,V_o,\mu/\rho,k_s,\omega)$

Eliminate the time dimension by combining V with other variables

$$[V] = L/T$$
$$[\omega] = T^{-1}$$
$$[F/\rho] = (F)/(FT^2/L^4) = L^4/T^2$$
$$[\mu/\rho] = (FT/L^2)/(FT^2/L^4) = L^2/T$$
$$F/(\rho V^2) = f(D,\mu/(\rho V),k_s,\omega/V)$$

Eliminate the length dimension by combining D with other variables

$$[D] = L$$
$$[F/(\rho V^2)] = F/((FT^2/L^4)(L^2/T^2) = L^2$$
$$[\mu/(\rho V)] = (FT/L^2)/[(FT^2/L^4)(L/T)] = L$$
$$[k_s] = L$$
$$[\omega/V] = T^{-1}/(L/T) = L^{-1}$$
$$\underline{\underline{F/(\rho V^2 D^2) = f(\mu/(\rho VD),k_s/D,D\omega/V)}}$$

8.9

$$F_D = f(V,\rho,B,\mu,u^2,L_x)$$
$$\text{where } [F_D] = F$$
$$[V] = L/T$$
$$[\rho] = FT^2/L^4$$
$$[B] = L$$
$$[\mu] = FT/L^2$$
$$[u'] = L/T$$
$$[L_x] = L$$

Combine ρ with the other variables to eliminate the F dimension

$$F_D/\rho = f(V,B,\mu/\rho,u',L_x)$$

Combine V with the other variables to eliminate the T dimension

$$[F_D/\rho] = F/(FT^2/L^4) = L^4/T^2$$
$$[\mu/\rho] = (FT/L^2)/(FT^2/L^4) = L^2/T$$
$$F_D/(\rho V^2) = f(B,\mu/\rho V,u'/V,L_x)$$

Combine B with the other variables to eliminate the L dimension

$$[F_D/(\rho/V^2)] = (L^4/T^2)/(L^2/T^2) = L^2$$
$$[\mu/\rho V] = (L^2/T)/(L/T) = L$$

Final combination:

$$\underline{\underline{F_D/(\rho V^2B^2) = f(\mu/(\rho VL),u'/V,L_x/B)}}$$

8.10 $\Delta p/\Delta \ell = f(\mu,V,D)$

$[\Delta p/\Delta \ell] = (F/L^2)/L$ where $[\mu] = FT/L^2$ and $[V] = L/T$

F: $(\Delta p/\Delta \ell)/\mu = f(V,D)$ where $[(\Delta p/\Delta \ell)/\mu] = (F/L^3)/(FT/L^2) = T^{-1}L^{-1}$

T: $(\Delta p/\Delta \ell)/(\mu V) = f(D)$ where $[(\Delta p/\Delta \ell)/(\mu V)] = L^{-2}$

L: $(\Delta p/\Delta \ell)D^2/(\mu V) = 1$ or $\underline{\underline{\Delta p = f(\mu V\Delta \ell/D^2)}}$

8.11 $\Delta p = f(D, n, Q, \rho)$ where $[\Delta p] = F/L^2$; $[D] = L$; $[n] = T^{-1}$;

$[Q] = L^3/T$ and $[\rho] = FT^2/L^4$

F: $\Delta p/\rho = f(D, n, Q)$

T: $\Delta p/\rho n^2 = f(D, Q/n)$

L: $\underline{\underline{\Delta p/\rho n^2 D^2 = f(Q/nD^3)}}$

8.12

$$f = f(p, R, \rho, k)$$

where $[f] = T^{-1}$

$[p] = F/L^2$

$[R] = L$

$[\rho] = FT^2/L^4$

$[k]$ is dimensionless

Eliminate the F dimension by combining p and ρ

$$f = f(p/\rho, R, k)$$

Eliminate the time dimension by combing p/ρ with f

$$[p/\rho] = (F/L^2)/(FT^2/L^4) = L^2/T^2$$

Then $[(p/\rho)^{1/2}] = L/T$

So $f/(p/\rho)^{1/2} = f(R, k)$

Eliminate the length dimension by combining R with f $(p/\rho)^{1/2}$

$$[f/(p/\rho)^{1/2}] = T^{-1}/(L/T) = L^{-1}$$

Then $fR/(p/\rho)^{1/2} = f(k)$

or $\underline{\underline{f\rho^{1/2}R/p^{1/2} = f(k)}}$

8.13

$$F = \lambda^a \rho^b D^c c^d$$
$$ML/T^2 = L^a (M/L^3)^b L^c (L/T)^d$$
$$= L^{a-3b+c+d} M^b T^{-d}$$

Equating powers of M, L and T, we have

$$T: d = 2$$
$$M: b = 1$$
$$L: 1 = a - 3 + c + d$$
$$1 = a - 3 + c + 2$$
$$a + c = 2$$
$$a = 2 - c$$

Therefore,

$$F = \lambda^{(2-c)} \rho D^c c^2$$
$$\underline{\underline{F/(\rho c^2 \lambda^2) = f(D/\lambda)}}$$

Another valid answer would be

$$\underline{\underline{F/(\rho c^2 D^2) = f(D/\lambda)}}$$

8.14 $V = f(\ell, \rho, \sigma)$

F: $V = f(\ell, \sigma/\rho)$ where $[\sigma/\rho] = (F/L)/(FT^2/L^4) = L^3/T^2$

T: $1 = f(\ell, \sigma/\rho V^2)$ where $[\sigma/\rho V^2] = L$

L: $1 = f(\sigma/\rho V^2 \ell)$ or $V = \underline{\underline{C\sqrt{\sigma/\rho\ell}}}$

8.15 $T = f(\mu, \omega, S, D)$

F: $T/\mu = f(\omega, S, D)$ where $[T/\mu] = FL/(FTL^{-2}) = L^3/T$

T: $T/\omega\mu = f(S, D)$

L: $\underline{\underline{T/\omega\mu D^3 = f(S/D)}}$

8.16 $h = f(t, \sigma, \rho, \gamma, \mu, d)$

F: $h = f(t, \sigma/\rho, \gamma/\rho, \mu/\rho, d)$ where $[\sigma/\rho] = L^3/T^2$

and $[\gamma/\rho] = L/T^2$; $[\mu/\rho] = L^2/T$

T: $h = f(\sigma t^2/\rho, \gamma t^2/\rho, \mu t/\rho, d)$

L: $\underline{\underline{h/d = f(\sigma t^2/\rho d^3, \gamma t^2/\rho d, \mu t/\rho d^2)}}$

or $\underline{\underline{h/d = f(\sigma t^2/\rho d^3, g t^2/d, \nu t/d^2)}}$

8.17 First, establish the dimensions of the variables:

$$[\Delta p] = F/L^2 = (ML/T^2)/L^2 = M/(T^2L)$$
$$[\rho] = M/L^3; [\mu] = FT/L^2 = (ML/T^2)(T)/L^2 = M/LT$$
$$[V] = L/T; [E_v] = F/L^2 = (ML/T^2)/L^2 = M/LT^2$$
$$[\sigma] = F/L = (ML/T^2)/L = M/T^2$$
$$[\Delta\gamma] = F/L^3 = (ML/T^2)/L^3 = M/L^2T^2$$

Assume

$$f(\rho, L, V, \mu, E_v, \sigma, \Delta\gamma) = \rho^a L^b V^c \mu^d E_V^e \sigma^f \Delta\gamma^g$$
$$\therefore [\Delta p] = M/T^2L = (M/L^3)^a L^b (L/T)^c (M/LT)^d (M/LT^2)^e (M/T^2)^f (M/L^2T^2)^g$$

Equating powers of M, L and T, we have

$$L: -3a + b + c - d - e - 2g = -1$$
$$M: a + d + e + f + g = 1$$
$$T: -c - d - 2e - 2f - 2g = -2$$
$$-3a + b + c = d + e + 2g - 1$$
$$a = 1 - d - e - f - g$$
$$c = 2 - d - 2e - 2f - 2g$$

Solving the above three equations for b yields

$$b = -d - f + g$$

$$\Delta p = \rho^{1-d-e-f-g} L^{-d-f+g} V^{2-d-2e-2f-2g} \mu^d E_V^e \sigma^f \Delta\gamma^g$$
$$\Delta p = \rho V^2 (\mu/\rho LV)^d (E_v/\rho V^2)^e (\sigma/\rho LV^2)^f (\Delta\gamma L/\rho V^2)^g$$
$$\text{or } \Delta p/\rho V^2 = (\mu/\rho LV)^d (E_v/\rho V^2)^e (\sigma/\rho LV^2)^f (\Delta\gamma L/\rho V^2)^g$$

Arbitrary values can be given to d, e, f and g and the combinations of variables will still be dimensionless. Therefore, let d = -1, e = -1/2, f = -1 and g = -1 to yield

$$\underline{\underline{\Delta p/\rho V^2 = f(\rho LV/\mu), V/\sqrt{E_v/\rho}, (\rho LV^2/\sigma, \rho V^2/\Delta\gamma)}}$$

8.18

$$e = f(Br,\ \sigma,\ E,\ V,\ d,\ \dot{M}_p,\ D)$$

where $[e] = M/(L^2T)$; $[Br]$ = dimensionless

$[E] = M/(LT^2)$; $[\sigma] = M/(LT^2)$

$[V] = L/T$; $[d] = L$; $[\dot{M}_p] = M/T$; $[D] = L$

$$\therefore \quad [e] = [E^\alpha \sigma^\beta V^\gamma d^\delta \dot{M}_p{}^\varepsilon D^\lambda]$$

$$M/(L^2T) = (M/(LT^2))^\alpha (M/(LT^2))^\beta (L/T)^\gamma L^\delta (M/T)^\varepsilon L^\lambda$$

M: $\quad 1 = \alpha + \beta + \varepsilon$

L: $\quad 2 = \alpha + \beta - \gamma - \delta - \lambda$

T: $\quad 1 = 2\alpha + 2\beta + \gamma + \varepsilon$

Use α, γ and ε as unknowns

$$\alpha + 0 + \varepsilon = 1 - \beta \qquad (1)$$

$$\alpha - \gamma + 0 = 2 - \beta + \delta + \lambda \qquad (2)$$

$$2\alpha + \gamma + \varepsilon = 1 - 2\beta \qquad (3)$$

(1): $\quad \alpha + \varepsilon = 1 - \beta$

(2) + (3): $\quad 3\alpha + \varepsilon = 3 - 3\beta + \delta + \lambda$

(2) + (3) - (1): $\quad 2\alpha = 2 - 2\beta + \delta + \lambda$

$$\alpha = 1 - \beta + (\delta + \lambda)/2$$

$$\varepsilon = -\alpha + 1 - \beta = -1 + \beta - ((\delta + \lambda)/2) + 1 - \beta = -(\delta + \lambda)/2$$

$$\gamma = \alpha - 2 + \beta - \delta - \lambda$$

$$= 1 - \beta + ((\delta + \lambda)/2) - 2 + \beta - (\delta + \lambda) = -1 - ((\delta + \lambda)/2)$$

$$e = f(E^{(1-\beta + ((\delta + \lambda)/2)} \sigma^\beta V^{-1 - ((\delta + \lambda)/2} d^\delta \dot{M}_p{}^{-((\delta + \lambda)/2)} D^\lambda,\ Br)$$

$$e = (E/V)\ f((\sigma/E)^\beta (Ed^2/(V\dot{M}_p))^{\delta/2} (ED^2/(\dot{M}_pV))^{\lambda/2},\ Br)$$

or $\underline{\underline{eV/E = f(\sigma/E,\ Ed^2/(V\dot{M}_p),\ ED^2/(\dot{M}_pV),\ Br)}}$

Alternate form:

$$\underline{\underline{eV/E = f(\sigma/E,\ Ed^2/V\dot{M}_p,\ d/D,\ Br)}}$$

8.19 $\Delta p = f(Q,\rho,\mu,D,d)$

F: $\Delta p/\rho = f(Q,\mu/\rho,D,d)$

where $[\Delta p/\rho] = (F/L^2)/(FT^2/L^4) = L^2/T^2$

$[\mu/\rho] = (FT/L^2)/(FT^2/L^4) = L^2/T$

T: $\Delta p/(\rho Q^2) = f(\mu/(Q\rho),D,d)$

where $[\Delta p/(\rho Q^2)] = (L^2/T^2)/(L^6/T^2) = L^{-4}$

$[\mu/(\rho Q)] = (L^2/T)/(L^3/T) = L^{-1}$

L: $\underline{\underline{\Delta p d^4/(\rho Q^2) = f(\mu d/(\rho Q),d/D)}}$ ← One possible solution.

8.20 **The viscous forces are relatively small.**

8.21 $Re_w = Re_0$

$V_w d/\nu_w = V_0 d/\nu_0$

$V_w = V_0 \nu_w/\nu_0 = 2\ \text{m/s}\ (10^{-6}/10^{-5}) = \underline{\underline{0.20\ \text{m/s}}}$

8.22

$$Re_5 = Re_{20}$$

$$V_5 D_5/\nu_5 = V_{20} D_{20}/\nu_{20}$$

$$V_5 = V_{20}\,(D_{20}/D_5)\,(\nu_5/\nu_{20})$$

$$= (4\text{ m/s})\,(20/5)\,(10^{-6})/((4)\,(10^{-6}))$$

$$\underline{\underline{V_5 = 4\text{ m/s}}}$$

8.23

$$Re_m = Re_p$$

$$V_m L_m/\nu_m = V_p L_p/\nu_p$$

$$V_m/V_p = (L_p/L_m)(\nu_m/\nu_p) \quad (1)$$

Multiply both sides of Eq. (1) by $A_m/A_p = L_m^2/L_p^2$:

$$(V_m A_m)/(V_p A_p) = (L_p/L_m) \times (1) \times L_m^2/L_p^2$$

$$Q_m/Q_p = L_m/L_p$$

$$\underline{\underline{Q_m/Q_p = 1/8}}$$

$$C_{p_m} = C_{p_p}$$

$$(\Delta p/\rho V^2)_m = (\Delta p/\rho V^2)_p$$

$$\Delta p_p = \Delta p_m (\rho_p/\rho_m)(V_p/V_m)^2$$

$$= \Delta p_m (1)(L_m/L_p)^2$$

$$= 300 \times (1/8)^2 = \underline{\underline{4.69\text{ kPa}}}$$

8.24

$$n = f(V, d, \rho, \mu)$$

F: $\quad n = f(V, d, \rho/\mu)$

where $[\rho/\mu] = (FT^2/L^4)/(FT/L^2) = T/L^2$

T: $\quad n/V = f(d, \rho V/\mu)$

where $[n/V] = (T^{-1})/(L/T) = L^{-1}$

$[\rho V/\mu] = (T/L^2)(L/T) = L^{-1}$

L: $\quad \underline{\underline{nd/V = f(\rho V d/\mu)}}$ ← One possible solution.

8.25 $Re_m = Re_p$

$V_m L_m/\nu_m = V_p L_p/\nu_p$; Assume $\nu_m = \nu_p$

$\underline{\underline{V_m = V_p L_p/L_m = 3\ V_p}}$

8.26 Given: $D_m = 1$ ft; $D_p = 3$ ft; $\nu_p = 1.58 \times 10^{-4}$ ft^2/sec;

$\nu_m = 1.22 \times 10^{-5}$ ft^2/sec; $V_m = 5$ ft/sec; $F_m = 20$ lb.

Reynolds model law must be applied, so:

$Re_m = Re_p$; $V_m D_m/\nu_m = V_p D_p/\nu_p$

or $V_p/V_m = (D_m/D_p)(\nu_p/\nu_m) = (1/3)(1.58 \times 10^{-4}/1.22 \times 10^{-5})$ (1)

Also $C_{p_m} = C_{p_p}$ for dynamic similitude; thus, $\Delta p_m/(\rho_m V_m^2/2) = \Delta p_p/(\rho_p V_p^2/2)$

$\Delta p_p/\Delta p_m = (\rho_p/\rho_m)(V_p^2/V_m^2)$

$F_p/F_m = (\Delta p_p A_p)/(\Delta p_m A_m) = (A_p/A_m)(\rho_p/\rho_m)(V_p^2/V_m^2)$ (2)

Combine Eq. (1) and (2)

$F_p/F_m = (\rho_p/\rho_m)(\nu_p/\nu_m)^2 = (0.00237/1.94)(1.58 \times 10^{-4}/1.22 \times 10^{-5})^2$

$= 0.2049$

$F_p = 20 \times 0.2049 = \underline{\underline{4.098\text{ lb}}}$ or $\underline{\underline{F_p = 18.23\text{ N}}}$

8.27 $Re_m = Re_p$

$V_m D_m/\nu_m = V_p D_p/\nu_p$

$D_m/D_p = (\nu_m/\nu_p)(V_p/V_m)$

$= ((1.66 \times 10^{-5})/(1.5 \times 10^{-3}))(100/10) = 0.111$

$\underline{\underline{D_m = 0.111\ D_p = 0.111\text{ in.}}}$

8.28 $Re_m = Re_p$; $(VD/\nu)_m = (VD/\nu)_p$

$(V_m/V_p) = (D_p/D_m)(\nu_m/\nu_p)$; $\nu_m/\nu_p = (V_m D_m/V_p D_p)$

$(\mu_m \rho_p/\mu_p \rho_m) = (V_m D_m/V_p D_p)$ or $\rho_m = \rho_p(\mu_m/\mu_p)(V_p/V_m)(D_p/D_m)$ (1)

8.28 (continued)

$M_m = M_p$; $(V/c)_m = (V/c)_p$

$(V_m/V_p) = c_m/c_p = ((\sqrt{kRT})_m/(\sqrt{kRT})_p) = \sqrt{T_m/T_p} = (298/283)^{1/2}$ (2)

Combining Eqs. (1) and (2):

$\rho_m = 1.23(1.83 \times 10^{-5}/1.76 \times 10^{-5})(283/298)^{1/2}(5) = \underline{\underline{6.23\ kg/m^3}}$

8.29 $Re_A = Re_W$

$V_A L_A/\nu_A = V_W L_W/\nu_W$; but $L_A/L_W = 1$

$\therefore$ $V_A/V_W = \nu_A/\nu_W \approx (1.6)(10^{-4})/(1.2)(10^{-5})$ (at 60°F)

$\underline{\underline{V_A/V_W > 1}}$

8.30 $Re_m = Re_p$

$V_m d_m \rho_m/\mu_m = V_p d_p \rho_p/\mu_p$

$V_m = V_p (d_p/d_m)(\rho_p/\rho_m)(\mu_m/\mu_p)$

$V_m = (10\ ft/s)(48/4)(1.75/1.94)((2.36 \times 10^{-5})/(4 \times 10^{-4}))$

$\underline{\underline{V_m = 6.39\ ft/s}}$

8.31 Following same procedure as in solution to Prob. 8.30

$V_{air} = 3(20/10)(1.6 \times 10^{-5}/1.00 \times 10^{-6}) = \underline{\underline{96\ m/s}}$

The pressure coefficients will be the same or:

$C_{p_a} = C_{p_w}$

$(\Delta p/(\rho V^2))_{air} = (\Delta p/(\rho V^2))_{water}$

or $\Delta p_a = \Delta p_w (\rho_a/\rho_w)(V_a^2/V_w^2)$

$\Delta p_a = 2.0\ kPa(1.17/998)(96/3)^2$

$\Delta p_a = 2.40\ kPa$

8.31 (continued)

The Δp_a = 2.40 kPa is for the pressure difference in an air pipe that is geometrically similar to the water pipe. Therefore, this air pipe would be half as long as the water pipe because the $D_a = 1/2\ D_w$. Consequently, the Δp_a as obtained from the pressure coefficient similarity relationship will have to be multiplied by two to obtain the Δp_a for an air pipe that is the same length as the water pipe:

$$\Delta p_a = 2 \times 2.40 \text{ kPa} = \underline{\underline{4.80 \text{ kPa}}}$$

8.32 Following the same basic procedure as in the solution to Prob. 8.30.

$$V_{air} = 8(1/1)(1.41 \times 10^{-5}/1.31 \times 10^{-6}) = \underline{\underline{86.1 \text{ m/s}}}$$

8.33

$$Fr_{moon} = Fr_{earth}$$

$$(V/\sqrt{gL})_m = (V/\sqrt{gL})_e$$

$$V_e/V_m = (g_e/g_m)^{0.5} (L_e/L_m)^{0.5}$$

$$= (5)^{0.5} (1)$$

$$Re_m = Re_e$$

$$(VL/\nu)_m = (VL/\nu)_e$$

$$\nu_e = (V_e/V_m)\ \nu_m = (5)^{0.5} (10^{-5}) \text{ m}^2/\text{s}$$

$$\underline{\underline{\nu_e = 2.24 \times 10^{-5} \text{ m}^2/\text{s}}}$$

8.34

$$Re_m = Re_p$$

$$V_m L_m/\nu_m = V_p L_p/\nu_p$$

$$V_m = (L_p/L_m)\ (\nu_m/\nu_p)\ V_p$$

$$= (10)(10^{-6}/((4)\ (10^{-5})))\ (10 \text{ m/s})$$

$$\underline{\underline{V_m = 2.50 \text{ m/s}}}$$

8.35 $Re_{prot.} = Re_{model}$

$V_{prot.} = V_{model}(L_{model}/L_{prot.})(\nu_{prot.}/\nu_{model})$

$V_{prot.} = 1(1/4)(10^{-5}/10^{-6}) = \underline{\underline{2.5\ m/s}}$

$C_{p_m} = C_{p_p}$

$(\Delta p/(\rho V^2))_m = (\Delta p/(\rho V^2))_p$

$\Delta p_p = \Delta p_m(\rho_p/\rho_m)\ (V_p/V_m)^2$

$= 2.0\ kPa(860/998)(2.5/1.0)^2$

$= \underline{\underline{10.8\ kPa}}$

8.36

$Re_{air} = Re_{water}$

$(VD\rho/\mu)_{air} = (VD\rho/\mu)_{water}$

$V_a = V_w(D_w/D_a)(\rho_w/\rho_a)(\mu_a/\mu_w)$

$\rho_w = 1{,}000\ kg/m^3$

$\rho_a = \rho_{a,std.atm.} \times (150\ kPa/101\ kPa)$

$= 1.20 \times (150/101) = 1.78\ kg/m^3$

$\mu_a = 1.81 \times 10^{-5} N{\cdot}S/m^2$

$\mu_w = 1.31 \times 10^{-3} N{\cdot}S/m^2$

Then $V_a = 1.5\ m/s\ (1{,}000/1.78)(1.81 \times 10^{-5}/1.31 \times 10^{-3})$

$V_a = \underline{\underline{11.6\ m/s}}$

$C_{p_w} = C_{p_a}$

$(\Delta p/\rho V^2)_w = (\Delta p/\rho V^2)_a$

$\Delta p_w = \Delta p_a(\rho_w/\rho_a)(V_w/V_a)^2$

$= 780 \times (1{,}000/1.78)(1.5/11.6)^2$

$= 7{,}330\ Pa = \underline{\underline{7.33\ kPa}}$

8.37

$$Re_{tunnel} = Re_{prototype}$$

$$V_{tunnel} = V_{prot.}(4/1)(\nu_{tunnel}/\nu_{prot.})$$

$$V_{tunnel} = 3(4/1)(1) = \underline{\underline{12\ m/s}}$$

$$C_{p_{tunnel}} = C_{p_{prototype}}$$

$$(\Delta p/\rho V^2)_{tunnel} = (\Delta p/\rho V^2)_{prototype}$$

$$(\Delta p_{tunnel}/\Delta p_{prot.}) = (\rho_{tunnel}/\rho_{prot.})(V^2_{tunnel}/V^2_{prot.})$$

Multiply both sides of Eq. by $A_{tunnel}/A_{prot.} = L_t^2/L_p^2$

$$(\Delta p \times A)_{tunnel}/(\Delta p \times A)_{prot.} = (\rho_{tunnel}/\rho_{prot.}) \times (V^2_{tunnel}/V^2_{prot.}) \times (L_t/L_p)^2$$

$$F_{tunnel}/F_{prot.} = (1/1)(4)^2(1/4)^2$$

$$F_{tunnel} = F_{prot.} = \underline{\underline{868\ N}}$$

8.38 $Re_m = Re_p$

$$(\rho VL/\mu)_m = (\rho VL/\mu)_p$$

$$\rho_m/\rho_p = (V_p/V_m)(L_p/L_m)(\mu_m/\mu_p)$$

$$= (30/300)(100)(1)$$

$$= 10$$

$$\underline{\underline{\rho_m = 10\rho_p = 0.024\ slugs/ft^3}}$$

$$F_m/F_p = (\Delta p_m/\Delta p_p)(A_m/A_p) \qquad (1)$$

$$\frac{C_{p,m}}{C_{p,p}} = \left(\frac{\Delta p_m}{\rho_m V_m^2}\right)\left(\frac{\rho_p V_p^2}{\Delta p_p}\right)$$

$$1 = \left(\frac{\Delta p_m}{\Delta p_p}\right)\left(\frac{\rho_p}{\rho_m}\right)\left(\frac{V_p^2}{V_m^2}\right) = \left(\frac{\Delta p_m}{\Delta p_p}\right)\left(\frac{1}{10}\right)\left(\frac{1}{10}\right)^2$$

8.38 (continued)

Then $\Delta p_m/\Delta p_p = 1{,}000$ (2)

solve Eqs. (1) and (2) for F_m/F_p

$F_m/F_p = 1{,}000\ A_m/A_p = 1{,}000(1/10^4) = 10^{-1}$

$\underline{\underline{F_p = F_m \times 10 = 500 \text{ lbf}}}$

8.39 $Re_m = Re_p$ or $(VD\rho/\mu)_m = (VD\rho/\mu)_p$

Then $V_m/V_p = (D_p/D_m)(\rho_p/\rho_m)(\mu_m/\mu_p)$

Multiply both sides of above equation by $A_m/A_p = (D_m/D_p)^2$

$(A_m/A_p)(V_m/V_p) = (D_p/D_m)(D_m/D_p)^2(\rho_p/\rho_m)(\mu_m/\mu_p)$

$$Q_m/Q_p = (D_m/D_p)(\rho_p/\rho_m)(\mu_m/\mu_p)$$

$$= (1/3)(0.82)(10^{-3}/(3 \times 10^{-3}))$$

$$Q_m/Q_p = 0.0911$$

or $Q_m = Q_p \times 0.0911$

$Q_m = 0.50 \times 0.0911 \text{ m}^3/\text{s} = \underline{\underline{0.0455 \text{ m}^3/\text{s}}}$

$C_p = \underline{\underline{1.07}}$

8.40 $C_{p_m} = C_{p_p}$; $(\Delta p/\rho V^2)_m = (\Delta p/\rho V^2)_p$

or $\Delta p_m/\Delta p_p = (\rho_m V_m^2)/(\rho_p V_p^2)$ (1)

Multiply both sides of Eq. (1) by $(A_m/A_p) \times (L_m/L_p) = (L_m/L_p)^3$

Obtain $\text{Mom.}_m/\text{Mom.}_p$: $\text{Mom.}_m/\text{Mom}_p = (\rho_m/\rho_p)(V_m/V_p)^2(L_m/L_p)^3$ (2)

Also $Re_m = Re_p$ or $V_m L_m/\nu_m = V_p L_p/\nu_p$

$V_m/V_p = (L_p/L_m)(\nu_m/\nu_p)$ (3)

Substitute Eq. (3) into Eq. (2) to obtain

$M_m/M_p = (\rho_m/\rho_p)(\nu_m/\nu_p)^2(L_m/L_p)$

$M_p = M_m(\rho_p/\rho_m)(\nu_p/\nu_m)^2(L_p/L_m) = 2(1{,}026/1{,}000)(1.4/1.31)^2(60) = \underline{\underline{141 \text{ N}\bullet\text{m}}}$

Also, $V_p = 10(1/60)(1.41/1.31) = \underline{\underline{0.179 \text{ m/s}}}$

8.41 $C_{p_m} = C_{p_p}$; $(\Delta p/\rho V^2)_m = (\Delta p/\rho V^2)_p$

$\Delta p_m/\Delta p_p = (\rho_m/\rho_p)\ (V_m^2/V_p^2)$

Multiply both sides of the above equation by $A_m/A_p = (L_m/L_p)^2$

$(\Delta p_m/\Delta p_p)(A_m/A_p) = (\rho_m/\rho_p)(V_m^2/V_p^2)(L_m^2/L_p^2) = F_m/F_p$ (1)

For dynamic similitude $Re_m = Re_p$ or $(VL\rho/\mu)_m = (VL\rho/\mu)_p$

or $(V_p/V_m)^2 = (L_m/L_p)^2(\rho_m/\rho_p)^2(\mu_p/\mu_m)^2$ (2)

Eliminating $(V_p/V_m)^2$ between Eq.(1) and Eq.(2) yields

$F_p/F_m = (\rho_m/\rho_p)(\mu_p/\mu_m)^2$

Then if the same fluid is used for model and prototype, we have

$F_p/F_m = 1$ or $F_p = F_m$; $\underline{\underline{F_p = 20\text{ kN}}}$

8.42 $\mathbf{M}_m = \mathbf{M}_p$

$V_m/c_m = V_p/c_p$; $V_m/V_p = c_m/c_p$ (1)

Also $Re_m = Re_p$; $V_mL_m\rho_m/\mu_m = V_pL_p\rho_p/\mu_p$

or $V_m/V_p = (L_p/L_m)(\rho_p/\rho_m)(\mu_m/\mu_p)$ (2)

Eliminate V_m/V_p between Eqs. (1) and (2) to obtain

$c_m/c_p = (L_p/L_m)(\rho_p/\rho_m)(\mu_m/\mu_p)$ (3)

But $c = \sqrt{E_v/\rho} = \sqrt{kp/\rho} = \sqrt{kp/(p/RT)} = \sqrt{kRT}$

Therefore $c_m/c_p = 1$

Then from Eq. (3) $1 = (10)(\rho_p/\rho_m)(1)$

or $\rho_m = 10\rho_p$ But $\rho = p/RT$

so $(p/RT)_m = 10(p/RT)_p$; $p_m = 10\ p_p = 10$ atm $= \underline{\underline{1.01\text{ MPa abs.}}}$

8.43 $Re_m = Re_p$

$V_m L_m \rho_m/\mu_m = V_p L_p \rho_p/\mu_p$; But $\rho_m/\mu_m = \rho_p/\mu_p$

so $V_m = V_p(L_p/L_m) = 80 \times 10 = 800$ km/hr = 222 m/s

$\mathbf{M} = V/c = 222/345 = 0.644$ Mach number effects would be important.

8.44 $M/Re = (V/c)(\mu/\rho VD) = (\mu)/(\rho cD)$

where $\rho = p/RT = (22)/(1{,}716 \times 393) = 3.26 \times 10^{-5}$

$c = 975$ ft/s

$\mu = 3.0 \times 10^{-7}$ lbf-s/ft^2

$M/Re = 3.0 \times 10^{-7}/(3.26 \times 10^{-5} \times 975 \times 2) = 4.72 \times 10^{-6} < 1$

Not rarefied.

8.45 $W/(Re)^{0.5} = (\rho_g V^2 d/\sigma)\ (\mu_g/(Vd\rho_g))^{0.5} = 0.5$

$= \rho_g^{0.5} V^{1.5} \mu^{0.5} d^{0.5}/\sigma = 0.5$; $\rho_g = 1.2$ kg/m^3

$d^{0.5} = 0.5\sigma/(\rho_g^{0.5} V^{1.5} \mu^{0.5}) = 0.5\ (\sigma^2/(\rho_g \mu V^3))^{0.5}$

$= (0.5)\ ((7.3 \times 10^{-2})^2/((1.2)\ (1.81)\ (10^{-5})\ (30^3)))^{0.5}$

$d = 0.00227$ m = 2.27 mm

8.46 $W = 6.0 = \rho DV^2/\sigma$; $\rho = 0.95$ kg/m^3

$D = 6\sigma/\rho V^2 = 6 \times 0.02/(0.95 \times (30)^2) = 1.40 \times 10^{-4}$m = 140 μm

---

8.47 $W = 6.0 = \rho DV^2/\sigma$; $\rho = 1.20$ kg/m^3; $\sigma = 0.073$ N/m (from Table A-5)

$D = 6\sigma/\rho V^2 = 6 \times 0.073/(1.2 \times (15)^2) = 1.62 \times 10^{-3}$m = 1.62 mm

---

8.48 $F_m = F_p$; $(V/\sqrt{gL})_m = (V/\sqrt{gL})_p$

or $V_m/V_p = \sqrt{g_m L_m/g_p L_p}$ (1)

$Re_m = Re_p$; $(VL/\nu)_m = (VL/\nu)_p$ or $V_m/V_p = (L_p/L_m)(\nu_m/\nu_p)$ (2)

Eliminate V_m/V_p between Eq's (1) and (2) to obtain:

$\sqrt{g_m L_m/g_p L_p} = (L_p/L_m)(\nu_m/\nu_p)$, but $g_m = g_p$

Therefore: $\nu_m/\nu_p = (L_m/L_p)^{3/2}$

8.49 $t_p/t_m = (L_p/L_m)^{1/2}$

Then wave period$_{prot}$ $= 1 \times (10)^{1/2} = \underline{\underline{3.16\text{ s}}}$

wave height$_{prot}$ $= 10\text{ cm} \times 10 = \underline{\underline{1\text{ m}}}$

8.50 $Fr_m = Fr_p$

$V_m/((g_m)(L_m))^{0.5} = V_p/((g_p)(L_p))^{0.5}$

$V_p/V_m = (L_p/L_m)^{0.5} = 5$

$V_p = (2.0)(5)\text{ m/s} = \underline{\underline{10\text{ m/s}}}$

$Q_p/Q_m = (V_p/V_m)(A_p/A_m)$

$= (5)(625)$

$Q_p = (5)(625)(0.10) = \underline{\underline{312.5\text{ m}^3/\text{s}}}$

8.51 Assume Froude model law applies, then following the procedure of solution to Prob. 8.50 we have:

$V_m = V_p\sqrt{L_m/L_p} = 100\sqrt{1/10} = \underline{\underline{31.6\text{ m/s}}}$

8.52 From solution to Prob. 8.50 we have:

$V_m/V_p = \sqrt{L_m/L_p}$ (1)

or for this case $V_m/V_p = \sqrt{1/25} = \underline{\underline{1/5}}$

Multiply both sides of Eq. (1) by $A_m/A_p = (L_m/L_p)^2$

$V_mA_m/V_pA_p = (L_m/L_p)^{1/2}(L_m/L_p)^2$

$Q_m/Q_p = (L_m/L_p)^{5/2}$ or for this case $Q_m/Q_p = (1/25)^{5/2} = \underline{\underline{1/3,125}}$

$Q_m = 3,000/3,125 = \underline{\underline{0.96\text{ m}^3/\text{s}}}$

8.53 $Fr_m = Fr_p$

$V_m/((g_m)(L_m))^{0.5} = V_p/((g_p)(L_p))^{0.5}$

$V_m = V_p(L_m/L_p)^{0.5} = V_p(1/8) = \underline{\underline{1.875\text{ ft/s}}}$

$d_m/d_p = 1/64$; $d_m = (1/64)\,d_p$

$= (1/64)(20) = \underline{\underline{0.312\text{ ft}}}$

8.54 Froude model law applies:

$$V_p = V_m\sqrt{L_p/L_m} = 7.87\sqrt{25} = \underline{\underline{39.3\ \text{ft/s}}}$$

From solution to P8-46:

$$Q_p/Q_m = (L_p/L_m)^{5/2};\ Q_p = 3.53 \times (25)^{5/2} = \underline{\underline{11{,}030\ \text{ft}^3/\text{s}}}$$

8.55 Froude model law applies:

$$V_p = V_m\sqrt{L_p/L_m} = 0.90\sqrt{10} = \underline{\underline{2.85\ \text{m/s}}}$$

$L_p/L_m = 10$; therefore, wave height$_{\text{prot.}}$ $= 10 \times 2.5$ cm $= \underline{\underline{25\ \text{cm}}}$

8.56 Froude model law applies:

$$V_p/V_m = \sqrt{L_p/L_m} \text{ or } (L_p/t_p)/(L_m/t_m) = (L_p/L_m)^{1/2}$$

Then $t_p/t_m = (L_p/L_m)(L_m/L_p)^{1/2}$

$$t_p/t_m = (L_p/L_m)^{1/2}$$

$$t_p = 1 \times \sqrt{25} = \underline{\underline{5\ \text{min}}}$$

Also $Q_p/Q_m = (L_p/L_m)^{5/2};\ Q_p = 0.10 \times (25)^{5/2} = \underline{\underline{312.5\ \text{m}^3/\text{s}}}$

8.57 $F_m = F_p$ or $(V/\sqrt{gL})_m = (V/\sqrt{gL})_p$

$$V_m/V_p = (L_m/L_p)^{1/2} \text{ because } g_m = g_p \quad (1)$$

$$(L_m/t_m)/(L_p/t_p) = (L_m/L_p)^{1/2}$$

or $t_m/t_p = (L_m/L_p)^{1/2}$ (2)

Then from Eq. (1) $V_m = V_p(L_m/L_p)^{1/2} = 3.0 \times (1/300)^{1/2} = \underline{\underline{0.173\ \text{m/s}}}$

From Eq. (2) $t_m = 12.5\ \text{hr}(1/300)^{1/2} = 0.722\ \text{hr} = \underline{\underline{43.3\ \text{min.}}}$

8.58 $C_{p_m} = C_{p_p}$; $(\Delta p/\rho V^2)_m = (\Delta p/\rho V^2)_p$

$\Delta p_m/\Delta p_p = (\rho_m/\rho_p)(V_m/V_p)^2$ (1)

Multiply both sides of Eq. (1) by $A_m/A_p = L_m^2/L_p^2$

$(\Delta p_m A_m)/(\Delta p_p A_p) = (\rho_m/\rho_p)(L_m/L_p)^2(V_m/V_p)^2$

Also from the Froude model law $V_m/V_p = \sqrt{L_m/L_p}$ (2)

Eliminating V_m/V_p from Eqs.(1) and (2) yields

$F_m/F_p = (\rho_m/\rho_p)(L_m/L_p)^2(L_m/L_p)$

$F_m/F_p = (\rho_m/\rho_p)(L_m/L_p)^3$

$F_p = F_m(\rho_p/\rho_m)(L_p/L_m)^3 = 80(1{,}026/1{,}000)(36)^3 = \underline{\underline{3.83\ \text{MN}}}$

8.59 $Q_m/Q_p = (L_m/L_p)^{5/2}$

$Q_m = 200 \times (1/20)^{5/2} = \underline{\underline{0.112\ \text{m}^3/\text{s}}}$

From solution to Prob. 8.58 we have:

$F_p = F_m(\rho_p/\rho_m)(L_p/L_m)^3 = 22(1/1)(20)^3 = \underline{\underline{176\ \text{kN}}}$

8.60 Check the scale ratio as dictated by Q_m/Q_p:

$Q_m/Q_p = 0.90/5{,}000 = (L_m/L_p)^{5/2}$

or $L_m/L_p = 0.0318$ (1)

Then with the scale ratio of Eq.(1):

$L_m = 0.0318 \times 1{,}200\ \text{m} = 38.1\ \text{m}$

$W_m = 0.0318 \times 300\ \text{m} = 9.53\ \text{m}$

Therefore, model will fit into the available space,

so use $\underline{\underline{L_m/L_p = 0.0318}}$

8.61 $V_m/\sqrt{g_m L_m} = V_p/\sqrt{g_p L_p}$

$V_p = V_m\sqrt{L_p}/\sqrt{L_m}$

$= (4\ \text{ft/s})(150/4)^{1/2}$

$\underline{\underline{V_p = 24.5\ \text{ft/s}}}$

8.62 Froude model law applies, so we follow the solution procedure of Prob. 8.58:

$V_m/V_p = \sqrt{L_m/L_p}$; $V_p = 4 \times \sqrt{25} = \underline{\underline{20\ \text{ft/s}}}$

$F_m/F_p = (L_m/L_p)^3$; $F_p = 2(25)^3 = \underline{\underline{31,250\ \text{lb}}}$

8.63 $Fr_m = Fr_p$

$V_m/(g_m L_m)^{0.5} = V_p/(g_p L_p)^{0.5}$; $V_p = V_m (L_p/L_m)^{0.5} = 12\ \text{m/s}$

$F_p = (12\text{N})(L_p/L_m)^3 = (12)(16)^3 = \underline{\underline{49,152\ \text{N}}}$

8.64 Assume $C_{p_m} = C_{p_p}$

$(\Delta p/(\rho V^2/2)_m = (\Delta p/(\rho V^2/2)_p$

$\Delta p_m/\Delta p_p = (\rho_m/\rho_p)(V_m^2/V_p^2)$

Assume $\rho_m = \rho_p$

$F_m/F_p = (\Delta p_m/\Delta p_p)(A_m/A_p) = (V_m/V_p)^2 (L_m/L_p)^2$

$(F_p/F_m) = (40/20)^2 (20)^2$

$\underline{\underline{F_p}} = (200\ \text{N})(4)(400) = 320,000\ \text{N} = \underline{\underline{320\ \text{kN}}}$

Choice (d) is the correct one.

8.65 $C_{p,model} = C_{p,prot.}$

Then $\Delta p_p/(1/2)\rho_p V_p^{\,2}) = C_{p_p} = C_{p_m}$

or $\Delta p_p = C_{p_m}((1/2)\rho_p V_p^{\,2}) = C_{p_m} \times (1/2) \times 1.25 \times (160{,}000/3{,}600)^2$

$p - p_0 = 1{,}234.6\ C_{p_m}$ but $p_0 = 0$ gage

so $p = 1{,}234.6\ C_{p_m}$ Pa

Extremes of pressure are therefore:

$p_{windward\ wall} = \underline{\underline{1.235\ kPa}}$

$p_{side\ wall} = 1{,}234.6 \times (-2.7) = \underline{\underline{-3.33\ kPa}}$

$p_{leeward\ wall} = 1{,}234.6 \times (-0.8) = \underline{\underline{-988\ Pa}}$

Lateral Force: $\Delta p_m/\Delta p_p = ((1/2)\rho_m V_m^{\,2})/((1/2)\rho_p V_p^{\,2})$ (1)

Multiply both sides of Eq. (1) by $A_m/A_p = L_m^2/L_p^2$

$(\Delta p_m A_m)/(\Delta p_p A_p) = (\rho_m/\rho_p)(V_m^2/V_p^2)(L_m^2/L_p^2) = F_m/F_p$

$F_p/F_m = (\rho_p/\rho_m)(V_p^2/V_m^2)(L_p^2/L_m^2)$

$F_p = 20(1.25/120)((160{,}000/3{,}600)^2/(20)^2)(250)^2$

$F_p = \underline{\underline{6.43\ MN}}$

CHAPTER NINE

9.1

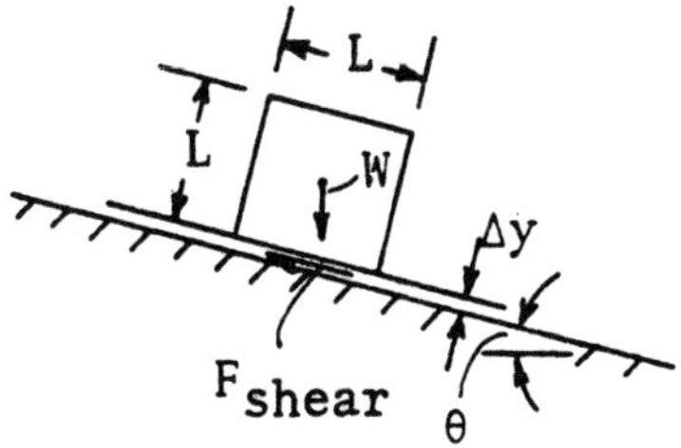

$$F_{shear} = W \sin\theta$$

$$\tau = F_{shear}/A_s = W \sin\theta/L^2 \qquad (1)$$

$$\tau = \mu dV/dy = \mu \times V/\Delta y \qquad (2)$$

or $$V = \tau\Delta y/\mu$$

from (1) $$V = (W \sin\theta/L^2)\ \Delta y/\mu \qquad (3)$$

$$V = (200 \sin 10°/0.30^2) \times 1 \times 10^{-4}/10^{-2}$$

$$V = \underline{\underline{3.86 \text{ m/s}}}$$

9.2 Same solution procedure applies as in Prob. 9.1. Then from Eq. (3) of solution to Prob. 9.1, we have

$$\mu = (W \sin\theta/L^2)\ \Delta y/V$$

$$\mu = (40 \times (5/13)/3^2) \times (0.02/12)/0.5$$

$$\mu = \underline{\underline{5.70 \times 10^{-3} \text{ lbf-s/ft}^2}}$$

9.3 Same type of solution procedure applies as in Prob. 9.1 and 9.2. Then

$$\mu = (15 \times (5/13)/1^2) \times 5 \times 10^{-4}/0.12$$

$$\mu = \underline{\underline{2.40 \times 10^{-2} \text{ N}\cdot\text{s/m}^2}}$$

9.4

a) Upper plate is moving to the left relative to the lower plate.

b) Minimum shear stress occurs where the maximum velocity occurs (where $du/dy = 0$).

9.5 A. True B. False C. False D. False E. True

9.6

a) By similar triangles $u/y = u_{max}/\Delta y$

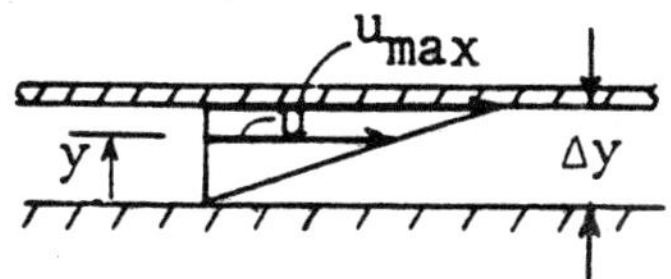

or $u = (u_{max}/\Delta y)y$

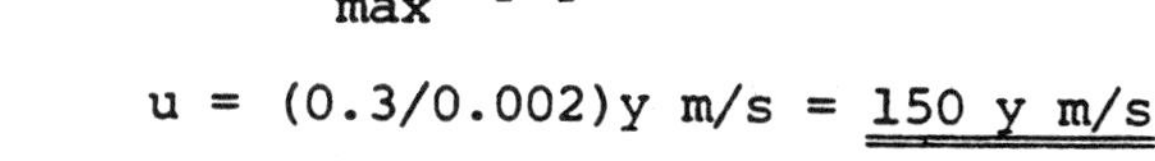

$$u = (0.3/0.002)y \text{ m/s} = \underline{\underline{150\ y \text{ m/s}}}$$

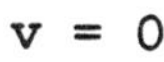

$$v = 0$$

9.6 (Cont'd)

b) For flow to be irrotational $\partial u/\partial y = \partial V/\partial x$

here $\partial u/\partial y = 150$ and $\partial V/\partial x = 0$

The equation is not satisfied; flow is rotational.

c) $\partial u/\partial x + \partial v/\partial y = 0$ (continuity equation)

$\partial u/\partial x = 0$ and $\partial v/\partial y = 0$ so continuity is satisfied.

d) Use the same formula as developed for solution to P 9-1, but $W \sin\theta = F_{shear}$.

Then $V = (F_s/(LW))\Delta y/\mu$

or $F_s = VLW\mu/\Delta y$

$F_s = 0.3 \times (1 \times 0.3) \times 3/0.002$

$F_s = \underline{135\ N}$

9.7 Valid statements are (c),(e).

9.8 The shear force is the same on the wire and tube wall; however, there is less area in shear on the wire so there will be a greater shear stress on the wire.

9.9 Assume a linear velocity distribution within the oil. The velocity distribution will appear as below:

Because the lower plate is moving at a constant speed, the shear stresses on the top and bottom of it will be the same, or

$\tau_1 = \tau_2$

$\mu_1 dV_1/t_1 = \mu_2 dV_2/t_2$

$\mu_1 \times (V - V_{lower})/t_1 = \mu_2 V_{lower}/t_2$

$V\mu_1/t_1 - \mu_1 V_{lower}/t_1 = \mu_2 V_{lower}/t_2$

$V_{lower}(\mu_2/t_2 + \mu_1/t_1) = V\mu_1/t_1$

$V_{lower} = \underline{V\mu_1/t_1/(\mu_2/t_2 + \mu_1/t_1)}$

9.10

$\tau = \mu dv/dy$

$\tau = \mu r\omega/\Delta y$

$dT = rdF$

$dT = r\tau dA$

$dT = r(\mu r\omega/\Delta y)\ 2\pi r dr$

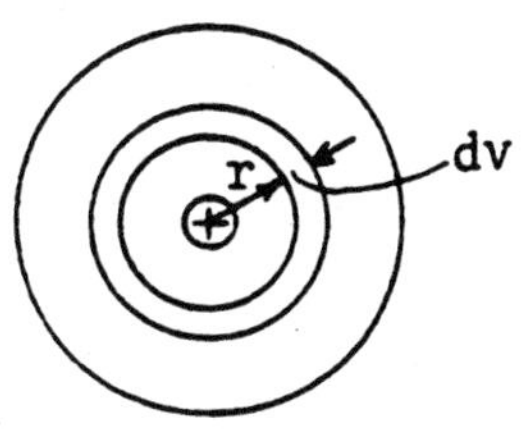

Plan View

Then $T = \int_0^{r} dT = \int_0^{r_0} (\mu\omega/\Delta y) 2\pi r^3 dr$

$T = (2\pi\mu\omega/\Delta y) r^4/4\Big|_0^{r_0} = 2\pi\mu\omega r_0^{\ 4}/(4\Delta y)$

For $\Delta y = 0.001$ ft; $r_0 = 6" = 0.50$ ft; $\omega = 180 \times 2\pi/60 = 6\pi$ rad/s

$\mu = 0.10$ lbf- s/ft^2

$T = (2\pi \times 0.10 \times 6\pi/0.001)(0.5^4/4)$

$\underline{\underline{T = 185 \text{ ft-lbf}}}$

9.11 The problem is the same type as Prob. 9.10; therefore,

$T = 2\pi\mu\omega r_0^4/(4\Delta y)$

where $r = 0.10$ m; $\Delta y = 2 \times 10^{-3}$ m; $\omega = 110$ rad/s; $\mu = 6\text{N}\cdot\text{s/m}^2$

$T = 2\pi \times 6 \times 110 \times 10^{-4}/(4 \times (2 \times 10^{-3})) = \underline{\underline{4.71 \text{ N}\cdot\text{m}}}$

9.12

$dT = (\mu u/s)\ dA \times r$

$= \mu r\omega \sin\beta 2\pi r^2 dr/(r\theta\sin\beta)$

$= 2\pi\mu\omega r^2 dr/\theta$

$T = (\mu\omega/\theta)(2\pi r^3/3)\Big]_0^{r_0} = \underline{\underline{(2/3)\pi r_0^3\mu\omega/\theta}}$

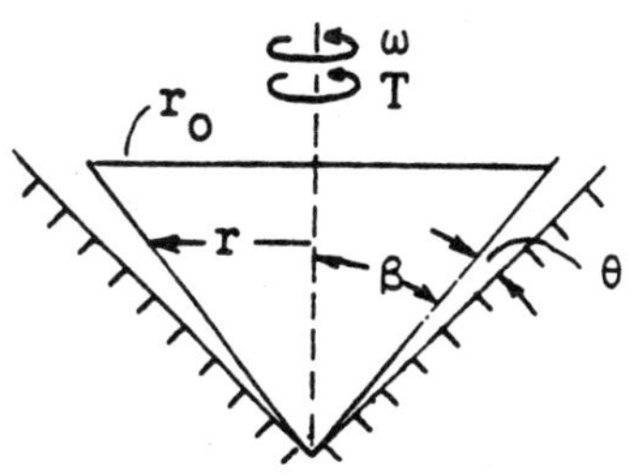

9.13

Velocity distribution:

Force $= \tau A$

$= \mu(dV/dy)A$

$= (0.62 \times (0.4/0.002) \times 1\times 2) \times 2$

$\underline{\underline{F = 496 \text{ N}}}$

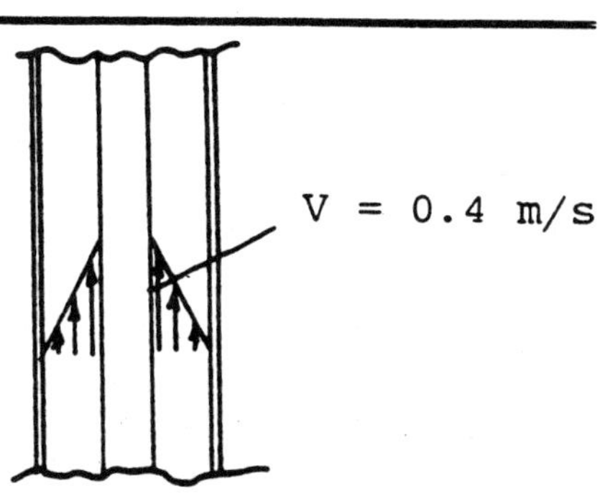

9.14 $\tau = \mu V/\delta$

$T = \tau A r$ where T = torque, A = bearing area = $2\pi r b$

$T = \tau 2\pi r b r = \tau 2\pi r^2 b$

$= (\mu V/\delta)(2\pi r^2 b)$ where $V = r\omega$

$= (\mu/\delta)(r\omega)(2\pi r^2 b)$

$= (\mu/\delta)(2\pi\omega) r^3 b$

$= (0.1/0.001)(2\pi)(200)(0.01)^3(0.01)$

$\underline{T = 12.6 \times 10^{-4}\ \text{N}\cdot\text{m}}$

9.15 Subscript s refers to inner cylinder.
Subscript 0 refers to outer cylinder.
ℓ is the length of the ring of fluid.

$T_s = \tau(2\pi r)(r\ell)$

$T_0 = \tau(2\pi r)(r\ell) + d/dr(\tau 2\pi r\cdot r\ell)\Delta r$

$T_s - T_0 = 0;\ d/dr(\tau 2\pi r^2 \ell)\Delta r = 0;\ d/dr(\tau r^2) = 0$

Then $\tau r^2 = C_1;\quad \tau = \mu r(d/dr)(v/r)$

So $\mu r^3(d/dr(V/r)) = C_1;\quad \mu(d/dr(V/r)) = C_1 r^{-3}$

Integrating, $\mu v/r = (-1/2)C_1 r^{-2} + C_2$

$v = 0$ at $r = r_0$ and $v = r_s\omega$ at $r = r_s$

$0 = (-1/2)C_1 r_0^{-2} + C_2;\quad \mu\omega = (-1/2)C_1 r_s^{-2} + C_2$

$C_1 = 2C_2 r_0^2;\quad \mu\omega = C_2(1 - r_0^2/r_s^2);\quad C_2 = \mu\omega/(1 - r_0^2/r_s^2)$

Then, $\tau_s = C_1 r_s^{-2} = 2C_2(r_0/r_s)^2 = 2\mu\omega r_0^2/(r_s^2 - r_0^2) = 2\mu\omega/((r_s^2/r_0^2) - 1)$

So $T_s/\ell = \tau 2\pi r_s^2 = 4\pi\mu\omega r_s^2/((r_s^2/r_0^2) - 1)$ (Torque on fluid)

Torque on shaft per unit length = $\underline{\underline{4\pi\mu\omega r_s^2/(1 - (r_s^2/r_0^2))}}$

9.16 $P = T\omega;\quad T = 4\pi\mu\omega\ell r_s^2/(1 - (r_s^2/r_0^2))$

$= 4\pi \times 0.1 \times (40)(0.01)^2\ 0.04/(1 - (1/1.1)^2) = 0.00116\ \text{N}\cdot\text{m}$

$P = 0.00116\ (40) = \underline{\underline{0.0463\ \text{W}}}$

9.17

$T = 0.6(0.02) = 0.012\ N\cdot m$

$\mu = T(1-r_s^2/r_0^2)/(4\pi\omega\ell r_s^2) = 0.012(1-2^2/2.25^2)/(4\pi(20)(2\pi/60)(0.1)(0.02)^2)$
$= \underline{\underline{2.39 N\cdot s/m^2}}$

9.18 $u = (g\sin\theta/2\nu)y(2d-y)$

u_{max} occurs at the liquid surface where $y = d$

so $u_{max} = (g\sin\theta/(2\nu))d^2$

where $\theta = 30°$, $\nu = 10^{-3} m^2/s$ and $d = 3.5 \times 10^{-3} m$

$u_{max} = (9.81 \times \sin 30°/(2 \times 10^{-3})) \times (3.5 \times 10^{-3})^2$

$u_{max} = 30 \times 10^{-3} m/s = \underline{\underline{0.030\ m/s}}$

$V = (gd^2\sin\theta)/(3\nu)$

$= 9.81 \times (3.5 \times 10^{-3})^2 \sin 30°/(3 \times 10^{-3})$

$V = \underline{\underline{0.020\ m/s}}$

$V = \underline{\underline{(2/3)\ u_{max}}}$

9.19 $Re = Vd/\nu = q/\nu$; $q = 200 \times 1.2 \times 10^{-3} = \underline{\underline{0.240\ cfs/ft}}$

$q = (1/3)\gamma d^3\sin\theta/\mu$; $d^3 = 3\mu q/\gamma\sin\theta = 3\nu q/g\sin\theta = 3 \times 1.2 \times 10^{-3}$

$\times\ 0.24/(32.2 \times 0.707) = 0.3795 \times 10^{-4}$

$d = \underline{\underline{0.0336\ ft = 0.403\ in.}}$

9.20 Total discharge per unit width of roof is:

$q = L \times 1 \times R_r$ (1)

where R_r = rainfall rate

but $q = (1/3)(\gamma/\mu)d^3\sin\theta$

or $d = (3q\mu/(\gamma\sin\theta))^{1/3}$ (2)

$d = (3LR_r\mu/(\gamma\sin\theta))^{1/3}$ (combined Eq's. (1) and (2))

9.20 (continued)

In this problem $L = 15$ ft; $R_r = 0.4$ in./hr. $= 9.26 \times 10^{-6}$ ft/s

$\mu = 2.73 \times 10^{-5}$ lb-s/ft^2 ; $\gamma = 62.4$ lbf/ft^3; $\theta = 10°$

Then $d = (3 \times 15 \times 9.26 \times 10^{-6} \times 2.73 \times 10^{-5}/(62.4 \times \sin 10°))^{1/3}$

$d = 1.02 \times 10^{-3}$ ft $= \underline{0.012 \text{ in.}}$; $\underline{V = 0.137 \text{ ft/s}}$ (using Eq. (9.4a))

9.21 $V = (gd^2/3\nu)\sin\theta$ from Eq. (9-4)

but $u = (g/2\nu)\sin\theta y(2d - y)$

or $u_{max} = (g/2\nu)\sin\theta(d^2) = (gd^2/2\nu)\sin\theta$

Then $V/u_{max} = [(gd^2/3\nu)\sin\theta]/[(gd^2/2\nu)\sin\theta] = 2/3$

or $\underline{V = (2/3)u_{max}}$

9.22 $u = -(\gamma/2\mu)(By-y^2)dh/ds$

u_{max} occurs at $y = B/2$

So $u_{max} = -(\gamma/2\mu)(B^2/2-B^2/4)dh/ds = -(\gamma/2\mu)(B^2/4)dh/ds$

$dp/ds = -1{,}200$ Pa/m ; $dh/ds = (1/\gamma)dp/ds$

$B = 2$mm $= 0.002$ m and $\mu = 10^{-1}$N•s/m^2

Then $u_{max} = -(\gamma/2\mu)(B^2/4)((1/\gamma)(-1{,}200))$

$= (B^2/8\mu)(1{,}200)$

$= (0.002^2/(8 \times 0.1))(1{,}200) = 0.006$ m/s

$= \underline{6.0 \text{ mm/s}}$

$F_s = \tau A = \mu(du/dy) \times 2 \times 1.5$

$\tau = \mu \times [-(\gamma/2\mu)(B-2y)dh/ds]$

but τ_{plate} occurs at $y = 0$

Thus $F_s = -\mu \times (\gamma/2\mu) \times B \times (-1{,}200/\gamma) \times 3 = (B/2) \times 1{,}200 \times 3$

$F_s = (0.002/2) \times 1{,}200 \times 3 = \underline{3.6 \text{ N}}$

9.23 From the solution to problem 9-22 we have

$u_{max} = -(\gamma B^2/8\mu)((1/\gamma)(dp/ds))$

so $u_{max} = -(0.01^2/(8 \times 10^{-3}))(-12) = \underline{0.150 \text{ ft/s}}$

9.24 $h_A = (p_A/\gamma) + z_A = (150/100) + 0 = 1.5$ (Assume $z_A = 0$)

$h_B = (p_B/\gamma) + z_B = (100/100) + 1 = 2$

$h_B > h_A$; therefore flow is from B to A: (downward)

9.25 Assume the flow will be laminar. Then from

Example 9.3: $q = (2/3)u_{max}B$

$u = (-\gamma/2\mu)(By - y^2)dh/ds$ (9.6a)

$u = u_{max}$ when $y = B/2$

Then $u_{max} = (-\gamma/2\mu)(B^2/2 - B^2/4)dh/ds$

$dh/ds = d/ds(p/\gamma + z)$

$= (1/\gamma)dp/ds + dz/ds$ ($dp/ds \to 0$)

$= -1$

$u_{max} = (-\gamma/2\mu)(B^2/4)(-1)$

$= (-12{,}300 \text{N/m}^3/(2\times6.2\times10^{-1}\text{N}\cdot\text{s/m}^2))(.01^2\text{m}^2/4)(-1)$

$= 0.248$ m/s

$q = (2/3)u_{max}B$

$= (2/3)(0.248 \text{ m/s})(0.01\text{m})$

$\underline{q = 0.00165 \text{ m}^2/\text{s}}$

Now check to see if the flow is laminar (Re < 1,000):

$Re = VB/\nu = (2/3)V_{max}B/\nu = qB/\nu$

$= (0.00165 \text{ m}^2/\text{s})(0.01 \text{ m})/(5.1 \times 10^{-4} \text{ m}^2/\text{s})$

$Re = 3.2 \leftarrow$ Laminar

Therefore our original assumption of laminar flow was correct.

9.26 From solution to Prob. 9-22, we have

$V_{max} = -(\gamma/2\mu)(B^2/4)\ dh/ds$

where $dh/ds = dh/dz = d/dz(p/\gamma + z)$

$= (1/\gamma)dp/dz + 1$

$= (1/(0.8 \times 62.4))\ (-8) + 1 = -0.16 + 1 = 0.840$

Then $V_{max} = -((0.8 \times 62.4)/(2 \times 10^{-3})\ (0.01^2/4)\ (0.840)$

V_{max} = $\underline{\underline{-0.524\ ft/s}}$ (flow is downward)

9.27 From solution to Prob. 9-22, we have

$V_{max} = -(\gamma/2\mu)(B^2/4)dh/ds$

where $dh/ds = dh/dz = d/dz(p/\gamma + z)$

$= (1/\gamma)dp/dz + 1$

$= (1/(0.85 \times 9{,}810))\ (-9{,}000) + 1$

$= -0.0793$

Then $V_{max} = -((0.85 \times 9{,}810)/(2 \times 0.1))\ (0.002^2/4)\ (-0.0793)$

= 0.00331 m/s = $\underline{\underline{3.31\ mm/s}}$ (flow is upward)

9.28 From solution to P9-22 we have

$V_{max} = -(\gamma/2\mu)(B^2/4)dh/ds$

where $dh/ds = dh/dz = d/dz(p/\gamma + z)$

$= (1/\gamma)dp/dz + 1$

$= (1/(0.8 \times 62.4))\ (-60)) + 1 = -0.202$

Then $V_{max} = -((0.8 \times 62.4)/(2 \times 0.001))\ (0.01^2/4)\ (-0.202)$

= $\underline{\underline{+0.126\ ft/s}}$ (flow is upward)

9.29 $\bar{V} = q/B = 0.00833/(0.10/12) = 1.00$ ft/s

$V_{max} = (3/2)\bar{V} = 1.50$ ft/s

From solution to P.9-22 $dh/ds = -8\mu V_{max}/(\gamma B^2)$

where $\mu = 2 \times 10^{-3}$ lbf•s/ft^2; $\gamma = 55.1$ lbf/ft^3

Then $dh/ds = -8 \times 2 \times 10^{-3} \times 1.50/(55.1 \times (0.1/12)^2) = -6.27$

But $dh/ds = (1/\gamma)dp/ds + dz/ds$ where $dz/ds = -0.866$

Then $-6.27 = (1/\gamma)dp/ds - 0.866$

$dp/ds = \gamma(-6.27 + 0.866) = \underline{\underline{-298 \text{ psf/ft}}}$

9.30 $\bar{V} = q/B = 24 \times 10^{-4}/(0.002) = 1.2$ m/s

$V_{max} = (3/2)\bar{V} = 1.8$ m/s

From solution to Prob. 9-22 $dh/ds = -8\mu V_{max}/(\gamma B^2)$

where $\mu = 0.1$ N•s/m^2 and $\gamma = 0.8 \times 9{,}810$ N/m^2

Then $dh/ds = -8 \times 0.1 \times 1.8/(0.8 \times 9{,}810 \times 0.002^2) = -45.87$

But $dh/ds = (1/\gamma)\ dp/ds + dz/ds$ where $dz/ds = -0.866$

Then $-45.87 = (1/\gamma)\ dp/ds - 0.866$

$dp/ds = \gamma(-45.87 + 0.866) = \underline{\underline{-353 \text{ kPa/m}}}$

9.31

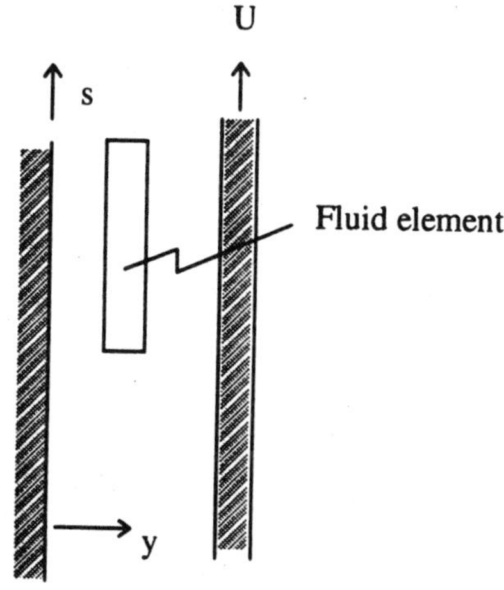

Consider the forces on the fluid element

9.31 (Cont'd)

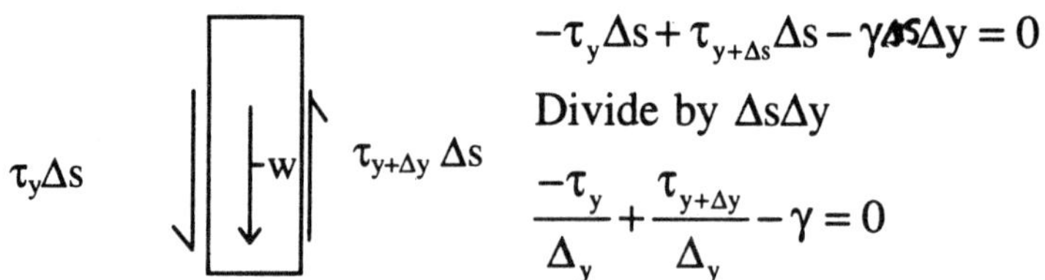

$$-\tau_y \Delta s + \tau_{y+\Delta s}\Delta s - \gamma \Delta s \Delta y = 0$$

Divide by $\Delta s \Delta y$

$$\frac{-\tau_y}{\Delta_y} + \frac{\tau_{y+\Delta y}}{\Delta_y} - \gamma = 0$$

Take the limit as Δy approaches zero

$$d\tau / dy = \gamma$$

But $$\tau = \mu du / dy$$

So $$\frac{d}{dy}(\mu du / dy) = \gamma$$

$$\mu du / dy = \gamma y + C_1$$

$$du / dy = \frac{\gamma}{\mu} y + C_1$$

Integrate again $$u = \frac{\gamma}{\mu}\frac{y^2}{2} + Cy_1 + C_2$$

Boundary Conditions: $$\text{At } y = 0, u = 0 \text{ and}$$
$$\text{At } v = L, u = U$$

Therefore, $$C_2 = 0 \text{ and } C_1 = \frac{U}{L} - \frac{\gamma}{\mu}\frac{L}{2}$$

$$\underline{\underline{u = \frac{\gamma}{\mu}\frac{y^2}{2} + \left(\frac{U}{L} + \frac{\gamma}{\mu}\frac{L}{2}\right) y}}$$

The discharge per unit dimension (normal to page) is given by

$$q = \int_0^L u dy$$

$$= \int_0^L \left[\frac{\gamma}{\mu}\frac{y^2}{2} + \left(\frac{U}{L} - \frac{\gamma}{\mu}\frac{L}{2}\right) y\right] dy$$

$$= \frac{\gamma}{\mu}\frac{y^3}{6} + \frac{Uy^2}{2L} - \frac{\gamma}{\mu}\frac{Ly^2}{4}\Big|_0^L$$

$$= \frac{\gamma}{\mu}\frac{L^3}{6} + \frac{UL}{2} - \frac{\gamma}{\mu}\frac{L^3}{4}$$

9.31 (Cont'd)

For zero discharge

$$\frac{UL}{2} = \frac{\gamma L^3}{4\mu} - \frac{\gamma}{\mu}\frac{L^3}{6}$$

or

$$U = \frac{1}{6}\frac{\gamma}{\mu}L^2$$

9.32 a) First consider the forces on an element of mud Δx long and y_0 deep as shown below.

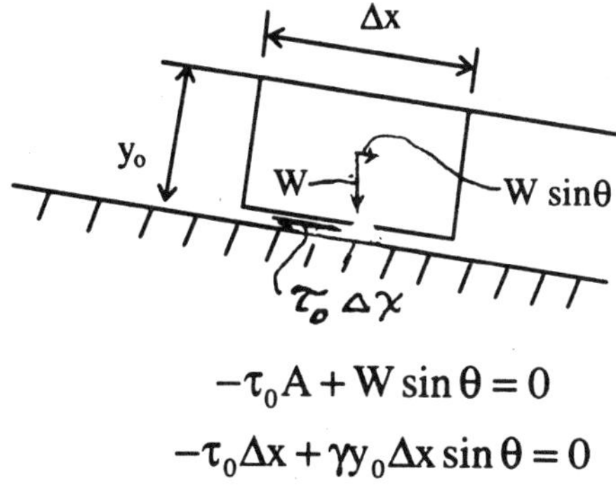

$$-\tau_0 A + W\sin\theta = 0$$
$$-\tau_0 \Delta x + \gamma y_0 \Delta x \sin\theta = 0$$
$$\tau_0 = \gamma y_0 \sin\theta$$

There will be no motion if $\gamma y_0 \sin\theta < \tau_0$

b. Determine the velocity field when there is flow. Consider forces on the element of mud shown below. Assume unit dimension normal to page.

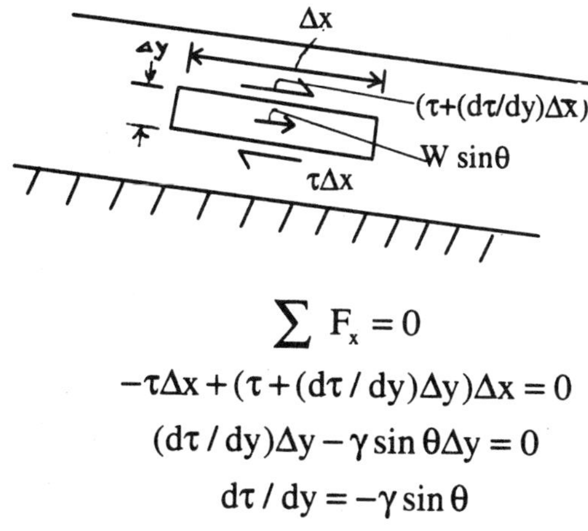

$$\sum F_x = 0$$
$$-\tau\Delta x + (\tau + (d\tau / dy)\Delta y)\Delta x = 0$$
$$(d\tau / dy)\Delta y - \gamma \sin\theta\Delta y = 0$$
$$d\tau / dy = -\gamma \sin\theta$$

9.32 (Cont'd)

$$\tau = -\int \gamma \sin\theta dy + C$$
$$= -\gamma \sin\theta y + C$$
$$\tau = 0 \text{ when } y = y_0 \therefore C = \gamma \sin\theta y_0$$

and
$$\tau = -\gamma \sin\theta y + \gamma \sin\theta y_0$$
$$\tau = \gamma \sin\theta (y_0 - y) \qquad (1)$$

But for the mud $\tau = \tau_0 + \eta du/dy$ (2)

Eliminate τ between equations (1) and (2)

$$\tau_0 + \eta du/dy = \gamma \sin\theta (y_0 - y)$$

or
$$du/dy = [\gamma \sin\theta (y_0 - y) - \tau_0]/\eta \qquad (3)$$

$$u = (1/\eta)[\gamma \sin\theta (y_0 y - y^2/2) - \tau_0 y] + C$$

when $y = 0 \; u = 0 \Rightarrow C = 0$

If $\tau < \tau_0$, $du/dy = 0$

Transition point is obtained from Eq. (3)

$$0 = (\gamma \sin\theta (y_0 - y) - \tau_0)$$
$$\tau_0 = \gamma \sin\theta (y_0 - y)$$
$$\tau_0 = \gamma \sin\theta y_0 - \gamma \sin\theta y$$

$$y = \frac{\gamma \sin\theta y_0 - \tau_0}{\gamma \sin\theta} \qquad (4)$$
$$y_{tr} = y_0 - (\tau_0 / \gamma \sin\theta) \qquad (5)$$

when $0 < y < y_{tr}$ $\tau > \tau_0$

and
$$u = [\gamma \sin\theta (y y_0 - y^2/2) - \tau_0 y]/\eta \qquad (6)$$

when $y_{tr} < y < y_0$ $\tau < \tau_0$ so $u = u_{max} = u_{tr}$. The velocity distrib. looks like this.

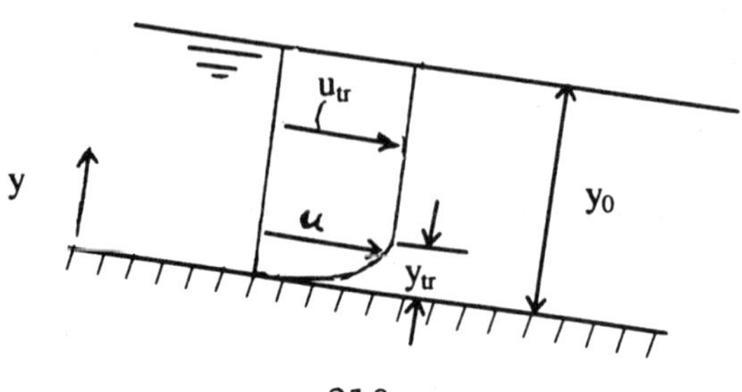

9.33 $q = (2/3)\ u_{max}B \quad ; \quad dh/ds = -1$

$u = -(\gamma/2\mu)\ (By - y^2)\ dh/ds$

and $u = u_{max}$ when $y = B/2$

$\therefore\ u_{max} = -(\gamma/2\mu)\ ((B^2/2) - (B^2 4))\ dh/ds$

$= -(\gamma/2\mu)\ (B^2/4)\ dh/ds$

Thus $q = (2/3)\ (-\gamma/2\mu)\ (B^2/4)\ (dh/ds)\ (B)$

or $Q = -(2/3)\ (\gamma/2\mu)\ (B^3/4)\ (dh/ds)\ (\pi D)$

$= (-2/3)\ (12{,}300/(2 \times 6.2 \times 10^{-1}))\ ((10^{-3})^3/4)\ (-1)\ (\pi \times .019)$

$Q = 9.87 \times 10^{-8}\ m^3/s$

9.34

$F = p_{avg.} \times A$

$= 1/2\ p_{max} \times A$

$= 1/2\ p_{max} \times 0.3m \times 1m$

or $p_{max} = 2F/0.3\ m^2 = 2 \times 50{,}000/0.30$

$= 333{,}333\ N/m^2$

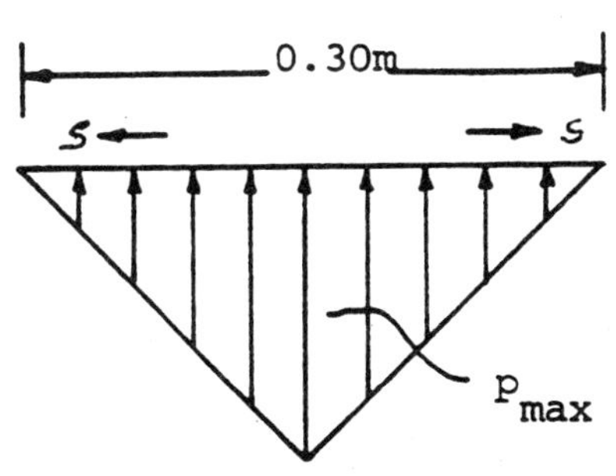

Then $dp/ds = -333{,}333\ N/m^2/0.15m = -2{,}222{,}222\ N/m^3$

For flow between walls where $\sin\theta = 0$, we have

$u_{max} = -(\gamma/2\mu)(B \times B/2 - B^2/4)(d/ds(p/\gamma))$

$u_{max} = -(B^2/8\mu)\ dp/ds$

$V_{avg} = 2/3\ u_{max}$

$= -(1/12)(B^2/\mu)\ dp/ds$

Then $q_{per\ side} = VB = -(1/12)(B^3/\mu)\ dp/ds$

and $q_{total} = 2VB = -(1/6)(B^3/\mu)\ dp/ds$

$= -(1/6) \times ((6\times10^{-4}m)^3/(20^{-1}N{\cdot}s/m^2)) \times 2{,}222{,}222\ N/m^3)$

$= 4.00 \times 10^{-4} m^3/s$

$q = \underline{1.44\ m^3/hr}$

9.37 $q = -(B^3/(12\mu))dp/dr$

$Q = 2\pi rq = \text{constant}$

$Q = 2\pi r\ (-B^3/(12\mu))dp/dr$

Separate variables:

$((-\pi B^3)/(6\mu))dp = Q(dr/r)$

Integrate:

$-\pi B^3 p/(6\mu) = Q \ln r + C$

Use boundary conditions to determine Δp: p_1 at r_1 and p_2 at r_2

$\therefore\ (-\pi B^3/(6\mu))p_1 = Q \ln r_1 + C$

$(-\pi B^3/(6\mu))p_2 = Q \ln r_2 + C$

$(p_2 - p_1)(\pi B^3/(6\mu)) = Q\ (\ln r_1 - \ln r_2)$

$\underline{\Delta p = (6\mu/(\pi B^3))\ Q \ln (r_1/r_2)}$

$\therefore$ For r_1=0.01m , r_2=0.1m , B=0.005m and Q=0.0003m^3/s

we have:

$\Delta p = ((6 \times 3.6 \times 10^{-2})/(\pi \times 0.005^3))(0.0003)\ln(0.01/0.1)$

$\Delta p = -380$ Pa

Pressure drop = $\underline{p_1 - p_2 = 380 \text{ Pa}}$

9.38 $\mathrm{Re} = U_0 x/\nu$

$x = \mathrm{Re}\nu/U_0 = 500{,}000 \times 1.22 \times 10^{-5}/6$

$x = 1.017$ ft

$\delta = 5x/\mathrm{Re}_x^{1/2} = 5 \times 1.017/(500{,}000)^{1/2}$

$\delta = 0.0072 \text{ ft} = \underline{\underline{0.086 \text{ in.}}}$

$\tau_0 = 0.332\mu(U_0/x)\mathrm{Re}_x^{1/2}$

$\tau_0 = 0.332 \times 2.36 \times 10^{-5}(6/1.017) \times (500{,}000)^{1/2} = \underline{\underline{0.0327 \text{ lbf/ft}^2}}$

9.39 $\delta/x = 5/Re_x^{1/2} = 5/(500{,}000)^{1/2} = 0.0071$

9.40 $F_{s,lam} = C_f A\rho U_0^2/2$

where $C_f = 1.33/Re^{1/2}$

$= 1.33/(500{,}000)^{1/2}$

$= 0.00188$

$F_{s,lam} = (0.00188 \times 3 \times 1.017) \times 1.94 \times 6^2/2$

$= \underline{\underline{0.200 \text{ lbf}}}$

9.41 At the edge of the boundary layer the shear stress, τ, is approximately zero. Therefore, $\tau/\tau_0 \approx 0$. Choice (a) is the correct one.

9.42 This is a turbulent boundary layer; therefore,

$$
\begin{aligned}
c_f &= 0.058\,Re_x^{-1/5} \\
Re_x &= Ux/\nu \\
&= 20 \times 1/(1.5 \times 10^{-5}) \\
&= 13.3 \times 10^5 \\
&= 1.33 \times 10^6 \\
Re_x^{-1/5} &= 0.0596 \\
c_f &= 0.058 \times 0.0596 \\
&= 0.00346 \\
F &= c_f A \times (\rho U^2/2) \\
&= 0.00346 \times (0.01)^2 (1.2)(20)^2/2 \\
&= \underline{\underline{8.30 \times 10^{-5}\,\text{N}}}
\end{aligned}
$$

9.43 $u/U_0 = (y/\delta)^{1/2} \quad \tau_0 = 1.66\ U_0\mu/\delta$

$$\tau_0 = \rho U_0^2\ d/dx\int_0^\delta (u/U_0(1 - u/U_0))dy \qquad (9\text{-}35)$$

$$= \rho U_0^2\ d/dx\int_0^\delta((y/\delta)^{1/2} - (y/\delta))\ dy$$

$$= \rho U_0^2\ d/dx[(2/3)(y/\delta)^{3/2} - 1/2(y/\delta)^2]_0^\delta\ \delta$$

$$1.66\ U_0\mu/\delta = (1/6)\rho U_0^2\ d\delta/dx$$

$$\delta\ d\delta/dx = 9.96\mu/(\rho U_0)$$

$$\delta^2/2 = 9.96\ \mu x/(\rho U_0) = 9.96\ x^2/Re_x$$

$$\delta = 4.46x/Re_x^{1/2} \qquad \text{Blasius} \quad \underline{\delta = 5x/Re^{1/2}}$$

9.44 $Re_x = Vx/\nu = 1 \times 1/10^{-5} = 100{,}000$

∴ The boundary layer is laminar. Use Fig. 9-5 to obtain u/U_0

$yRe_x^{0.5}/x = 0.001\ (10^5)^{0.5}/1 = 0.316$

Then from Fig. 9.5 $u/U_0 \approx 0.11$; $u = 0.11$ m/s

9.45 $Re_L = 3 \times 10^5$; $C_f = 1.33/Re_L^{0.5} = 0.00243$

$F_s = C_f\ BL\rho U^2/2 = .00243 \times 1 \times 3 \times 1{,}000 \times 1^2/2 = \underline{3.65\ N}$

9.46 $Re_x = Ux/\nu = 6 \times 1/10^{-4}$ (laminar boundary layer)

$yRe_x^{0.5}/x = (0.010)\ (6 \times 10^4)^{0.5}/1 = 2.45$

$u/U_0 = 0.72$ (From Fig. 9.5)

$\underline{u = 6 \times 0.72 = 4.3\ m/s}$

9.47 $Re_x = 1 \times 1/10^{-4} = 10^4$ (laminar B.L.)

$yRe_x^{0.5}/x = 0.10 \times 10^2/1) = 10$ (outside boundary layer)

Therefore $\underline{u = U_0 = 1\ m/s}$

9.48 $Re_x = 500{,}000;\ U_0x/\nu = 500{,}000$

$x = 500{,}000\ \nu/U_0 = 500{,}000 \times 1.31 \times 10^{-6}/2 = \underline{\underline{0.327\ m}}$

$\delta = 5x/Re_x^{1/2} = 5 \times 0.327/(500{,}000)^{1/2} = 2.31 \times 10^{-3}m = 2.31\ mm$

$\tau_0 = 0.332\mu\ (U_0/x)Re_x^{1/2}$

$\tau_0 = 0.332 \times 1.31 \times 10^{-3}(2/0.327) \times (500{,}000)^{1/2} = \underline{\underline{1.88\ N/m^2}}$

9.49 $F_s = 0.664\ B\mu U_0 Re_L^{1/2} = 0.664 \times 1 \times 1.31 \times 10^{-3} \times 2 \times (500{,}000)^{1/2}$

$F_s = \underline{\underline{1.23\ N}}$

$F_{s_{total}} = C_f A\rho U_0^2/2;\ Re_L = U_0 \times 1/\nu = 2 \times 1/(1.31 \times 10^{-6}) = \underline{\underline{1.53 \times 10^6}}$

C_f 0.0031 from Fig. 9-13

$F_{s_{total}} = 0.0031 \times 1 \times 500 \times 4 = \underline{\underline{6.20\ N}}$

$F_{s_{lam.}}/F_{s_{total}} = 1.23/6.20 = 0.198;\ \underline{\underline{19.8\%}}$

9.50 Find Re: At 30°C $\nu = 1.6 \times 10^{-5} m^3/s$

$\rho = 1.17\ kg/m^3$

$U_0 = (200km/hr)(1{,}000m/km)/(3{,}600s/hr)$

$U_0 = 55.56\ m/s$

Then $Re = U_0L/\nu = (55.56)(2)/(1.6 \times 10^{-5}) = 6.9 \times 10^6$

From Fig. 9.13, the flow is combined laminar and turbulent

$C_f = (0.074/Re^{0.2}) - (1700/Re) = 0.00292$

Drag: $F_s = C_f BL\rho U_0^2/2$

Wing has 2 sides $\therefore F_{s,wing} = 2C_f BL\rho U_0^2/2$

$= \not{2}(0.00292)(11)(2)(1.17)(55.56)^2/\not{2}$

a) $\underline{\underline{F_{s,wing} = 232\ N}}$

b) Power = FU_0

$\underline{\underline{\text{Power} = (232)(55.56) = 12.89\ kW}}$

9.50 (continued)

c) Critical laminar Re = $5 \times 10^5 = U_0x/\nu$

$x_{cr} = 5 \times 10^5\ \nu/U_0$

$x_{cr} = (5 \times 10^5)(1.6 \times 10^{-5})/55.56$

$x_{cr} = 14$ cm

d) If all of boundary layer is turbulent then

$C_f = 0.074/Re^{0.2}$

$C_f = 0.00317$

Then $F_{tripped\ B.L}/F_{normal} = 0.00317/0.00292 = 1.086$

Change in drag with tripped B.L. is 8.6% increase

9.51 $T = 20°C$; $\rho = 998\ kg/m3$; $\nu = 10^{-6}\ m^2/s$; $\tau = 0.1\ N/m^2$

$u_* = (\tau_0/\rho)^{0.5} = (0.1/998)^{0.5} = 0.01$ m/s

$u_*y/\nu = (0.01)\ (0.01)/(10^{-6}) = 10^2$

From Fig. 9-9 for $u_*y/\nu = 100$ it is seen that Eq. 9-27 applies

$u/u_* = 5.75 \log (yu_*/\nu) + 5.56$

$= 5.75 \log (100) + 5.56 = 17.06$

$u = u_*\ (17.06) = 0.01\ (17.06) = 0.1706$ m/s

9.52 $Re_L = U_0L/\nu = 0.20 \times 1.5/(10^{-6}) = 3.0 \times 10^5$

Re_L is less than 500,000; therefore, laminar boundary layer

$\delta = 5x/Re_x^{1/2} = 5 \times 1.5/(3.0 \times 10^5)^{1/2} = 0.0137m = 13.7$ mm

$C_f = 1.33/Re_L^{1/2} = 1.33/(3.0 \times 10^5)^{1/2} = 0.00243$

$F_s = C_fA\rho U_0^2/2 = 0.00243 \times 1.0 \times 1.5 \times 2 \times 1{,}000 \times 0.20^2/2 = 0.146$ N

9.53 $R_L = U_0 L\rho/\mu = 20 \times 2 \times 1.5/10^{-5} = 6 \times 10^6$

$$C_f = (0.074/Re_L^{1/5}) - (1{,}700/Re_L) = 0.00298$$

$$F_s = C_f(2BL)\rho U_0^2/2$$

$$= 0.00298 \times (2 \times 2 \times 1.5) \times 20^2/2$$

$$= \underline{\underline{3.58 \text{ N}}}$$

$$Re_{lm} = 6 \times 10^6 \times (1/2) = 3 \times 10^6$$

$$c_f = 0.058/(3 \times 10^6)^{1/5} = 0.00294$$

$$\tau_0 = c_f \rho U_0^2/2 = 0.00294 \times 1.5 \times 20^2/2 = 0.881 \text{ N/m}^2$$

but $\tau_0 = \mu\, du/dy$ or $du/dy = 0.881/10^{-5}$

$$= \underline{\underline{8.81 \times 10^4 \text{s}^{-1}}}$$

9.54 $u/u_* = 8.74(yu_*/\nu)^{1/7}$ (9-38)

but at the outer limit of the boundary layer $u = U_0$ and $y = \delta$

so $U_0/u_* = 8.74(\delta u_*/\nu)^{1/7}$

or $u_*^{8/7} = U_0\nu^{1/7}/(8.74\delta^{1/7})$

$$(u_*^{8/7})^{14/8} = U_*^2 = (U_0\nu^{1/7}/(8.74\delta^{1/7}))^{14/8}$$

$$u_*^2 = \tau_0/\rho = \underline{\underline{0.0225\, U_0^2(\nu/U_0\delta)^{1/4}}}$$

9.55 $Re_{turb} = 5 \times 10^5 = Uc/\nu$

$$U = (5 \times 10^5)\nu/c = (5 \times 10^5)(1.58 \times 10^{-4})/0.5 = 158 \text{ ft/s}$$

$$C_f = 1.33/(5 \times 10^5)^{0.5} = 0.00188$$

$$F = C_f(\rho U^2/2)A = (0.00188)((0.00237)(158)^2/(2))(2)(3)(0.5)$$

$$= \underline{\underline{0.167 \text{ lbf}}}$$

9.56

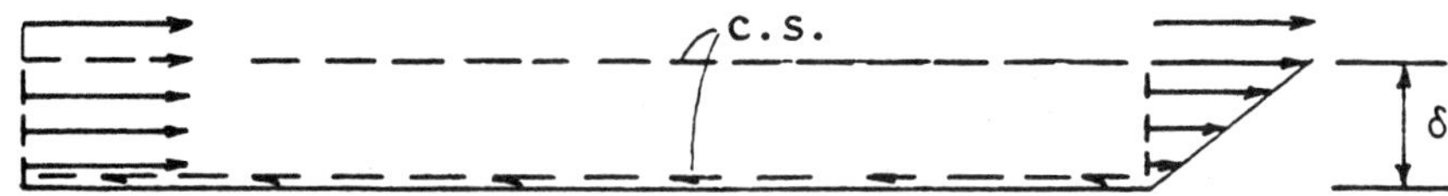

Apply the momentum equation to the c.v. shown above.

$$\Sigma F_x = \int_{c.v.} V_x \rho \underline{V} \cdot d\underline{A}$$

$$F_{s,\text{plate on c.v.}} = -\rho V_1^2 \delta + \int \rho V_2^2 dA + \rho V_1 q_{top}$$

where $V_2 = (V_{max}/\delta) y = V_1 y/\delta$

$$q_{top} = V_1 \delta - \int_0^\delta V_2 dy = V_1 \delta - \int_0^\delta V_1 y/\delta dy$$

$$q_{top} = V_1 \delta - V_1 y^2/2\delta \Big|_0^\delta = V_1 \delta - 0.5 V_1 \delta = 0.5 V_1 \delta$$

Then $F_s = -\rho V_1^2 \delta + \int_0^\delta \rho (V_1 y/\delta)^2 dy + 0.5 \rho\ V_1^2 \delta$

$$= -\rho V_1^2 \delta + \rho V_1^2 \delta/3 + 0.5 \rho V_1^2 \delta$$

$$= \rho V_1^2 \delta(-1 + (1/3) + (1/2)) = -0.1667 \rho V_1^2 \delta$$

For $V_1 = 40$ m/s, $\rho = 1.2$ kg/m^3 and $\delta = 3 \times 10^{-3}$m we have

$$F_s = -0.1667 \times 1.2 \times 40^2 \times 3 \times 10^{-3} = -0.960\text{N}$$

or the skin friction drag on top side of plate is $F_s = \underline{+0.960 \text{ N}}$

The shear stress at the downstream end of plate is

$$\tau_0 = \mu dV/dy = 1.8 \times 10^{-5} \times 40/(3 \times 10^{-3}) = \underline{\underline{0.24 \text{ N/m}^2}}$$

9.57

$$\tau_0/\rho = U_0^2\ d/dx \int_0^\delta (y/\delta)^{1/7} [1 - (y/\delta)^{1/7}] dy$$

$$= U_0^2 d/dx \int_0^\delta [(1/\delta^{1/7}) y^{1/7} dy - (1/\delta^{2/7}) y^{2/7} dy]$$

$$= U_0^2\ d/dx [(1/\delta^{1/7})(y^{8/7}/(8/7)) - (1/\delta^{2/7})(y^{9/7}/(9/7))]_0^\delta$$

$$= U_0^2\ d/dx[(7/8)\delta - (7/9)\delta] = \underline{\underline{(7/72) U_0^2 d\delta/dx}}$$

9.58

$$\dot{m} = \int_0^\delta \rho u dy = \int_{\delta^*}^\delta \rho_\infty U_\infty dy = \rho_\infty U_\infty\ (\delta - \delta^*)$$

$$\rho_\infty U_\infty \delta^* = \rho_\infty U_\infty \delta - \int_0^\delta \rho u dy$$

$$= \rho_\infty U_\infty \int_0^\delta (1 - (\rho u)/\rho_\infty U_\infty)\ dy$$

$$\therefore\ \underline{\underline{\delta^* = \int_0^\delta (1 - (\rho u)/(\rho_\infty U_\infty))\ dy}}$$

9.59 The streamlines will be displaced a distance $\delta^* = q_{defect}/V_1$

where $q_{defect} = \int_0^\delta (V_1 - V_2)dy = \int_0^\delta (V_1 - V_1 y/\delta)dy$

Then $\delta^* = [\int_0^\delta (V_1 - V_1 y/\delta)dy]/V_1 = \int_0^\delta (1 - y/\delta)dy = \delta - \delta/2 = \underline{\underline{\delta/2}}$

9.60 $F_s = C_f BL\rho U_0^2/2$

where $C_f = (0.455/(\log_{10} Re_L)^{2.58}) - (1{,}700/Re_L)$

$Re_{L,20} = 30 \times 10/10^{-6} = 3 \times 10^8$

$Re_{L,10} = 10^8$

Then $C_{f,30} = (0.455/(\log_{10}(3 \times 10^8))^{2.58} - (1{,}700/(2 \times 10^8))$

$= 0.00184$

$C_{f,10} = 0.00213$

Then $F_{s,20}/F_{s,10} = (0.00184/0.00213) \times 3$

$= \underline{\underline{2.59}}$

9.61 $F_s = C_f A\rho U_0^2/2$

$R_{e_L} = V_0 L/\nu = 30 \times 30/(1.4 \times 10^{-5})$

$R_{e_L} = 6.42 \times 10^7$

Then from Fig. 9-13 $C_f = 0.0022$; $\rho = 1.25$ kg/m^3

$F_s = 0.0022 \times 2 \times 30 \times 1.5 \times 1.25 \times 30^2/2 = 111.4$ N

$P = FV = 111.4 \times 30 = \underline{\underline{3.34 \text{ kW}}}$

9.62 $\Sigma F_z = 0$

$T + F_s + F_{Buoy.} - W = 0$

$T = W - F_s - F_{Buoy.}$ (1)

where $W = 200$ N

$F_{Buoy.} = \forall\gamma_{water} = 0.003 \times 3 \times 9{,}810 = 88.3$ N

$F_s = C_f A\rho U_0^2/2$; $R_{e_L} = VL/\nu = 2 \times 1/(1.31 \times 10^{-6}) = 1.53 \times 10^6$

Therefore, from Fig. 9-13, $C_f = 0.0033$

Then $F_s = 0.0033 \times 2 \times 3 \times 1{,}000 \times 4/2 = \underline{\underline{39.6 \text{ N}}}$

Solving Eq. (1): $T = 200 - 39.6 - 88.3 = \underline{\underline{72.1 \text{ N}}}$

9.63

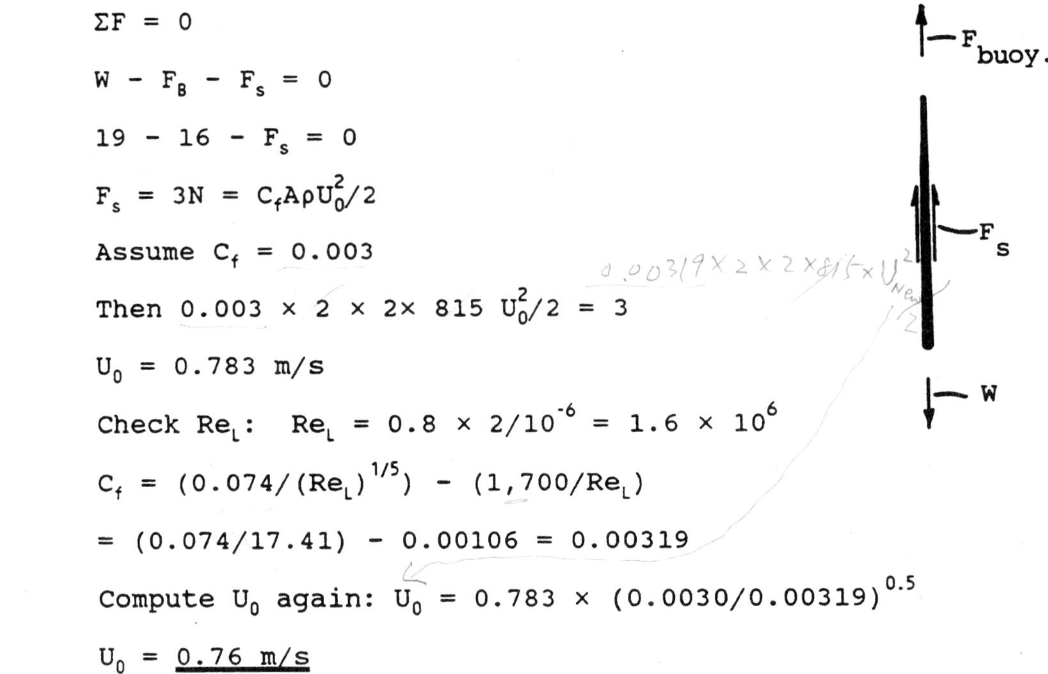

$\Sigma F = 0$

$W - F_B - F_s = 0$

$19 - 16 - F_s = 0$

$F_s = 3N = C_f A\rho U_0^2/2$

Assume $C_f = 0.003$

Then $0.003 \times 2 \times 2\times 815\ U_0^2/2 = 3$

$U_0 = 0.783$ m/s

Check Re_L: $Re_L = 0.8 \times 2/10^{-6} = 1.6 \times 10^6$

$C_f = (0.074/(Re_L)^{1/5}) - (1{,}700/Re_L)$

$= (0.074/17.41) - 0.00106 = 0.00319$

Compute U_0 again: $U_0 = 0.783 \times (0.0030/0.00319)^{0.5}$

$U_0 = \underline{0.76\text{ m/s}}$

9.64 $\delta' = 5\nu/u_*$; $\nu = 10^{-6} m^2/s$

where $u_* = (\tau_0/\rho)^{0.5}$

and $\tau_0 = c_f \rho U_0^2/2$

$\tau_0/\rho = (0.058/Re_x^{0.2})U_0^2/2$

$Re_x = U_0 x/\nu$

$= (5)(1)/10^{-6} = 5 \times 10^6$; $Re_x^{0.2} = 21.87$

Then $\tau_0/\rho = (0.058/21.87)(25/2)$

$\tau_0/\rho = 0.0332\ m^2/s^2$

$u_* = (\tau_0/\rho)^{0.5} = 0.1822$ m/s

Finally $\delta' = 5\nu/u_* = (5)(10^{-6})/(0.1822)$

$\underline{\delta' = 27.4 \times 10^{-6}\text{ m}}$

Roughness element size of 100 microns is about 4 times greater than the thickness of the viscous sublayer; therefore, <u>it would definitely affect the skin friction coefficient</u>.

9.65 Wgt. $= F_s = C_f \rho (U_0^2/2) A$; $\rho = 1.2$ kg/m^3 ; $\nu = 1.51 \times 10^{-5}$m^2/s^2

$C_f = 0.074/Re^{0.2}$

Wgt. $= 3 = \not{2}(0.074/(U_0 \times 0.1/(1.51\times10^{-5}))^{0.2})(1.2)(U_0^2/\not{2})(1\times0.1)$

Solving for U_0 yields $\underline{U_0 = 67.6 \text{ m/s}}$.

9.66 $Re_L = U_0 L/\nu = 50 \times (44/30) \times 8/(1.22 \times 10^{-5}) = 4.8 \times 10^7$

$C_f = 0.0024$ (from Fig. 9-13)

Then $F_D = C_f A \rho U_0^2/2 = 0.0024 \times 32 \times 1.94 \times (73.33)^2/2 = 401$ lbs.

$P = FV = 401 \times 73.33 = 29{,}405$ ft-lbf/sec $= \underline{53.5 \text{ hp}}$

9.67 $F_s = C_f A_s \rho U_0^2/2$; Assume turbulent boundary layer

where $A_s = \pi DL = \pi \times 0.025 \times 2.65 = 0.208$ m^2;

$Re_L = U_0 L/\nu = 30 \times 2.65/(1.51 \times 10^{-5})$

$Re_L = 5.3 \times 10^6$ Then from Fig. 9-13, $C_f = 0.0032$; $\rho = 1.20$ kg/m^3

Then $F_s = 0.0032 \times 0.208 \times 1.2 \times 30^2/2 = \underline{0.360 \text{ N}}$

$F = m/a$ or $a = F/m = 0.360/(8.0/(9.81)) = 0.441$ m/s^2

Deceleration = 0.441 m/s

With tailwind or headwind C_f will still be about the same value:

$C_f \approx 0.0032$

Then $F_{s,headwind} = 0.360 \times (35/30)^2 = \underline{0.490 \text{ N}}$

$F_{s,tailwind} = 0.360 \times (25/30)^2 = \underline{0.250 \text{ N}}$

Maximum distance: As a first approximation, assume no drag or lift. So for maximum distance, the original line of flight (from release point) will be at 45° with the horizontal--this is obtained from basic mechanics. Also, from basic mechanics:

$y = -gt^2/2 + V_0 t \sin\theta$ and $x = V_0 t \cos\theta$

or upon eliminating t from the above with Y = 0, we get

$x = 2V_0^2 \sin\theta\cos\theta/g = 2 \times 32^2 \times 0.707^2/9.81 = \underline{104.4 \text{ m}}$

Then $t = x/V_0\cos\theta = 104.4/(32 \times 0.707) = 4.61$ s

Then total change in velocity over 4.6s $\approx 4.6 \times a_s = 4.6 \times (-0.44) = 2$m/s

Avg. $V = (32 + 30)/2 = 31$ m/s

Then, better estimate of distance of throw is: $x = 31^2/9.81 = \underline{98.0 \text{ m}}$

9.68 $F_s = C_f A_s \rho V_0^2/2$; $Re_L = 1.7 \times 50/(1.31 \times 10^{-6}) = 6.49 \times 10^7$

$C_f = 0.0022$ from Fig. 9.13:

$F_s = 0.0022 \times \pi \times 0.5 \times 50 \times 1{,}000 \times 1.7^2/2 = \underline{\underline{250\ N}}$

9.69 $F_s = C_f A \rho U_0{}^2/2$

$Re_L = U_0 L/\nu = (100{,}000/3{,}600) \times 150/(1.41 \times 10^{-5})$

$R_{e_{L_{100}}} = 2.95 \times 10^8$; $R_{e_{L_{200}}} = 5.9 \times 10^8$

$C_{f_{100}} = 0.0018$; $C_{f_{200}} = 0.0017$

Then $F_{s_{100}} = 0.0018 \times 10 \times 150 \times 1.25 \times (100{,}000/3{,}600)^2/2$

$F_{s_{100}} = \underline{\underline{1{,}302\ N}}$; $P_{100} = 1{,}302 \times (100{,}000/3{,}600) = \underline{\underline{36.2\ kW}}$

$F_{s_{200}} = \underline{\underline{4{,}919\ N}}$; $P_{200} = 4{,}919 \times (200{,}000/3{,}600) = \underline{\underline{273\ kW}}$

9.70 $Re_x = Ux/\nu = (30)\ (100)/(1.22 \times 10^{-5}) = 2.46 \times 10^8$

$c_f = (2 \log Re_x - 0.65)^{-2.3} = (2 \log (2.46 \times 10^8) - 0.65)^{-2.3}$

$= 0.00167$

$\tau_0 = c_f\ \rho U_0{}^2/2 = (0.00167)\ (1.94)\ (30^2)/2 = 1.456\ lbf/ft^2$

$u_* = (\tau_0/\rho)^{0.5} = (1.456/1.94)^{0.5} = 0.866\ ft/s$

$\delta/x = 0.079\ Re_x{}^{-0.085} = 0.01529$

$\delta = (0.01529)\ (100) = 1.529\ ft$

$\delta/2 = 0.764\ ft$

From Fig. 9-11 at $y/\delta = 0.50$ $(U_0 - u)/u_* \approx 3.0$

Then $(30 - u)/0.866 = 3.0$

$\underline{\underline{u_{\delta/2} = 27.4\ ft/s}}$

9.71 $F_s = C_f A_s \rho U^2{}_0/2$ where $C_f = f(Re_L)$

$Re_L = U_0 L/\nu = (30)\ (600)/(1.41 \times 10^{-5}) = 1.28 \times 10^9$

$C_f = (0.455/(\log Re_L)^{2.58}) - (1{,}700/Re_L) = 0.00152$

$F_s = (0.00152)\ (50{,}000)\ (1.94)\ (30)^2/2 = \underline{\underline{66{,}350\ lbf}}$

9.72 $F_s = C_f B L \rho V^2{}_0/2$; $Re_L = VL/\nu = 10 \times 208/(1.2 \times 10^{-5}) = 1.73 \times 10^8$

$C_f = (0.455/(\log Re_L)^{2.58}) - (1700/(1.73 \times 10^8)) = 0.00196$

$F_s = (0.00196)(44)(208)(1.94/2)(10^2) = \underline{\underline{1{,}740 \text{ lbf}}}$

9.73 $R_{e_L} = U_0 L/\nu = (15 \times 0.515) \times 325/(1.4 \times 10^{-6}) = 1.79 \times 10^9$

From formula: $C_f = 0.455/(\log_{10} Re_L)^{2.58} = 0.00146$

Then $F_s = C_f A \rho U_0{}^2/2$

$F_s = 0.00146 \times 325(48 + 38) \times 1{,}026 \times (15 \times 0.515)^2/2 = \underline{\underline{1.250 \text{ MN}}}$

$P = 1.250 \times 10^6 \times 15 \times 0.515 = \underline{\underline{9.66 \text{ MW}}}$

Get δ at $x = 300$ m

$Re_{300} = U_0 x/\nu = 15 \times 0.515 \times 300/(1.4 \times 10^{-6})$

$= 1.66 \times 10^9$

$\delta/x = 0.079 \, Re_x^{-0.085}$

$\delta = 300 \text{ m } (0.079 \times (1.66 \times 10^9)^{-0.085})$

$\delta = \underline{\underline{3.90 \text{ m}}}$

9.74

$Fr_m = Fr_p \qquad L_m/L_p = 1/100$

$V_m \backslash (gL_m)^{0.5} = V_p/(gL_p)^{0.5}$

$V_m/V_p = (L_m/L_p)^{0.5} = 1/10$; $V_m^2/V_p^2 = 1/100$

$V_m = (1/10)(30 \text{ ft/s}) = 3 \text{ ft/s}$

Viscous drag on model:

$Re_L = VL/\nu = (3)(5)/(1.22 \times 10^{-5}) = 1.23 \times 10^6$

$C_f = (0.074/(1.23 \times 10^6)^{0.2}) - (1700/(1.23 \times 10^6)) = 0.003098$

$F_{s,m} = C_f(1/2)\rho V^2 A = (0.003098)(1/2)(1.94)(3^2)(2.5)$

$= 0.0676$ lbf

$\therefore \; F_{wave,m} = 0.1 - 0.0676 = 0.0324$ lbf

9.74 (continued)

Assume, for scaling up wave drag, that

$(C_p)_m = (C_p)_p$

$(\Delta p/(\rho V^2/2))_m = (\Delta p/(\rho V^2/2))_p$

$\Delta p_m/\Delta p_p = (\rho_m/\rho_p)(V_m^2/V_p^2)$

But $F_m=/F_p = (\Delta p_m/\Delta p_p)(A_m/A_p) = (\rho_m/\rho_p)(V_m^2/V_p^2)(A_m/A_p)$

$= (\rho_m/\rho_p)(L_m/L_p)^3 = (1.94/1.99)(1/100)^3$

$F_p = F_m(1.99/1.94)(100)^3 = 0.0324(1.99/1.94)(10^6)$

$\underline{F_p = 3.32 \times 10^4 \text{ lbf}}$

9.75 $\nu_m = 1.00 \times 10^{-6} m^2/s$ at 20°C

$\nu_p = 1.31 \times 10^{-6} m^2/s$ at 0°C

$V_m = 1.45$ m/s $V_p = (Lp/Lm)^{1/2} \times V_m = \sqrt{30} \times 1.45 = \underline{7.94 \text{ m/s}}$

$Re_m = 1.45(250/30)/(1.00 \times 10^{-6}) = 1.2 \times 10^7$

$Re_p = 7.94(250)/1.31 \times 10^{-6} = 1.52 \times 10^9$

$C_f = 0.455/(\log Re)^{2.58} - 1{,}700/Re$

$C_{fm} = 0.455/(\text{Log}[1.2 \times 10^7])^{2.58} - 1{,}700/1.2 \times 10^7 = 0.00277$

$C_{fp} = 0.455/(\text{Log}[1.52 \times 10^9])^{2.58} - (1{,}700/1.52 \times 10^9) = 0.00149$

$F_{sm} = C_{fm} A\rho\, V^2/2 = 0.00277(8{,}800/30^2)998 \times 1.45^2/2 = \underline{28.42 \text{ N}}$

$F_{wave_m} = 38.00 - 28.42 = \underline{9.58 \text{ N}}$

$F_{wave_p} = (\rho_p/\rho_m)(L_p/L_m)^3\ F_{wave_m} = (1{,}026/1{,}000)30^3(9.58) = 266{,}000N$

$F_{sp} = C_{f_p} A\rho V^2/2 = 0.00149(8{,}800)1{,}026 \times 7.94^2/2 = 424{,}000N$

$F_p = F_{wave_p} + F_{sp} = 266{,}000 + 424{,}000 = \underline{690{,}000 \text{ N}}$

9.76 Minimum τ_0 occurs where C_f is minimum. Two points to check: (1) where Re_x is highest; i.e. $Re_x = Re_L$; (2) end of laminar sublayer where c_f reaches minimum value for the laminar part.

(1) $Re_L = V_0 L/\nu = 20 \times 4/10^{-6} = 8 \times 10^7$

$c_f \approx 0.058/Re_x^{1/5} = 0.00152$

(2) $Re_x = 5 \times 10^5$ (end of laminar boundary layer)

$c_f = 0.664/Re_x^{1/2} = \underline{\underline{0.00094}} \leftarrow$ minimum

So $\tau_{0_{min}} = c_{f_{min}} \rho V_0^2/2 = 0.00094 \times 998 \times 20^2/2 = \underline{\underline{188 \text{ N/m}^2}}$

9.77 $Re_L = VL/\nu = 44(4)/(1.2(10^{-5})) = 144(10^5) = 1.44(10^7)$

$C_f = 0.0027$, Fig. (9-13)

Then $F_D/1\text{ski} = 0.0027(4)(1/2)(1.94)(44^2/2) = 10.135$ lb

$F_D/2\text{ ski} = 20.27$ lb

H.P. = 20.27(44)/550 = $\underline{\underline{\text{1.62 horsepower}}}$

9.78 $Re_L = U_0 L/\nu = 10 \times 80/(1.4 \times 10^{-6})$

$Re_L = 5.7 \times 10^8$; $C_f = 0.0017$ from Fig. 9-13.

Then $F_D = C_f A \rho U_0^2/2 = 0.0017 \times 1{,}500 \times 1{,}026 \times 10^2/2 = \underline{\underline{130.8 \text{ kN}}}$

$\delta/x = C_f(0.980 \log_{10} Re_x - 0.732)$

$= 0.0017(0.980 \times 8.79 - 0.732)$

$\delta/x = 0.0133$

or $\delta = 80 \times 0.0133 = \underline{\underline{1.07 \text{ m}}}$

CHAPTER TEN

10.1 a. (3) b. (1) c. (2) d. (1) e. (3) f. (2)

10.2

Write the energy equation from the 0 elevation to the 10-ft elevation:

$p_0/\gamma + V_0{}^2/2g + z_0 = p_{10}/\gamma + V_{10}^2/2g + z_{10} + h_L$

$200{,}000/10{,}000 + 0 = 110{,}000/10{,}000 + 10 + h_L$

$$h_L = 20 - 11 - 10 = -1\,\text{m}$$

Because h_L is negative, the flow must be downward.

$32\mu LV/\gamma D^2 = h_L = 1$ m

$V = \gamma D^2/(32\mu L)$

$= 10{,}000 \times 0.01^2/(32 \times 3.0 \times 10^{-3} \times 10)$

$V = 1.04$ m/s

10.3 Valid statements are (a), (d) and (e).

10.4

$V = Q/A = 0.20/((\pi/4) \times (1/6)^2) = 9.167$ ft/sec

$Re = VD\rho/\mu = 9.167 \times (1/6) \times 0.97 \times 1.94/10^{-2} = 288$, laminar

$\Delta p = 32\mu LV/D^2 = 32 \times 10^{-2} \times 100 \times 9.167/(144 \times (1/6)^2) = 73.3$ psi/100 ft

---

10.5 Write the energy equation from the point of highest elevation to a point of 10 meters below. First, check Re:

$Re = VD\rho/\mu = 2 \times 0.01 \times 1{,}000/0.05 = 400$, laminar

$p_1/\gamma + V_1^2/2g + z_1 = p_2/\gamma + V_2^2/2g + z_2 + h_L$

10.5 (continued)

$$600{,}000/(9.81 \times 1{,}000) + 10 = p_2/\gamma + 0 + 32\mu LV/\gamma D^2$$

$$p_2/\gamma = 600{,}000/\gamma + 10 - 32 \times 0.05 \times 10 \times 2/(\gamma(0.01)^2)$$

$$p_2 = 600{,}000 + 10 \times 9{,}810 - 320{,}000 = 378 \text{ kPa}$$

10.6 $Re = VD\rho/\mu = (1)(0.01)(1{,}000)/10^{-1} = 100$ (laminar)

Because the flow is laminar the velocity distribution will be parabolic. $\tau/\tau_0 = 0.80$.

10.7

$\mu = 3.8 \times 10^{-1}\ \text{N}\cdot\text{s/m}^2$; $\nu = 2.2 \times 10^{-4}\text{m}^2/\text{s}$

$V = Q/A = 8 \times 10^{-6}/((\pi/4) \times 0.030^2) = 0.0113$ m/s

$Re = VD/\nu = 0.0113 \times 0.030/(2.2 \times 10^{-4}) = 1.54$ (laminar)

Then $\Delta p_f = 32\mu LV/D^2 = 32 \times 0.38 \times 10 \times 0.0113/(0.030)^2 = \underline{1{,}527 \text{ Pa}/10 \text{ m}}$

10.8 Write energy equation from reservoir liquid surface to outlet of pipe. Assume laminar flow so $\alpha=2$.

$$p_1/\gamma + V_1^2/2g + z_1 = p_2/\gamma + 2V^2/2g + z_2 + 32\mu LV/(\gamma D^2)$$

$$0 + 0 + 0.50 = 0 + V^2/g + 32\mu LV/(\gamma D^2)$$

$$V^2/g + 32\mu LV/(\gamma D^2) - 0.50 = 0$$

$$V^2/32.2 + 32(4\times10^{-5})(10)V/(0.80\times62.4\times(1/48)^2) - 0.50 = 0$$

$$V^2 + 19.02V - 16.1 = 0$$

Solving the above quadratic equation for V yields:

$\underline{V = 0.812 \text{ ft/s}}$

Check Reynolds number to see if flow is indeed laminar

$$Re = VD\rho/\mu = 0.812 \times (1/48)(1.94 \times 0.8)/(4 \times 10^{-5})$$

$Re = 656$ (Laminar)

$Q = VA = 0.812 \times (\pi/4)(1/48)^2 = \underline{2.77 \times 10^{-4} \text{ cfs}}$

10.9

$V = Q/A = 2 \times 10^{-3}/(\pi/4 \times (0.05)^2)$

$V = 1.019$ m/s; $Re = VD\rho/\mu = 1.019 \times 0.05 \times 940/0.048 = 997$

Write energy equation: $p_1/\gamma + V_1^2/2g + z_1 = p_2/\gamma + V_2^2/2g + z_2 + 32\mu LV/\gamma D^2$

$p_1 - p_2 = 32\mu LV/D^2 = 32 \times 0.048 \times 10 \quad \times 1.019/(0.05)^2$

$p_1 - p_2 = \underline{\underline{6.26 \text{ kPa}}}$

10.10 **Write energy equation from (1) to (2):**

$p_1/\gamma + z_1 + V^2{}_1/2g + h_p = p_2/\gamma + V^2{}_2/2g + z_2 + h_L$

$h_p = h_L = f(L/D)\ (V^2/2g)$

$V = Q/A = 7.85 \times 10^{-4}/((\pi/4)\ (0.01)^2) = 10$ m/s

$Re = VD/\nu = (10)\ (0.01)/(7.6 \times 10^{-5}) = 1{,}316$ (Laminar flow)

$f = 64/Re = 64/1{,}316 = 0.0486$

$h_p = 0.0486\ (6/0.01)\ (10^2/((2)\ (9.81)) = 149$ m

$P = \gamma Q h_p = (8{,}630)\ (7.85 \times 10^{-4})\ (149) = \underline{\underline{1{,}008 \text{ Watts}}}$

10.11 $V = Q/A = 0.0157/((\pi/4)(0.1^2)) = 2.0$ ft/s

$Re = VD/\nu = (2)(0.10)/(0.0057) = 35.1$ (Laminar)

$d/ds\ (p + \gamma z) = 32\mu V/D^2$ Eq. (10.13)

$-dp/ds - \gamma dz/ds = (32)(10^{-2})(2)/0.1^2)$

$-dp/ds - \gamma(-0.5) = 64$

$dp/ds = 0.5\gamma - 64$

$= (0.5)\ (0.9)\ (62.4) - 64$

$= 28.08 - 64 = \underline{\underline{-35.9 \text{ psf/ft}}}$

10.12 First check Re: $Re = VD\rho/\mu = 0.1 \times 0.1 \times 800/0.01 = 800$

Therefore, the flow is laminar

$V_{max} = 2V = \underline{\underline{20 \text{ cm/s}}}$

10.12 (continued)

$f = 64/Re = 64/800 = \underline{\underline{0.080}}$

$u_*/V = \sqrt{f/8}$; $u_* = \sqrt{0.08/8} \times 0.1$

$= \underline{\underline{0.010 \text{ m/s}}}$

$\tau_0 = \rho u_*^2 = 800 \times 10^{-4} = 0.08 \text{ N/m}^2$

Get $\tau_{r=0.025}$ by proportions:

$0.025/0.05 = \tau/\tau_0$; $\tau = 0.50\ \tau_0$

$\tau = 0.50 \times 0.080 = \underline{\underline{0.040 \text{ N/m}^2}}$

10.13 $Re = VD\rho/\mu = (Q/A)D/\nu = 4QD/(\pi D^2\nu) = 4Q/(\pi D\nu)$

$Re = 4 \times 0.03/(\pi \times 0.2 \times 2.37 \times 10^{-6}) = \underline{\underline{80{,}585}}$;

Turbulent

10.14 $h_{L,1}/h_{L,2} = (f_1/f_2)((L/D)_1/(L/D_2))\ (V_1^2/V_2^2)$

$= (D_2/D_1)\ (V_1^2/V_2^2)$

$V_1A_1 = V_2A_2$

$V_1/V_2 = A_2/A_1 = (D_2/D_1)^2$

$(V_1/V_2)^2 = (D_2/D_1)^4$

$h_{L,1}/h_{L,2} = (D_2/D_1)\ (D_2/D_1)^4 = (D_2/D_1)^5$

$= 2^5 = \underline{\underline{32}}$

Correct choice of answers is (d).

10.15 $Re = VD/\nu = 2 \times 1/(5.3 \times 10^{-3}) = 377$; laminar

$h_L/L = 32\mu\overline{V}/\gamma D^2 = dh/ds$

$V = (\gamma/4\mu)(r_0^2 - r^2)(32\mu\overline{V}/(\gamma D^2))$

$V = (1/4)(r_0^2 - r^2)(32\overline{V}/4r_0^2)$

$V = (32/16)\overline{V}(1 - (r/r_0)^2) = [1 - (r/r_0)^2]\, 2\,\overline{V} = 4[1 - (r/r_0)^2]$

10.15 (continued)

r	r/r_0	V
0	0	4 ft/s
1"	1/6	3.89
2"	1/3	3.56
3"	1/2	3.00
4"	2/3	2.22
5"	5/6	1.22
6"	1	0

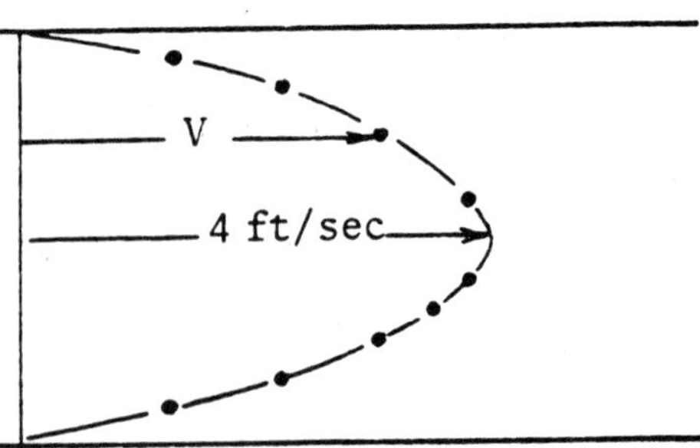

10.16 $\mu = 0.62\ \text{N·s/m}^2 \qquad \gamma = 12{,}300\ \text{N/m}^3$

Write energy equation from surface in funnel to outlet

$$p_1/\gamma + V_1^2/2g + z_1 = p_2/\gamma + V_2^2/2g + z_2 + h_L$$

$$0 + 0 + 0.30 = 0 + 2V_2^2/2g + 0 + 32\mu L V_2/(\gamma D^2)$$

$$(0.30) \times (g) = V_2^2 + V_2((g)(32)(0.62)(0.2)/(12{,}300 \times .01^2)$$

$$2.943 = V_2^2 + 31.69\ V_2$$

$$V_2^2 + 31.69\ V_2 - 2.943 = 0$$

$$V_2 = (-31.69 \pm (31.69^2 + (4)(2.943))^{0.5})/2 = \underline{0.093\ \text{m/s}}$$

10.17

Assume laminar flow, so $\Delta p_f = 32\mu LV/D^2$ or $\Delta p_f = 32\mu LQ/((\pi/4) \times D^4)$

Then $D^4 = 128\mu LQ/(\pi\Delta p_f) = 128 \times 8.5 \times 10^{-3} \times 2{,}640 \times 0.1/(\pi \times 10 \times 144)$

$D^4 = 0.0635$; $D = 0.502$ ft

A 6-in. pipe would almost fit exactly, or 8-in. pipe could be used if one wanted to be conservative. Check Re on 6" pipe:

$V = Q/A = 0.1/((\pi/4) \times 0.5^2) = 0.51$ ft/sec.

$Re = VD\rho/\mu = 0.51 \times 0.5 \times 0.85 \times 1.94/(8.5 \times 10^{-3}) = \underline{\underline{47}}$ <u>laminar</u>

10.18 Mercury at 20°C:

$\mu = 1.5 \times 10^{-3}\ \text{N·s/m}^2$, $\nu = 1.2 \times 10^{-7}\ \text{m}^2/\text{s}$, $\gamma = 133{,}000\ \text{N/m}^3$

Energy Eq.: $\cancel{p_1}^{0}/\gamma + \cancel{\alpha V_1^2/2g} + z_1 = \cancel{p_2}^{0}/\gamma + \cancel{\alpha V_2^2/2g} + z_2 + h_f$

$z_1 - z_2 = L = h_f$

$\cancel{L} = 32\mu \cancel{L} V/\gamma D^2$

10.18 (continued)

$D^2 = 32\mu V/\gamma$ (1)

$Re = VD/\nu = 2{,}000$

$V = 2{,}000\ \nu/D = 2{,}000 \times 1.2 \times 10^{-7}/D$

$V = 2.4 \times 10^{-4}/D$ (2)

Solve Equations (1) and (2) for D:

$D^2 = 32\mu(2.4 \times 10^{-4})/\gamma D$

$D^3 = 32(1.5 \times 10^{-3})(2.4 \times 10^{-4})/133{,}000$

$\underline{D = 4.42 \times 10^{-4} m}$

10.19 $Re = VD/\nu = 0.40 \times 0.04/5.3 \times 10^{-3} = \underline{3.02}$ Laminar

From solution to P10-16: $dh/ds = -32\mu V/(\gamma D^2)$

$dh/ds = -32 \times 6.2 \times 10^{-1} \times 0.40/(12{,}300 \times (0.04)^2)$

$d/ds(p/\gamma + z) = -0.403$ or $(1/\gamma)dp/ds + dz/ds = 0.403$

Because flow is downward $dz/dz = -1$

Then: $dp/ds = \gamma[1 - 0.403] = \underline{7.34\ kPa/m;}$ pressure increases downward.

Shear stress: $\tau = \gamma(r/2)[-dh/ds]$, [from Eq. (10-3)]

or $\tau = 12{,}300(r/2) \times 0.403$

Then at $r = 0$, $\tau = 0$ and $\tau_{wall} = \tau_0 = 12{,}300(0.02/2) \times 0.403$

$\tau_{wall} = \underline{49.6\ N/m^2}$

10.20

The design might have a physical configuration as shown below. The design should be based upon solving Eq.10.17 ($h_f = 32\ \mu LV/(\gamma D^2)$) for the viscosity μ. Since this is for laminar flow, the size of pipe and depth of liquid in the tank should be such that laminar flow will be assured ($R_e < 1{,}000$). For the design

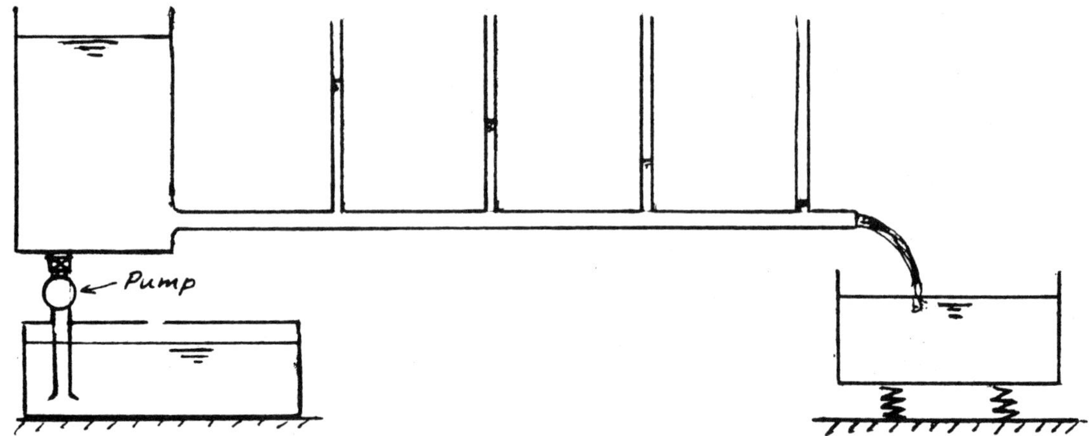

suggested here, the following measurements, conditions, and calculations would have to be made:

A. Measure tube diameter by some means.

B. Measure γ or measure temperature and get γ from a handbook.

C. Establish steady flow by having a steady supply source (pump liquid from a reservoir).

D. Measure Q. This could be done by weighing amount of flow for a given period of time or by some other means.

E. Measure h_f/L by the slope of the piezometric head line as obtained from piezometers. This could also be obtained by measuring Δp along the tube by means of pressure gages or pressure transducers from which h_f/L could be calculated.

F. Solve for μ with Eq. 10.17.

10.21 Since the velocity distribution is parabolic, the flow is laminar.

Then $\Delta p_f = 32\mu LV/D^2$; $\nu = \mu/\rho = \Delta p_f D^2/(32LV\rho)$

$\nu = 16 \times 1^2/(32 \times 100 \times 3/2 \times 0.9 \times 1.94) = \underline{\underline{0.00191\ ft^2/s}}$

10.22 As in Prob. 10.21

$\nu = \Delta p_f D^2/(32LV\rho) = 1{,}900 \times (0.3)^2/(32 \times 100 \times 0.75 \times 800)$

$= \underline{\underline{8.91 \times 10^{-5} m^2/s}}$

10.23 Check Re: $Re_{20°} = VD/\nu = 0.10 \times 0.005/10^{-6} = 500$ So flow is laminar.

Then $\Delta p = 32\mu LV/D^2$

Assume linear variation in μ

Thus $\mu_{avg.} = \mu_{25°} = 8.91 \times 10^{-4} N{\cdot}s/m^2$

and $\Delta p = 32 \times 8.91 \times 10^{-4} \times 5 \times 0.10/(0.005)^2 = \underline{\underline{570\ Pa}}$

10.24

The flow is downward (from right to left).

$\Delta h = ((13.55 - 0.8)/0.8) \times \text{deflection} = 5.312$ ft of oil $= h_f$

Then $h_f = f(L/D)(V^2/2g)$; $f = 5.312 \times ((1/6)/(30)) \times 2 \times 32.2/5^2) = 0.076$

Assume flow is laminar. Then $h_f = 32\mu LV/\gamma D^2$

or $\mu = h_f\gamma D^2/(32LV)$ and $Re = VD\rho/\mu$ (1)

So $Re = VD\rho/[h_f\gamma D^2/(32LV)] = 32LV^2/(gh_f D)$

$Re = 32 \times 30 \times 5^2/(32.2 \times 5.312 \times (1/6)) = 842$ $\underline{\underline{\text{laminar}}}$

From Eq.(1): $\mu = 5.312 \times 0.8 \times 62.4 \times (1/6)^2/(32 \times 30 \times 5)$

$\underline{\underline{\mu = 1.53 \times 10^{-3}\ lb\text{-}sec/ft^2}}$

10.25

The flow is downward (from right to left).

$\Delta h = [(13.55 - 0.8)/0.8] \times \text{deflection} = 1.59 \text{ m} = h_f$

$h_f = f(L/D)(V^2/2g)$; $f = 1.59 \times (0.05/10) \times 2 \times 9.81/1.2^2 = \underline{\underline{0.108}}$

Assume flow is laminar. Then as in P10-24

$Re = 32LV^2/(gh_fD) = 32 \times 10 \times 1.2^2/(9.81 \times 1.59 \times 0.05) = 590$, laminar

Then $\mu = h_f\gamma D^2/(32LV) = 1.59 \times 9.81 \times 800 \times 0.05^2/(32 \times 10 \times 1.2)$

$= \underline{\underline{0.0812 \text{ N}\cdot\text{s/m}^2}}$

10.26 $h_f/L = 2 = (f/D)(V^2/2g) = (f/0.03)(1/(2 \times 9.81))$

$f = 1.177$

Assume laminar flow: $f = 64/Re$

or $Re = 64/1.177 = 54.4$, laminar

Indeed, the flow is laminar and it will be laminar if the flow rate is doubled. The head loss varies directly with V (and Q); therefore, the head loss will also be doubled when the flow rate is doubled.

10.27 For a straight pipe there is no momentum change, so the forces on the column of fluid in the pipe are zero, or

$\tau_0\pi DL = -\Delta pA$ (for horizontal pipe)

$\tau_0\pi DL = \gamma h_f A$

$\tau_0\pi DL = \gamma f(L/D)(V^2/2g)(\pi D^2/4) = f\pi D\rho V^2/8$

Thus, $\tau_0 = f\rho V^2/8 = 0.017 \times 0.82 \times 1.94 \times 6^2/8 = \underline{\underline{0.122 \text{ psf}}}$

10.28 $Re_{oil} = VD\rho/\mu = (1)(900)(0.1)/(10^{-1}) = 900$; $V_{max} = 2\bar{V}$

$Re_{gas} = (1.0)(0.1)(1)/(10^{-5}) = 10^4$; $V_{max} < 2\bar{V}$

$(V_{max,oil}/V_{max,gas}) > 1$; case (a)

10.29 $u_* \delta_N'/\nu = 11.6;\quad \delta_N' = 11.6\nu/u_*$

From solution to P10-27 $\tau_0 = (f/8)\rho V^2$

Thus $u_* = \sqrt{\tau_0/\rho} = V\sqrt{f/8}$. Also $Re = VD/\nu$ or $\nu = VD/Re$

So, $\delta_N' = 11.6(VD/Re)/(V\sqrt{f/8})$. From Fig. 10-8 $f = \underline{\underline{0.018}}$

So $\delta_N' = 11.6(0.12/100{,}000)/\sqrt{0.018/8} = 2.93 \times 10^{-4} m = \underline{\underline{0.293\ mm}}$

10.30 Shear stress decreases with x near the entrance; therefore, case (a) is correct choice.

10.31 $Re = 4Q/(\pi D\nu) = 4 \times 2 \times 10^5/(\pi \times (10/12) \times 1.06) = 2.9 \times 10^5$

From Fig. 10.8 $\underline{\underline{f = 0.0145}}$

10.32 $Re = 4 \times 0.06 \times 10^6/(\pi \times 0.25 \times 1.31) = 2.33 \times 10^5$

From Fig. 10.8 $f = \underline{\underline{0.015}}$

10.33

$V = Q/A = 0.012 \times 4/(\pi \times 0.03^2) = 16.98$ m/s

$\mu = 1.81 \times 10^{-5} N\cdot s/m^2;\quad \rho = p/RT = 110{,}000/(287 \times 293) = 1.31\ kg/m^3$

$Re = VD\rho/\mu = 16.98 \times 0.03 \times 1.31/(1.81 \times 10^{-5}) = 3.7 \times 10^4$

Then from Fig. 10-8 $f = 0.0225;\ \Delta p = (0.0225 \times 16.98^2 \times 1.31/(0.03 \times 2)$

$= \underline{\underline{142\ N/m^2/m}}$

10.34 Write energy equation from one standpipe to the other:

$$p_1/\gamma + V_1^2/2g + z_1 = p_2/\gamma + V_2^2/2g + z_2 + h_L$$

$$p_1/\gamma + z_1 = p_2/\gamma + z_2 + h_L$$

$\Delta h = ((p_1/\gamma) + z_1)) - ((p_2/\gamma) + z_2) = h_L$

$Re = VD/\nu = (0.6)(0.02)/(5.1 \times 10^{-4}) = 23.5$ (Laminar)

$$\Delta h = h_1 - h_2 = 32\mu LV/(\gamma D^2)$$

$\Delta h = (32)(6.2 \times 10^{-1})(1)(0.6)/(12{,}300 \times 0.02^2)$

$\underline{\underline{\Delta h = 2.42\ m}}$

10.35

$\Delta p = f(L/D)\rho V^2/2 \quad V = Q/A = 25 \times 4/(60 \times \pi \times (1/12)^2) = 76.4$ ft/s

$\rho = p/(RT) = 15 \times 144/(1{,}716 \times 540) = 0.00233$ slugs/ft^3

$\mu = 3.85 \times 10^{-7}$ lbf-s/ft^2 ; $Re = VD\rho/\mu = 76.4 \times (1/12) \times 0.00233$

$/(3.85 \times 10^{-7}) = 3.9 \times 10^4$

Then $f = 0.022$

$\Delta p = 0.022 \times 1 \times 12 \times 76.4^2 \times 0.00233/2 =$ 1.8 psf/ft

---

10.36

$h_f = f(L/D)\,(V^2/2g)$

$0.30 = f(1/0.01)\,(2^2/(2 \times 9.81))$; $f = 0.0147$

$Re = 2.7 \times 10^5$ (from Fig. 10-8)

$\nu = VD/Re = (2)\,(0.01)/(2.7 \times 10^5) = 7.4 \times 10^{-8}$ m^2/s

---

10.37 $\Delta h = h_f = 0.80(2.5 - 1) = 1.2$ ft of water

$h_f = f(L/D)V^2/2g;\ f = 1.2 \times (0.05/4) \times 2 \times 9.81/3^2 = 0.033$

10.38

$Re = VD/\nu = 3(0.3)/(1.31 \times 10^{-6}) = 6.87 \times 10^5$

$k_s/D = 0.00026/0.30 = 0.00087$

From Fig. 10-8 $f = 0.020$

$u/u_* = 5.75 \log(y/k_s) + 8.5$

where $u_* = V\sqrt{f/8} = 3\sqrt{0.020/8} = 0.15$ m/s

Then $u = 0.15[5.75 \log(y/0.00026) + 8.5]$

y(m) →	0.02	0.04	0.06	0.10	0.15
u(m/s)→	2.90	3.16	3.31	3.50	3.66

10.39

$Re = Vd/\nu = (1)\,(0.10)/(10^{-4}) = 10^3$ (laminar)

$f = 64/Re = 64/1{,}000 = 0.064$ (case a).

10.40 $V = Q/A = 0.002/[(\pi/4) \times (0.06)^2] = 0.707$ m/s

$Re = VD/\nu = 0.707 \times 0.06/10^{-6} = 4.24 \times 10^4$

From Fig. 10-8 $\underline{f = 0.021}$

10.41 **Write the energy equation from front of train to outlet of tunnel.**

$$p_1/\gamma + V_1^2/2g + z_1 = p_2/\gamma + V_2^2/2g + z_2 + h_L$$

$$p_1/\gamma + V_1^2/2g = 0 + 0 + 0 + V_2^2/2g + f(L/D)\,V_2^2/2g$$

$$p_1/\gamma = f(L/D)\,V^2/2g$$

Assume: $k_s = 0.05$ ft; $T = 60^\circ F$; $\gamma = 0.0764$ lb/ft^3

$k_s/D = 0.05/10 = 0.005$; $Re = VD/\nu = (50)(10)/(1.58 \times 10^{-4}) = 3.2 \times 10^6$

$f = 0.030$ (from Fig. 10-8)

$p_1 = \gamma f(L/D)(V^2/2g)$

$= (0.0764)(0.03)(2{,}500/10)(50^2/(64.4))$

$p_1 = 22.24$ psfg

Now write energy equation from outside entrance to rear of train

$$p_3/\gamma + V_3^2/2g + z_3 = p_4/\gamma + V_4^2/2g + z_4 + h_L$$

$$0 + 0 + 0 = p_4/\gamma + V_4^2/2g + 0 + (K_e + f(L/D))V^2/2g$$

$$p_4/\gamma = (V^2/2g)(1.5 + f(L/D))$$

$$= -(50^2/2g)(1.5 + 0.03\,(2{,}500/10))$$

$$p_4 = -\gamma(349.4) = -26.69 \text{ psf}$$

$$\Delta p = p_1 - p_4 = 22.24 - (-26.69) = \underline{48.93 \text{ psf}}$$

$$P = FV = (\Delta pA)(50) = (48.93)(\pi/4)(10^2)(50)$$

$$= 192{,}158 \text{ ft-lbf/s} = \underline{349 \text{ hp}}$$

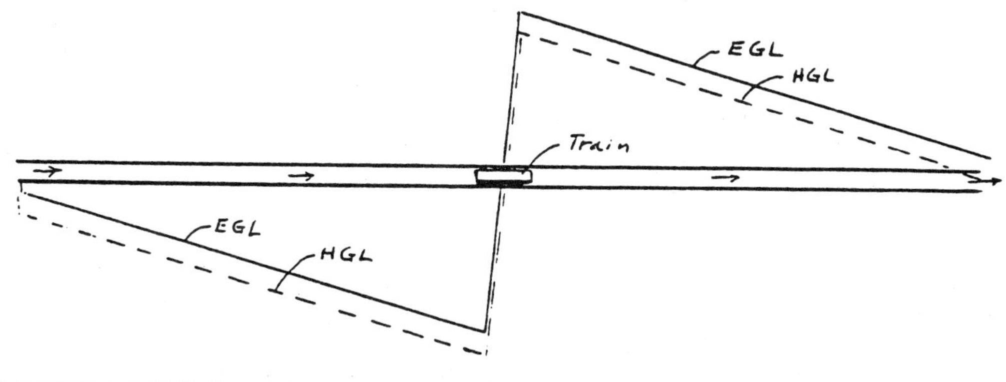

10.42

$$d_{tube} = 3/16\,\text{in.} = 0.01562\,\text{ft}$$
$$L_{tube} = 50\,\text{in.}$$
$$\text{Assume } \nu = 1.2\times10^{-5}\,\text{ft}^2/\text{s}$$

Write the energy equation from the jug to the cylinder

$$p_j/\gamma + V_j^2/2g + z_j = p_c/\gamma + V_c^2/2g + z_c + \sum h_L \qquad (1)$$

Assume the bend in the tube has a large radius so that the bend loss is negligible. Also assume that the entrance loss coefficient is equal to 0.5. It would actually be somewhat larger than 0.5 but this should yield a reasonable approximation. Therefore

$$\sum h_L = (0.5 + fL/D + K_E)V^2/2g$$

The exit loss coefficient, K_E, is equal to 1.0. Therefore, Eq. 1 becomes.

$$\Delta z = z_j - z_c = (V^2/2g)(1.5 + fL/D)$$
$$\text{or } V = \sqrt{2g\Delta z/(1.5+fL/D)} \qquad (1)$$
$$= \sqrt{2g\Delta z/(1.5+f\times267)}$$

First assume f=0.03 and initial Δz=(21-2.5)/12=1.54ft.
Then

$$V = \sqrt{(2g)(1.54)/(1.5+8)} = 3.23\,\text{ft/s}$$
$$Re = VD/\nu = 3.23\times.01562/(1.2\times10^{-5}) = 4205$$
$$f = 0.038 \text{ (From Fig. 10.8)}$$

Try again

$$V = \sqrt{(2g)(1.54)/(1.5+10.15)} = 2.92\,\text{ft/s}$$
$$Re = 3800 \text{ and } f = 0.040$$

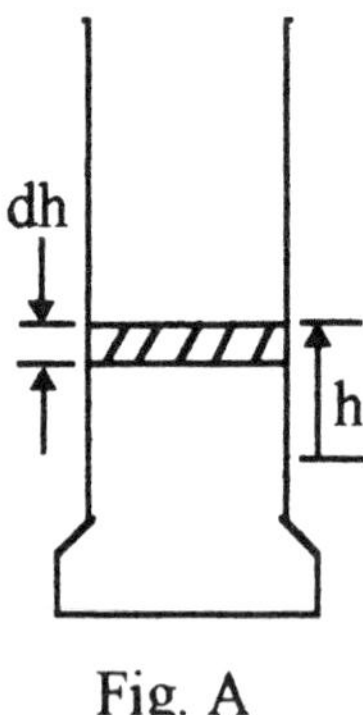

Fig. A

Use f=0.040 for final solution. As a simplifying assumption assume that as the cylinder fills the level of water in the jug has negligible change. As the cylinder is being filled one can visualize (see Fig. A) that in time dt a volume of water equal to Qdt will enter the cylinder and that volume in the cylinder can be expressed a $A_c$dh. Or in equation form:

$$Qdt = A_c dh$$
$$dt = (A_c/Q)dh$$

But $Q = V_tA_t$ so $dt = ((A_c/A_t)/V)dh$ (2)

Substitute V of Eq. (1) into Eq. (2):

10.42 (cont'd)

$$dt = (A_c / A_t) / (2g\Delta z / (1.5 + 267f))^{1/2} dh$$

$$\forall_c = .500 \text{liter} = 0.01766 \text{ft}^3$$

$$\text{or } 0.01766 = A_c \times (11.5\text{in.}/12); A_c = 0.01842 \text{ft}^2$$

$$A_{tube} = (\pi / 4)((3/16)/12)^2 = 0.0001917 \text{ft}^2; A_c / A_t = 96.1$$

$$dt = 96.1 / (2g\Delta z / (1.5 + 10.7))^{1/2} dh$$

Let h be measured from the level where the cylinder is 2in full. Then

$$\Delta z = ((21\text{in} - 2.5\text{in})/12) - h$$

$$\Delta z = 1.542 - h$$

$$\text{or } dt = (96.1 / (2g(1.54 - h)/(12.2)^{1/2})dh$$

$$dt = 41.8 / (1.54 - h)^{1/2})dh$$

$$dt = -41.8(1.54 - h)^{-1/2})(-dh)$$

Integrate:

$$t = -41.8(1.54 - h)^{1/2} / (1/2)\Big|_0^h$$

$$= -83.6(1.54 - h)\Big|_0^{0.75}$$

$$= -83.6[(0.79)^{1/2} - (1.54)^{1/2}]$$

$$= -83.6(0.889 - 1.241)$$

$$= \underline{29.4 \text{s}}$$

Possible problems with this solution: The Reynolds number is very close to the point where laminar flow will occur and this would be an unstable condition. The flow might alternate between turbulent and laminar flow.

10.43

Write the energy equation between water surfaces of the reservoirs:

$$p_1/\gamma + V_1^2/2g + z_1 = p_2/\gamma + V_2^2/2g + z_2 + \Sigma h_L$$

$$0 + 0 + z_1 = 0 + 0 + 100 + \Sigma h_L$$

where $\Sigma h_L = (K_e + 2K_b + K_E + fL/D)(V^2/2g)$

$$K_e = 0.50 \;;\; K_b = 0.40 \text{(assumed)}; \; K_E = 1.0; \; fL/D = 0.025 \times 430/1 = 10.75$$

$$V = Q/A = 10.0/((\pi/4) \times 1^2) = 12.73 \text{ ft/s}$$

Then $z_1 = 100 + (0.5 + 2 \times 0.40 + 1.0 + 10.75)(12.73^2)/64.4 = \underline{\underline{133 \text{ ft}}}$

The point of minimum pressure will occur just downstream of the first bend as shown by the hydraulic grade line (below).

10.43 (Cont'd)

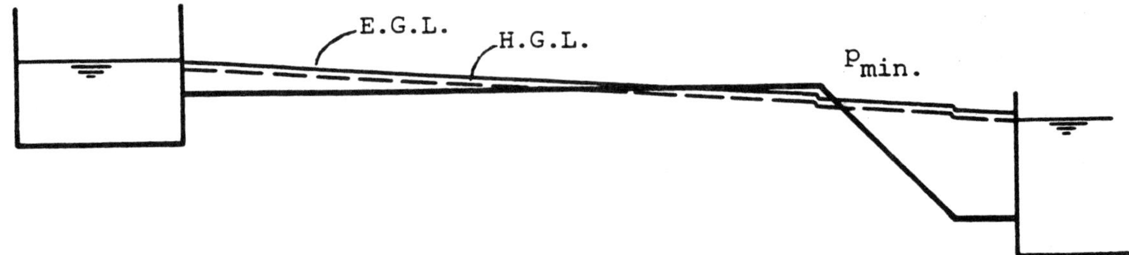

To determine the magnitude of the minimum pressure, write the energy equation from the upstream reservoir to just downstream of bend:

$$z_1 = z_b + p_b/\gamma + V^2/2g + (fL/D)V^2/2g + K_e V^2/2g + K_b V^2/2g$$

$$p_b/\gamma = 133 - 110.70 - (12.73^2/64.4)(1.9+0.025 \times 300/1) = -1.35 \text{ ft}$$

$$p_B = -1.35 \times 62.4 = \underline{\underline{-84 \text{ psfg}}} = \underline{\underline{-0.59 \text{ psig}}}$$

$$Re = VD/\nu = 12.73 \times 1/(1.41 \times 10^{-5}) = 9.0 \times 10^5$$

With an f of 0.025 at a Reynolds number of 9×10^5 we read a value of 0.0025 (approx) for k_s/D from Fig. 10-8. Then from the table on Fig. 10-8 or Fig. 10-9 the pipe appears to be fairly rough concrete pipe.

10.44 Write the energy equation from the reservoir water surface to the jet at the end of the pipe.

$$p_1/\gamma + V_1^2/2g + z_1 = p_2/\gamma + V_2^2/2g + z_2 + h_T + \sum h_L$$

$$0 + 0 + z_1 = 0 + V_2^2/2g + z_2 + h_T + (K_e + fL/D)V^2/2g$$

$$z_1 - z_2 = h_T + (1 + 0.5 + fL/D)V^2/2g$$

$$100\text{ft} = h_T + (1.5 + fL/D)V^2/2g$$

But

$$V = Q/A = 4/((\pi/4)1^2) = 5.093\text{ft/s}; \; V^2/2g = 0.403\text{ft}$$

$$\nu = 1.06 \times 10^{-5}\text{ft/s then } Re = VD/\nu = 4.80 \times 10^5$$

From Fig. 10.8 f=0.0150 for $k_s/D = 0.0016$

Then

$$100\text{ft} = h_T + (1.5 + 0.0150 \times 1000/1)(0.403)$$

$$h_T = (100 - 6.65)\text{ft}$$

$$P = Q\gamma h_T \times \text{eff.}$$

$$= 4 \times 62.4 \times 93.35 \times 0.80$$

$$= 18{,}640\text{ft} \cdot \text{lb/s}$$

$$= \underline{\underline{33.89 \text{ horsepower}}}$$

10.45 $Re = VD\rho/\mu = (0.1)\ (0.1)\ (800)/(10^{-2}) = 800$

Because Re < 2,000 the flow is laminar.

Therefore, a) $V_{max} = 2\bar{V} = 2 \times 0.1\ m/s = \underline{\underline{0.2\ m/s}}$

b) $f = 64/Re = 64/800 = 0.080$

c) $(u_*/V) = (f/8)^{0.5} = (0.08/8)^{0.5} = 0.10$

$u_* = (0.1)\ (0.1) = \underline{\underline{0.01\ m/s}}$

d) $(\tau_0/\rho)^{0.5} = 0.01 \Rightarrow \tau_0/\rho = 10^{-4}$

$\tau_0 = (10^{-4})\ (800) = 0.08\ N/m^2$

$\tau_{25mm} = 0.5\ \tau_0 = \underline{\underline{0.04\ N/m^2}}$

e) Yes it will be doubled because flow will still be laminar and head loss is linear with V (and Q) in laminar range.

10.46 The valid statements are: a, b, d. For cases c & e:

$Re = VD/\nu = (1)\ (1)/(10^{-6}) = 10^6$

The flow at 1 m/s is in the turbulent range; therefore, the head loss will be more than doubled with a doubling of the velocity.

10.47 a) Pumps are at A and C

b) A contraction, such as a Venturi meter or orifice, must be at B.

c)

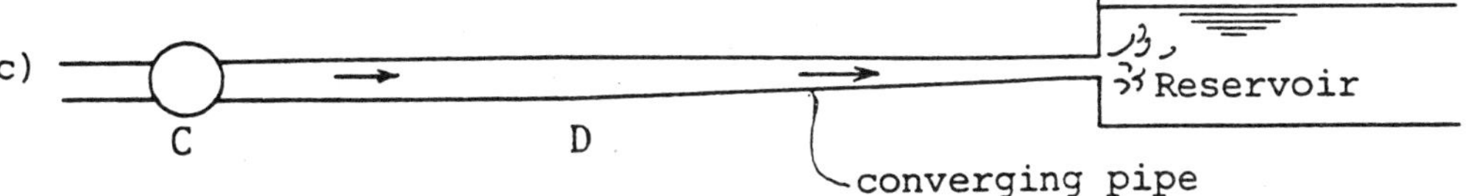

d) Other information:
(1) Flow is from left to right
(2) The pipe between AC is smaller than before or directly after it.
(3) The pipe between BC is probably rougher than AB.

10.48 $V = Q/A = 0.04/((\pi/4) \times 0.15^2) = 2.26\ m/s$

$Re = VD/\nu = 2.26 \times 0.15/(10^{-6}) = 3.4 \times 10^5$

$k_s/D = 0.002$ (Fig. 10-9) and $f = 0.024$ (Fig. 10-8)

From Eq. (10-21) $\tau_0 = f\rho V^2/8$

Thus, $\tau_0 = 0.024 \times 998 \times 2.26^2/8 = \underline{\underline{15.3\ N/m^2}}$

10.48 (continued)

Assume linear shear stress variation; thus,

$\tau_1 = (6.5/7.5) \times \tau_0 = \underline{\underline{13.25\ \text{N/m}^2}}$

Assume logarithmic velocity distribution. Thus, $u/u_* = 5.75\log(y/k_s) + 8.5$

$u_* = \sqrt{\tau_0/\rho} = 0.124$ m/s; $u = 0.124[5.75\log(0.01/(0.002 \times 0.15)) + 8.5] = \underline{\underline{2.14\ \text{m/s}}}$

10.49 One possibility is shown below:

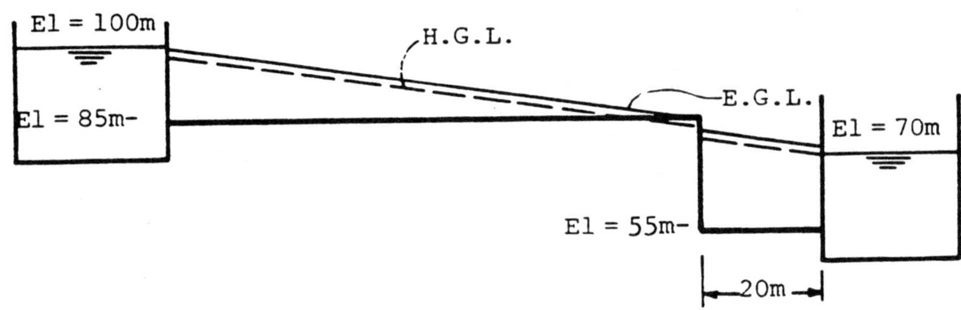

Assume that the pipe dia = 0.50 m, assume K_b = 0.20, and f = 0.015

Then $100 - 70 = (0.5 + 2 \times 0.20 + 1 + 0.015 \times 130/0.5)V^2/2g$

$V^2/2g = 5.17$

The minimum pressure will occur just downstream of the first bend and its magnitude will be as follows:

$p_{min.}/\gamma = 100 - 85 - (0.5 + 0.20 + (0.015 \times 80/0.5) + 1)\ V^2/2g$

$= -6.20$ m

$p_m = 6.20 \times 9{,}810 = \underline{-60.8\ \text{kPa gage}}$

10.50 $Re = 4Q/(\pi D\nu) = 4 \times 0.02/(\pi \times 0.10 \times 10^{-6}) = 2.55 \times 10^5$

$k_s/D = 0.0005$ (Fig. 10-9) and $f = 0.0185$ (Fig. 10-8)

Then $h_f = f(L/D)V^2/2g$ where $V = 0.02/((\pi/4) \times 0.1^2) = 2.546$ m/s

$h_f = 0.0185 \times (80/0.10) \times 2.546^2/(2 \times 9.81) = 4.89$ m

Write energy Eq. from pump to point 80 m higher:

$p_1/\gamma + V_1^2/2g + z_1 = p_2/\gamma + V_2^2/2g + z_2 + h_f$; $V_1 = V_2$

$1.6 \times 10^6/9{,}790 + V_1^2/2g = p_2/\gamma + V_2^2/2g + 80 + 4.89$; $p_2 = \underline{\underline{769\ \text{kPa}}}$

10.51 Write the energy equation from the water surface in the tank to the outlet of the pipe.

$$p_1/\gamma + V_1^2/2g + z_1 = p_2/\gamma + V_2^2/2g + z_2 + \sum h_L$$

$$0+0+12 = 0+V_2^2/2g+0+(K_e+fL/D)V^2/2g$$

$$12 = V_2^2/2g(1+K_e+fL/D) \qquad (1)$$

K_e=0.50 from Table 10.2

Get f by trial and error using Fig. 10.8. From Fig. 10.9 $k_s/D = 0.006$. With this value of k_s/D assume f = 0.034 (it will never be less than 0.032 as seen in Fig. 10.8). Plugging these values for K_e and f into Eq. (1) we have:

$$12 = V_2^2/2g(1.5+(.034\times 8/(1/12)))$$

$$V_2^2 = 2g\times 12/(4.764)$$

$$V_2 = 12.74\text{ft}/\text{s}$$

$$\text{Re} = VD/\nu = 12.74\times(1/12)/1.22\times 10^{-5} = 8.70\times 10^4; f = 0.033$$

Solve for V_2 again $\underline{\underline{V_2 = 12.87\text{ft}/\text{s}}}$

10.52 Assume the pipe is galvanized iron so $k_s = 0.15\text{mm} = 0.015\text{cm}$ (Fig. 10.8). $k_s/D = .015/2 = .0075$. Then f=0.035 (from Fig. 10.8). Assume K_b = 0.9 and K_e = 0.5 (Table 10.2). Writing the energy equation from the water surface in the tank to the pipe outlet we have:

$$p_1/\gamma + V_1^2/2g + z_1 = p_2/\gamma + V_2^2/2g + z_2 + \sum h_L$$

$$0+0+5 = 0+V_2^2/2g+0+(K_e+2K_b+fL/D)V_2^2/2g$$

$$5 = (V_2^2/2g)(1+0.5+2\times 0.9+.035\times 10/.02)$$

$$5 = (V_2^2/2g)(20.8)$$

$$V_2 = 2.17\text{m}/\text{s}$$

Check Re and new f: Assume $\nu = 10^{-6}\text{m}^2/\text{s}$; Re = VD/$\nu$ = $2.17\times 0.02/10^6 = 4.34\times 10^4$
f = 0.036 (Fig. 10.8)

With new f:

$$\underline{\underline{V_2 = 2.15\text{m}/\text{s}}}$$

$$h = V^2/2g = (2.15)^2/(2\times 9.81)$$

$$\underline{\underline{= 0.24\text{m} = 24\text{cm}}}$$

10.53

$$Re = VD/\nu; D = 1/3 ft; \nu = 1.22 \times 10^{-5} ft^2/s$$

$$V = Q/A = 1.0/((\pi/4)D^2)$$
$$= 1.0/((\pi/4)(1/3)^2)$$
$$= 11.46 ft/s$$

Then

$$Re = 11.46 \times (1/3)/(1.22 \times 10^{-5}) = 3.13 \times 10^5$$
$$k_s = 1.5 \times 10^{-4} ft \text{ and } k_s/D = 4.5 \times 10^{-4}$$

From Fig. 10.8 f = 0.0165 and fL/D = 0.0165 × 300/(1/3) = 14.86. Write the energy equation from water surface A to water surface B

$$p_A/\gamma + V_A^2/2g + z_A + h_p = p_2/\gamma + V_2^2/2g + z_2 + \sum h_L$$
$$0 + 0 + 0 + h_p = (10 \times 144/62.4) + 0 + (K_e + K_E + fL/D)V^2/2g$$

Assume $K_e = 0.03$ (from Table 10.2)

$$h_p = 23.08 + (0.03 + 1 + 14.86)(11.46^2/64.4)$$
$$= 55.48 ft$$

$$P = Q\gamma h_p/eff$$
$$= 1.0 \times 62.4 \times 55.48/0.85$$
$$= \underline{\underline{4{,}073 ft \cdot lbf/s}}$$
$$= \underline{\underline{7.41 \text{ horsepower}}}$$

10.54 Write the energy equation from the reservoir water surface to the tank water surface. The head losses will be due to entrance, pipe resistance, and exit.

Let $K_e = 0.5$ (Table 10.2), $K_E = 1.0$.

Energy equation:

$$p_1/\gamma + V_1^2/2g + z_1 + h_p = p_2/\gamma + V_2^2/2g + z_2 + \sum h_L$$
$$0 + 0 + z_1 + h_p = 0 + 0 + z_2 + (K_e + fL/D + K_E)V^2/2g \quad (1)$$
$$h_p = (z_2 - z_1) + (0.5 + (0.015 \times 30/0.9) + 1.0)V^2/2g$$
$$h_p = h + (2.0)V^2/2g$$

But the head supplied by the pump is $h_o(1 - (Q^2/Q_{max}^2))$ so Eq. (1) becomes

$$h_o(1 - (Q^2/Q_{max}^2)) = h + V^2/g$$
$$50(1 - Q^2/4) = h + Q^2/(gA^2)$$
$$50 - 12.5Q^2 = h + Q^2/(gA^2)$$

10.54 (cont'd)

However, $A = (\pi/4)D^2 = (\pi/4)(0.9^2) = 0.636\text{m}^2$ and $g = 9.81\text{m/s}^2$ so

$$50 - 12.5Q^2 = h + 0.752Q^2$$

$$50 - h = 12.752Q^2 \qquad (2)$$

$$\sqrt{50-h} = 3.571Q$$

It should be noted that as the tank fills a volume of water given by Qdt can be expressed as $A_{tank}dh$: $A_{tank}\,dh = Qdt$ or

$$Q = A_{tank}\,dh/dt \qquad (3)$$

Eliminated Q between Eq's. (2) and (3) to obtain

$$\sqrt{50-h} = 3.571A_{tank}dh/dt$$

$$\text{or } dt = 3.571A_{tank}(50-h)^{-\frac{1}{2}}dh \qquad (4)$$

Integrate Eq. (4)

$$t = -3.571A_{tank}(50-h)^{1/2}/(1/2) + C$$

$$\text{when } h = 0\ t = 0 \text{ so } C = (2)(3.571)A_{tank}(50)^{1/2}$$

$$t = 2 \times 3.571A_{tank}(50^{1/2} - (50-h)^{1/2});\ A_t = 100\text{m}^2$$

$$t = 714.2(7.071 - (50-h)^{1/2})$$

$$\text{when } h = 40\text{m}$$

$$t = 714.2(7.071 - (10)^{1/2})$$

$$= 2{,}792\text{s}$$

$$= \underline{\underline{46.5\,\text{minutes}}}$$

10.55 $Re = VD/\nu = 4 \times 0.03/(2 \times 10^{-6}) = 6 \times 10^4$

$f_{lam} = 64/Re = 64/(6 \times 10^4)$; $f_{turb} = 0.020$ (Fig. 10-8)

Then $h_{f_{lam}}/h_{f_{turb}} = 64/((6 \times 10^4) \times (0.020)) = \underline{\underline{0.0533}}$

10.56 $Re = 4Q/\pi D\nu = 4 \times 0.02/(\pi \times (4/12) \times (1.22 \times 10^{-5}))$

$= 6.3 \times 10^3$

$k_s/D = 0.0025$ (Fig. 10.9)

Then from Fig. 10.8, $\underline{\underline{f = 0.038}}$

10.57 $Re = 4Q/(\pi D\nu) = 4(0.75)10^3/(\pi(1/2)3.33) = 570$, laminar ;

$\mu = \rho\nu = 0.005$ lbf-s/ft^2

$h_f = 32\mu LV/(\gamma D^2) = 32(5 \times 10^{-3})1500(0.75)/(1.5 \times 32.2 \times (\pi/4)(1/2)^4)$

$= \underline{\underline{75.9 \text{ ft}}}$

10.58

$Re = VD/\nu = 4Q/(\pi D\nu) = 4 \times 0.03/(\pi \times 0.15 \times (10^{-2}/820))$

$Re = 2.09 \times 10^4$ (turbulent)

$k_s = 4.6 \times 10^{-5}$m $\quad k_s/D = 4.6 \times 10^{-5}/0.15 = 3.1 \times 10^{-4}$

$f = 0.027$ from Fig. 10-8 in text; $V = Q/A = 0.03/(\pi \times 0.15^2/4) = 1.698$m/s

Then $h_f = f(L/D)(V^2/2g) = 0.027(1{,}000/0.15)(1.698^2/(2 \times 9.81)) = 26.4$ m

Energy Eq.: $p_A/\gamma + V_A^2/2g + z_A = p_B/\gamma + V_B^2/2g + z_B + h_f$

$p_A = 0.82 \times 9{,}810[(250{,}000/(0.82 \times 9{,}810)) + 20 + 26.4] = \underline{\underline{623 \text{ kPa}}}$

10.59 Galvanized iron pipe: $k_s = 0.15$ mm (from Fig. 10.8)

$k_s/D = 0.15/26 = 5.8 \times 10^{-3}$

$= 0.0058$; Neglect entrance loss

Assume $f = 0.032$ (first estimate from Fig. 10.8)

Energy Eq: Write it from reservoir water surface to outlet

$p_1/\gamma + V_1^2/2g + z_1 = p_2/\gamma + V_2^2/2g + z_2 + h_L$

10.59 (continued)

$$0 + 0 + 10 = 0 + V_2^2/2g + 0 + K_V V^2/2g + (fL/D)V^2/2g$$

$$10 = (V_2^2/2g)(1 + 5 + (0.032 \times 2.6/.026))$$

$$10 = 9.2\ V^2/2g$$

$$V = 4.62\ m/s$$

Check Re: $Re = VD/\nu$

$$= (4.62)(0.026)/(10^{-6})$$

$$Re = 1.20 \times 10^5$$

With $Re = 1.2 \times 10^5$ and $k_s/D = 0.0058$ we find from Fig. 10.8 that f indeed equals 0.032.

Thus $\underline{\underline{V = 4.62\ m/s}}$

Note: When valves are tested to evaluate K_V the pressure taps are usually connected to pipes both upstream and downstream of the valve. Therefore, the head loss in this problem may not actually be $5V^2/2g$. Also, the velocity exiting the valve will probably be highly non-uniform; therefore, this solution should be considered a gross approximation only.

10.60 $h_f = \Delta(p/\gamma + z) = (-20 \times 144/62.4) + 30 = -16.2$ ft

Therefore, flow is from B to A

$$Re\, f^{1/2} = (D^{3/2}/\nu)(2gh_f/L)^{1/2} = (2^{3/2}/(1.41 \times 10^{-5})$$

$$\times (64.4 \times 16.2/(3 \times 5{,}280))^{1/2} = 5.14 \times 10^4$$

$k_s/D = 0.0004$ then $f = 0.0175$

$$V = \sqrt{h_f 2gD/fL} = \sqrt{(16.2 \times 64.4 \times 2)/(0.0175 \times 3 \times 5{,}280)} = 2.74\ ft/s$$

$Q = VA = 2.74 \times (\pi/4) \times 2^2 = \underline{\underline{8.60\ cfs}}$

10.61 $V = Q/A = 0.10/((\pi/4) \times 0.15^2) = 5.66$ m/s; $V^2/2g = 1.63$ m

$k_s/D = 0.0003$ (Fig. 10-9); $Re = VD/\nu = 5.66 \times 0.15/(1.3 \times 10^{-6})$

$$= 6.5 \times 10^5$$

10.61 (continued)

Then f = 0.016 (Fig. 10-8). Now write the energy equation between the two reservoirs:

$$p_1/\gamma + V_1^2/2g + z_1 + h_p = p_2/\gamma + V_2^2/2g + z_2 + \Sigma h_L$$

$$h_p = z_2 - z_1 + V^2/2g(K_e + f(L/D) + K_0) = 13 - 10 + 1.63$$

$$\times (0.1 + 0.016 \times 80/(0.15) + 1) = 3 + 15.7 = 18.7 \text{ m}$$

$$P = Q\gamma h_p = 0.10 \times 9{,}810 \times 18.7 = 18{,}345 \text{ W} = \underline{\underline{18.3 \text{ kW}}}$$

10.62 Write energy equation from upstream reservoir water surface to downstream water surface.

$$p_1/\gamma + V_1^2/2g + z_1 = p_2/\gamma + V_2^2/2g + z_2 + h_L$$

$$0 + 0 + z_1 = 0 + 0 + z_2 + h_f$$

$$100 \text{ m} = (fL/D)V^2/2g$$

$k_s/D = 0.00035$ (Fig. 10.9)

Assume f = 0.016 (from Fig. 10.8)

Then $100 \text{ m} = (0.016 \times 10{,}000/1)V^2/2g$

$V = (100\ (2g)/(160))^{1/2} = 3.50$ m/s

$Re = VD/\nu = (3.50)(1)/(1.31 \times 10^{-6}) = 2.67 \times 10^6$

Check f from Fig. 10.8: f = 0.0155 solve again:

V = 3.55 m/s

$\underline{Q = VA = (3.55)(\pi/4)D^2 = 2.79 \text{ m}^3/\text{s}}$

Riveted Steel: $k_s/D = 0.001$ f = 0.0198

$Q_{R.S}/Q_c = \sqrt{0.0155/0.0198} = 0.885$

$\underline{\underline{Q_{R.S} = 2.47 \text{ m}^3/\text{s}}}$

Pump Power: $h_p = (z_1 - z_2) + h_L$

$= 100 \text{ m} + 100\ (2.8/2.79)^2$

= 201 m

$\underline{\underline{P = Q\gamma h_p = (2.8)(9{,}810)(201) = 5.52 \text{ MW}}}$

10.63

$$p_1/\gamma + V_1^2/2g + z_1 = p_2/\gamma + V_2^2/2g + z_2 + h_f$$

$$150{,}000/(900 \times 9.81) + \cancel{V_1^2/2g} + 0 = 120{,}000/(900 \times 9.81) + \cancel{V_2^2/2g} + 3 + h_f$$

$$h_f = 0.398 \text{ m}$$

$$((D^{3/2})/(\nu)) \times (2gh_f/L)^{1/2} = ((0.08)^{3/2}/10^{-6}) \times (2 \times 9.81 \times 0.398/30)^{1/2}$$

$$= 1.15 \times 10^4$$

$k_s/D = 1.5 \times 10^{-4}/0.08 = 1.9 \times 10^{-3}$; $f = 0.026$ (from Fig. 10-8)

Then $h_f = f(L/D)(V^2/2g)$; $V = \sqrt{(h_f/f)(D/L)2g}$

$$V = \sqrt{(0.398/0.026)(0.08/30) \times 2 \times 9.81} = 0.895 \text{ m/s}$$

$$Q = VA = 0.895 \times (\pi/4) \times (0.08)^2 = \underline{\underline{4.50 \times 10^{-3} \text{m}^3/\text{s}}}$$

10.64 $p_1/\gamma + V_1^2/2g + z_1 + h_p = p_2/\gamma + V_2^2/2g + z_2 + \Sigma h_L$

$$0 + 0 + 100 + h_p = 0 + 0 + 112 + V^2/2g(K_e + fL/D + K_0)$$

$$h_p = 12 + (V^2/2g)(0.03 + fL/D + 1$$

Here $V = Q/A = 0.20/((\pi/4) \times 0.30^2) = 2.83$ m/s; $V^2/2g = 0.408$ m

$$Re = VD/\nu = 2.83 \times 0.30/(10^{-5}) = 8.5 \times 10^4$$

$k_s/D = 4.6 \times 10^{-5}/0.3 = 1.5 \times 10^{-4}$; $f = 0.019$ (from Fig. 10.8)

Then $h_p = 12 + 0.408(0.03 + (0.019 \times 150/0.3) + 1.0) = 16.3$ m

Finally $P = Q\gamma h_p = 0.20 \times (940 \times 9.81) \times 16.3 = 2.67 \times 10^4 \text{W} = \underline{\underline{30.1 \text{ kW}}}$

10.65

$$Re = VD/\nu = 3 \times 0.40/10^{-5} = 1.2 \times 10^5$$

The flow is turbulent and obviously the conduit is very rough ($f = 0.06$); therefore, one would expect f to be virtually constant. Thus, $h_f \alpha V^2$, so if the velocity is doubled, the head loss will be quadrupled.

10.66 Write the energy equation from the water surface in the upper reservoir to the water surface in the lower reservoir:

$$p_1/\gamma + V_1^2/2g + z_1 = p_2/\gamma + V_2^2/2g + z_2 + \Sigma h_L$$

$$0 + 0 + 80 = 0 + 0 + 40 + (K_e + 2K_v + K_E + fL/D)V^2/2g$$

$$80 = 40 + (0.5 + 2 \times 0.2 + 1.0 + f \times 200/1)V^2/2g$$

$k_s/D = 0.0009$ (from Fig. 10-9); Assume $f = 0.020$

Then $V^2/2g = (80-40)/(0.5 + 0.4 + 1.0 + 4.0) = 6.78$ ft

$$V = 20.9 \text{ ft/s}; \quad Re = VD/\nu = 20.9 \times 1/(1.22 \times 10^{-5}) = 1.7 \times 10^6$$

Check f: $f = 0.0195$ (from Fig. 10-8)

Better estimate for V : $V^2/2g = 40/(1.9 + 3.9)$; $V = 21.07$ ft/s

$$Q = VA = 21.07 \times (\pi/4) \times 1^2 = \underline{\underline{16.55 \text{ cfs}}}$$

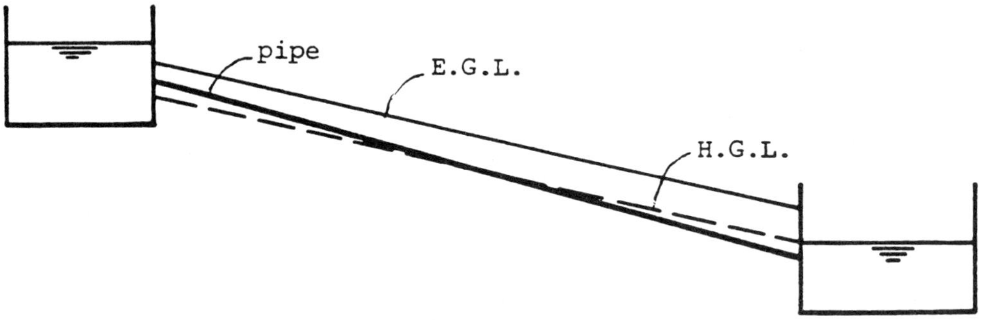

10.67

$$p_1/\gamma + V_1^2/2g + z_1 = p_2/\gamma + V_2^2/2g + z_2 + \Sigma h_L$$

$$0 + 0 + z_1 = 0 + 0 + z_2 + \Sigma h_L$$

$$20 = (V^2/2g)(K_e + f(L/D) + K_0) = V^2/2g(0.5 + f(L/D) + 1.0)$$

$$[(\pi/4)^2 \times 2g \times 20/Q^2] = [1.5 + f(L/D)] \times D^{-4}; \; 7.94 D^4 = (1.5 + fL/D)$$

For first trial assume $f = 0.02$ and neglect minor losses:

$$D^5 = 0.02 \times 2 \times 5{,}280/7.94; \quad D = 1.93 \text{ ft}$$

Then $V = Q/A = 10/((\pi/4) \times (1.93)^2) = 3.43$ ft/sec

and $Re = 3.43 \times 1.93/(1.2 \times 10^{-5}) = 5.5 \times 10^5$; $k_s/D = 0.005$; $f = 0.0175$

2nd trial: $D^5 = 0.0175 \times 2 \times 5{,}280/7.94$; $D = 1.88 \text{ ft} = \underline{\underline{22.5 \text{ in.}}}$

Use next commercial size larger; $\underline{\underline{D = 24 \text{ in.}}}$

10.68 $h_f = 1$ ft ; $L = 1{,}000$ ft ; $Q = 300\ ft^3/s$; $\nu = 1.22 \times 10^{-5} ft^2/s$

First assume $f = 0.015$

Then $h_f = (fL/D)\ V^2/2g$

$= (0.015 \times 1{,}000/D)(Q^2/((\pi/4)^2 D^4)/2g)$

1 ft $= 33{,}984/D^5$; $D = 8.06$ ft ; $k_s/D = 0.00002$

Now get better estimate of f: $Re = 4Q/(\pi D \nu) = 3.9 \times 10^6$

Then $f \approx 0.010$

Compute D again: $1 = 22{,}656/D^5$; $D = 7.43$ ft = 89 in.

Use 90 in. pipe.

10.69 $h_f = f(L/D)V^2/2g = f(L/D)(Q^2/(2gA^2)) = f(L/D)(Q^2/(2g(\pi/4)^2 \times D^4))$

$= fLQ^2/(2g(\pi/4)^2 D^5)$, or $D = [8fLQ^2/(g\pi^2 h_f)]^{1/5}$

Assume $f = 0.015$

$D = [8 \times 0.015 \times 1{,}000 \times 0.1^2/(9.81 \times \pi^2 \times 30)]^{1/5} = 0.21$

Then $k_s/D = 0.0002$ and $Re = 4\ Q/(\pi D \nu) = 4 \times 0.1/(\pi \times 0.21 \times 10^{-5})$

$= 6 \times 10^4$ so $f = 0.022$

Try again: $D = (0.022/0.015)^{1/5} \times 0.21 = 0.227$ m = 22.7 cm

Use next commercial size larger; D = 23 cm

Still assume $h_L \approx 30$ m/1,000 m

Then $P = Q\gamma h_f = 0.1 \times 0.93 \times 9{,}810 \times 30 = 27{,}370W =$ 27.4 kW/km

---

10.70 $Q = 15$ cfs ; $L = 3 \times 5{,}280$ ft ; $h_L = 30$ ft

Energy Eq.: $30 = (K_e + K_E + fL/D)(Q^2/A^2)/2g$

$K_e = 0.5$; $K_E = 1.0$; Assume $f = 0.015$

Then $30 = (1.5 + 0.015 \times 3 \times 5{,}280/D)(Q^2/((\pi/4)^2 D^4)/2g$

$30 = (1.5 + 237.6/D)(15^2/(0.617D^4)/64.4$

$30 = (1.5 + 237.6/D)(5.66/D^4)$

10.70 (continued)

Solving: $D = 2.15$ ft $Re = 4Q/(\pi D\nu) = 7.3 \times 10^5$ for $T = 60°F$

$k_s/D = 0.00007$ (for commercial pipe)

$f = 0.0135$

Solve again: $30 = (1.5 + 214/D)(5.66/D^4)$

$D = 2.10$ ft $= 25.2$ in. Use 26 in. steel pipe. (one possibility)

10.71 First you might consider how to physically hold the disk in the pipe. One way to do this might be to secure the disk to a rod and then secure the rod to streamlined vanes in the pipe such as shown below. The vanes would be attached to the pipe.

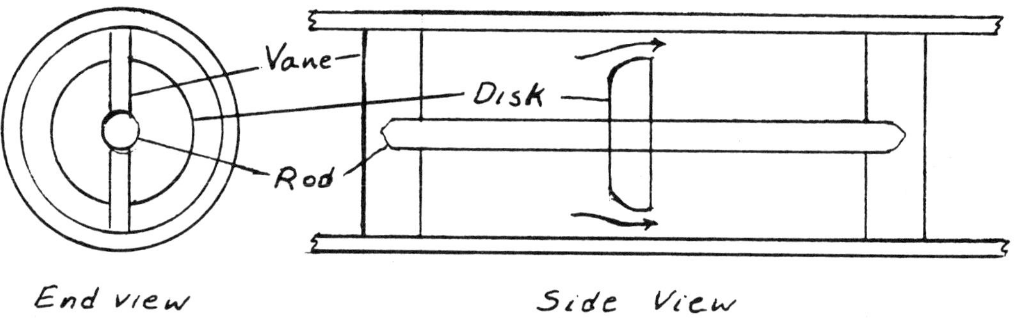

To establish cavitation around the disk, the pressure in the water at this section will have to be equal to the vapor pressure of the water. The designer will have to decide upon the pipe layout in in which the disk is located. It might be something like shown below. By writing the energy equation from the disk section to the pipe outlet one can determine the velocity required at the disk to create vapor pressure at that section.

10.71 (Cont'd)

This calculation will also establish the disk size relative to the pipe diameter. Once these calculations are made, one can calculate the required

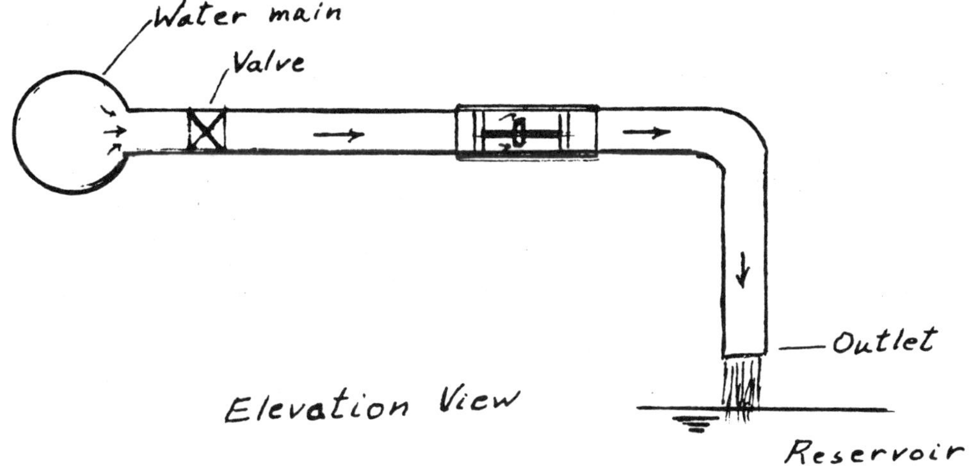

discharge, etc. Once that calculation is made, one can see if there is enough pressure in the water main to yield that discharge with the control valve wide open. If not, re-design the system. If it is OK, then different settings of the control valve will yield different degrees of cavitation.

10.72 Write energy equation from water surface in reservoir to the outlet:

$p_1/\gamma + V_1^2/2g + z_1 = p_2/\gamma + V_2^2/2g + z_2 + h_L$

$0 + 0 + 120 = 0 + V^2/2g + 80 + (K_e + f(L/D))V^2/2g$; $K_e = 0.5$

$(V^2/2g)\ (1.5 + f(L/D)) = 40$ ft

$k_s/D = 0.0008$ (Fig. 10-9) ; $f \approx 0.019$ (guess from Fig. 10-8)

$(V^2/2g)\ (1.5 + (.019 \times 100/0.5)) = 40$; $V = 22.05$ ft/s

Try again: $R_e = VD/\nu = (22.05)\ (0.5)/(1.41 \times 10^{-5}) = 7.8 \times 10^5$

$f = 0.019$ Thus orig. guess was OK ; $V = 22.05$ ft/s

$Q = VA = (22.05)\ (\pi/4)\ (0.5^2) = \underline{\underline{4.33\text{ cfs}}}$

10.73 Write energy equation from water surface in reservoir to the outlet.

$$p_1/\gamma + V_1^2/2g + z_1 = p_2/\gamma + V_2^2/2g + z_2 + \Sigma h_L \quad ; \quad V = Q/A = 50 \text{ ft/s}$$

Assume $K_e = 0.10$; $R_e = VD/\nu = (50)(2)/(1.41 \times 10^{-5}) = 7.1 \times 10^6$

$$0 + 0 + 600 = 0 + V_2^2/2g + 200 + (K_e + f(L/D))V^2/2g$$

$$400 = (V^2/2g)(1.10 + f(1,200/2))$$

$$400 = (50^2/64.4)(1.10 + 600f) \rightarrow \underline{\underline{f = 0.0153}}$$

$$k_s/D = 0.00035 \text{ (From Fig. 10-8)} \rightarrow \underline{\underline{k_s = 0.00070 \text{ ft}}}$$

10.74 $\dot{m} = 50$ kg/s so $\dot{m}$/tube = 0.50 kg/s

and Q/tube = 0.50/860 = 5.8139×10^{-4} m^3/s

$V = Q/A = 5.8139 \times 10^{-4}/((\pi/4) \times (2 \times 10^{-2})^2) = 1.851$ m/s

$Re = VD\rho/\mu = 1.851 \times 0.02 \times 860/(1.35 \times 10^{-4}) = 2.35 \times 10^5$

$k_s/D \simeq 0.006$ (Fig. 10-9) and $f = 0.034$ (Fig. 10-8)

Then $h_f = f(L/D)V^2/2g = 0.034(5/0.02) \times (1.851^2/2 \times 9.81) = 1.48$ m

a) $P = Q\gamma h_f = 5.8139 \times 10^{-4} \times 860 \times 9.81 \times 1.48 \times 100 = \underline{\underline{728 \text{ W}}}$

b) $k_s/D = 0.5/16 = 0.031$ so $f = 0.058$ (Fig. 10-8)

$P = 728 \times (0.058/0.034) \times (20/16)^4 = \underline{\underline{3.03 \text{KW}}}$

10.75

$$p_1/\gamma + V_1^2/2g + z_1 + h_p = p_2/\gamma + V_2^2/2g + z_2 + h_L$$

$V_1 = V_2$; $p_1 = p_2$; $z_2 - z_1 = 0.8$ m

average temperature = 50°C; $\nu = 0.58 \times 10^{-6}$ m^2/s

$V = Q/A = 3 \times 10^{-4}/(\pi/4(0.02)^2) = 0.955$

$Re = VD/\nu = 0.955(0.02)/(0.58 \times 10^{-6}) = 3.3 \times 10^4$; $f = 0.023$

$h_L = (fL/D + 19 K_b)V^2/2g = (0.023(20)/0.02 + 19 \times 0.7)0.955^2/(2(9.81))$

$= 1.69$ m

Note: Examination of the data given indicates that the tubing in the exchanger has an $r/d \approx 1$. Assuming smooth bends of 180°, $K_b \approx 0.7$;

$h_p = z_2 - z_1 + h_L = 0.8 + 1.69 = 2.49$ m

$P = \gamma h_p Q = 9,685 (2.49)\ 3 \times 10^{-4} = \underline{7.23 \text{ W}}$

10.76

a) $k_s/D = 0.004$ Assume $f = 0.028$ and $r/d \approx 2 \rightarrow K_b \simeq 0.2$

$p_1/\gamma + z_1 + V_1^2/2g = p_2/\gamma + z_2 + V_2^2/2g + \Sigma h_L$

$100 = 64 + (V^2/2g)(1 + 0.5 + K_b + f \times L/D)$

$= 64 + (V^2/2g)(1 + 0.5 + 0.2 + 0.028 \times 100/1)$

$36 = (V^2/2g)(4.5)$

$V^2 = 72g/4.5 = 515 \rightarrow V = 22.7$

$Re = 22.7(1)/(1.22 \times 10^{-5}) = 1.9 \times 10^6$; $f = 0.028$

$Q = 22.7(\pi/4)\ 1^2$ $\underline{\underline{17.8\text{ cfs}}}$

b) $V^2/2g = 36/4.5 = 8.0$ ft

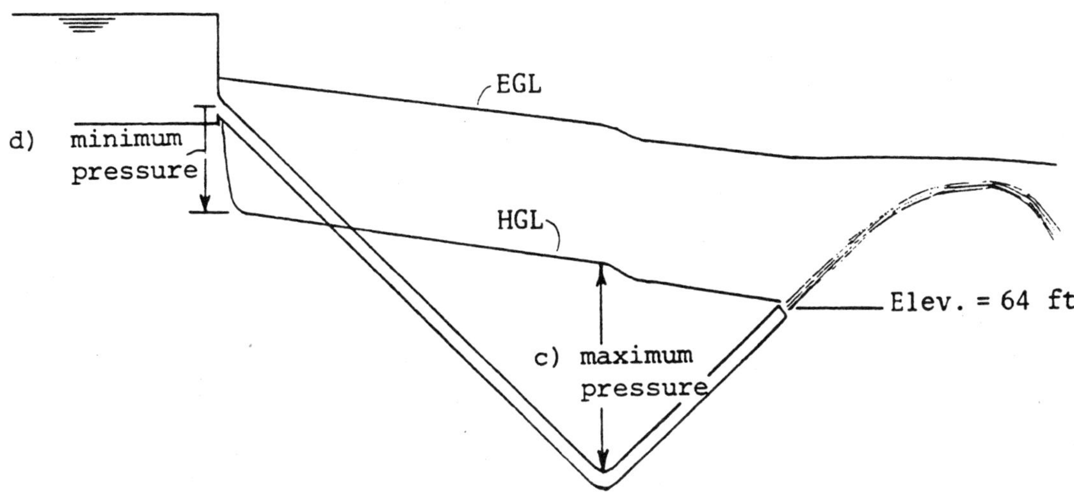

e) $p_{min}/\gamma = 100 - 95 - (V^2/2g)(1 + 0.5) = 5 - 8(1.5) = -7$ ft

$p_{min} = -7(62.4) = -437$ psfg $= \underline{\underline{-3.03\text{ psig}}}$

$p_{max}/\gamma + V_m^2/2g + z_m = p_2/\gamma + z_2 + V_2^2/2g + \Sigma h_L$

$p_{max}/\gamma = 64 - 44 + 8.0(0.2 + 0.028(28/1)) = 27.9$ ft

$p_{max} = 27.9(62.4) = 1{,}739$ psfg $= \underline{\underline{12.1\text{ psig}}}$

10.77

$Q = 0.1 \text{ gpm} = 2.23 \times 10^{-4} \text{cfs}$

$d_1 = (1/4)(1/12) = 0.0208 \text{ ft}; \; d_2 = (1/32)(1/12) = 0.0026 \text{ ft}$

$d_2/d_1 = (1/32)/(1/4) = 0.125;$ Assume $T = 50°F$; sp.gr. = 0.68

$\gamma = 62.4(0.68) = 42.4 \text{ lbf/ft}^3$; From Fig. A-2, $\nu = 5.5 \times 10^{-6} \text{ft}^2/\text{sec}$

$V_1 = Q/A = 2.23 \times 10^{-4}/(\pi/4(1/48)^2) = 0.653; \; V_1^2/2g = 0.00663$

$V_2 = (32/4)^2 \times 0.653 = 4\;\;8; \; V_2^2/2g = 27.15 \text{ ft}$

$Re_1 = V_1 D_1/\nu = 0.653(0.0208)/(5.5 \times 10^{-6}) = 2{,}475; \; f \approx 0.044$

$p_1 = 14.7 \text{ psia}; \; z_2 - z_1 = 2 \text{ ft}; \; p_2 = 14.0 \text{ psia}$

$h_L = (fL/D + 5K_b)\, V_1^2/2g + K_e\, V_2^2/2g$

$= (0.044(10)/0.0208 + 5 \times 0.21)0.00663 = 0.15 \text{ ft}$

$h_p = (p_2 - p_1)/\gamma + z_2 - z_1 + V_2^2/2g + h_L$

$= (14.0 - 14.7)144/42.4 + 2 + 27.15 + 0.15 = 26.9 \text{ ft}$

$P = \gamma h_p Q/(550e) = 42.4(26.9)0.000223/(550 \times 0.8) = \underline{\underline{5.76 \times 10^{-4} \text{hp}}}$

10.78

First find Q for valve wide open. Assume valve is a gate valve.

$p_1/\gamma + V_1^2/2g + z_1 = p_2/\gamma + V_2^2/2g + z_2 + \Sigma h_L$

$2 = 0 + 0 + 0 + (V^2/2g)(0.5 + 0.9 + 0.2 + 0.9 + 1 + fL/D)$

$V^2 = 4g/(3.5 + fL/D)$; Assume $f = 0.015$

Then $V = [4 \times 9.81/(3.5 + 0.015 \times 14/0.1)]^{1/2} = 2.65 \text{ m/s}$

$k_s/D = 0.0005; \; Re = 2.65 \times 0.10/(1.3 \times 10^{-6}) = 2.0 \times 10^5; \; f = 0.019$

Then $V = [4 \times 9.81/(3.5 + 0.019 \times 14/0.10)]^{1/2} = 2.52 \text{ m/s}$

$Re = 2.0 \times 10^5 \times 2.52/2.65 = 1.9 \times 10^5$; O.K.

for $(1/2)Q$: $Re = 9.5 \times 10^4; \; f = 0.021; \; V = 1.26 \text{ m/s}$

So $V^2 = 1.588 = 4 \times 9.81/(3.3 + K_v + 0.021 \times 14/0.1)$

$3.3 + K_v + 2.94 = 24.7; \; \underline{\underline{K_v = 18.5}}$

10.79 $p_1/\gamma + V_1^2/2g + z_1 = p_2/\gamma + V_2^2/2g + z_2 + h_f$

$(300{,}000/9{,}810) + 0 = (60{,}000/9{,}810) + 10 + h_f;\ h_f = 14.46$ m

$f(L/D)(Q^2/A^2)/2g = 14.46$

$f(L/D)[Q^2/((\pi/4)D^2)^2/2g] = 14.46$

$(4^2 fLQ^2/\pi^2)/2gD^5 = 14.46;\quad D = [(8/14.46)fLQ^2/(\pi^2 g)]^{1/5}$

Assume f = 0.020; Then

$D = [(8/14.46) \times 0.02 \times 140 \times (0.025)^2/(\pi^2 \times 9.81)]^{1/5} = 0.100$

Then $k_s/D = 0.002$ and f = 0.024; Try again:

$D = 0.100 \times (0.024/0.020)^{1/5} = \underline{\underline{0.104\text{ m}}}$

Use a <u>12-cm pipe</u>

10.80

$p_1/\gamma + z_1 + V_1^2/2g = p_2/\gamma + z_2 + V_2^2/2g + \Sigma h_L$

$11 = \Sigma h_L = (V_1^2/2g)(Ke + 3K_{b1} + f_1 \times 45/1) + (V_2^2/2g)(K_c + 2K_{b2} + K_0 + f_2 \times 30/(1/2))$

$K_e = 0.5;\ K_0 = 1.0;$ From Table 10-2, $K_{b1} = 0.35;\ K_{b2} = 0.16;\ K_c = 0.39$

Assume $f_1 = 0.015;\ f_2 = 0.016$

$11 \times 2g = V_1^2(0.5 + 3 \times 0.35 + 0.015(45)) + V_2^2(0.39 + 2 \times 0.16 + 1.0 + 0.016(60))$

$708 = V_1^2(2.23) + V_2^2(2.67) = Q^2(2.23/((\pi/4)^2(1)^4) + 2.67/((\pi/4)^2(1/2)^4)) = Q^2(72.9)$

$Q^2 = 708/72.9 = 9.71;\quad Q = 3.12$ cfs

$Re = 4Q/(\pi D\nu)\ ;\ Re_1 = 4(3.12)/(\pi(1.22 \times 10^{-5})) = 3.3 \times 10^5$

$k_s/D = 0.00015;\quad f = 0.016;\ Re_2 = 6.5\ 10^5;\ k_s/D = 0.0003;\ f = 0.016$

So $\underline{\underline{Q = 3.1\text{ cfs}}}$

10.81

Using a pipe diameter of 10 cm,

$p_1/\gamma + z_1 + V_1^2/2g = p_2/\gamma + z_2 + V_2^2/2g + \Sigma h_L$

$0 + 12 + 0 = 0 + 0 + (V^2/2g)(1 + K_e + K_v + 4K_b + f \times L/D)$

$24g = V^2(1 + 0.5 + 10 + 4(0.9) + 0.025 \times 1{,}000/(0.10))$ (assuming f = 0.025)

10.81 (continued)

$V^2 = 24g/265.1 = 0.888\ m^2/s^2$

$V = 0.942 m/s;\ Q = VA = 0.942(\pi/4)(0.10)^2 = 0.0074\ m^3/s;$

$Re = 7 \times 10^4$ and $f \simeq 0.025$

$p_A/\gamma + z_A + V^2/2g = p_2/\gamma + z_2 + V^2/2g + \Sigma h_L$

$p_A/\gamma + 15 = V^2/2g(2K_b + f \times L/D)$

$p_A/\gamma = (0.888/2g)(2 \times 0.9 + 0.025 \times 500/0.10) - 15 = -9.26\ m$

$p_A = 9{,}810 \times (-9.26) = \underline{\underline{-90.8\ kPa}}$

Note: This is not a good installation because the pressure at A is near cavitation level.

10.82 Assume that your vacuum cleaner is a tank type (the type with a hose). One might consider a setup as shown below:

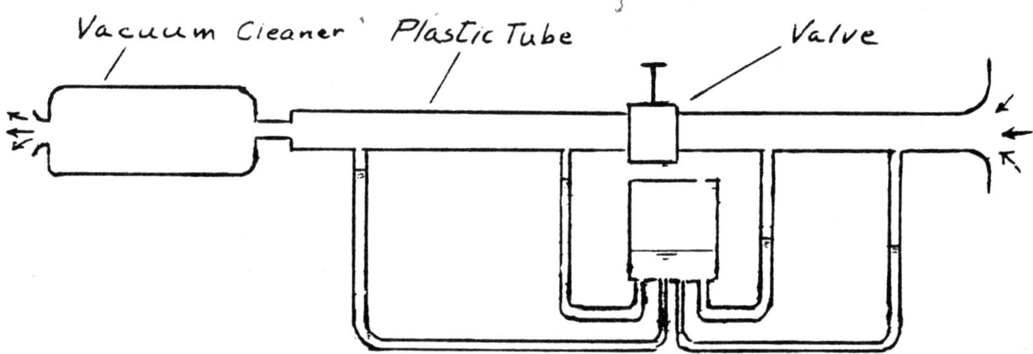

By measuring the height of liquid in each piezometer one may calculate the pressure at each section. One could also measure the pressure at each section by a pressure gage or pressure transducer. Once the pressures at all sections are obtained, one can plot

10.82 (Cont'd)

pressure vs. distance along the tube. It might appear as shown below:

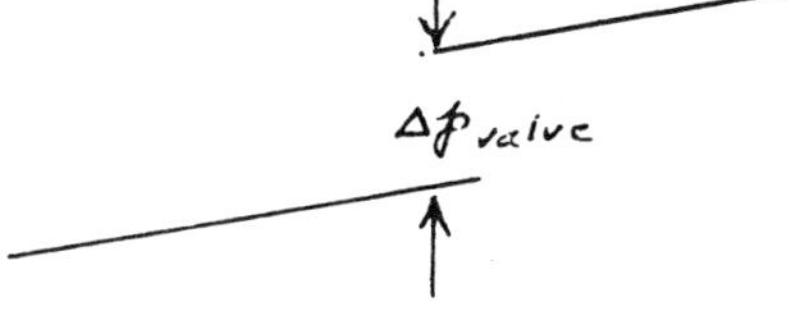

By extending the slopes of p vs. distance one can determine the Δp for the valve.

The air density can be determined by measuring the temperature and then picking ρ from the appropriate table or by solving for it by the equation of state. The discharge can be determined by taking a pitot tube velocity traverse across the section of the pipe and integrating that to obtain Q. Once the discharge is determined, the mean velocity can be calculated from $V = Q/A$. Then the loss coefficient can be calculated using Eq. (10.29):

$$\gamma h_L = K\gamma V^2/2g$$

$$\Delta p_f = K\rho V^2/2$$

or $$K = 2\Delta p/\rho V^2$$

10.84 Write the energy equation from the water surface in the lower reservoir to the water surface in the upper reservoir.

$$p_1/\gamma + V_1^2/2g + z_1 + h_p = p_2/\gamma + V_2^2/2g + z_2 + \Sigma h_L$$

$$0 + 0 + 200\text{m} + h_p = 0 + 0 + 235\text{m} + (V^2/2g)(K_e + K_b + K_E + fL/D)$$

$$V = Q/A = 0.314/((\pi/4) \times 0.2^2) = 10.0 \text{ m/s}; \; V^2/2g = 5.10 \text{ m}$$

$$K_e = 0.03; K_b = 0.35; K_E = 1.0; \; Re = VD/\nu = 10 \times 0.20/10^{-6} = 2 \times 10^6$$

$$k_s/D \simeq 0.00025 \quad \text{so} \quad f = 0.015 \quad fL/D = 0.015 \times 140/0.2 = 10.50$$

$$h_p = 235 - 200 + 5.10(0.03 + 0.35 + 1 + 10.5) = 95.6 \text{ m}$$

$$P = Q\gamma h_p = 0.314 \times 9,790 \times 95.6 = \underline{\underline{294 \text{ kW}}}$$

10.85 From the solution to Prob. 10.84: $h_p = 35 + 12.35\ V^2/2g$

$h_p = 35 + 12.35[(Q/((\pi/4) \times 0.2^2))^2/2g] = 35 + 638Q^2$

System data computed and shown below:

$Q(m^3s)$ →	0.05	0.10	0.15	0.20
$h_p(m)$ →	36.6	41.4	49.3	60.5

Then, plotting the system curve on the pump performance curve of Fig. 10-17 yields $Q = 0.17\ m^3/s$ for the operating point.

10.86

For the system to operate as a pump, the increase in head produced by the jet must be greater than 9 ft (the difference in elevation between the lower and upper reservoir). Consider the head change between a section just to the right of the jet and far to the right of it with zero flow in the lower pipe. Determine this head change by applying the momentum equation:

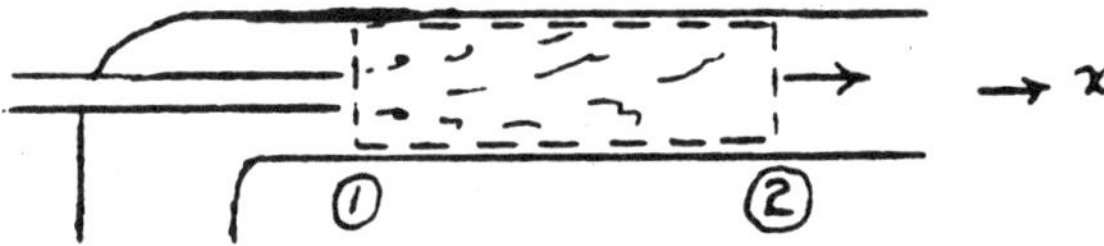

$V_1 = 60$ ft/s $\qquad Q = V_1A_1 = 2.94$ cfs

$V_2 = Q/A_2 = (60)\ (\pi/4)\ (3^2)/((\pi/4)\ (12^2))$

$V_2 = 60\ (3^2/12^2) = 3.75$ ft/s

$\Sigma F_x = \Sigma V_x \rho \mathbf{V}\cdot\mathbf{A}$

$p_1A_1 - p_2A_2 = (60)(1.94)(-60 \times (\pi/4)(1/4)^2) + (3.75)(1.94)(3.75 \times (\pi/4)(1^2))$

$A(p_1 - p_2) = 1.94\ (-176.7 + 11.04)$

$p_2 - p_1 = 409.2$ psf

$h_2 - h_1 = (409.2\ \text{lbf/ft}^2)/(62.4\ \text{lb/ft}^3) = 6.56$ ft

The change in head of 6.56 ft. is not enough to overcome the static head of 9.0 ft.; therefore, the system will not act as a pump.

10.87 $p_1/\gamma + V_1^2/2g + z_1 + h_p = p_2/\gamma + V_2^2/2g + z_2 + \Sigma h_L$

$0 + 0 + 10 + h_p = 0 + 0 + 20 + V_2^2/2g(K_e + fL/D + k_0)$

$h_p = 10 + (Q^2/(2gA^2))(0.1 + 0.02 \times 1{,}000/(10/12) + 1)$

$A = (\pi/4) \times (10/12)^2 = 0.545\ \text{ft}^2$

10.87 (continued)

$h_p = 10 + 1.31\ Q^2_{cfs}$ but $Q_{cfs} = 449$ gpm

$h_p = 10 + 1.31\ Q^2_{gpm}/(449)^2$

$h_p = 10 + 6.51 \times 10^{-6} Q^2_{gpm}$

Q →	1,000	2,000	3,000
h →	16.5	36.0	68.6

Plotting this on pump curve figure yields $Q \simeq$ 2,950 gpm

---

10.88 h_p = 20 ft - 10 ft = 10 ft

Then from the pump curve for Prob. 10.87 we get Q = 4,700 gpm.

10.89

$p_1/\gamma + V_1^2/2g + z_1 + h_p = p_2/\gamma + V_2^2/2g + z_2 + \Sigma h_L$

$0 + 0 + 100 + h_p = 0 + V_2^2/2g + 150 + V_2^2/2g\,(0.03 + fL/D)$

$V_2 = Q/A_p = 15/((\pi/4) \times 1.5^2) = 8.49$ m/s

$Re = VD/\nu = 8.49 \times 1.5/(1.31 \times 10^{-6}) = 9.72 \times 10^6;\ k_s/D = 0.000035$

$f = 0.0105$ (from Fig. 10-8)

Then $h_p = 140 - 100 + V_2^2/2g\,(1.03 + 0.0105 \times 300/1.5) =$

$40 + 11.5 = 51.5$ m

$P = Q\gamma h_p = 15 \times 9{,}810 \times 51.5 =$ 7.58 MW

10.90 $k_s/D_{15} = 0.1/150 = 0.00067;\ k_s/D_{30} = 0.1/300 = 0.00033$

$V_{15} = Q/A_{15} = 0.1/((\pi/4) \times 0.15^2) = 5.659$ m/s; $V_{30} = 1.415$ m/s

$Re_{15} = VD/\nu = 5.659 \times 0.15/10^{-6} = 8.49 \times 10^5;\ Re_{30} = 1.415 \times 0.3/10^{-6}$

$= 4.24 \times 10^5$

From Fig. 10-8 $f_{15} = 0.0185;\ f_{30} = 0.0165$

Energy Eq: $z_1 - z_2 = \Sigma h_L$

10.90 (continued)

$$z_1 - z_2 = (V_{15}^2/2g)(0.5 + 0.0185 \times 50/0.15) + (V_{30}^2/2g)(1 + 0.0165 \times 160/0.30)$$
$$+ (V_{15} - V_{30})^2/2g$$

$$z_1 - z_2 = (5.659^2/(2 \times 9.81))(6.67) + (1.415^2/(2 \times 9.81))(9.80)$$
$$+ (5.659 - 1.415)^2/(2 \times 9.81)$$

$$z_1 - z_2 = 1.632(6.67) + 1.00 + 0.918 = \underline{\underline{12.80\text{ m}}}$$

10.91 $K_e = 0.5$ and $K_E = 1.0$ (From Table 10.2) $20°C = 9/5 \times 20 + 32 = 68°F$; $\nu = 1.1 \times 10^{-5}$ ft^2/s. Write the energy equation from the water surface in the tank at the left to the water surface in the tank on the right.

$$p_1/\gamma + V_1^2/2g + z_1 = p_2/\gamma + V_2^2/2g + z_2 + \sum h_L$$

$$z_1 = z_2 + (K_e + f_1L_1/D_1)V_1^2/2g + (V_1 - V_2)^2/2g + ((f_2L_2/D_2) + K_E)V_2^2/2g$$

where

$$V_1 = Q/A_1 = Q/((\pi/4)(1/2)^2) = 20.37\text{ft}/\text{s}$$

$$\text{Re}_1 = 20.37 \times (1/2)/(1.1 \times 10^{-5}) = 9.3 \times 10^6$$

$$V_1^2/2g = 6.44\text{ft}$$

$$V_2 = V_1/4 = 5.09\text{ft}/\text{s}; \text{Re}_2 = 5.09 \times 1/1.1 = 4.6 \times 10^6$$

$$V_2^2/2g = 0.403\text{ft}$$

$$k_s/D_1 = 4 \times 10^{-4}/0.5 = 8 \times 10^{-5}; f_1 = 0.012 \text{ (From Fig. 10.8)}$$

$$k_s/D_2 = 4 \times 10^{-4}; f_2 = 0.011 \text{ (From Fig. 10.8)}$$

$$z_1 - z_2 = h = (0.5 + .012 \times 150/(1/2))6.44 + (20.37 - 5.09)^2/64.4$$
$$+ ((0.011 \times 500/1) + 1)0.403$$
$$= 26.4 + 3.63 + 5.90$$
$$= \underline{\underline{30.0\text{ft}}}$$

10.92 Write energy equation from reservoir water surface to pipe outlet:

$$p_1/\gamma + V_1^2/2g + z_1 = p_2/\gamma + V_2^2/2g + z_2 + \Sigma h_L$$

$$0 + 0 + 100\text{ ft} = 0 + V_2^2/2g + 64 + (V^2/2g)(K_e + K_v + fL/D)$$

$K_e = 0.50$; $K_v = 5.6$; assume $f = 0.015$ for first trial

Then $(V^2/2g)(0.5 + 5.6 + 1 + 0.015 \times 300/1) = 36$

$V = 14.1$ ft/s $\quad$ $\text{Re} = VD/\nu = 14.1 \times 1/10^{-4} = 1.4 \times 10^5$

$k_s/D = 0.00015$ $\quad$ so $f \simeq 0.0175$

Second trial: $V = 13.7$ ft/s; $\text{Re} = 1.37 \times 10^5$; $f = 0.0175$

$Q = VA = 13.7 \times (\pi/4) \times 1^2 = \underline{\underline{10.8\text{ ft}^3/\text{s}}}$

10.93 $p_1/\gamma + z_1 + V_1^2/2g + h_p = p_2/\gamma + V_2^2/2g + z_2 + \Sigma h_L$

$0 + 30 + 0 + h_p = 0 + 60 + (V^2/2g)(1 + 0.5 + 4K_b + fL/D)$

$V = Q/A = 2.0/((\pi/4) \times (1/2)^2) = 10.18$ ft/sec; $V^2/2g = 1.611$ ft

$r/d = 12/6 = 2 \rightarrow K_b = 0.19$; $Re = 4Q/(\pi D\nu) = 4 \times 2/(\pi \times (1/2) \times 1.22 \times 10^{-5})$

$= 4.17 \times 10^5 \rightarrow f = 0.0135$

$h_p = 30 + 1.611(1 + 0.5 + 4 \times 0.19 + 0.0135 \times 1{,}700/(1/2)) = 107.6$ ft

$P = Q\gamma h_p/550 =$ 24.4 horsepower

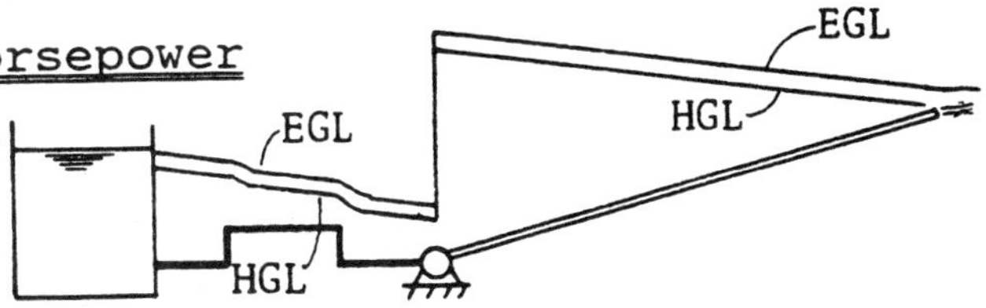

$p_m/\gamma + z_m = z_2 + h_L$

$p_m = \gamma[(z_2 - z_m) + h_L]$

$p_m = 62.4[(60 - 35) + 0.0135 \times (600/0.5) \times 1.611]$

$p_m = 3{,}189$ psf = 22.1 psig

10.94

$p_1/\gamma + V_1^2/2g + z_1 + h_p = p_2/\gamma + V_2^2/2g + z_2 + \Sigma h_L$

$0 + 0 + 20 + h_p = 0 + 0 + 40 + V^2/2g(K_e + 2K_b + K_0 + fL/D)$

$h_p = 20 + V^2/2g(0.5 + 2 \times 0.19 + 1 + fL/D)$

$V = Q/A = 1/((\pi/4 \times 0.6^2) = 3.54$ m/s; $V^2/2g = 0.639$ m

$Re = VD/\nu = 3.54 \times 0.6/(5 \times 10^{-5}) = 4.25 \times 10^4$; $k_s/D = 0.00009$

$f = 0.021$ (from Fig. 10-8);

$h_p = 20 + 0.639(0.5 + 0.38 + 1 + 6.65) = 25.4$ m

Then $P = Q\gamma h_p = 1 \times 0.94 \times 9{,}810 \times 25.4/0.70 =$ 335 KW

10.95

$p_1/\gamma + V_1^2/2g + z_1 = p_2/\gamma + V_2^2/2g + z_2 + \Sigma h_L$

$0 + 0 + z_1 = 0 + 0 + 12 + (V_{30}^2/2g)(0.5 + fL/D) + (V_{15}^2/2g)(K_C + f(L/D) + 1.0)$

$V_{30} = Q/A_{30} = 0.15/((\pi/4) \times 0.30^2) = 2.12$ m/s; $V_{30}^2/2g = 0.229$ m

10.95 (Cont'd)

$V_{15} = 4V_{30} = 8.488$ m/s; $V_{15}^2/2g = 3.67$ m; $D_2/D_1 = 15/30 = 0.5 \rightarrow K_c = 0.37$

Then $z_1 = 12 + 0.229[0.5 + 0.02 \times (20/0.3)] + 3.67[0.37 + 0.02(10/0.15) + 1.0]$

z_1 <u>22.3 m</u>

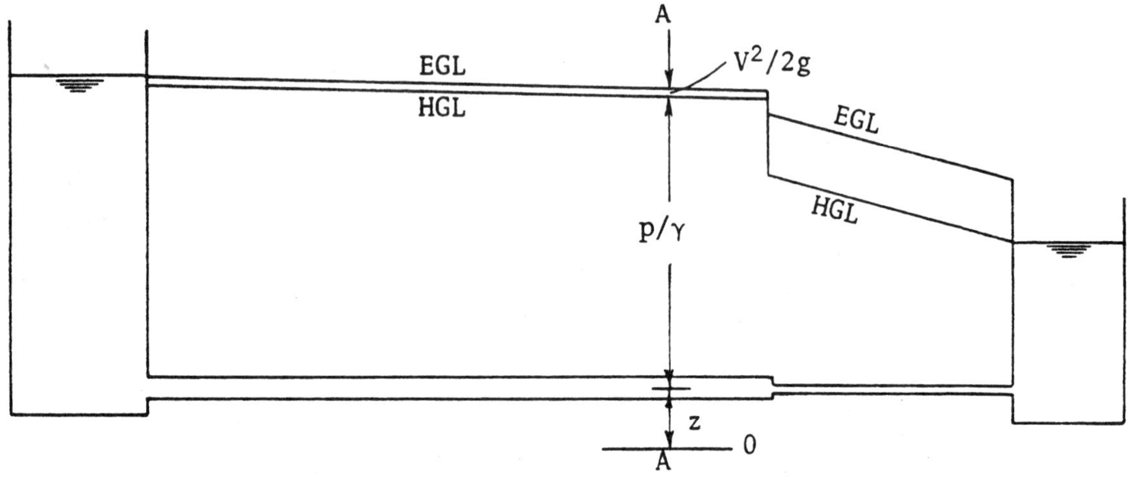

10.96

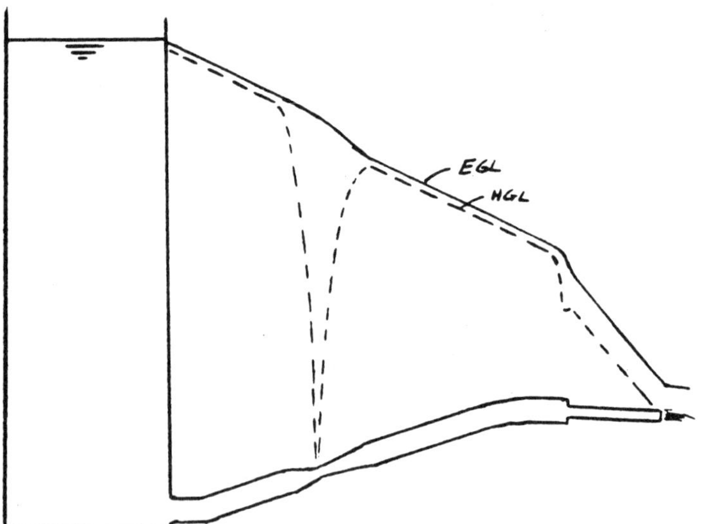

Cavitation could occur in the venturi throat section or just downstream of the abrupt contraction (where there will be a contraction of the flow area).

10.97 $p_1/\gamma + z_1 + V_1^2/2g = p_2/\gamma + z_2 + V_2^2/2g + \Sigma h_L$

$p_1/\gamma + 20 + 0 = 0 + 90 + 0 + V^2/2g(0.5 + 2K_b + K_v + f\ (L/D) + 1)$

$k_b = 0.9$, $K_v = 10$ from Table 10-2

$V = Q/A = (45/449)/((\pi/4)(2/12)^2) = 4.59$; $V^2/2g = 4.59^2/64.4 = 0.328$

10.97 (Cont'd)

$Re = 4.59(2/12)/(1.41 \times 10^{-5}) = 5.4 \times 10^4 \quad k_s/D = 0.003 \quad f = 0.029$

$p_1 = \gamma[70 + 0.328(0.5 + 2 \times 0.9 + 10 + (0.029 \times 240/(2/12)) + 1.0)]$

$= 62.4\ (88.1) = 5{,}497$ psfg $=$ 38.2 psig

10.98

$k_s/D_{20} = 0.0015$; $k_s/D_{15} = 0.0018$ (Fig. 10-9)

$V_{20} = Q/A_{20} = 0.03/((\pi/4) \times 0.20^2) = 0.955$ m/s; $Q/A_{15} = 1.697$ m/s

$Re_{20} = VD/\nu = 0.955 \times 0.2/(1.3 \times 10^{-6}) = 1.5 \times 10^5$

$Re_{15} = 1.697 \times 0.15/1.3 \times 10^{-6} = 1.9 \times 10^5$

From Fig. 10-8: $f_{20} = 0.022$; $f_{15} = 0.024$

$z_1 = z_2 + \Sigma h_L$; $z_1 = 100 + \Sigma h_L$

$z_1 = 110 + V_{20}^2/2g(0.5 + 0.022 \times 100/0.2 + 0.19) + V_{15}^2/2g[(0.024 \times 150/0.15)$

$+ 1.0 + 0.19)] = 110 + 0.0465(11.5) + 0.1468(25.19)$

$= 110 + 0.535 + 3.70 =$ 114.2 m

10.99

One possible design given below:

$L \approx 300 + 50 + 50 = 400$ m; $K_b = 0.19$

$50 = \Sigma h_L = V^2/2g(K_e + 2K_b + f(L/D) + 1.0) = V^2/2g(1.88 + f(L/D))$

$50 = [Q^2/(2gA^2)](f(L/D) + 1.88) = [2.5^2/(2 \times 9.81 \times A^2)]((400\ f/D) + 1.88)$

Assume $f = 0.015$. Then $50 = [0.318/((\pi/4)^2 \times D^4)](0.015 \times (400/D)) + 1.88)$

Solving, one gets $D \approx 0.59$ m $= 59$ cm. Try commercial size $D = 60$ cm.

Then $V_{60} = 2.5/((\pi/4) \times 0.6^2) = 8.84$ m/s

$Re = 8.8 \times 0.6/10^{-6} = 5.3 \times 10^6$; $k_s/D = 0.0001$ and $f \approx 0.013$

10.99 (Cont'd)

Since $f = 0.13$ is less than originally assumed f, the design is conservative. So use D = 60 cm and L ≈ 400 m.

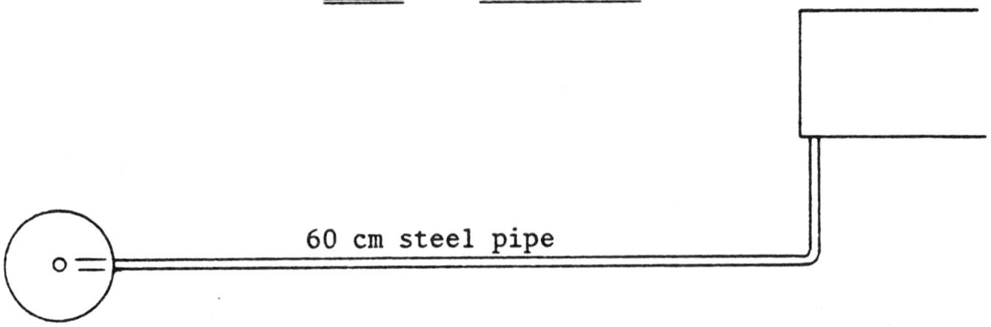

60 cm steel pipe

10.100

First write the energy equation from the reservoir to the tank and assume that the same pipe configuration as used in the solution to P10-99 is used. Also a pump, two open gate valves, and two bends will be in the pipe system. Assume steel pipe will be used.

Assume $L \simeq 400$ ft.

$$p_1/\gamma + V_1^2/2g + z_1 + h_p = p_2/\gamma + V_2^2/2g + z_2 + \Sigma h_L$$

$$0 + 0 + 450 + h_p = 0 + 0 + 500 + (V^2/2g)(K_e + 2K_b + 2K_v + K_E + fL/D)$$

Assume $V \simeq 2$ m/s; $A = Q/V = 1.0/2 = 0.50\text{m}^2$

$A = (\pi/4)D^2 = 0.50$ or $D = .799$ m Choose a pipe size of 0.80m

Then $V = Q/A = 1.0/((\pi/4) \times 0.8^2) = 1.99$ m/s and $V^2/2g = 0.202$m

$k_s/D = 0.00006$; $Re = VD/\nu = 1.6/10^{-6} = 1.6 \times 10^6$

Then $f = 0.012$ (from Fig. 10-8)

$$h_p = 50 + (V^2/2g)(0.5 + 2 \times 0.2 + 2 \times 0.19 + 1.0 + 0.012 \times 400/1)$$

$$= 50 + 1.43 = 51.43\text{m}$$

$$P = Q\gamma h_p = 2.0 \times 9{,}810 \times 51.43$$

$$= 1.01 \text{ MW}$$

Design will include 0.80m steel pipe and a pump with output of 1.01 MW

Note: An infinite number of other designs is possible. Also, a design solution would include the economics of the problem to achieve the desired result at minimum cost.

10.101 Because you want to design equipment to illustrate cavitation, it would be desirable to make the flow restriction device from clear plastic so that one may observe the formation of cavitation bubbles. The design calculation for pressure and discharge would be the same as given for Prob. 10.71.

10.102 The equipment for the momentum experiment could be shown below:

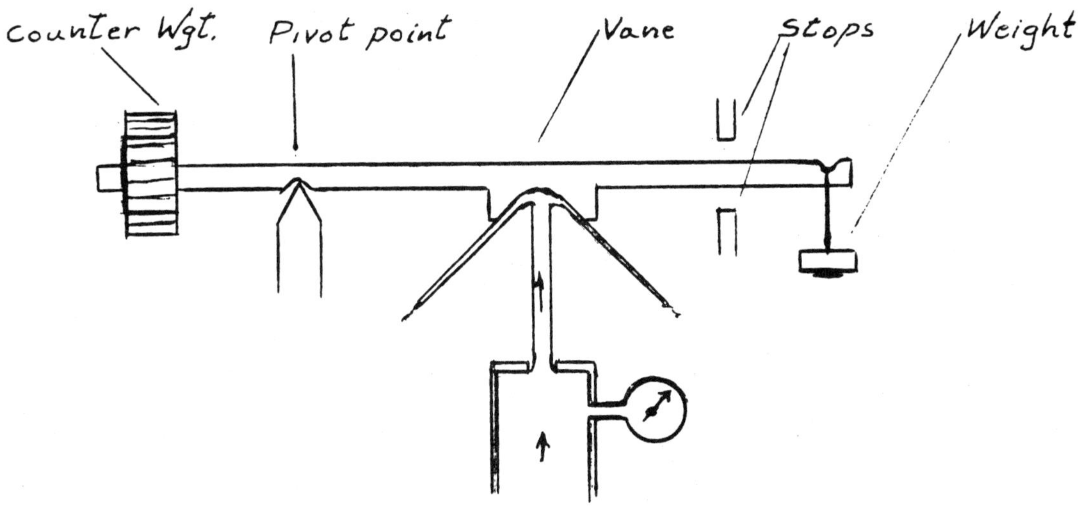

Necessary measurements and calculations:

A. Discharge. This could be done by using weight scales and tank to weigh the flow of water that has occurred over a given period of time.

B. The velocity in the jet could be measured by means of a stagnation tube or solving for the velocity by using Bernoulli's equation given the pressure in the nozzle from which the jet issues.

C. Initially set the counter balance so that the beam is in its horizontal equilibrium position. By opening a valve establish the jet of water. Apply

10.102 (Cont'd)

necessary weight at the end of the beam balance to bring the beam back to horizontal equilibrium. By calculation (using moment summation) determine the force that the jet is exerting on the vane. Compare this force with the calculated force from the momentum equation (using measured Q, V, and vane angle).

10.103

$$h_{LA} = h_{LB}$$

$$0.2V_A^2/2g = 10V_B^2/2g \qquad (1)$$

$$V_A = \sqrt{50}V_B$$

$$Q_B/Q_A = V_BA_B/V_AA_A$$

$$= V_BA_B/V_A((1/4)A_B) \qquad (2)$$

$$Q_B/Q_A = 4V_B/V_A$$

Solve Eqs. (1) and (2) for Q_B/Q_A:

$$Q_B/Q_A = 4 \times V_B/\sqrt{50}V_B$$

$$\underline{\underline{= 0.566}}$$

10.104

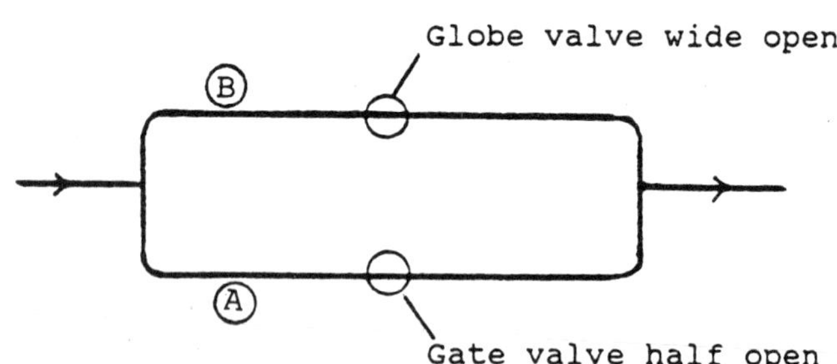

$\Sigma h_{LB} = \Sigma h_{LA}$

$h_{L,globe} + 2h_{L,elbow} = h_{L,gate} + 2h_{L,elbow}$

$10V_B^2/2g + 2(0.9V_B^2/2g) = 5.6V_A^2/2g + 2\ (0.9V_A^2/2g)$

$11.8V_B^2/2g = 7.4V_A^2/2g$

$\underline{V_A/V_B = 1.26}$

10.105 $V_1/V_2 = [(f_2/f_1)(L_2/L_1)(D_1/D_2)]^{1/2}$

Initially assume $f_1 = f_2$

Then $V_1/V_2 = [(1,500/1,000)(0.50/0.40)]^{1/2} = 1.369$; $V_1 = 1.369\ V_2$

$V_1A_1 + V_2A_2 = 1$

$1.369\,V_2 \times (\pi/4) \times 0.5^2 + V_2 \times (\pi/4) \times 0.4^2 = 1$

$V_2 = 2.535$ m/s; Then $V_1 = 1.369 \times 2.535 = 3.47$ m/s

$Q_1 = V_1A_1 = 3.47(\pi/4) \times 0.5^2 = 0.68\ m^3/s$; $Q_2 = 0.32 m^3/s$

10.106 $h_{f,1} = h_{f,2}$; Let length of pipe 2 be 4L

Then $f(L/D)\ (V_1^2/2g) = f(4L/D)\ (V_2^2/2g)$

$$V_1^2 = 4V_2^2 \rightarrow V_1 = 2V_2$$

Thus $Q_1 = 2Q_2 = 2$ cfs

10.107

$$h_{p,A} = h_{f,B} = h_{f,C}$$

$$f(L/D)\ (V^2/2g)_A = f(L/D)\ (V^2/2g)_B = f(L/D)\ (V^2/2g)_C$$

$$0.012\ (6,000/1.5)V_A^2 = 0.02\ (2,000/.5)V_B^2 = .015\ (5,000)V_C^2$$

$$48V_A^2 = 80V_B^2 = 75V_C^2$$

Therefore, V_A will have the greatest velocity.

10.108 $(V_1/V_2) = [(f_2/f_1)(L_2/L_1)(D_1/D_2)]^{1/2}$

Let pipe 1 be large pipe and pipe 2 be smaller pipe

Then $(V_1/V_2) = [(0.013/0.01)(L/3L)(2D/D)]^{1/2} = 0.931$

$(Q_1/Q_2) = (V_1/V_2)(A_1/A_2) = 0.931 \times (2D/D)^2 = 3.72$

$(Q_{large}/Q_{small}) = 3.72$

10.109

$Q_{18} + Q_{12} = 14$ cfs

$h_{L_{18}} = h_{L_{12}}$; $f_{18}(L_{18}/D_{18})(V_{18}^2/2g) = f_{12}(L_{12}/D_{12})(V_{12}^2/2g)$

$f_{18} = 0.018 = f_{12}$ so $L_{18}Q_{18}^2/D_{18}^5 = L_{12}Q_{12}^2/D_{12}^5$

$Q_{18}^2 = (D_{18}/D_{12})^5(L_{12}/L_{18})Q_{12}^2 = (18/12)^5(2{,}000/6{,}000)Q_{12}^2 = 2.53Q_{12}^2$

$Q_{18} = 1.59\ Q_{12}$; $1.59\ Q_{12} + Q_{12} = 14$

$2.59\ Q_{12} = 14$; $\underline{\underline{Q_{12} = 5.4 \text{ cfs}}}$

$Q_{18} = 1.59\ Q_{12} = 1.59(5.4) = \underline{\underline{8.6 \text{ cfs}}}$

$V_{12} = 5.4/((\pi/4(1)^2) = 6.88$ $V_{18} = 8.6/((\pi/4)(18/12)^2) = 4.87$

$h_{L_{12}} = 0.018((2{,}000)/1)(6.88)^2/64.4 = 26.5$

$h_{L_{18}} = 0.018(6{,}000/1.5)(4.87^2/64.4) = 26.5$

Thus, $h_{L_{A-B}} = \underline{\underline{26.5 \text{ ft}}}$

10.110

$Q = Q_{14} + Q_{12} + Q_{16}$

$20 = V_{14} \times (\pi/4) \times (14/12)^2 + V_{12} \times (\pi/4) \times 1^2 + V_{16} \times (\pi/4) \times (16/12)^2$; (1)

Also, $h_{f_{14}} = h_{f_{12}} = h_{f_{16}}$ and assuming f = 0.03 for all pipes

$(3{,}000/14)\ V_{14}^2 = (2{,}000/12)\ V_{12}^2 = (3{,}000/16)\ V_{16}^2$ (2)

$V_{14}^2 = 0.778\ V_{12}^2 = 0.875\ V_{16}^2$

From Eq(1) $20 = 1.069\ V_{14} + 0.890\ V_{14} + 1.49\ V_{14}$; $V_{14} = 5.79$ ft/s

and $V_{12} = 1.134\ V_{14} = 6.56$ ft/s; $V_{16} = 6.19$ ft/s

$\underline{\underline{Q_{12} = 5.15 \text{ ft}^3\text{/sec}}}$; $\underline{\underline{Q_{14} = 6.19 \text{ ft}^3\text{/s}}}$; $\underline{\underline{Q_{16} = 8.64 \text{ ft}^3\text{/s}}}$

$V_{24} = Q/A_{24} = 20/(\pi/4 \times 2^2) = 6.37$ ft/s;

$V_{30} = 4.07$ ft/s

$h_{LAB} = (0.03/64.4)[(2{,}000/2.00)(6.37)^2 + (2{,}000/1) \times (6.56)^2$

$+ (3{,}000/(30/12) \times (4.07)^2] = \underline{\underline{68.3 \text{ ft}}}$

10.111

Call pipe A-B pipe 1
Call pipe ACB pipe 2

Then $h_{f,1} + h_p = h_{f,2}$; $d_s/D = .0001$

Assume $f_1 = f_2 = 0.013$ (guess from Fig. 10-8)

$$f(L_1/D_1)\ (V_1^2/2g) + h_p = f(L_2/D_2)\ (V_2^2/2g)$$

$$0.013(2{,}000/0.5)\ (V_1^2/2g) + h_p = 0.013(6{,}000/0.5)\ (V_2^2/2g)$$

$$2.65\ V_1^2 + h_p = 7.951\ V_2^2 \qquad (1)$$

Continuity: $(V_1 + V_2)A = 0.60\ m^3/s$

$$V_1 + V_2 = 0.60/A = 0.6/((\pi/4)\ (0.5^2)) = 3.0558$$

$$V_1 = 3.0558 - V_2 \qquad (2)$$

By iteration (Eqs. (1), (2) and pump curve) one can solve for the division of flow:

$Q_1 = 0.27\ m^3/s$ $\qquad Q_2 = 0.33\ m^3/s$

10.112

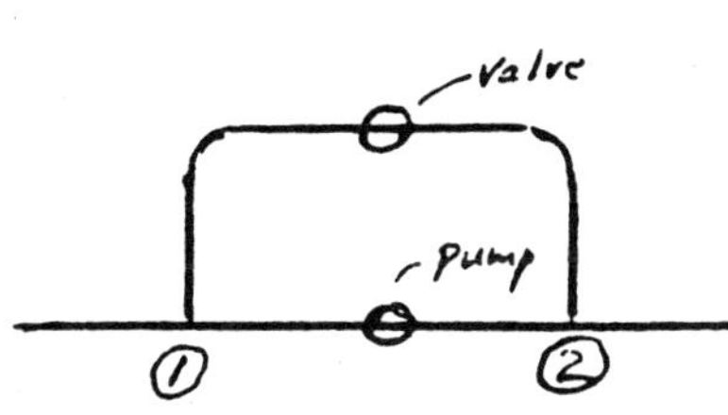

$$Q_p = Q_v + 0.2$$

$$(p_2 - p_1)/\gamma = h_p$$

$$A = (\pi/4)\ (0.1^2) = 0.00785\ m^2$$

$$K_v V_v^2/2g = K_v Q_v^2/(2gA^2) = h_p = 100 - 100(Q_v + 0.2)$$

$$(0.2)\ (Q_v^2)/(2 \times 9.81 \times (0.00785)^2) = 100 - 100\ Q_v - 20$$

$$165\ Q_v^2 + 100\ Q_v - 80 = 0 \leftarrow \text{Solve by quadratic formula}$$

$Q_v = 0.456\ m^3/s$; $Q_p = 0.456 + 0.2 = 0.656\ m^3/s$

10.113 $\quad R_h = A/P$

$R_{h,A} = (A/P)_A = 16/16 = 1$

$R_{h,W} = (A/P)_W = 8/8 = 1$ $\qquad \therefore\ R_{h,A} = R_{h,W}$

The correct choice is (a).

10.114

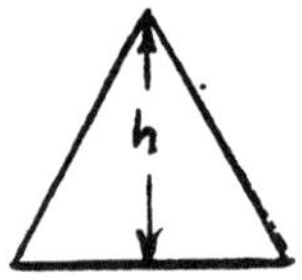

$h = (6 \text{ in})(\cos 30°) = 5.20$ 	$T = 60°F$

$A = (6)(5.20)/2 = 15.6 \text{ in}^2 = 0.108 \text{ ft}^2$ 	$\nu = 1.58 \times 10^{-4} \text{ ft}^2/\text{s}$

$R_h = A/P = 15.6 \text{ in}^2/(3 \times 6) = 0.867 \text{ in.}$ 	$\rho = 0.00237 \text{ slug/ft}^3$

$4R_h = 3.47 \text{ in.} = 0.289 \text{ ft.}$

$k_s/4R_h = 0.0005 \text{ ft}/0.289 \text{ ft.} = 0.00173$

$Re = (V)(4R_h)/\nu = (10)(0.289)/(1.58 \times 10^{-4}) = 1.83 \times 10^4$

$f = 0.030$ (from Fig. 10-8)

$\Delta p_f = (f(L/4R_h)(\rho V^2/2)$

$\Delta p_f = 0.030\,(100/0.289)(0.00237 \times 10^2/2)$

$\Delta p_f = \underline{1.23 \text{ lbf/ft}^2}$

10.115

$Q = (1.49/n)A\,R_h^{2/3}\,S^{1/2}$

$Q_A/Q_B = R_h^{2/3}{}_{,A}/R_h^{2/3}{}_{,B} = (R_{h,A}/R_{h,B})^{2/3}$

where $R_{h,A} = 50/20 = 2.5$; $R_{h,B} = 50/(3 \times 7.07) = 2.36$

$R_{h,A} > R_{h,B}$ $\therefore \underline{Q_A > Q_B}$. The correct choice is (c).

10.116 $A = 0.15 \text{ m}^2$; $P = 2.30$; $R = A/P = 0.0652 \text{ m}$; $4R = 0.261 \text{ m}$

Assume $k_s = .15 \text{ mm} = 1.5 \times 10^{-4} \text{ m}$

$k_s/4R = 1.5 \times 10^{-4}/0.261 = 5.7 \times 10^{-4}$;

$V = Q/A = 5/0.15 = 33.3 \text{ m/s}$

And $Re = V \times 4R/\nu = 33.3 \times 0.261/(1.46 \times 10^{-5}) = 5.95 \times 10^5$

Then $f = 0.018$ (from Fig. 10.8)

$h_f = f(L/D)(V^2/2g) =$

$0.018 \times (100/0.26)(33.3^2/(2 \times 9.81)) = 391 \text{ m}$

$P = Q\gamma h_f = 5 \times 12.0 \times 391 = \underline{23.5 \text{ kW}}$

10.117 $\Delta h_{before} = \Delta h_{after}$

$\therefore h_{f,before} = h_{f,after}$

$(f_b L/4R_b) V_b^2/2g = (f_a L/4R_a) V_a^2/2g$

$R_b = A_b/P_b = 2/6 = 0.333$ ft

$R_a = A_a/P_a = 1.4/6 = 0.233$ ft

$V_a^2/V_b^2 = R_a/R_b = 0.70$

$\underline{V_{after}/V_{before} = 0.84}$

10.118 $f = f(Re, k_s/4R)$ $R = A/P = 0.7m^2/3.4m = 0.206m$

$R_e = V(4R)/\nu = 10 \times 4 \times .206 \times 10^6 = 8.2 \times 10^6$

$k_s/4R = 10^{-3}m/0.824m = 1.2 \times 10^{-3} = .0012$

From Fig. 10.8: $\underline{f \approx 0.020}$ Choice (b) is correct one.

10.119

Use Manning's formula

$Q = (1/n)\ A\ R_h^{2/3}\ S_0^{1/2}$

$n = 0.012$ (assume)

$A = (1)\ (2)/2 = 1m^2$

$R_h = A/P$

$= 1/2(1^2 + 1^2)^{0.5}$

$= 0.35$ m

$Q = (1/0.012)\ (1)\ (0.35)^{2/3}\ (0.0013)^{0.5}$

$\underline{\underline{Q = 1.49\ m^3/s}}$

10.120 Assume $k_s = 30cm$; $R=A/P\approx 2.21m$; $k_s/4R=0.034$; $f\approx 0.060$ (from Fig. 10.8)

$C = \sqrt{8g/f} = 36.2\ m^{1/2}s^{-1}$

$\underline{Q = CA\sqrt{RS} = 347\ m^3/s}$

10.121

Assume $k_s = 10^{-3}$m; $A = 4.5\ m^2$ and $P = 6$m. Then $R = A/P = 0.75$ m

and $4R = 3$ m; $k_s/4R = 0.333 \times 10^{-3}$m; Assume $f = 0.016$

$h_f/L = fV^2/(2g4R)$ or $V = \sqrt{(8g/f)RS} = 1.92$ m/s

$Re = 1.92 \times 3/(1.31 \times 10^{-6}) = 4.4 \times 10^6$; $f = 0.015$

Then $V = 1.92 \times \sqrt{0.016/0.015} = 1.98$ m/s

Finally, $Q = 1.98 \times 4.5 = \underline{\underline{8.73\ m^3/s}}$

10.122 $R = A/P = 4 \times 12/(12 + 2 \times 4) = 2.4$; assume $k_s = 0.003$

$k_s/(4R) = 0.003/(4 \times 2.4) = 0.00031$

$Re\ f^{1/2} = ((4R)^{3/2}/\nu)(2gS)^{1/2} = ((4 \times 2.4)^{3/2}/(1.22 \times 10^{-5}))$

$\times\ (2g \times 10/8{,}000)^{1/2} = 6.9 \times 10^5$; $f = 0.015$

$V = \sqrt{8gRS/f} = \sqrt{8g(2.4)6/(0.015(8{,}000))} = 5.56$; $Q = 5.56(4)12 = \underline{\underline{267\text{ cfs}}}$

Alternate solution:

$Q = (1.49/n)AR^{2/3}S^{1/2}$ Assume $n = 0.015$

$= (1.49/0.015)\ 4 \times 12(2.4)^{2/3}(6/8{,}000)^{1/2} = \underline{\underline{234\text{ cfs}}}$

10.123 $Q = 100$ cfs; $S = 0.001$; $n = 0.015$

$Q = (1.49/n)AR^{0.667}S_0^{0.5}$

or $Qn/(1.49S^{0.5}) = AR^{0.667}$

$31.84 = AR^{0.667}$

$31.84 = (by)(by/(b + 2y))^{0.667}$

For different values of b one can compute *y* and the area *by*. These solutions were made and the results are shown in the Table and Fig. below.

b (ft)	y(ft)	A (ft^2)	y/b
2	16.5	33.0	8.2
4	6.0	24.0	1.5
6	3.8	22.5	0.63
8	2.8	22.4	0.35
10	2.3	23.3	0.23
15	1.7	25.5	0.11

10.123 (continued)

y/b: 1.0, 0.8, 0.6, 0.4, 0.2, 0.0

A (ft²): 20, 25

Minimum A at y/b=0.5 (Verified)

10.124

$$Q = (1.49/n)AR^{2/3}S^{1/2}$$

$R = A/P = (\pi D^2/8)/(\pi D/2) = D/4$; $A = \pi D^2/8$

Assume $n = 0.013$

Then $Q = (1.49/0.013) \times (\pi D^2/8)(D/4)^{2/3} \times (0.9/1{,}000)^{1/2}$

$$\underline{\underline{Q = 21.6 \text{ ft}^3/\text{s}}}$$

Alternate solution:

$Q = CA\sqrt{RS}$ where $C = \sqrt{8g/f}$; Assume $k_s = 10^{-3}$ft

$k_s/4R = 10^{-3}/(4 \times D/4) = 2.5 \times 10^{-4}$; Assume $f = 0.016$

Then $C = \sqrt{8 \times 32.2/0.016} = 127$

$Q = 127 \times (\pi \times 4^2/8)\sqrt{1 \times 0.9/1{,}000} = 23.9$ ft^3/s; $V = 3.80$ ft/s

$Re = V \times 4R/\nu = 3.80 \times 4 \times 1/(1.2 \times 10^{-5}) = 1.3 \times 10^6$; $f = 0.015$

Solve again: $C = \sqrt{8g/0.015} = 131$; $\underline{\underline{Q = 24.7 \text{ ft}^3/\text{s}}}$

10.125 $Q = (1.49/n)\ A\ R_h^{0.667}S_0^{0.5}$; Assume n=0.012

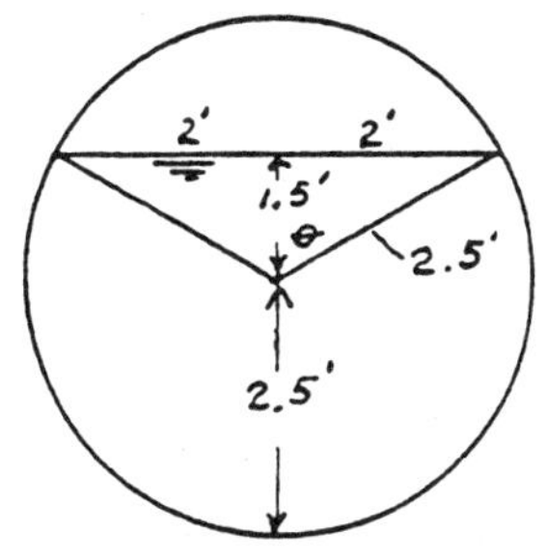

$\cos\theta = 1.5\text{ft}/2.5\text{ft}$; $\theta = 53.13°$

$A = \pi r^2((360° - 2\times 53.13°)/360) + 0.5\times 4\text{ft}\times 1.5\text{ft}$

$A = 16.84 \text{ ft}^2$

10.125 (continued)

$P = \pi D((360° - 2\times 53.13°)/360) = 11.071$ ft

$R_h = A/P = 1.521$ ft ; $R_h^{0.667} = 1.323$

Then $Q = (1.49/0.012)(16.84)(1.323)(0.001)^{0.5}$

$Q = 87.5$ cfs

10.126

$R = A/P = (10 + 12)6/(10 + 6\sqrt{5} \times 2) = 132/36.8 = 3.58$

Assume $k_s = 0.003$; $(k_s/4R) = 0.003/(4 \times 3.58) = 0.00021$

$Re\, f^{1/2} = ((4R)^{3/2}/\nu)(2gS)^{1/2} = [(4 \times 3.58)^{3/2}(2g/2{,}000)^{1/2}/(1.41 \times 10^{-5})]$

$= 6.9 \times 10^5$ Thus, $f = 0.014$

$V = \sqrt{8gRS/f} = \sqrt{8g \times 3.58/(0.014(2{,}000))} = 5.74$ fps

$Q = VA = 5.74(132) = 758$ cfs

Alternate method, assuming $n = 0.015$

$V = (1.49/n)R^{2/3}S^{1/2} = (1.49/0.015)(3.58)^{2/3}(1/2{,}000)^{1/2} = 5.18$ fps

$Q = 5.18(132) = 684$ cfs

10.127 $Q = (1.49/n)\, AR^{0.667}S_0^{0.5}$; Assume $n = 0.012$

$1000 = (1.49/0.012)(10d+d^2)((10d+d^2)/(10+2\sqrt{2}d))^{0.667}(1/500)^{0.5}$

$1000 = (1.49/0.012)(10d+d^2)^{5/3}(10+2\sqrt{2}d)^{-0.667}(0.0447)$

Solving for d yields $d = 5.3$ ft

10.128 $Q = (1.49/n)\, AR^{2/3}S^{1/2}$; assume $n = 0.012$

$A = 10 \times 5 + 5^2$, $P = 10 + 2\sqrt{5^2 + 5^2} = 24.14$ ft

$R = A/P = 75/24.14 = 3.107$ ft

Then $Q = (1.49/0.012)(75)(3.107)^{2/3}(4/5280)^{1/2} = 546$ cfs

---

10.129 $Q = (1/n)AR^{2/3}S^{1/2}$; assume n = 0.015

$25 = (1.0/0.015)4d(4d/(4 + 2d))^{2/3} \times 0.004^{1/2}$

Solve by trial and error: d = 1.6 m

10.130 $Q = (1.49/n)AR^{2/3}S^{1/2}$; assume n = 0.012

$600 = (1.49/0.012)\ 12d\ (12d/(12 + 2d))^{2/3} \times (10/8{,}000)^{1/2}$

Solve by trial-and-error: d = 5.6 ft

10.131

$Q = (1.49/n)\ A\ R_h^{2/3}\ S_0^{1/2}$ Assume n = 0.015

$3{,}000 = ((1.49)/(0.015))(10d + 2d^2)((10d + 2d^2)/(10 + 2\sqrt{5}d))^{2/3}\ (0.001)^{1/2}$

$955 = (10d + 2d^2)\ ((10d + 2d^2)/(10 + 2\sqrt{5}d))^{2/3}$

Solve for d by trial and error

d = 10.1 ft

10.132 Divide the channel into parts 1, main, & 2.

With these divisions we compute A_1, A_{main} & A_2 and also R_1, R_{main} & R_2. The results of these computations along with $R^{2/3}$ values shown in tabular form below:

Channel Part	$A(ft^2)$	P(ft)	$R_h = A/P(ft)$	$R_h^{2/3}(ft)^{2/3}$
1	1218	208.5	5.842	3.244
main	5600	228.3	24.53	8.443
2	1218	208.5	5.842	3.244

Assume: $n_1 = n_2 = 0.035$ (from Table 10.4)

$n_{main} = 0.025$ (from Table 10.3)

10.132 (continued)

Then using Eq. 10.46 we have:

$Q = 1.49\sqrt{0.0015}[(2(1218 \times 3.244)/.035) + (5600 \times 8.443/.025)]$

$\underline{\underline{Q = 122,170\ ft^3/s}}$

10.133 For best hydraulic section, the shape will be a half hexagon as depicted below

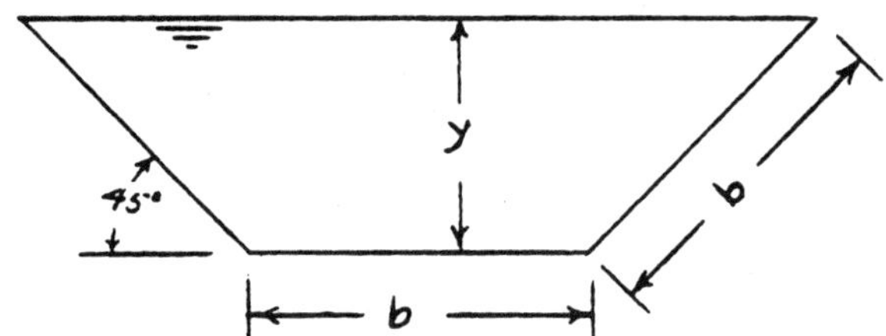

Use the Manning equation to solve this problem:

$Q = (1.49/n)AR_h^{0.667}S_0^{0.5}$

Assume n = 0.015 (concrete, wood forms unfinished - Table 10.3)

Then $900 = (1.49/0.015)AR_h^{0.667}(0.002)^{0.5}$

$AR_h^{0.667} = 202.6$

But $A = by + y^2$ where $y = b \cos 45° = 0.7071b$

$A = 0.7071b^2 + 0.50b^2 = 1.2071b^2$

$R_h = A/P = 1.2071b^2/3b = 0.4024b$

Thus $AR_h^{0.667} = 202.6 = 1.2071b^2(0.4024b)^{0.667}$

$b^{2.667} = 308$; $\underline{\underline{b = 8.57\ ft}}$

10.134 $Q = 50{,}750$ cfs ; $S = 0.0040$

Eq. 10.46: $Q = (1.49/n)[A_1R_1^{2/3} + A_2R_2^{2/3}]S^{1/2}$

Let A_1 & R_1 denote the main channel

A_2 & R_2 denote the total overbank channel

$A_1 = 10(200) = 2000\ ft^2$, $P_1 = 200 + 2(5) = 210$

$R_1 = 2000/210 = 9.523$

$R_1^{2/3} = 4.493$

$A_2 = 2(200)5 = 2000\ ft^2$, $P_2 = 200 + 200 + 2(5) = 410$

$R_2 = 2000/410 = 4.878$, $R_2^{2/3} = 2.876$

$n = (1.49/50750)[2000(4.493) + 2000(2.876)](.0040)^{1/2}$

$\underline{\underline{n = 0.433(.0632)\ 0.027}}$

10.135 $Q = 1.49\ [A_1R_1^{2/3}/n_1 + A_2R_2^{2/3}/n_2]S^{1/2}$

$50750 = 1.49\ [2000(4.493)/0.02 + 2000(2.876)/n_2](.0040)^{1/2}$

$538542 = 449300 + 5752/n_2$

$n_2 = 5752/(538542 - 449300)$

$= \underline{\underline{0.0645}}$

$V_1 = (1.49/n_1)R_1^{2/3}S^{1/2}$

$= (1.49/0.02)(4.493)(0.06325)$

$Q_1 = V_1A_1 = 21.17 \times 2{,}000 = \underline{\underline{42{,}343\ cfs}}$

$V_2 = (1.49/0.0984)(2.879)(0.06325)$

$= \underline{\underline{2.757\ ft/s}}$

$Q_2 = V_2A_2 = 2.757 \times 2{,}000 = \underline{\underline{5{,}515\ cfs}}$

10.136 $Q = 45{,}400,\ S = 0.004,\ n_1 = 0.025,\ n_2 = 0.07$

Assume y > 10 but < 18

Main Channel: $A_1 = 102.5(10) + [105+(y-10)/8](y-10)$

$= 1025 + (103.75+0.13y)(y-10)$

$P_1 = 100 + (10^2+2.5^2)^{1/2} + [y^2+(y/4)^2]^{1/2}$

$= 100 + 10.3 + 1.03y$

$= 110.3 + 1.03y$

Overbank: $A_2 = 200(y-10),\ P_2 = 200 + y-10 = 190 + y$

$$45400/(1.49)(.004)^{1/2} = [1025 + (103.75 + .13y)(y - 10)][(1025 + (103.75 + .13y)(y - 10))/(110.3 + 1.03y)]^{2/3}/.025 + 200(y - 10)[200(y - 10)/(190 + y)]^{2/3}/0.07$$

Solving for y by trial and error yields: y = 17 ft

Thus for Main Channel: y = 17.0 ft

$$Q(\text{Main}) = (1.49/.025)(.004)^{1/2}[1025 + (103.75 + .13y)(y - 10)][(1025 + (103.75 + 13y)(y - 10))/(110.3 + 1.03y)]^{2/3} = 3.769\ (10176) = \underline{38350\ \text{cfs}}$$

$Q(\text{Overbank}) = 45400 - 38350 = \underline{7050\ \text{cfs}}$

10.137 $S = 0.0009,\ Q = 100{,}000$ cfs

From Table 10.4: $n(\text{overbank}) = 0.06 = n_2$

From Table 10.3: $n(\text{main channel}) = 0.025 = n_1$

The solution for y must be made differently for y < 20 and y > 20. Initially we assume y < 20.

$A_1 = (500 + y)y,\ P_1 = 500 + 2.828y$

Eq. 10.45: $Q = 100{,}000 = (1.49/0.025)(500 + y)y[(500 + y)y/(500 + 2.828y)]^{2/3}(0.0009)^{1/2}$

Solve for y by trial and error yields $\underline{y = 16.9\ \text{ft}}$

10.138 An assumption is made for the discharge in all pipes making certain that the continuity equation is satisfied at each junction. Figure A shows the network with assumed flows.

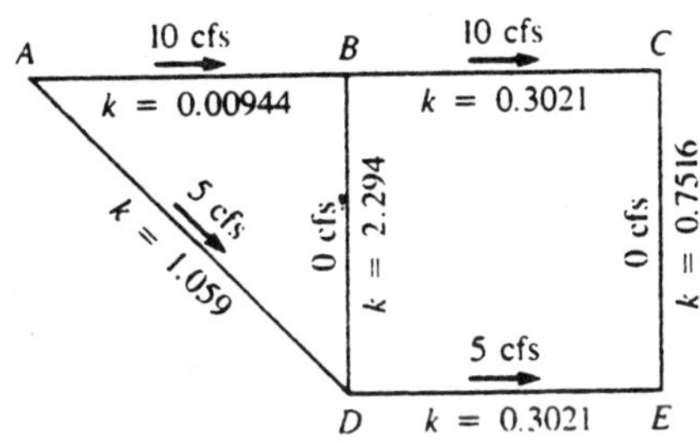

Figure A

The Darcy-Weisbach equation is used for computing the head loss; therefore, we have

$$h_f = f\left(\frac{L}{D}\right)\left(\frac{V^2}{2g}\right)$$

$$= 8\left(\frac{fL}{gD^5\pi^2}\right)Q^2$$

$$= kQ^2$$

where $k = 8\left(\dfrac{fL}{gD^5\pi^2}\right)$.

The loss coefficient, k, for each pipe is computed and shown in Fig. A. Next, the flow corrections for each loop are calculated as shown in the accompanying table. since $n = 2$ (exponent on Q), $nkQ^{n-1} = 2kQ$. When the corrections obtained in the table are applied to the two loops, we get the pipe discharges shown in Fig. B. Then with additional iterations, we get the final distribution of flow as shown in Fig. C. Finally, the pressures at the load points are calculated.

Loop *ABC*

Pipe	$h_f = kQ^2$	$2kQ$
AB	+0.944	0.189
AD	−26.475	10.590
BD	0	0
$\sum kQ_c^2 - \sum kQ_{cc}^2 =$	−25.53	$\sum 2kQ =$ 10.78

$\Delta Q = -25.53/10.78 = -2.40$ cfs

Loop *BCDE*

Pipe	h_f	$2kQ$
BC	+30.21	6.042
BD	0	0
CE	0	0
DE	− 7.55	3.02
	+22.66	9.062

$\Delta Q = 22.66/9.062 = 2.50$ cfs

10.138 (continued)

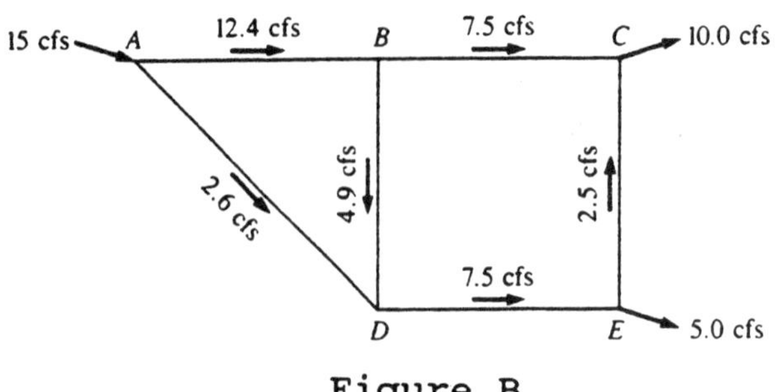

Figure B

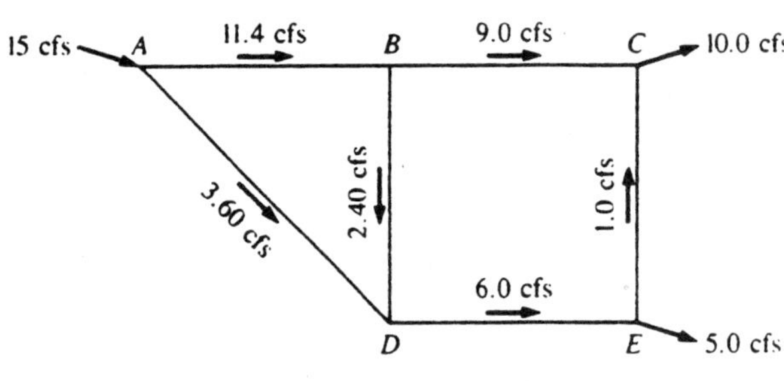

Figure C

$$
\begin{aligned}
p_C &= p_A - \gamma(k_{AB}Q_{AB}^2 + k_{BC}Q_{BC}^2) \\
&= 60 \text{ psi} \times 144 \text{ psf/psi} - 62.4(0.00944 \times 11.4^2 + 0.3021 \times 9.0^2) \\
&= 8640 \text{ psf} - 1603 \text{ psf} \\
&= 7037 \text{ psf} \\
&= 48.9 \text{ psi}
\end{aligned}
$$

$$
\begin{aligned}
p_E &= 8640 - \gamma(k_{AD}Q_{AD}^2 + k_{DE}Q_{DE}^2) \\
&= 8640 - 62.4(1.059 \times 3.5^2 + 0.3021 \times 6^2) \\
&= 7105 \text{ psf} \\
&= 49.3 \text{ psi}
\end{aligned}
$$

10.139 Assume that the equipment will have a maximum weight of 1,000 lbf and assume that the platform itself weighs 200 lbf. Assume that the platform will be square and be 5 ft on a side. The plan and elevation view are shown below:

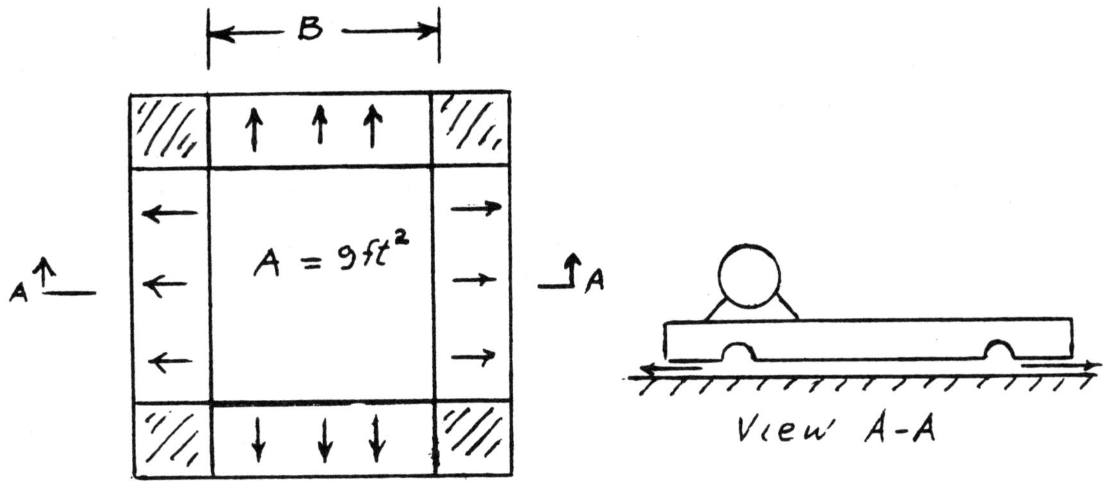

Assume that a plenum 1 ft inside the perimeter of the platform will be the source of air for the underside of the platform.

Now develop the relationship for pressure distribution from plenum to edge of platform. The flow situation is shown below.

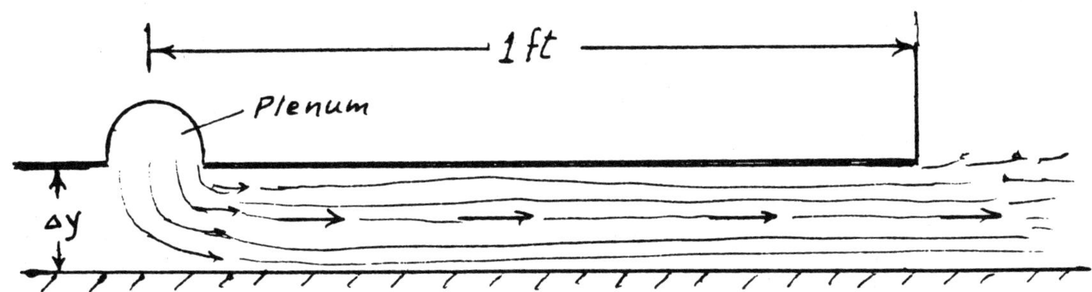

10.139 (Cont'd)

Determine the h_f from the plenum to the edge of the platform:

$$h_f = f(L/4R)\ V^2/2g$$

Assume $\quad f = 0.02;\ R = A/P = \Delta y \cancel{B}/(2\cancel{B}) = \Delta y/2$

$$L = 1 \text{ ft}$$

$$h_f = (0.02 \times 1/(\Delta y/2))V^2/2g$$

$$= (0.02 \times \cancel{2}/(\Delta y))V^2/(\cancel{2}g)$$

$$= .02V^2/(\Delta y g)$$

Multiply both sides by γ

$$\Delta p_f = \gamma h_f = (0.02/\Delta y)\ \rho V^2$$

Assume $\quad \rho = 0.0023 \text{ slugs/ft}^3$

Then $\quad \Delta p_f = (0.02/\Delta y)(.0023)V^2$

$$\Delta p_f = (46V^2/\Delta y) \times 10^{-6}$$

$$p_{avg.} \text{ (over 4 ft}^2 \text{ area)} = (23\ V^2/\Delta y) \times 10^{-6}$$

Also determine the Δp due to the change in momentum as the flow discharges from the plenum.

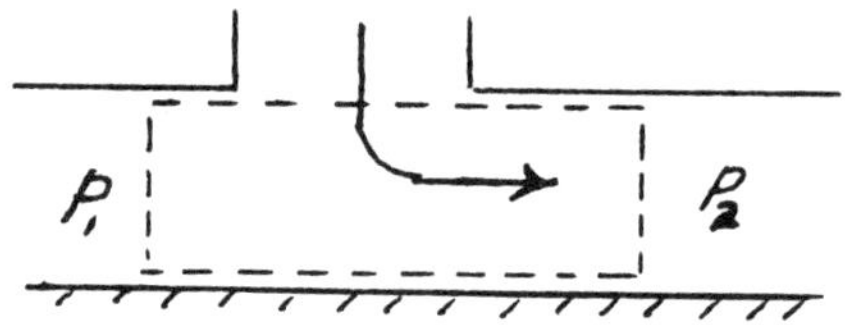

Apply the momentum equation in the x-direction:

$$\Sigma F_x = \Sigma V_x \rho \mathbf{V} \cdot \mathbf{A}$$

$$B\Delta y(p_1 - p_2) = V\rho(VB\Delta y)$$

$$\Delta p_{mom.} = \rho V^2$$

The pressure within the 9 ft^2 interior area of the platform will be $\Delta p_{mom.} + \Delta p_f = V^2(.0023 + (46/\Delta y)\times 10^{-6})$

10.139 (Cont'd)

The pressure force on platform is given by

$$F = 9\ ft^2 \times (\Delta p_{mom.} + \Delta p_f) + \Delta p_{f,avg.} \times 12\ ft^2$$

$$F = 9 \times V^2[.0023 + (46/\Delta y) \times 10^{-6})]$$

$$+ 12\ V^2[(23V^2/\Delta y) \times 10^{-6}]$$

$$F = V^2[9 \times .0023 + (9 \times 46/\Delta y) \times 10^{-6}$$

$$+ 12 \times 23 \times 10^{-6}/\Delta y]$$

$$F = V^2[9 \times .0023 + 690 \times 10^{-6}/\Delta y]$$

Let $\Delta y = 1/8$ in. $= 0.01042$ ft

$$F = V^2\ [9 \times .0023 + 690 \times 10^{-6}/.01042]$$

$$= V^2[0.0207 + .0662]$$

$$F = .0869\ V^2$$

$$1200 = .0869\ V^2$$

$$V^2 = 13,809\ ft^2/s^2$$

$$V = 117.5\ ft/s$$

$$Q = 117.5 \times \Delta y \times 12 = 14.69\ ft^3/s$$

$$\Delta p = V^2(.0023 + 46 \times 10^{-6}/\Delta y)$$

$$= V^2(.0023 + 46 \times 10^{-6}/0.01042)$$

$$= V^2(.0023 + .00441)$$

$$= 92.7\ psf$$

$$Power = Q\Delta p/550$$

$$= 14.69 \times 92.7/550$$

$$= 2.48\ hp$$

Assume 50% efficiency for blower, so required power $\approx$ 5 horsepower.

Blower could be driven by gasoline engine and also *be* located on the platform.

CHAPTER ELEVEN

11.1

Force normal to plate = $\Delta p_{average} \times A$

$= (1.5\ \rho V_0^2/2) \times b \times 1$ (for unit length of plate)

F_D = Force parallel to free stream direction = $F_{normal} \cos 60°$

$= (1.5\rho V_0^2/2) \times b \times 1/2 \quad (1)$

But $F_D = C_D A_p\ \rho V_0^2/2$

$= C_D b \sin 30° \rho V_0^2/2 = C_D \times 1/2 \times \rho V_0^2/2 \quad (2)$

Eliminate F_D between Eqs. (1) and (2) to yield

$C_D b \times 1/2 \times \rho V_0^2/2 = (1.5\ \rho V_0^2/2) \times b \times 1/2$

$\underline{\underline{C_D = 1.5}}$

11.2 **Flow is from the N.E. direction.**

11.3 The force contributing to drag on the downstream face is

$F_D = 0.4\ A_p \rho V^2/2$

The force on each side face is $F_s = ((1.0 + 0)/2) A_p \rho V^2/2$

Then the drag force on one side is

$F_s \sin \alpha = 0.5\ A_p \rho V_0^2/2 \times 0.5$

The total drag is:

Drag $= 2((0.5\ A_p \rho V_0^2/2) \times 0.5) + 0.4\ A_p \rho V_0^2/2 = C_D A_p \rho V_0^2/2$

Solving for C_D one gets $\underline{C_D = 0.9}$

11.4 Similar analysis as for solution of Prob. 11.3:

$C_D A_p \rho V_0^2/2 = 0.8\ A_p \rho V_0^2/2 + 2\ (\ (1.10\ A_p \rho V_0^2/2) \times 0.5)$

Solving: $\underline{C_D = 1.90}$

11.5 The relevant equation is

$$F_D = C_D A_p \rho V^2 / 2$$

$$\text{or } C_D = F_D / (A_p \rho V^2 / 2)$$

Thus F_D, A_p, and V will have to be measured. The air density ρ can be obtained by measuring the air temperature with a thermometer and the air pressure with a barometer and solving for ρ by the equation of state.

You will need to decide how to position the sphere in the wind tunnel so that its support does not have an influence on flow past the sphere. One possible setup might be as shown below.

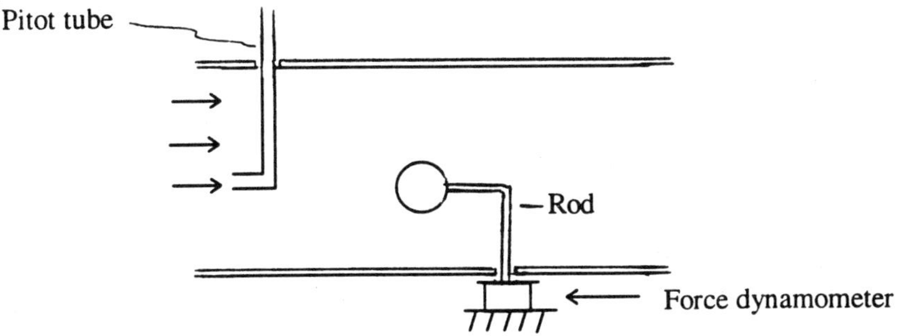

The sphere is attached to a rod and the rod in turn is attached to a force dynamometer as shown. Of course the rod itself will produce drag, however; its drag can be minimized by enclosing the vertical part of the rod in a streamlined housing. The horizontal part of the rod would have negligible drag because much of it would be within the low velocity wake of the sphere and the drag would be skin friction drag which is very small. The air velocity approaching the sphere could be measured by a pitot tube inserted into the wind tunnel. It would be removed when the drag of the sphere is being measured. The projected area of the sphere would be obtained by measuring the sphere diameter and then calculating the area.

pressure transducer placed outside the wind tunnel. Blockage effects could also be addressed in the design of this experiment.

Another design consideration that could be addressed is size of sphere. It should be large enough to get measurable drag readings but not so large as to produce significant blockage.

11.6 Much of the design for determining the drag coefficient of the square rod would be the same as for determining the drag coefficient of a sphere (Prob. 11.5). However, it would be very difficult to measure the drag of a square rod supported like a sphere (as suggested in the design of problem 11.5) because end effects of the rod as well as the velocity distribution across the tunnel would present serious problems. A better way to determine the drag would be to measure the pressure distribution on the upstream and downstream faces of the rod and evaluate the force on the rod by integrating the pressure over these faces. The pressure could be measured by drilling small holes through the faces and connecting a tube to each hole inside the square rods. The pressure in each tube could be measured by a pressure transducer placed outside the wind tunnel. Blockage effects could also be addressed in the design of this experiment.

11.7 $Re = VD/\nu = 30 \times 3/(1.5 \times 10^{-5}) = 6 \times 10^6$

From Fig. 11.5 $C_D \approx 0.60$; $\rho \approx 1.20\ kg/m^3$

$F_D = C_D A_p \rho V_0^2/2 = 0.60 \times (3 \times 90) \times 1.20 \times (30)^2/2 = 87.5\ kN$

The overturning moment is

$F_D \times H/2$ or $M = 87.5\ kN \times 90/2 = \underline{\underline{3.93\ MN\cdot m}}$

11.8 $Re = VD/\nu = 25 \times 0.10/(1.5 \times 10^{-5}) = 1.67 \times 10^5$

From Fig. 11-5: $C_D = 0.95$; $\rho \approx 1.20\ kg/m^3$

$$\text{Moment} = F_D H/2 = C_D A_p \rho (V_0^2/2) \times H/2$$

$$\text{Moment} = 0.95 \times 0.10 \times ((35)^2/2) \times 1.2 \times (25)^2/2 = \underline{\underline{21.8\ kN \cdot m}}$$

11.9 The cup, sphere or disk should probably be located at the center of the pipe (as shown below) because the greatest velocity of the air stream in the pipe will be at the center.

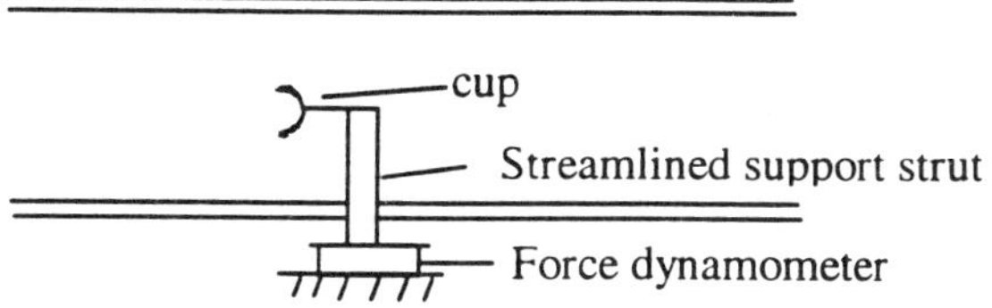

You want to correlate V and Q with the force acting on your device. First, neglecting the drag of the support device, the drag is given as

$$F_D = C_D A_p \rho V_0^2 / 2$$
$$\text{or } V_0 = (2F_D / (C_D A_p \rho))^{1/2}$$

you can measure temperature, barometric pressure, and gage pressure in the pipe. Therefore, with these quantities the air density can be calculated by the equation of state. Knowing the diameter of the cup, sphere or disk you can calculate A_p. Assume that C_D will be obtained from Table 11.1 or Fig. 11.11. Then the other quantity that is needed to estimate V_0 is the drag F_D. This can be measured by a force dynamometer as indicated on the sketch of the device. However, the support strut will have some drag so that should be considered in the calculations. Another possibility is to minimize the drag of the support strut by designing a housing to fit around, but be separate from the vertical part of the strut thus eliminating most of the drag on the strut. This was also suggested for Problem 11.5.

Once the centerline velocity is determined it can be related to the mean velocity in the pipe by Table 10.1 from which the flow rate can be calculated. For example, if the Reynolds number is about 10^5 then $\overline{V}/V_{max} \approx 0.82$ (from Table 10.1) and

$$Q = \overline{V}A$$
$$Q = 0.82V_{max}A$$

There may be some uncertainty about C_D as well as the drag of the support rod; therefore, the device will be more reliable if it is calibrated. This can be done as follows. For a given flow make a pitot-tube-velocity-traverse across the pipe from which Q can be calculated. Also for the given run measure the force on the force dynamometer. Then plot F vs Q. Do this for several runs so that a curve of F vs. Q is developed (calibration completed).

11.10 $\rho = 0.00237$ slugs/ft^3 ; $\nu = 1.58 \times 10^{-4}$ ft^2/s ; V = 140 mph = 205 ft/s

$Re = V_0 d/\nu = 205 \times 250/(1.58 \times 10^{-4}) = 3.2 \times 10^8$

$C_D \approx 0.70$ (Fig. 11.5--extrapolated)

Then $F_D = C_D A_p \rho V^2/2$

$= 0.70 \times 250 \times 350 \times 0.00237 \times 205^2/2$

$F_D = \underline{\underline{3.05 \times 10^6 \text{ lbf}}}$

11.11 $dF_D = C_D(dr)d\ \rho V^2_{rel.}/2$ where $V_{rel.} = r\omega$

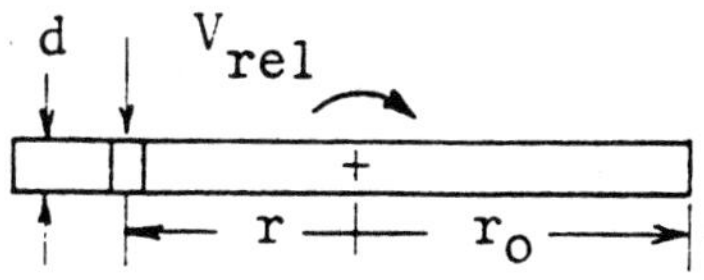

Then $dT = rdF_D = C_D d\rho(V^2_{rel.}/2)rdr$

$T_{total} = 2\int_0^{r_0} dT = 2\int_0^{r_0} C_D d\rho((r\omega)^2/2)rdr$

$T_{total} = C_D d\rho\omega^2\int_0^{r_0} r^3 dr = C_D d\rho\omega^2 r_0^4/4$ but $r_0 = L/2$

so $T_{total} = C_D d\rho\omega^2 L^4/64$ or $P = T\omega = C_D d\rho\omega^3 L^4/64$

Then for the given conditions:

$P = 1.2 \times 0.02 \times 1.2 \times (100)^3 \times 1^4/64 =$ 450 W

11.12 $V_0 = 30$ m/s Re $= 6 \times 10^6$ S = 0.25 (from Fig. 11-10)

$S = nd/V_0$ or $n = SV_0/d$

$n = 0.25 \times 30/3 =$ 2.5 Hz

11.13 $V_0 = 25$ m/s Re $= 1.7 \times 10^5$ S $\simeq$ 0.21 (from Fig. 11-10)

$S = nd/V_0$ or $n = SV_0/d$ $n = 0.21 \times 25/0.1 =$ 52.5 Hz

---

11.14 $\nu = 1.58 \times 10^{-4}$ ft²/s; $\rho = 0.00237$ slugs/ft³; $V_0 = 60$ mph $= 88$ ft/s

$Re = V_0 b/\nu = 88 \times 10/(1.58 \times 10^{-4}) = 5.6 \times 10^6$; $C_D = 1.19$ (Table 11-1)

Then $F_D = C_D A_p \rho V_0^2/2$

$= 1.19 \times 300 \times 0.00237 \times 88^2/2 =$ 3,276 lbf

11.15 $R_e = VL/\nu = (100)(4)/(1.58 \times 10^{-4}) = 2.5 \times 10^6$

$C_D = 1.18$ (Table 11-1)

$F_D = C_D A_p \rho V_0^2/2$

$F_D = (1.18)\ (4 \times 4)\ (0.00237)\ (100^2)/2 =$ 224 lbf

11.16 $F_{edge} = 2\ C_f A\rho V^2/2;\ \ F_{normal} = C_D A\rho V^2/2$

Then $F_{normal}/F_{edge} = C_D/2C_f;\ \ Re = Re_L = VB/\nu = 1 \times 2/(1.3 \times 10^{-6})$

$= 1.5 \times 10^6;\ C_f = 0.0032$ (from Fig. 9-13);

$C_D = 1.18$ (from Table 11-1);

$F_{normal}/F_{edge} = 1.18/(2 \times 0.0032) = \underline{\underline{184}}$

11.17 $C_D = 1.17$ (Table 11.1)

$F_D = C_D A_p \rho V^2/2 = 1.17 \times (\pi/4) \times 1^2 \times 1{,}000 \times 4^2/2 = \underline{\underline{7.35\ \text{kN}}}$

11.18 $\rho = 1.25\ \text{kg/m}^3;\ C_D = 1.17$ (Table 11.1)

$F_D = C_D A_p \rho V^2/2$

$= 1.17 \times (\pi/4) \times 6^2 \times 1.25 \times 30^2/2 = \underline{\underline{18{,}608\ \text{N}}}$

$= \underline{\underline{18.61\ \text{kN}}}$

11.19 $\rho \approx 1.23\ \text{kg/m}^3;\ C_D = 1.18$ (Table 11.1)

Then $M = 3 \times F_D = 3 \times C_D A_p \rho V^2/2 = 3 \times 1.18 \times 2 \times 1.23 \times 35^2 = \underline{\underline{10.67\ \text{kN}\cdot\text{m}}}$

11.20 Assume T = 20°C; $C_D = 1.20$ (Table 11-1)

Then $F_D = C_D A_p \rho V^2/2 = 1.2 \times 1.83 \times 0.46 \times 1.2 \times 20^2/2 = 242\ \text{N}$

Power = FV = 242 x 20 = $\underline{\underline{4.85\ \text{kW}}}$

11.21 Assume C_D will be like that for a rectangular plate: $\ell/b = 1.5/0.2 = 7.5$

Then $C_D \approx 1.25$ (Table 11-1). V = 105 km/hr. = 29.17 m/s

$\Delta P = C_D A_p (\rho V^2/2)V$; Assume $\rho = 1.20\ \text{kg/m}^3$

Then $\Delta P = 1.25 \times 1.5 \times 0.2 \times 1.2 \times 29.17^2/2 \times 80{,}000/3{,}600 = \underline{\underline{4.25\ \text{kW}}}$

11.22 The energy required per distance of travel = F x distance.

Thus, the energy, E, per unit distance, L, is: E = F

or $E = \mu \times Wgt + C_D A_p \rho V^2/2 = 0.02 \times 3{,}000 + 0.4 \times 20 \times (0.00237/2)V^2$

For V = 55 mph = 80.67 ft/sec: $E_{55} = 121.7$ ft-lbf

For V = 65 mph = 95.33 ft/sec: $E_{65} = 146.2$ ft-lbf

Then energy savings = (146.2 - 121.7)/146.2 = 0.167; $\underline{\underline{16.7\%}}$

11.23

$$F_D + F_r = w \times \sin 6^D$$
where F_D = Drag Force
F_r = Rolling friction
W = **Weight of Car**

$$C_D A_p \rho V^2 / 2 + W \times 0.02 = W \times \sin 7^\circ$$

$$V^2 = 2W(\sin 6^\circ - 0.02)/(C_D A_p \rho)$$

$$V^2 = 2 \times 2500(0.105 - 0.02)/(0.32 \times 25 \times 0.0024)$$

$$= 22{,}135 \text{ft}^2/\text{s}^2$$

$$\underline{\underline{V = 148.8 \text{ft/s} = 101 \text{mph}}}$$

11.24 P = FV

Where $F = F_D + F_{\text{rolling frict.}}$

$F_D = C_D A_p \rho V_0^2/2 = (0.4)(2)(1.2)(40)^2/2 = 768\text{N}$

$F_{r.f.} = 0.02 \text{ Wgt.} = 0.02 \times 10{,}000\text{N} = 200\text{N}$

$\underline{\underline{P = (768 + 200)30 = 29.04 \text{ kW}}}$

11.25 $\rho = p/RT = 96{,}000/(287 \times (273 + 20)) = 1.14 \text{ kg/m}^3$

Assume C_D is like a rectangular plate: $C_D \approx 1.20$

Then $F_D = C_D A_p \rho V^2/2 = 1.2 \times 1.83 \times 0.30 \times 1.14 \times 30^2/2 = \underline{\underline{338 \text{ N}}}$

Note: F_D will depend upon C_D and dimensions assumed.

11.26 Assume T = 10°C; $\rho = 1.25 \text{ kg/m}^3$

Take moments about one wheel for impending tipping. $\Sigma M = 0$

$W \times (1.44/2) - F_D \times ((3.2/2) + 0.91) = 0$

$F_D = (190{,}000 \times 1.44/2)/2.51 = 54{,}500 \text{ N} = C_D A_p \rho V^2/2$

Assume $C_D = 1.20$ (Table 11-1)

Then $V^2 = 54{,}500 \times 2/(1.2 \times 12.5 \times 3.2 \times 1.25)$; $\underline{\underline{V = 42.6 \text{ m/s}}}$

F_{wind} W W

11.27 Consider force balance parallel to direction of motion of the bicyclist:

$\Sigma F = 0$

$+F_{Wgt.\ comp.} - F_D - F_{rolling\ resist.} = 0$

$(Wgt)\sin 10° - C_D A_p \rho V^2_{Rel.}/2 - 0.02\ W\cos 10° = 0$

$W\sin 10° - 0.5 \times 0.5 \times 1.2\ V^2_R/2 - .02W\cos 10° = 0$

$M = W/g = 80\ kg$

$W = 80g = 784.8\ N$

$W\sin 10° = 136.3\ N$

$W\cos 10° = 772.9\ N$

Then $136.3 - 0.15\ V_R^2 - .02 \times 772.9 = 0$

$V_R = 28.4\ m/s = V_{bicycle} + 5\ m/s$

Note: 5 m/s is the head wind so the relative speed is $V_{bicycle} + 5$.

$\underline{V_{bicycle} = 23.4\ m/s}$

11.28 $P = F_D V = C_D A_p \rho (V_R^2/2) V \qquad V_R = (V + 5)$

$100 = 0.3 \times 0.5 \times 1.2(V + 5)^2 \times V/2;\quad \underline{V = 7.3\ m/s}$

11.29 $P = FV = (\mu_{roll} Mg + C_D A_p \rho V_0^2/2)V$

$P = \mu_{roll} Mg\ V + C_D A_p \rho V_0^3/2$

For this problem, $\mu = 0.10$, $M = 800$ kg, $A_p = 4m^2$, $\rho = 1.2\ kg/m^3$,

and $P = 80{,}000$ W

Then $80{,}000 = 0.1 \times 800 \times 9.81\ V + C_D \times 4 \times (1.2/2)V^3$

$80{,}000 = 784.8\ V + 2.40\ C_D V^3$

Solving with $C_D = 0.30$ (roof closed) one gets $\underline{V = 40.6\ m/s}$

Solving with $C_D = 0.42$ (roof open) one gets $\underline{V = 37.0\ m/s}$

11.30 Assume gas consumption is proportional to power. Then gas consumption = CF_DV where V is the speed of the automobile and F_D is the total drag of the auto (including rolling friction).

$$F_D = C_DA_p\rho V_0^2/2 + 0.1\ mg$$

$$= 0.3 \times 2 \times 1.2\ V_0^2/2 + 0.1 \times 500 \times 9.81$$

$$= 0.360\ V_0^2 + 490.5\ N$$

$$V_{0,\text{still air}} = (90{,}000/3{,}600) = 25.0\ m/s$$

Then $F_{D,\text{still air}} = 0.36 \times 25^2 + 490.5 = 715.5\ N$

$$P_{\text{still air}} = C \times 715.5 \times 25 = 17.89\ CkW$$

$$P_{\text{head wind}} = C \times 17{,}890 \times 1.20 = (0.36\ V_0^2 + 490.5)(25) \times C$$

$$V_0 = \text{headwind} + 25 = 32\ m/s;\ \text{headwind} = \underline{\underline{7\ m/s}}$$

11.31 $(F_D + F_r)V = \text{Power}$ $\qquad C_D = 0.60$ $\qquad$ 60 mph = 88 ft/s

$$F_r = (P/V) - F_D = (P/V) - C_DA_p\rho V^2/2$$

$$= ((40)(550)/88) - (0.60)(30)(0.00237)(88^2)/2$$

$$= 250 - 165 = 85\ lbf$$

"Souped up" version:

$$(F_D + 85)V = (220)(550)$$

$$((C_DA_p\rho V^2/2) + 85)V = (220)(550)$$

$$(C_DA_p\rho V^3/2) + 85V = (220)(550)$$

$$0.0213V^3 + 85V - 121{,}000 = 0$$

Solve for V: $\underline{\underline{V = 171.0\ ft/s = 116.6\ mph}}$

11.32 $F_D = C_DA_p\rho V^2/2$; Assume $\rho = 1.2\ kg/m^3$

$$F_D = C_D \times 8.36 \times 1.2 \times (93{,}000/3{,}600)^2/2$$

$$F_{D_{\text{reduction}}} = 0.25 \times 0.78 \times 8.36 \times 1.2(93{,}000/3{,}600)^2/2$$

$$F_{D_{\text{reduction}}} = \underline{\underline{653\ N}}$$

11.33 $Re = V_0 d/\nu = (25)(100)/(1.3 \times 10^{-4}) = 2.3 \times 10^7$

$C_D = 0.05$ (from Fig. 11.11; extrapolated)

$F_D = C_D A_p \rho V_0^2/2 = (0.05)(\pi/4)(100^2)(0.07/32.2)(25^2)/2$

$= \underline{\underline{267 \text{ lbf}}}$

$P = F_D V_0 = (267)(25) = \underline{\underline{6,670 \text{ ft-lbf/s} = 12.1 \text{ hp}}}$

11.34 $P = FV;\quad P = C_D \times 8.36 \times 1.2\, V^3/2 + 450\, V$

At 80 km/hr: $P_{w/o\ vanes} = .78 \times 8.36 \times 1.2\, V^3/2 + 450\, V$

$= 52.9$ kW

$P_{with\ vanes} = 42.2$ kW } 20.2% savings

At 100 km/hr:

$P_{w/o\ vanes} = 96.4$ kW

$P_{with\ vanes} = 75.4$ kW } 21.8% savings

The above savings assume that the fuel savings are directly proportional to power savings.

11.35 $F_{D_{form}} = C_D A_p \rho V_0{}^2/2;$ Assume $\rho = 1.25\ \text{kg/m}^3$

$F_{D_{form}} = 0.80 \times 9 \times 1.25 \times V_0^2/2 = 4.5\, V_0^2$

$F_{D_{skin}} = C_f A \rho V_0^2/2;\quad Re_L = VL/\nu = V \times 150/(1.41 \times 10^{-5})$

$Re_{L,100} = (100,000/3,600) \times 150/(1.41 \times 10^{-5}) = 2.9 \times 10^8$

$Re_{L,200} = 5.8 \times 10^8$

$C_{f,100} = 0.0018;\quad C_{f,200} = 0.0017$ (from Fig. 9-13)

V = 100 km/hr	V = 200 km/hr
$F_{D,form,100} = 3,472$ N	$F_{D,form,200} = 13,889$ N
$F_{D,skin,100} = 1,302$ N	$F_{D,skin,200} = 4,919$ N
$F_{bearing} = 3,000$ N	$F_{bearing} = 3,000$ N
$F_{total} = 7,774$ N	$F_{total} = 21,807$ N
45% form, 17% skin, 38% bearing	64% form, 22% skin, 14% bearing

11.36 Stoke's law is the equation of drag for a sphere for a Reynolds number less than 0.5:

$$F_D = 3\pi\mu V_0 d$$
$$\text{or } \mu = F_D / (3\pi V_0 d)$$

One can use this equation to determine the viscosity of a liquid by measuring the fall velocity of a sphere in a liquid. Thus one needs a container to hold the liquid (for instance a long tube vertically oriented). The spheres could be ball bearings, glass or plastic spheres. Then one needs to measure the time of fall between two points. This could be done by measuring the time it takes for the sphere to drop from one level to a lower level. The diameter of the sphere could be easily measured by a micrometer and the drag force, F_D, would be given by

$$F_D = Wgt_{sphere} - F_{buoy.of\ sphere}$$

If the specific weight of the material of the sphere is known then the weight of the sphere can be calculated. Or one could actually weigh the sphere on an analytic balance scale. The buoyant force can be calculated if one knows the specific weight of the liquid. If necessary the specific weight of the liquid could be measured with a hydrometer.

To obtain a reasonable degree of accuracy the experiment should be designed so that a reasonable length of time (not too short) elapses for the sphere to drop from one level to the other. This could be assured by

11.36 (Cont'd)

choosing a sphere that will yield a fairly low velocity of fall which could be achieved by choosing to use a small sphere over a large one or by using a sphere that is near the specific weight of the liquid (for instance, plastic vs. steel).

Other items that should or could be addressed in the design are:

A. Blockage effects if tube diameter is too small.

B. Ways of releasing sphere and retrieving it.

C. Possibly automating the measurement of time of fall of sphere.

D. Making sure the test is always within Stoke's law range (Re<0.5)

E. Making sure the elapsed time of fall does not include the time when the sphere is accelerating

---

11.37 $F_{buoy.} = \forall \times \gamma_{oil}$

$= (4/3)\pi \times (1/2)^3 \times 0.85 \times 62.4$

$= 27.77$ lbf

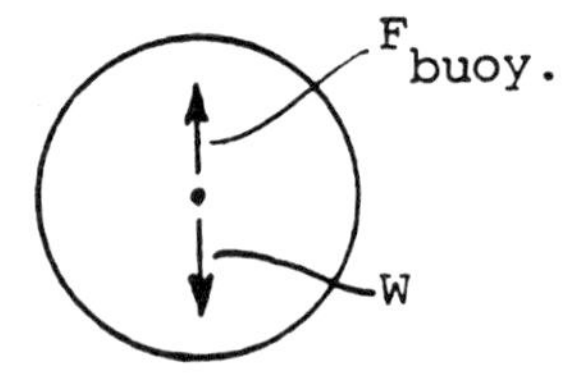

Net force producing motion $= -W + F_{buoy.}$

$= -27.0 + 27.77$ lbf

$= 0.77$ lbf upward

Assume laminar flow.

Then $F_D = 0.77 = 3\pi\mu DV_0$; $V_0 = 0.77/(3\pi D\mu)$

$V_0 = 0.77/(3\pi \times 1 \times 1)$

$V_0 = \underline{\underline{0.082 \text{ ft/s upward}}}$

$Re = V_0 d\rho/\mu = 0.082 \times 1 \times 1.94 \times 0.85/1$

$= 0.14 \leftarrow$ laminar--Assumption O.K.

11.38 $Re = VD\rho/\mu = 0.03 \times 0.02 \times 900/0.096 = 5.63$; Then $C_D \simeq 7.0$ (Fig.11.11)

$\Sigma F = 0 = -F_D - W + F_{buoy.}$; $F_D = F_{buoy.} - W$

$C_D A_p \rho V_0^2/2 = \forall(\gamma_{oil} - \gamma_{sphere})$; $\forall = (4/3)\pi r^3 = 4.19 \times 10^{-6} m^3$

$7 \times \pi \times 0.01^2 \times 900 \times 0.03^2/2 = 4.19 \times 10^{-6}(900\ g - g\rho_{sphere})$

$\rho_{sphere} = 878.3\ kg/m^3$; $\gamma_{sphere} = 8{,}616\ N/m^3$

11.39 Because the viscosity is large, it is expected that the sphere will fall according to Stokes' law.

Thus, $F_D = 3\pi\mu V_0 D = 3\pi\nu\rho V_0 D$

where $F_D = (1/6)\pi D^3(\gamma_{sphere} - \gamma_{oil})$

$= (1/6)\pi(0.0025)^3 \times 9{,}810(1.2 - 0.95) = 2.00 \times 10^{-5} N$

Then $V_0 = F_D/(3\pi\mu\rho D) = 2.0 \times 10^{-5}/(3\pi \times 10^{-4} \times 950 \times 0.0025)$

$V_0 = 0.0089\ m/s = 8.9\ mm/s$

Check Re: $Re = V_0 D/\nu = 0.0089 \times 0.0025/10^{-4} = 0.22$

Within Stokes' range so solution is correct.

11.40

$F_D = C_1 \gamma_{liq.} D^3 = C_2 D^3$ (1)

Also $F_D = C_D A_p \rho V^2/2 = C_3 D^2 V^2$ (2)

Eliminating F_D between Eq's (1) and (2) yields

$V^2 = C_4 D$ or $V = \sqrt{C_4 D}$

As the bubble rises it will expand because the pressure decreases with an increase in elevation; thus, the bubble will accelerate as it moves upward. The drag will be form drag because there is no solid surface to the bubble for viscous shear stress to act on.

11.41

$F_D = (1/6)\pi D^3(\gamma_{sphere} - \gamma_{fluid}) = (1/6)\pi(0.05)^3(9{,}810)(0.66 - 0.20)$

$F_D = 0.295$ N; Assume Stokes' law applies.

Then $0.295 = 3\pi\mu V_0 D$; $\mu = 0.295/(3\pi(0.005)(0.05)) = 125\ \text{N}\cdot\text{s/m}^2$

Check Re: $Re = VD\rho/\mu = 0.005 \times 0.05 \times 1{,}000 \times 0.66/125 = 1.32 \times 10^{-3}$

So the velocity is well into Stokes' range (Re < 0.5)

Thus, $\nu = \mu/\rho = 125\ \text{N}\cdot\text{s/m}^2/(660\ \text{kg/m}^3) = \underline{\underline{0.189\ \text{m}^2/\text{s}}}$

11.42

Assume $T_{air} = 60°F$; $\rho_{air} = 0.00237\ \text{slugs/ft}^3$; $\nu_{air} = 1.58 \times 10^{-4}\text{ft}^2/\text{sec}$;

$\mu_{air} = 3.74 \times 10^{-7}\text{lbf-sec/ft}^2$

$F_D = 3\pi\mu V_0 D$; $(1/6)\pi D^3\gamma_{water} = 3\pi\mu_{air}V_0 D$

$D^2\gamma_{water} = 18\ \mu_{air}V_0$ (1)

Also $VD/\nu = 0.5$; $V = 0.5\ \nu_{air}/D$ (2)

Solve Eq's. (1) & (2) for D:

$D^3 = 9\ \mu^2_{air}/(\rho_{air}\gamma_{water}) = 9 \times (3.74 \times 10^{-7})/(0.00237 \times 62.4)$

$= 8.51 \times 10^{-12}\text{ft}^3$

$D = 2.042 \times 10^{-4}\text{ft} = 0.000204\ \text{ft} = \underline{\underline{0.0024\ \text{in.}}}$

11.43 $\rho = p/RT = 96{,}000/(287 \times 273) = 1.23\ \text{kg/m}^3$

$F_D = \forall \times 6{,}000 = C_D A_p \rho V^2/2$ Assume $C_D = 0.5$

$(1/6)\pi d^3 \times 6{,}000 = 0.5 \times (\pi d^2/4) \times 1.23\ V^2/2$

$V = \sqrt{d \times 1{,}000 \times 16/1.23} = \sqrt{10 \times 16/1.23} = 11.4\ \text{m/s}$

$Re = 11.4 \times 0.01/(1.3 \times 10^{-5}) = 8.8 \times 10^3$; $C_D = 0.4$ (Fig. 11-11)

Recompute V: $V = 11.4 \times (0.5/0.4)^{1/2} = \underline{\underline{12.7\ \text{m/s}}}$

11.44

$$Wgt_{air} = (4/3)\pi r^3 \gamma_{rock}$$

$$45 = 4/3\pi r^3 \gamma_{rock} \quad (1)$$

$$F_{buoy} = (4/3)\pi r^3 \gamma_{water} = (45 - 25) = (4/3)\pi r^3 \gamma_{water} = (4/3)\pi r^3 \times 9{,}790 \quad (2)$$

Solve Eq's (1) & (2) for γ and r: $\gamma = 22{,}030\ \text{N/m}^3$; $r = 0.0787$ m

$F_D = C_D A_p \rho V_0^2/2$ or $V_0^2 = 2F_D/(C_D A_p \rho)$; $V_0^2 = 2 \times 25/(C_D \times 0.01947 \times 998)$

$$V_0 = 1.604/\sqrt{C_D}$$

Assume $C_D = 0.4$; also $A_p = \pi r^2 = 0.01947\ \text{m}^2$

Then $V_0 = 2.54$ m/s and $Re = (VD/\nu) = 2.54(2)(0.0787)/10^{-6} = 4 \times 10^5$

Try $C_D = 0.09$, $V_0 = 5.35$ m/s, $Re = 8.4 \times 10^5$

Try $C_D = 0.1$, $V_0 = 5.07$ m/s, $Re = 8 \times 10^5$ O.K.; $\underline{V = 5.07\ \text{m/s}}$

11.45

$$F_D = C_D A_p \rho V_0^2 / 2 = Ma$$

$$\text{then } a = C_D A_p \rho V_0^2 / (2M)$$

$$\text{where } M = 20{,}000 / 32.2 = 621.1 \text{ slugs}$$

$$C_D = 1.20 \text{ (Table 11.1)}$$

$$A_p = (\pi / 4)D^2 = 113.1 \text{ft}^2$$

$$\text{Then } a = 1.20 \times 113.1 \times 0.0024 \times 200^2 / (2 \times 621.1)$$

$$\underline{\text{deceleration} = 10.5 \text{ft} / \text{s}^2}$$

11.46 $F_D = C_D A_p \rho V_0^2/2$; $C_D = 1.20$ (Table 11.1)

$$V_0 = \sqrt{2F_D/(C_D A_p \rho)}$$

$$= \sqrt{2 \times 900/(1.2 \times (\pi/4) \times 49 \times 1.2)} = \underline{5.70\ \text{m/s}}$$

11.47 $F_{buoy} = \forall\gamma_{water} = 0.80 \times (\pi/4) \times 0.20^2 \times 9{,}810 = 246.5$ N

Then motive force $= F_{buoy} - Wgt = 246.5 - 200.0 = -46.5\ \text{N} = F_D$

$C_D = 0.87$ (Table 11-1) Then $46.5 = C_D A_p \rho V_0^2/2$

$$V_0 = \sqrt{(46.50 \times 2)/(0.87 \times (\pi/4) \times 0.2^2 \times 1{,}000)} = \underline{1.84\ \text{m/s upward}}$$

11.48 $F_D = C_D A_p \rho V_0^2/2$; $C_D = 0.81$ (From Table 11.1)

$A_p = (2)(L \cos 45°)(L) = 1.414L^2$

$F_D = Wgt - F_{buoy}$

$= 19.8 - 9{,}810L^3 = 19.8 - 9{,}810 \times (10^{-1})^3 = 10$ N

$10 = (0.81)(1.414 \times 10^{-2})(1{,}000)(V_c^2)/2$

$\underline{V_0 = 1.32 \text{ m/s}}$

11.49

$V_0 = (2F_D/(C_D A \rho))^{1/2}$

$F_{net} = F_D - W_{balloon} - W_{helium} + F_{buoy} = 0$

$F_D = +0.05 - (1/6)\pi D^3(\gamma_{air} - \gamma_{He})$

$= +0.05 - (1/6)\pi \times (0.30)^3\, 9.81(\rho_{air} - \rho_{He})$

Assume T = 15°C so $\rho_{air} = 1.22 \text{ kg/m}^3$; $\rho_{He} = 0.169 \text{ kg/m}^3$

$F_D = +0.05 - (1/6)\pi(0.30)^3 \times 9.81(1.22 - 0.169) = -0.099$ N

Assume $C_D \approx 0.40$ Then $V_0 = ((2 \times 0.099/(0.40 \times (\pi/4) \times 0.3^2 \times 1.3))^{1/2} = 2.32$m/s

Check Re and C_D: $Re = VD/\nu = 2.32 \times 0.3/(1.46 \times 10^{-5}) = 5 \times 10^4$

$C_D \approx 0.40$ O.K.; so V_0 = $\underline{2.32 \text{ m/s upward}}$

11.50 As in solution to Prob. 11.49:

$F_D = +W_{balloon} + W_{He} - F_{buoy}$

$F_D = +0.01 - (1/6)\pi \times 1^3(\gamma_{air} - \gamma_{air} \times 1{,}716/12{,}419)$

$F_D = +0.01 - (1/6)\pi \times 1^3 \times 0.0764(1 - 0.138)$

$F_D = +0.010 - 0.0345 = 0\ 0.0245$ lbs

$V_0 = \sqrt{2F_D/(C_D A_p \rho)} = \sqrt{2 \times 0.0245/((\pi/4) \times 0.00237\, C_D)} = \sqrt{26.3/C_D}$

Assume $C_D = 0.40$ Then $V_0 = \sqrt{26.3/0.4} = 8.1$ ft/s upward

$Re = VD/\nu = 8.1 \times 1/(1.58 \times 10^{-4}) = 5.2 \times 10^4$; $C_D = 0.50$

Again: $V_0 = \sqrt{26.3/0.5}$ = $\underline{7.25 \text{ ft/s}}$

11.51 $\Sigma F_y = 0;\ +T - Wgt - Drag + F_{buoy.} = 0$

$T = Wgt + Drag - F_{buoy.}$

$T = (\pi/4) \times 0.3^2 \times 0.3\ (15{,}000 - 9{,}810) + C_D(\pi/4) \times 0.3^2$

$\times 1{,}000 \times 1.5^2/2;\quad C_D = 0.90$ (Table 11-1)

Then $T = 110 + 71.6 = \underline{\underline{181.6\ N}}$

11.52 $V_0 = [(\gamma_s - \gamma_w)(4/3)D/(C_D\rho_w)]^{1/2};\quad$ Assume $C_D = 0.5$

$V_0 = [62.4(2.94 - 1)(4/3) \times (1/(4 \times 12))/(0.5 \times 1.94)]^{1/2}$

$V_0 = 1.86$ ft/s; $Re = 1.86 \times (1/48)/10^{-5} = 3.8 \times 10^3$

$C_D = 0.4$; Recompute V: $V_0 = 1.86 \times (0.5/0.4)^{1/2} = \underline{\underline{2.08\ ft/s}}$

11.53 $F_D = 15 - 9{,}810 \times (1/6)\pi D^3 = 15 - 9{,}810 \times (1/6)\pi \times 0.15^3 = 2.336$ N

Buoyant force is greater than weight, so ball will rise.

$2.336 = C_D(\pi D^2/4) \times 1{,}000\ V^2/2;\quad$ Assume $C_D = 0.4$

$V = \sqrt{2.336 \times 8/(\pi C_D \times 1{,}000 \times 0.15^2)} = 0.514/\sqrt{C_D}$

$V = 0.813$ m/s; $Re = VD/\nu = 0.813 \times 0.15/(1.3 \times 10^{-6}) = 9 \times 10^4$

$C_D = 0.48$; $V = 0.514/\sqrt{0.48} = \underline{\underline{0.742\ m/s\ upward}}$

11.54

Net $F = 0 = -W_{balloon} - W_{He} + F_{buoy.} + F_D$

$F_D = +3 - (1/6)\pi D^3(\gamma_{air} - \gamma_{He})$

$= +3 - (1/6)\pi \times 2^3 \times \gamma_{air}(1 - 287/2{,}077)$

$= +3 - (1/6)\pi \times 8 \times 1.225(1 - 0.138)$

$= +3 - 4.422 = -1.422$ N

Then $F_D = C_D A_p \rho V_0^2/2$; Then $V_0 = \sqrt{1.422 \times 2/((\pi/4) \times 2^2 \times 1.225\ C_D)} = \sqrt{0.739/C_D}$

Assume $C_D = 0.4$ then $V_0 = \sqrt{0.739/0.4} = 1.36$ m/s

$Re = VD/\nu = 1.36 \times 2/(1.46 \times 10^{-5}) = 1.86 \times 10^5$; $C_D = 0.42$

Try again: $V_0 = \sqrt{0.739/0.42} = \underline{\underline{1.33\ m/s\ upward}}$

11.55 $F_D = Wgt \qquad M = 1.0$

$C_D A_p \rho V_0^2/2 = wgt. = mg$

$C_D A_p k p M^2/2 = mg \quad ; \quad C_D = 0.80$ from Fig. 11-12

$\therefore \quad A_p = (mg)(2)/(C_D k p M^2)$

$\pi D^2/4 = (2500\pi D^3/6)(9.81)(2)/((0.8)(20 \times 10^3)(1.4)(1))$

$D = 0.69$ m

11.56 $F_D = C_D A_p \rho V_0^2/2$; Assume $C_D = 0.50$

$(\gamma_s - \gamma_w)\pi d^3/6 = C_D(\pi/4)d^2 \times 998 \quad V_0^2/2$

$\gamma_s = (93.56/d) + \gamma_w$

Now determine values of γ_s for different d values.
Results are shown below for a C_D of 0.50.

d(cm)	10	15	20
γ_s (N/m³)	10,725	10,413	10,238

$Re = VD/\nu = 0.5 \times 0.1/10^{-6} = 5 \times 10^4$

$C_D = 0.5$ O.K.

11.57 $F_L = C_L A_p \rho V_0^2/2 \quad ; \quad r\omega/V_0 = (0.1)(60)/3 = 2$

$C_L = 0.40$ (from Fig. 11-17)

$F_L = (0.40)(\pi/4)(0.2^2)(1.94)(3^2)/2$

$F_L = 0.11$ lbf

11.58 It will "break" toward the a) North

11.59 Assume T = 70°F; then $\rho = 0.0023$ slugs/ft³; $V_0 = 85$ mph $= 125$ ft/s

$r\omega/V_0 = (9/(12 \times 2\pi)) \times 35 \times 2\pi/125 = 0.21$

Then from Fig. 11-17 $C_L = 3 \times 0.05 = 0.15$

$F_L = C_L A \rho V_0^2/2 = 0.15 \times (9/12\pi)^2 \times (\pi/4) \times 0.0023 \times 125^2/2) = $ 0.121 lbf

Deflection will be $s = 1/2\ at^2$ where a is acceleration

$a = F/M$ and $t = L/V_0 = 60/125 = 0.48$ s

$a = F/M = 0.100/((5/16)/(32.2)) = 12.4$ ft/s²

Then deflection $= (1/2) \times 12.4 \times 0.48^2 = 1.43$ ft

---

11.60 Correct choice is force vector a.

11.61

Range: Before corn is popped, it should not be thrown out by the air,

so let $V_{max} = (2F_D/C_D A_p \rho_{air})^{1/2}$ $\rho = 0.83$ kg/m^3 (calculated)

where F_D = weight of unpopped corn = $0.15 \times 10^{-3} \times 9.81$ N

$A_p = (\pi/4) \times (0.006)^2 m^2$; $\rho_{air,start\ up} \approx 0.83 kg/m^3$; Assume $C_D \simeq 0.4$

Then $V_{max} = [2 \times 0.15 \times 10^{-3} \times 9.81/(0.40 \times (\pi/4)(0.006)^2 \times 1.2)]^{1/2} = \underline{18 \text{ m/s}}$

Check Re and C_D: $Re = VD/\nu = 14.7 \times 0.006/(2.8 \times 10^{-5}) = 3 \times 10^3$

$C_D \simeq 0.4$ so solution for V_{max} is O.K.

For minimum velocity let popped corn be suspended by stream of air. Assume only that diameter changes.

So $V_{min} = V_{max} \times (A_U/A_P)^{1/2} = V_{max}(D_U/D_P)$

where D_P = diameter of popped corn

and D_U = diameter of unpopped corn

$V_{min} \simeq (6/18) \times V_{max} = \underline{6 \text{ m/s}}$

11.62

An American flag is 1.9 times as long as it is high.

Thus $A = 6^2 \times 1.9 = 68.4$ ft^2

Assume T = 60°F $\rho = 0.00237$ slugs/ft^3

V = 100 mph = 147 ft/s

Compute drag of flag:

$F_D = C_D A \rho V_0^2/2$

$= 0.14 \times 68.4 \times 0.00237 \times 147^2/2$

$F_D = 244$ lbf

Make the flag pole of steel using one size for the top half and a larger size for the bottom half. To start the determination of d for the top half, assume that the pipe diameter is 6 in.

Then $F_{on\ pipe} = C_D A_p \rho V_0^2/2$ $Re = VD/\nu = 147 \times 0.5/(1.58 \times 10^{-4})$

$= 4.7 \times 10^5$

11.62 (continued)

With an Re of 4.7×10^5 C_D may be as low as 0.3 (Fig. 11-5); however, for conservative design purposes, assume $C_D = 1.0$.

Then $F_{pipe} = 1 \times 50 \times 0.5 \times 0.00237 \times 147^2/2 = 640$ lbf

$M = 244 \times 50 \times 12 + 640 \times 25 \times 12 = 338{,}450$ in.-lbf

$I/c = M/s$; assume allowable $s = 30{,}000$ psi

$I/c = 338{,}450/30{,}000 = 11.28$ in^3

From handbook it is found that a 6 in. double extra-strength pipe will be adequate.

Bottom half: $F_{flag} = 244$ lbf

Assume bottom pipe will be 12 in. in diameter.

$F_{6in.pipe} = 640$ lbf

$F_{12in.pipe} = 1 \times 50 \times 1 \times 0.00237 \times 147^2/2 = 1{,}280$ lbf

$M = 12(244 \times 100 + 640 \times 75 + 1{,}280 \times 25) = 1{,}253{,}000$ in.-lbf

$M_s = 41.8$ in.$^3 = I/c$

Handbook shows that 12 in. extra-strength pipe should be adequate.
Note: Many other designs are possible.

11.63 Force normal to plate will be based upon the $C_{p,net}$ where $C_{p,net}$ is the average net Cp producing a normal pressure on the plate. For example, at the leading edge of the plate the $C_{p,net} = 2.0 + 1.0 = 3.0$. Thus, for the entire plate the average net $C_p = 1.5$.

Then $$F_{\text{normal to plate}} = C_{p,net} A_{plate} \rho V_0^2/2$$

$$= 1.5\, A_{plate} \rho V_0^2/2$$

The force normal to $V_0 = F_L = (F_{\text{normal to plate}})(\cos 30°)$

$$C_L S \rho V_0^2/2 = (1.5)(A_{plate})(\rho V_0^2/2) \cos 30°$$

$C_L = 1.5 \cos 30° = 1.30$ (Based upon plan form area)

However, if C_L is to be based upon projected area where

$A_{proj} = A_{plate} \sin 30°$ then

$C_L = 2.60$ (Based upon projected area)

11.64 Assume $C_L \approx 0.60$ (Fig. 11-23)

$$F_L = C_L A \rho V_0^2/2$$

$$2{,}000 = (0.60)(4\ell)(0.0024)(200^2)/2$$

$$\ell = 17.4 \text{ ft}$$

$$\ell/c = 17.4/4 = 4.34 \rightarrow C_L = 0.59 \text{ (Fig. 11-23)}$$

Calculate ℓ again: $\ell = 17.4\,(0.60/0.59) = \underline{\underline{17.7 \text{ ft}}}$

11.65 Use Fig. 11-23 for characteristics; $\ell/c = 4$ so $C_L = 0.55$

$F_L = C_L A \rho V_0^2/2$; $10{,}000 = 0.55 \times 4\,c^2 \times (1.94/2) \times 3{,}600$

$c^2 = 1.30$ ft; $c = 1.14$ ft; $\ell = 4\,c = 4.56$ ft

Use a foil $\underline{\underline{1.14 \text{ ft wide} \times 4.56 \text{ ft long}}}$

11.66 $\rho = p/RT$ where $T = (-67 + 460) = 393°R$

$\rho = 475.2/(1716 \times 393)$ $\quad$ 3.3 psia = 475.2 psfa

$\rho = 0.000705 \text{ slug/ft}^3$

$$F_L = C_L S \rho V_0^2/2$$

$$S = 2F_L/(C_L \rho V_0^2)$$

$$= 2 \times 10{,}000/(0.2 \times 0.000705 \times 600^2)$$

$$\underline{\underline{S = 394 \text{ ft}^2}}$$

11.67 Correct choice is $\underline{\underline{(d)}}$ because C_L increases with increase in aspect ratio.

11.68

$$C_{Di} = C_L^2/(\pi(b^2/S)) \tag{11.19}$$

In the equation for the induced drag coefficient (see above) the only variable for a given airplane is C_L; therefore, one must determine if C_L varies for the given conditions. If the airplane is in level flight the lift force must be constant. Because $F_L = C_L A \rho V^2/2$ it is obvious that C_L must decrease with increasing V. This would be accomplished by decreasing the angle of attack. If C_L decreases, then Eq. (11.19) shows that C_{di} also must decrease. The correct answer is $\underline{(b)}$.

11.69

$$W/S = \tfrac{1}{2}\rho C_L V^2$$

$$\text{or } C_L = (2/\rho)(1/V^2)(W/S)$$

$$P = F_D V = (C_{Do} + (C_L^2/\pi\Lambda)(1/2)\rho V^3 S$$

$$P = \tfrac{1}{2}\rho V^3 S C_{Do} + (4/\rho^2)(1/V^4)(W^2/S^2)(1/(\pi\Lambda))(\tfrac{1}{2}\rho V^3 S)$$

$$P = \left[\tfrac{1}{2}\rho V^3 C_{Do} + (2/\rho)(1/(\pi\Lambda V)(W^2/S^2)\right]S$$

$$dP/dV = ((3/2)\rho V^2 C_{Do} - (2/\rho)(1/(\pi\Lambda V^2))(W/S)^2)S$$

for minimum power dP/dV = 0 and

$$(3/2)\rho V^2 C_{Do} = (2/\rho)(1/(\pi\Lambda V^2)(W/S)^2$$

$$V = \left[\tfrac{4}{3}(W/S)^2(1/(\pi\Lambda\rho^2 C_{Do}))\right]^{1/4}$$

for $\rho = 1$, $\Lambda = 10$, $W/S = 500$ and $C_{do} = 0.2$

$$V = \left[\tfrac{4}{3}(500^2)(1/(\pi \times 10 \times 1^2 \times 0.02))\right]^{1/4}$$

$$= 27.0 \text{m/s}$$

11.70 Take the stream tube between sections 1 and 2 as a control volume and apply the momentum equation

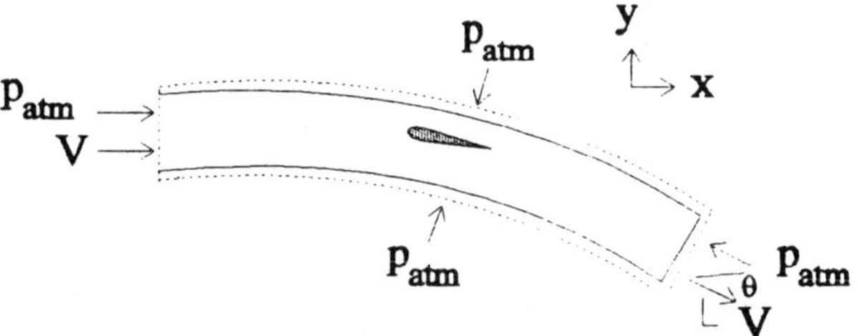

$$\sum F_y = \sum_{cs} V_y \rho \underline{V} \cdot \underline{A} + d/dt \int_{cv} V_y \rho d\forall$$

The flow is steady: d/dt( ) = 0. Also $V_1 = V_2 = V$. The only F_y is the force of the wing on the fluid in the control volume:

$$F_y = (-V\sin\theta)\rho VA$$

$$= -\rho V^2 A \sin\theta$$

But the fluid acting on the wing in the y direction is the lift F_L and it is the negative of F_y. Or

$$F_L = \rho V^2 A \sin\theta \qquad (1)$$

$$C_L = 2F_L/(\rho V^2 S) \qquad (2)$$

Eliminate F_L between Eqs. (1) and (2)

$$C_L = 2\rho V^2 A \sin\theta/(\rho V^2 S)$$

$$C_L = 2A\sin\theta/S$$

$$= 2(\pi/4)b^2 \sin\theta/S$$

$$C_L = (\pi/2)\sin\theta(b^2/S)$$

11.70 (Cont'd)

But $\sin\theta \approx \theta$ for small angles. Therefore,

$$C_L = (\pi/2)\theta(b^2/S)$$

$$\underline{\text{or } \theta = 2C_L/(\pi b^2/S)} \qquad (3)$$

$$F_{Di} = F_L(\theta/2)$$

$$C_{Di}\rho V^2 S/2 = (C_L \rho V^2 S/2)(\theta/2) \qquad (4)$$

Eliminate θ between Eqs. (3) and (4)

$$C_{Di}\rho V^2 S/2 = (C_L \rho V^2 S/2)(C_L/(\pi b^2/S))$$

$$\underline{\underline{C_{Di} = C_L^2/(\pi\Lambda)}}$$

11.71 $C_{L_{max}} = 1.40$ hence the C_L at stall; wgt = $C_L A\rho V_0^2/2$

for landing Wgt = $1.2\ A\rho V_L^2/2$

for stall Wgt = $1.4\ A\rho V_s^2/2$ (1)

But $V_L = V_s + 5$, so Wgt = $1.2\ A\rho(V_s + 5)^2/2$ (2)

Solve Eq's (1) and (2) for V_s

$1.2\ (V_s + 5)^2 = 1.4\ V_s^2$

$\underline{\underline{V_s = 62.4 \text{ m/s}}}$

$\underline{\underline{V_L = V_s + 5 = 67.4 \text{ m/s}}}$

11.72

Calculate p and then ρ: $p = p_0[(T_0 - \alpha(z - z_0))/T_0]^{g/\alpha R}$

$p = 101.3[(288 - (6.5 \times 10^{-3})(3{,}000))/288]^{(9.81/(6.5 \times 10^{-3} \times 287))} = 70.1$ kPa

$T = 288 - 6.5 \times 10^{-3} \times 3{,}000 = 268.5$ K

Then $\rho = p/RT = 70{,}100/(287 \times 268.5) = 0.910$ kg/m^3

$C_L = (F_L/A)/(\rho V_0^2/2) = (1{,}000 \times 9.81/20)/(0.91 \times 60^2/2) = 0.299$

Then $C_{D_i} = C_L^2/(\pi(b^2/S)) = 0.299^2/(\pi(14^2/20)) = 0.0029$

Then the total drag coefficient = $C_{D_i} + 0.01 = \underline{\underline{0.0129}}$

Total wing drag = $0.0129 \times 20 \times 0.910 \times 60^2/2 = \underline{\underline{423 \text{ N}}}$

Power = $60 \times 423 = \underline{\underline{25.4 \text{ kW}}}$

11.73

Cavitation will start at point where C_p is minimum, or in this case, where $C_p = -1.95$. Also $p_0 = 0.70 \times 9{,}810$ Pa gage

$$C_p = (p - p_0)/(\rho V_0^2/2)$$

and for cavitation $p = p_{vapor} = 1{,}230$ Pa abs.

$$p_0 = 0.7 \times 9{,}810 + 101{,}300 \text{ Pa abs.}$$

So $-1.95 = [1{,}230 - (0.7 \times 9{,}810 + 101{,}300)]/(1{,}000\ V_0^2/2)$

$$V_0 = \underline{\underline{10.5 \text{ m/s}}}$$

Lift: By approximating the C_p diagrams by triangles, it is found that $C_{p_{avg.}}$ on the top of the lifting vane is approx. -1.0 and

$C_{p_{avg.,bottom}} \approx +0.45$

Thus, $\Delta C_{p_{avg.}} \approx 1.45$

Then $F_{L/length} = 1.45 \times 0.20 \times 1{,}000 \times (10.5)^2/2$

$$F_{L/length} = \underline{\underline{16{,}000 \text{ N/m}}}$$

11.74 The correct choice is $\underline{\underline{(b)}}$.

11.75

$$C_D/C_L = (C_{D_0}/C_L) + (C_L/(\pi\Lambda))$$

$$d/dC_L(C_D/C_L) = (-C_{D_0}/C_L^2) + (1/(\pi\Lambda)) = 0; \quad \underline{\underline{C_L = \sqrt{\pi\Lambda C_{D_0}}}}$$

$$C_D = C_{D_0} + \pi\Lambda C_{D_0}/(\pi\Lambda) = 2\,C_{D_0}$$

Then $\underline{\underline{C_L/C_D = (1/2)\sqrt{\pi\Lambda/C_{D_0}}}}$

11.76

$$\ell = 1{,}000/(\sin 1.7°) = 33{,}708 \text{ m}$$

$$\text{Lift} = \text{Wgt} = (1/2)\rho V^2 C_L S$$

$$200 \times 9.81 = 0.5 \times 1.2 \times V^2 \times 1.0 \times 20;\ V = 12.8 \text{ m/s}$$

Then $t = 33{,}708 \text{ m}/(12.8 \text{ m/s}) = \underline{\underline{2{,}633 \text{ s} = 43.9 \text{ min}}}$

11.77 $L = C_L S \rho V_0^2/2$

$L/S = C_L \rho V_0^2/2$

$\rho V_0^2/2 = (2{,}000/0.4) = 5{,}000\ N/m^2$

At $C_L = 0.40$, $C_D = 0.02$

$\therefore$ $F_D = C_D S \rho V_0^2/2 = (0.02)\ (5{,}000)\ (10) = \underline{\underline{1{,}000\ N}}$

11.78 $S = 200\ ft^2$ Wgt. = 400 lbf

$Wgt. = C_L S \rho V_0^2/2$

$C_L = (wgt)/(S\rho V_0^2/2) = (400)/((200)\ (0.002)\ (50^2)/2) = 0.80$

$C_D = 0.06$ (from Fig. 11-23) ; $\alpha = 7°$

$F_D = C_D S \rho V_0^2/2 = (0.06)\ (200)\ (0.002)\ (50^2)/2 = \underline{\underline{30\ lbf}}$

12.1 $c = \sqrt{kRT} = \sqrt{(1.31)(518)(293)} = \underline{446 \text{ m/s}}$

12.2 $c = \sqrt{kRT} = \sqrt{(1.66)(2,077)(273 + 50)} = \underline{1,055 \text{ m/s}}$

12.3 $c = \sqrt{kRT} = \sqrt{(1.41)(24,677)(460 + 75)} = \underline{4,315 \text{ ft/sec}}$

12.4 $c_{He} = \sqrt{(kR)_{He}T} = \sqrt{(1.66)(2,077)(288)} = 996.5 \text{ m/s}$

$c_{N_2} = \sqrt{(kR)_{N_2}T} = \sqrt{(1.40)(297)(288)} = 346.0 \text{ m/s}$

$c_{He} - c_{N_2} = \underline{650.5 \text{ m/s}}$

12.5 $c^2 = \partial p/\partial \rho; \quad p = \rho RT$

If isothermal, T = const. $\therefore \partial p/\partial \rho = RT$

$\therefore \quad c^2 = RT$

$\underline{c = \sqrt{RT}}$

12.6 $p - p_0 = E_v \ln(\rho/\rho_0)$

$c^2 = \partial p/\partial \rho = E_v/\rho$

$\therefore \quad \underline{c = \sqrt{E_v/\rho}}$

$c = \sqrt{2.20 \times 10^9/10^3} = \underline{1,483 \text{ m/s}}$

12.7 Total temperature will develop at exposed surface.

At $M_1 = 2 \quad T/T_t = 0.5556$ (from Table A.1)

$T_t = (273 - 30)/0.5556 = 437\text{K} = \underline{164\text{C}}$

Airspeed behind the shock:

$M_2 = 0.5774$, $T_2/T_1 = 1.688$ (From Table A.1 at $M_1 = 2$)

Then $V_2 = M_2 c_2$

$c_2 = (kRT_2)^{0.5}$

12.7 (continued)

where k = 1.4

R = 287

$T_2 = (273 - 30)(1.688) = 410k$

$c_2 = (1.4(287)(410))^{0.5} = 406$ m/s

$V_2 = 0.5774\ (406) = 234$ m/s = 844 km/hr.

12.8 $T/T_t = 0.5556$ at M=2.0 (From Table A.1)

$T_t = (1/0.5556)(273) = 491k = 218°C$

12.9 At 10,000 m $c = \sqrt{(1.40)(287)(229)} = 303.3$ m/s

a) $V = (1.5)(303.3)(3{,}600/1{,}000) = 1{,}638$ km/hr

b) $T_t = 229(1 + ((1.4 - 1)/2) \times 1.5^2) = 332$ K = 59°C

c) $p_t = (30.5)(1 + 0.2 \times 1.5^2)^{(1.4/(1.4-1))} = 112$ kPa

d) M = 1; V = c

$V = (303.3)(3{,}600/1{,}000) = 1{,}092$ km/hr

12.10 $c = \sqrt{kRT} = \sqrt{(1.4)(287)(288)} = 340.2$ m/s

V = 800 km/hr = 222.2 m/s; M = 222.2/340.2 = 0.653

c at altitude $= \sqrt{(1.4)(287)(233)} = 306.0$ m/s

V = 199.8 m/s = 719.3 km/hr

12.11 $q = (k/2)pM^2 = (1.4/2)(30)\ (0.9)^2 = 17.01$ kPa

$W = C_L qS = L$

$W = L/S = C_L q = (0.05)(17.01) = 0.850$ kPa = 850 Pa

12.12 $c = \sqrt{kRT} = \sqrt{(1.4)(287)(293)} = 343$ m/s

$M = 250/343 = 0.729$

$T_t = (293)(1 + 0.2 \times (0.729)^2) = 293 \times 1.106 = 324$ K = $\underline{\underline{51°C}}$

$p_t = (200)(1.106)^{3.5} = \underline{\underline{284.6 \text{ kPa}}}$

12.13 $T_t = T(1 + ((k-1)/2)M^2$

$T = 283/(1 + 0.2 \times 0.5^2) = 270$ K

$p = 340/(1.05)^{3.5} = 287$ kPa

$c = [(1.4)(287)(270)]^{1/2} = 329.4$ m/s

$V = (0.5)(329.4) = 164.7$ m/s

$\rho = p/RT = 287 \times 10^3/(287 \times 270) = 3.70$ kg/m^3

$\dot{m} = (164.7)(3.70)(0.0065) = \underline{\underline{3.96 \text{ kg/s}}}$

12.14 $p_t = 300$ kPa; $T_t = 200°C = 473$ K

$T = 473/(1 + 0.2 \times 0.9^2) = 473/1.162 = \underline{\underline{407 \text{ K}}}$

$p = 300/(1.162)^{3.5} = \underline{\underline{177.4 \text{ kPa}}}$

$c = [(1.4)(260)(407)]^{1/2} = 384.9$ m/s

$V = (0.9)(384.9) = \underline{\underline{346.4 \text{ m/s}}}$

12.15 $T_t = 300$ K; $T = 50$ K

$T_0/T = 1 + ((k-1)/2)M^2$

$300/50 = 6 = 1 + 0.2\ M^2$; $\underline{\underline{M = 5}}$

12.16 $T_t = 20°C = 293$ K

$P_t = 500$ kPa

$V = 300$ m/s

$c_pT + V^2/2 = c_pT_0$

$\therefore T = T_t - V^2/(2c_p) = 293 - (300)^2/((2)(14,223)) = \underline{\underline{289.8 \text{ K}}}$

$c = \sqrt{kRT} = \sqrt{(1.41)(4,127)(289.8)} = 1,299$ m/s

12.16 (continued)

$M = 300/1,299 = \underline{\underline{0.231}}$

$p = 500/[1 + (0.41/2) \times 0.231^2]^{(1.41/0.41)} = \underline{\underline{481.6 \text{ kPa}}}$

$\rho = p/RT = (481.6)(10^3)/(4,127 \times 289.8) = 0.403 \text{ kg/m}^3$

$\dot{m} = \rho AV = (0.403)(0.02)^2(\pi/4)(300) = \underline{\underline{0.038 \text{ kg/s}}}$

12.17 $M = 2; \quad p_t = 600 \text{ kPa}; \quad F_D = C_D(1/2)\rho U^2 A$

$p = p_t/[1 + ((k-1)/2)M^2]^{k/(k-1)} = 600/[1 + 0.2(2)^2]^{3.5} = 76.7 \text{ kPa}$

$(1/2)\rho U^2 = kpM^2/2 = 1.4 \times 76.7 \times 2^2/2 = 214.8 \text{ kPa}$

$F_D = (0.95)(214.8 \times 10^3)(0.01)^2(\pi/4) = \underline{\underline{16.0 \text{ N}}}$

12.18 $C_p = (p_t - p)/\rho U^2/2 = (p_t - p)/kpM^2/2 = (2/kM^2)[(p_t/p) - 1]$

$= \underline{\underline{(2/kM^2)[(1 + (k-1/2)M^2)^{(k/(k-1))} - 1]}}$

$C_p(2) = \underline{\underline{2.43}}$

$C_p(4) = \underline{\underline{13.47}}$

$\underline{\underline{C_{p_{inc.}} = 1.0}}$

12.19 $p_t/p = [1 + (k-1)M^2/2]^{k/(k-1)}$

$M = \sqrt{(2/(k-1))[(p_t/p)^{(k-1)/k} - 1]}$

$p_t/p = 1 + \varepsilon; (p_t/p)^{(k-1)/k} = (1 + \varepsilon)^{(k-1)/k} = 1 + ((k-1)/k)\varepsilon + O(\varepsilon^2)$

$(p_t/p)^{(k-1)/k} - 1 \simeq ((k-1)/k)\varepsilon + O(\varepsilon^2)$

$M = [(2/(k-1))((k-1)/k)\varepsilon]^{1/2}$ neglecting higher order terms

$\underline{\underline{M = [(2/k)((p_t)/p) - 1)]^{1/2} \text{ as } \varepsilon \to 0}}$

12.20

$V_1 = 500$ m/s; $T_1 = -40°C = 233$ K; $p_1 = 70$ kPa

$c_1 = \sqrt{kRT} = \sqrt{(1.4)(297)(233)} = 311$ m/s

$M_1 = 500/311 = 1.61$

$M_2^2 = [(k-1)M_1^2 + 2]/[2kM_1^2 - (k-1)] = [(0.4)(1.61)^2+2]/[(2)(1.4)(1.61)^2 - 0.4]$

$M_2 = \underline{\underline{0.665}}$

$p_2 = p_1(1 + k_1M_1^2)/[(1 + k_1M_2^2)] = (70)(1 + 1.4 \times 1.61^2)/(1+1.4 \times 0.665^2)$

$= \underline{\underline{200 \text{ kPa}}}$

$T_2 = T_1(1 + ((k-1)/2)M_1^2)/(1 + ((k-1)/2)M_2^2)$

$= 233\ [1 + 0.2 \times 1.61^2]/[1+0.2 \times 0.665^2] = 325 \text{ K} = \underline{\underline{52°C}}$

$\Delta s = R\ell n[(p_1/p_2)(T_2/T_1)^{k/(k-1)}]$

$= R[\ell n(p_1/p_2) + (k/(k-1))\ell n(T_2/T_1)]$

$= 297\ [\ell n(70/200) + 3.5\ \ell n(325/233)] = \underline{\underline{34.1 \text{ J/kg K}}}$

12.21

$M = 3$; $T = 45°F = 505°R$; $p = 30$ psia

$M_2^2 = [(k-1)M_1^2 + 2]/[2kM_1^2 - (k-1)]$; $M_2 = \underline{\underline{0.475}}$

$(T_2/T_1) = [1 + ((k-1)/2)M_1^2]/[1 + ((k-1)/2)M_2^2]$

$= (1 + (0.2)(9))/(1 + (0.2)(0.475)^2) = 2.679$

$T_2 = 505 \times 2.679 = 1353°R = \underline{\underline{893°F}}$

$p_2/p_1 = (1 + kM_1^2)/(1 + kM_2^2) = (1 + 1.4 \times 9)/(1 + 1.4 \times (0.475)^2) = 10.33$

$p_2 = (10.33)(30) = \underline{\underline{310 \text{ psia}}}$

12.22 $p_{t_2} = 150$ kPa; $P_1 = 40$ kPa $\qquad P_{t_2}/P_1 = 3.75$

$p_{t_2}/p_1 = 3 = (p_{t_2}/p_{t_1})(p_{t_1}/p_1)$

12.22 (continued)

Using compressible flow tables:

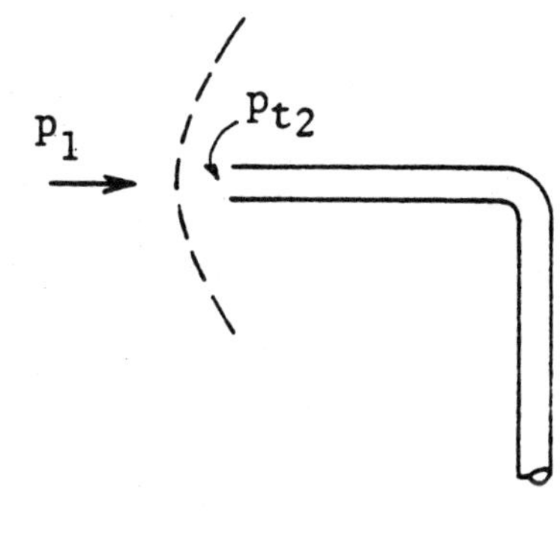

M	p_{t_2}/p_{t_1}	p_1/p_{t_1}	p_{t_2}/p_1
1.60	0.8952	0.2353	3.80
1.50	0.9278	0.2724	3.40
1.40	0.9582	0.3142	3.04
1.35	0.9697	0.3370	2.87

Therefore, interpolating, $M = \underline{\underline{1.59}}$

12.23

$M = 2$; $p = 100$ kPa, abs; $T = 20°C = 293$ K; $k = 1.31$

$M_2^2 = [(k-1)M_1^2 + 2]/\{2kM_1^2 - (k-1)\} = ((0.31)(4)+2)/((2)(1.31)(4)-0.31)=0.3186$

$M_2 = 0.564$

$p_2/p_1 = (1 + kM_1^2)/(1 + kM_2^2) = (1 + 1.31 \times 4)/(1 + 1.31 \times 0.3186) = 4.40$

$\underline{\underline{p_2 = 440 \text{ kPa, abs}}}$

$T_2/T_1 = [1 + ((k-1)/2)M_1^2]/[1 + ((k-1)/2)M_2^2] = 1.54$

$T_2 = (293)(1.54) = \underline{451 \text{ K}} = \underline{178°C}$

$\rho_2 = p_2/(RT_2) = (440)(10^3)/((518)(451)) = \underline{\underline{1.88 \text{ kg/m}^3}}$

12.24

$M_2 = 0.8$; $k = 1.66$; $R = 2{,}077$ J/kgK; $T_2 = 100°C = 373$ K

$M_1^2 = [(k-1)M_2^2 + 2]/[2kM_2^2 - (k-1)] = 1.653$; $M_1 = 1.28$

$T_1/T_2 = [1+((k-1)/2)M_2^2]/[1 + ((k-1)/2)M_1^2] = 0.786$; $T_1 = (0.786)(373) = 293$K

$c_1 = (1.66 \times 2{,}077 \times 293)^{1/2} = 1005$ m/s

$V_1 = (1005)(1.28) = 1286$ m/s

12.25 $M_2^2 = ((k-1)M_1^2 + 2)/(2kM_1^2 - (k-1))$

Because $M_1 >> 1$, $(k-1)M_1^2 >> 2$

$2kM_1^2 >> (k-1)$

So in limit $M_2^2 \rightarrow ((k-1)M_1^2)/2kM_1^2 = (k-1)/2k$

$\therefore \quad M_2 \rightarrow \sqrt{(k-1)/2k}$

$\rho_2/\rho_1 = (p_2/p_1)(T_1/T_2)$

$= ((1 + kM_1^2)/(1 + kM_2^2))(1+((k-1)/2)M_2^2)/(1+((k-1)/2)M_1^2)$

in limit $M_2^2 \rightarrow (k-1)/2k$ and $M_1 \rightarrow \infty$

$\therefore \rho_2/\rho_1 \rightarrow [(kM_1^2)/((k-1)/2)M_1^2][(1 + (k-1)^2/4k)/(1 + k(k-1)/2k)]$

$\rho_2/\rho_1 \rightarrow (k+1)/(k-1)$

$M_2(\text{air}) = 0.378$

$\rho_2/\rho_1(\text{air}) = 6.0$

12.26 $M_2^2 = [(k-1)M_1^2 + 2]/[2kM_1^2 - (k-1)]$

$= [(k-1)(1+\varepsilon)+2]/[2k(1+\varepsilon) - (k-1)] = [k+1+(k-1)\varepsilon]/[k+1+2k\varepsilon]$

$= [1+(k-1)\varepsilon/(k+1)]/[1+(2k\varepsilon)/(k+1)]$

$\simeq [1+(k-1)\varepsilon/(k+1)][1-(2k\varepsilon)/(k+1)]$

$\approx 1+(k-1-2k)\varepsilon/(k+1)$

$\approx 1-\varepsilon$

$\approx 1-(M_1^2-1)$

$\approx \underline{\underline{2-M_1^2}}$

M_1	M_2	M_2(Table A-1)
1.0	1.0	1.0
1.05	0.947	0.953
1.1	0.889	0.912
1.2	0.748	0.842

12.27

$A_T = 3 \text{ cm}^2 = 3 \times 10^{-4} \text{m}^2$

$p_t = 300$ kPa; $T_t = 20°\text{C} = 293$ K

$p_b = 90$ kPa

$p_b/p_t = 90/300 = 0.3$

Because $p_b/p_t < 0.528$, sonic flow at exit.

$\therefore \dot{m} = 0.685\, p_t A_*/\sqrt{RT_t} = (0.685)(3 \times 10^5)(3 \times 10^{-4})/\sqrt{(287)(293)} = \underline{\underline{0.212 \text{ kg/s}}}$

12.28 $A_T = 3\ cm^2 = 3 \times 10^{-4} m^2$

$A_p = 12\ cm^2 = 12 \times 10^{-4} m^2$

$p_t = 150$ kPa; $T_t = 303$ K

$p_b = 100$ kPa; $k = 1.31$; $R = 518$ J/kgK

$p_b/p_t = 100/150 = 0.667$; $p_*/p_t\big|_{methane} = (2/(k+1))^{k/(k-1)} = 0.544$

$p_b > p_*$, subsonic flow at exit

1) $M_e = \sqrt{(2/(k-1))[(p_t/p_b)^{(k-1)/k} - 1]} = \sqrt{6.45[(1.5)^{0.2366} - 1]} = 0.806$

$T_e = 303\ K/(1+(0.31/2)\times(0.806)^2) = 275$ K

$c_e = \sqrt{(1.31)(518)(275)} = 432$ m/s

$\rho_e = p_b/(RT_e) = 100 \times 10^3/(518 \times 275) = 0.702\ kg/m^3$

$\dot{m} = \rho_e V_e A_T = (0.702)(0.806)(432)(3 \times 10^{-4}) = \underline{\underline{0.0733\ kg/s}}$

2) Assuming Bernoulli's equation is valid, $p_t - p_b = (1/2)\rho V_e^2$

$V_e = \sqrt{2(150-100)10^3/0.702} = 377$ m/s

$\dot{m} = (0.702)(377)(3 \times 10^{-4}) = \underline{\underline{0.0794\ kg/s}}$

Error = 8.3% (too high)

12.29 $A_e = 5\ cm^2$; $\dot{m} = 0.25$ kg/s; $T_e = 10°C$; $p = 100$ kPa

$c_e = \sqrt{kRT_e} = \sqrt{(1.4)(287)(283)} = 337$ m/s

Assuming sonic flow at exit and exhausting to 100 kPa, one finds

$\rho_e = p/RT_e = 100 \times 10^3/(287)(283) = 1.23\ kg/m^3$

$\dot{m} = (1.23)(5 \times 10^{-4})(337) = 0.207$ kg/s

Because the mass flow is too low, flow must exit sonically at pressure higher than the back pressure.

$\therefore\ \rho_e = \dot{m}/(c_e A_e) = 0.25/((337)(5 \times 10^{-4})) = 1.48\ kg/m^3$

$\therefore\ p_e = \rho_e RT_e = 1.20 \times 10^5$ Pa

$\therefore\ p_t/p_e = ((k+1)/2)^{k/(k-1)} = (1.2)^{3.5} = 1.893$

$\therefore\ p_t = 2.27 \times 10^5\ Pa = \underline{\underline{227\ kPa}}$

12.30 a) p_t = 130 kPa; Helium k = 1.66

If sonic at exit, $P_* = [2/(k+1)]^{k/(k-1)} p_t$ = 0.487 x 130 kPa = 63.3 kPa

∴ Flow must exit subsonically

$M_e^2 = (2/(k-1))[(p_t/p_b)^{(k-1)/k} - 1]$

$= 3.03[(130/100)^{0.4} - 1] = 0.335$

$M_e = 0.579$

∴ $T_e = T_t/(1+((k-1)/2)M^2) = 301/(1+(1/3)(0.335)) = 271$ K

$\rho_e = 100 \times 10^3/[(2{,}077)(271)] = 0.178$ kg/m^3

$\dot{m} = \rho_e A_e V_e = (0.178)(12 \times 10^{-4})(0.579)\sqrt{(1.66)(2{,}077)(271)}$

$\dot{m} = 0.120$ kg/s

b) p_t = 350 kPa

∴ $p_* = (0.487)(350) = 170$ kPa

∴ Flow exits sonically

$\dot{m} = 0.727\, p_t A_*/\sqrt{RT_t} = (0.727)(350)10^3(12 \times 10^{-4})/\sqrt{2{,}077 \times 301}$

$\dot{m} = 0.386$ kg/s

12.31 U = 50 m/s; T = 600°C = 873 K; p = 100 kPa; R = 287 J/kgK; k = 1.4

$\rho = p/RT = 100 \times 10^3/(287)(873) = 0.399$ kg/m^3

$\dot{m} = (0.399)(50)(\pi/4)(4 \times 10^{-3})^2 = 0.000251$ kg/s

$M = 50/\sqrt{(1.4)(287)(873)} = 0.0844$

$p_t = (100)[1+(0.2)(0.0844)^2]^{3.5} = 100.5$ kPa

$T_t = 874$ K

If sonic flow at constriciton, then

$\dot{m} = 0.685(100.5 \times 10^3)(\pi/4)(2 \times 10^{-3})^2/\sqrt{(287)(873)} = 0.000432$ kg/s

∴ flow must be subsonic at constriciton.

Solution must be found iteratively.

12.31 (continued)

Assume M at constriction:

$\rho_e = \rho_t(1+((k-1)/2)M^2)^{(-1/(k-1))} = \rho_t(1+0.2M^2)^{-2.5}$

$c_e = c_t(1+((k-1)/2)M^2)^{-1/2} = c_t(1+0.2M^2)^{-0.5}$

$\dot{m} = \rho_e A_e c_e M_e = A_e M_e \rho_t c_t(1+0.2M^2)^{-3}$

$\rho_t = (0.399)[1+(0.2)(0.0844)^2]^{2.5} = 0.400\ kg/m^3$

$c_t = \sqrt{(1.4)(287)(874)} = 593\ m/s$

$\therefore\ \dot{m} = 7.45 \times 10^{-4} M(1+0.2M^2)^{-3}$

M	$\dot{m} \times 10^4$
0.5	3.22
0.4	2.71
0.35	2.42
0.36	2.48
0.365	2.51 (correct flow rate)

$\therefore\ p_b = (100.5)(1+0.2 \times 0.365^2)^{-3.5} = \underline{\underline{91.6\ kPa}}$

12.32 M = 2.5; p = 1.5 psia; T = -10°F; k = 1.4

$A/A_* = (1/M)[(1+((k-1)/2)M^2)/((k+1)/2)]^{(k+1)/(2(k-1))}$

$= (1/2.5)[(1+0.2 \times 2.5^2)/1.2]^3 = \underline{\underline{2.64}}$

From Table A1, $p/p_t = 0.0585$; $T/T_t = 0.444$

$\therefore\ p_t = 1.5\ psia/0.0585 = \underline{\underline{25.6\ psia}}$

$T_t = 450°R/0.444 = \underline{\underline{1{,}013°R}} = \underline{\underline{553°F}}$

12.33 $p_e = 30\ kPa$; $k = 1.4$; $R = 297\ J/kgK$

$p_t = 10^6 Pa = 1{,}000\ kPa$; $\dot{m} = 5\ kg/s$; $T_t = 550\ K$

$M_e = \sqrt{(2/(k-1))[(p_t/p_e)^{(k-1)/k}-1]} = \sqrt{5[(1{,}000/30)^{0.286}-1]} = 2.94$

$A_e/A_* = (1/2.94)[(1+(0.2)(2.94)^2)/1.2]^3 = \underline{\underline{4.00}}$

$\dot{m} = 0.685\ p_t A_T/\sqrt{RT_t}$

$A_T = \dot{m}\sqrt{RT_t}/(0.685 \times p_t) = 5 \times \sqrt{(297)(550)}/((0.685)(10^6)) =$

$= \underline{\underline{0.00295\ m^2}} = \underline{\underline{29.5\ cm^2}}$

12.34 $A/A_* = 4;\ p_t = 1.3\text{ MPa} = 1.3 \times 10^6\text{ Pa};\ p_b = 35\text{ kPa};\ k = 1.4$

From Table A1: $M_e \approx 2.94 \Rightarrow p_e/p_t \approx 0.030$

$\therefore p_e = 39\text{ kPa}$

$\therefore p_e > p_b$ (under expanded)

12.35 $p = p_t$ in reservoir because V=0 in reservoir

$p/p_t = 0.1278$ for $A/A_* = 1.688$ and $M_2 = 2$ (Table A.1)

$p_t = p/0.1278 = 100/0.1278 = \underline{782.5\text{ kPa}}$

Throat conditions for M=1:

$p/p_t = 0.5283 \qquad T/T_t = 0.8333$

$p = 0.5283(782.5) = 413.4\text{ kPa} \quad T = 0.8333(17+273) = 242\text{K} = -31°\text{C}$

Conditions for $p_t = 700$ kPa:

$p/p_t = 0.1278$

$p = 0.1278(700) = 89.5\text{ kPa} \Rightarrow 89.5\text{ kPa} < 100\text{ kPa}$

overexpanded exit condition

p_t for normal shock at exit:

Assume shock exists at M=2; we know $p_2 = 100$ kPa.

From Table A.1: $p_2/p_1 = 4.5$

$p_1 = p_2/4.5 = 22.2\text{ kPa}$

$p/p_t = 0.1278$

$p_t = p/0.1278 = 22.2/0.1278 = \underline{173.9\text{ kPa}}$

12.36 $q = (k/2)pM^2 = (k/2)p_t[1+((k-1)/2)M^2]^{-k/(k-1)}M^2$

$\ell nq = \ell n(kp_t/2) - (k/(k-1))\ell n(1+((k-1)/2)M^2) + 2\ell nM$

$(\partial/\partial M)\ell nq = (1/q)(\partial q/\partial M) = (-k/(k-1))[1/(1+((k-1)/2)M^2)][(k-1)M] + 2/M$

$0 = [-kM\ ]/[1+((k-1)/2)M^2] + (2/M) = [-kM^2+2+(k-1)M^2]/[(1+((k-1)/2)M^2)M]$

$0 = 2-M^2 \rightarrow \underline{M = \sqrt{2}}$

$A/A_* = (1/M)[1+((k-1)/2)M^2]/[(k+1)/2]^{(k+1)/2(k-1)}$

$= (1/\sqrt{2})[(1+0.2(2))/1.2]^3 = \underline{1.123}$

12.37 $p_b = 20$ kPa; $A/A_* = 4$; $p_t = 1.2$ MPa

$T_t = 3{,}000°C = 3{,}273$ K; $k = 1.2$; $R = 400$ J/kgK; $A_* = 100\ cm^2 = 10^{-2} m^2$

$A/A_* = (1/M_e)((1+0.1 \times M_e^2)/1.1)^{5.5} = 4$

Solve for M by iteration:

M_e	A/A_*
3.0	6.73
2.5	3.42
2.7	4.45
2.6	3.90
2.62	4.00

1) $\therefore M_e = \underline{\underline{2.62}}$

$p_e/p_t = (1+0.1 \times 2.62^2)^{-6} = 0.0434$

$\therefore p_e = (0.0434)(1.2 \times 10^6) = \underline{\underline{52.1 \times 10^3 \text{ Pa}}}$

$T_e/T_t = (1+0.1 \times 2.62^2)^{-1} = 0.593$

$T_e = (3{,}273 \times 0.593) = 1{,}941$ K

$\rho_e = p_e/(RT_e) = (52.1 \times 10^3)/(400 \times 1{,}941) = \underline{\underline{0.0671 \text{ kg/m}^3}}$

$c_e = \sqrt{(1.2 \times 400 \times 1{,}941)} = 965$ m/s

$V_e = (965)(2.62) = \underline{\underline{2{,}528 \text{ m/s}}}$

2) $\dot{m} = \rho_e A_e V_e = (0.0671)(4)(10^{-2})(2{,}528) = \underline{\underline{6.78 \text{ kg/s}}}$

3) $T = (6.78)(2{,}528) + (52.1-20) \times 10^3 \times 4 \times 10^{-2} = \underline{\underline{18.42 \text{ kN}}}$

4) $p_t = 20/0.0434 = 461$ kPa

$\dot{m} = (20/52.1)(6.78) = 2.60$ kg/s

$T = (2.60)(2{,}528) = \underline{\underline{6.57 \text{ kN}}}$

12.38 $p_b = 100$ kPa; $p_t = 1.8$ MPa; $T_t = 3{,}300$ K; $k = 1.2$; $R = 400$ J/kgK

$A_* = 10\ cm^2 = 10^{-3} m^2$

$p_t/p_e = (1+((k-1)/2)M^2)^{k/(k-1)} = (1+0.1\ M_e^2)^6$

$M_e = \sqrt{10[(p_t/p_e)^{1/6}-1]} = \sqrt{10[(1{,}800/100)^{1/6}-1]} = 2.49$

$A_e/A_* = (1/M_e)[(1+0.1\ M_e^2)/1.1]^{5.5} = \underline{\underline{3.38}}$

$T_e = 3{,}300/(1+(0.1)(2.49)^2) = 2{,}037$ K

12.38 (continued)

$$\rho_e = 100 \times 10^3/(400 \times 2{,}037) = 0.123 \text{ kg/m}^3$$

$$c_e = \sqrt{(1.2)(400)(2{,}037)} = 989 \text{ m/s}$$

$$\dot{m} = \rho_e A_e V_e = (0.123)(3.38)(10^{-3})(989)(2.49) = 1.024 \text{ kg/s}$$

1) $T = (1.024)(989)(2.49) = \underline{\underline{2{,}522 \text{ N}}}$

$$A_e/A_* = (0.9)(3.38) = 3.042$$

$$3.042 = (1/M_e)((1+0.1\,M_e^2)/1.1)^{5.5}$$

Solve by iteration:

M_e	A/A_*
2.0	1.88
2.2	2.36
2.3	2.65
2.4	3.010
2.42	3.088
2.41	3.049

$$p_e/p_t = (1+0.1\,M_e^2)^{-6} = 0.0641$$

$$p_e = (0.0641)(1.8 \times 10^6) = 115 \text{ kPa}$$

$$T_e = 3{,}300/(1+0.1 \times 2.41^2) = 2{,}087 \text{ K}$$

$$c_e = \sqrt{(1.2)(400)(2{,}087)} = 1{,}001 \text{ m/s}$$

$$T = (1.024)(1{,}001)(2.41) + (115 - 100) \times 10^3 \times 3.042 \times 10^{-3}$$

2) $\quad = \underline{\underline{2{,}516 \text{ N}}}$

12.39 $A_e/A_T = 4$; $k = 1.4$; $p_t = 200$ kPa; $p_b = 100$ kPa; $p_b/p_t = 0.5$

Solution by iteration:

(1) Choose M

(2) Determine A/A*

(3) Find $p_{t_2}/p_{t_1} = A_{*1}/A_{*2}$

(4) $(A_e/A_*)_2 = 4(A_{*1}/A_{*2})$

(5) Find M_e

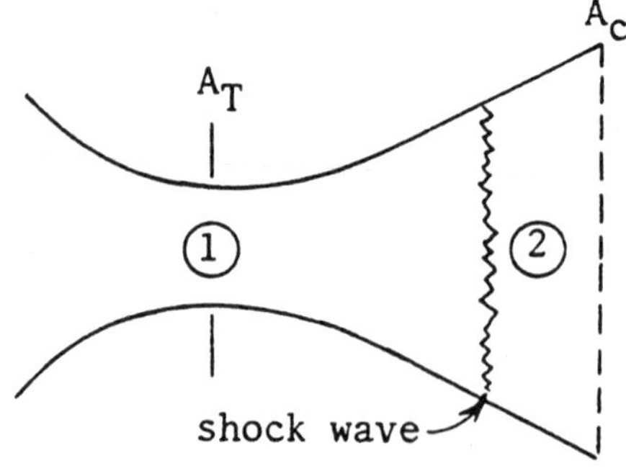

(6) $p_e/p_{t_1} = (p_e/p_{t_2})(p_{t_2}/p_{t_1})$ and converge on $p_e/p_{t_1} = 0.5$

12.39 (continued)

M	A/A_*	p_{t_2}/p_{t_1}	$(A_e/A_*)_2$	M_e	p_e/p_{t_1}	
2	1.69	0.721	2.88	0.206	0.7	
2.5	2.63	0.499	2.00	0.305	0.468	
2.4	2.40	0.540	2.16	0.28	0.511	$\therefore A/A_* = \underline{\underline{2.46}}$
2.43	2.47	0.527	2.11	0.287	0.497	
2.425	2.46	0.530	2.12	0.285	0.50	

12.40

Use same iteration scheme as problem 12-39 but with $k = 1.2$ to find A/A_* of shock:

$p_b/p_t = 100/250 = 0.4$ $\qquad$ $A_e/A_T = (8/4)^2 = 4$

M	A/A_*	p_{t_2}/p_{t_1}	$(A_e/A_*)_2$	M_e	p_e/p_{t_1}	
2.0	1.88	0.671	2.68	0.227	0.568	
2.4	3.01	0.463	1.85	0.341	0.432	
2.5	3.42	0.416	1.65	0.385	0.380	$\therefore A/A_* = 3.25$
2.46	3.25	0.434	1.74	0.366	0.400	

From geometry: $d = d_t + 2\,x \tan 15°$

$$d/d_t = 1 + (2x/d_t)\tan 15°$$

$$A/A_* = (d/d_t)^2 = 3.25$$

$$= [1 + (2x/d_t)(0.268)]^2$$

$$= [1 + (0.536x/d_t)]^2$$

$$\therefore x/d_t = 1.498$$

$$x = (1.498)(4) = \underline{\underline{5.99\text{ cm}}}$$

12.41

$A_s/A_* = 4;\ k = 1.41$

$$4 = (1/M)((1 + 0.205 \times M^2)/1.205)^{2.939}$$

Solve iteratively for M:

M	A/A_*
2	1.68
3	4.16
2.9	3.79
2.95	3.97
2.96	4.01
2.957	4.00

12.41 (continued)

$M_1 = 2.957$

$M_2^2 = ((k-1)M_1^2+2)/(2kM_1^2-(k-1))$

$M_2 = 0.480$

$p_2/p_1 = (1+kM_1^2)/(1+kM_2^2) = 10.06$

$p_t/p\big|_1 = (1+((k-1)/2)M_1^2)^{k/(k-1)} = 34.2$

$p_t/p\big|_2 = 1.172$

$p_{t_2}/p_{t_1} = (p_{t_2}/p_2)(p_2/p_1)(p_1/p_{t_1}) = 0.345$

$\Delta s = R\ell n(p_{t_1}/p_{t_2}) = 4{,}127\ \ell n(1/0.345) = \underline{\underline{4{,}392\ J/kgK}}$

12.42 $M = 2.1$ from Table A-1 $A/A_* = 1.837$; $p/p_t = 0.1094$

$A_* = 100/1.837 = 54.4$; $p_t = 65/0.1094 = 594$ kPa

$A_2/A_* = 75/54.4 = 1.379 \rightarrow p_2/p_t = 0.1904 \rightarrow p_2 = 0.1904(594) = 113$ kPa

after shock, $M_2 = 0.630$; $p_2 = 3.377(113) = 382$ kPa

$A_2/A_* = (1/M)((1+0.2M^2)/1.2)^3 = 1.155$; $p_t/p_2 = (1+0.2M^2)^{3.5} = 1.307$

$A_* = 75/1.155 = 64.9$; $p_t = 382(1.307) = \underline{\underline{499\ kPa}}$

$A_3/A_* = 120/64.9 = 1.849$; from Table A.1, $M_3 = \underline{\underline{0.336}}$

$p_3/p_t = 0.9245$; $p_3 = 0.9245(499) = \underline{\underline{461\ kPa}}$

12.43 Assume $M_e = 1$; $p_e = 100$ kPa; $T_e = 373(0.8333) = 311$ K

$c_e = \sqrt{1.4(287)311} = 353$ m/s; $\rho_e = 100 \times 10^3/(287 \times 311) = 1.12\ kg/m^3$

$A = \dot{m}/(\rho V) = 0.2/(1.12 \times 353) = 5.06 \times 10^{-4} m^2 = 5.06\ cm^2$

$D = ((4/\pi)A)^{1/2} = 2.54$ cm

$Re = (353 \times 0.0254)/(1.7 \times 10^{-5}) = 5.3 \times 10^5 \rightarrow f = 0.0132$

$f\Delta x/D = (0.0132 \times 10)/0.0254 = 5.20$ from Fig. 12.19 $M_1 = 0.302$

from Fig. 12.20 $p/p_* = 3.6$

12.43 (continued)

$p_1 = 100(3.6) = 360$ kPa > 240 kPa

$\therefore$ Case B

Solve by iteration.

M_e	T_e	c_e	V_e	ρ_e	$A \times 10^4$	$Re \times 10^{-5}$	M_1	p_1/p_e
0.8	331	365	292	1.054	6.51	4.54	0.314	2.55
0.7	340	369	259	1.026	7.54	4.11	0.322	2.18

By interpolation, for $p_1/p_e = 2.4$, $M_e = 0.76$

$T_e = 334$; $c_e = 367$; $V_e = 279$; $\rho_e = 1.042$

$A = 6.89 \times 10^{-4}\ m^2$; $D = 0.0296$ m = 2.96 cm

12.44 $M_1 = 0.3$; $A/A_* = 2.0351$; $A_* = 200/2.0351 = 98.3\ cm^2$

$p/p_t = 0.9395$; $p_t = 400/0.9395 = 426$ kPa; $A_s/A_* = 120/98.3 = 1.2208$

By interpolation from Table A-1:

$M_{s1} = 1.562$; $p_1/p_t = 0.2490 \rightarrow p_1 = 0.249(426) = 106$ kPa

$M_{s2} = 0.680$; $p_{s2}/p_1 = 2.679 \rightarrow p_{s2} = 2.679(106) = 284$ kPa

$A_s/A_{*2} = 1.1097 \rightarrow A_{*2} = 120/1.1097 = 108\ cm^2$

$p_{s2}/p_{t2} = 0.7338$; $p_{t2} = 284/0.7338 = 387$ kPa

$A_2/A_{*2} = 140/108 = 1.296 \rightarrow M_2 = 0.525$

$p_2/p_{t2} = 0.8288$; $p_2 = 0.8288(387) = 321$ kPa

12.45 $V_1 = 150$ ft/sec; $R = 1{,}716$ ft-lbf/slug; $d = 1$ in.; $p = 30$ psia;

$T = 67°F = 527°R$

$c = \sqrt{(1.4)(1{,}716)(527)} = 1{,}125$ ft/sec

$M_1 = 150/1{,}125 = 0.133$

$\rho = p/RT = (30 \times 144)/(1{,}716 \times 527) = 0.00478\ slug/ft^3$

$\mu = 3.8 \times 10^{-7}\ lbf\text{-}sec/ft^2$

$Re = (150 \times 1/12 \times 0.00478)/(3.8 \times 10^{-7}) = 1.57 \times 10^5$

12.45 (continued)

From Figs. 10-8 and 10-9, f = 0.025

$\bar{f}(x_* - x_M)/D = (1-M^2)/kM^2 + ((k+1)/2k)\ln[(k+1)M^2/(2+(k-1)M^2)] = 36.4$

$\therefore\ x_* - x_M = L = (36.4)(D/\bar{f}) = (36.4 \times 1/12)/0.025 = \underline{\underline{121.3\ \text{ft}}}$

$p_M/p_* = 8.2$ from Eq. (12.79)

$\therefore\ p_* = 30/8.2 = \underline{\underline{3.66\ \text{psia}}}$

12.46 $M_e = 0.8$; $d = 3\ \text{cm} = 3 \times 10^{-2}\text{m}$; $T_t = 373\ \text{k}$; $R = 287\ \text{J/kgK}$

$P_a = 100\ \text{kPa}$; brass-tube

$T_e = 373/(1+0.2 \times 0.8^2) = 331\ \text{K}$

$c_e = \sqrt{(1.4)(287)(331)} = 365\ \text{m/s}$

$V_e = (0.8)(365) = 292\ \text{m/s}$

$\mu_e = 2.03 \times 10^{-5}\text{N}\cdot\text{s/cm}^2$

$\rho_e = (100 \times 10^3)/(287 \times 331) = 1.053\ \text{kg/m}^3$

$Re = (292)(1.053)(3 \times 10^{-2})/(2.03 \times 10^{-5}) = 4.54 \times 10^5$

from Figs. 10-8 and 10-9, f = 0.0145

$\bar{f}(x_* - x_{0.8})/D = 0.1$

$\bar{f}(x_* - x_{0.2})/D = 14.5$

$\therefore\ \bar{f}(x_{0.8} - x_{0.2})/D = 14.4 = \bar{f}L/D$

$\therefore\ L = (14.4)(3 \times 10^{-2})/0.0145 = \underline{\underline{29.8\ \text{m}}}$

12.47 By Eq. (12-75)

$M = 0.2\quad \bar{f}(x_* - x_{0.2})/D = 14.53$

$M = 0.6\quad \bar{f}(x_* - x_{0.6})/D = 0.49$

$\bar{f}(x_{0.6} - x_{0.2})/D = 14.04$

$\bar{f} = 14.04(0.5)/(20 \times 12) = \underline{\underline{0.0293}}$

12.48 O_2 in 2.5 cm iron pipe; 10 m long

$p_b = 100$ kPa; $k = 1.4$; $R = 260$ J/kgK

$p_1 = 300$ kPa; $T_t = 293$ K; $\dot{m} = ?$

Assume sonic flow at exit.

$T_e = 293/1.2 = 244 = -29°C$

$c_e = V_e = \sqrt{(1.4)(260)(244)} = 298$ m/s

$\nu_e \simeq 1 \times 10^{-5} m^2/s$ (Fig. A3)

$Re = (298 \times 2.5 \times 10^{-2})/(1 \times 10^{-5}) = 7.45 \times 10^5$

From Figs. 10-8 and 10-9, f = 0.024

$\therefore$ $f(x_*-x_M)/D = (10 \times 0.024)/0.025 = 9.6$

$\therefore$ M at entrance = 0.235 (from Fig. 12-19)

$\therefore$ $p_M/p_* = 4.6$

$\therefore$ $p_1 = 460$ kPa > 300 kPa

$\therefore$ flow must be subsonic at exit. $p_e/p_1 = 100/300 = 0.333$

Use iterative procedure:

M_1	$\frac{f(x_*-x_M)}{D}$	Re x 10^5	f	fL/D	$\frac{f(x_*-x_e)}{D}$	M_e	p_e/p_1
0.20	14.5	6.34	0.024	9.6	4.9	0.31	0.641
0.22	11.6	6.97	0.024	9.6	2.0	0.42	0.516
0.23	10.4	7.30	0.024	9.6	0.8	0.54	0.416
0.232	10.2	7.34	0.024	9.6	0.6	0.57	0.396
0.234	10.0	7.38	0.024	9.6	0.4	0.62	0.366
0.2345	9.9	7.40	0.024	9.6	0.3	0.65	0.348

For M_1 near 0.234, $p_M/p_* = 4.65$, $p_e/p_* = (p_M/p_*)(p_e/p_M)$

$\therefore$ $p_e/p_* = (4.65)(0.333) = 1.55$

which corresponds to $M_e = 0.68$

$\therefore$ $T_e = 293/(1+(0.2)(0.68)^2) = 268$ K

$c_e = \sqrt{(1.4)(260)(268)} = 312$ m/s

$V_e = 212$ m/s

$\rho_e = 10^5/(260 \times 268) = 1.435$ kg/m^3

$\therefore$ $\dot{m} = (1.435)(212)(\pi/4)(0.025)^2 = \underline{\underline{0.149 \text{ kg/s}}}$

12.49 From Prob. 12.48, we know flow at exit must be sonic since

$p_1 > 460$ kPa. Iterative solution:

Assume f = 0.025

$\bar{f}(x_* - x_M)/D = 10$

M = 0.23

$T_t = 293/(1+0.2(0.23)^2) = 290$ K

$c_1 = \sqrt{(1.4)(290)(260)} = 325$ m/s

$\rho_1 = (500 \times 10^3)/(260 \times 290) = 6.63$ kg/m^3

$\mu_1 = 1.79 \times 10^{-5}$ N•s/m^2 (assuming μ not a function of pressure)

$\therefore$ Re = $(0.23)(325)(6.63)(2.5 \times 10^{-2})/(1.79 \times 10^{-5}) = 6.9 \times 10^5$

$\therefore$ f = 0.024 from Figs. 10-8 and 10-9

Try f = 0.024

$f(x_* - x_M)/D = 9.6$; M = 0.235; $T_t \simeq 290$

$c_1 = 325$; $\rho_1 = 6.63$; $\mu_1 \simeq 1.79 \times 10^{-5}$; Re = 7×10^5

gives same f: f = 0.024

For M = 0.235 $p_M/p_x = 4.64$

$\therefore$ $p_* = 107.8$ kPa

$T_e = 293/1.2 = 244$

$c_e = 298$

$\rho_e = (107.8 \times 10^3)/(260 \times 244) = 1.70$ kg/m^3

$\therefore$ $\dot{m} = (1.70)(298)(\pi/4)(0.025)^2 = \underline{0.248 \text{ kg/s}}$

12.50

Assume $M_e = 1$; $p_e = 7$ psia

$T_e = 560(0.8333) = 467°R$; $c_e = \sqrt{1.4(1,776)467} = 1,077$ ft/s

$\rho_e = 7(144)32.2/(1,776 \times 467) = 0.039$ lbm/ft^3

$A = \dot{m}/(\rho V) = 0.06/(0.039 \times 1,077) = 1.43 \times 10^{-3}$ ft^2; D = 0.0425 ft = 0.51 in.

Re = $(1,077)(0.0425)(0.039)/(1.36 \times 10^{-7} \times 32.2) = 4.1 \times 10^5$

12.50 (continued)

$k_s/D = 0.0117; \quad f = 0.040$

$f\Delta x/D = (0.04 \times 10)/0.0426 = 9.40$ from Fig. 12-19 $M_1 = 0.24$

from Fig. 12-20 $p_1/p_* = 4.54; \quad p_1 = 31.8$ psia < 45 psia $\therefore$ Case D

$M = 1$ at exit and $p_e > 7$ psia

Solve by iteration:

M_1	T_1	V_1	ρ_1	D	$Re \times 10^{-5}$	f	M_1	p_e
0.24	553	281	0.212	0.0358	1.62	0.040	0.223	9.16
0.223	554	262	0.212	0.0371	1.56	0.040	0.223	9.16

D = 0.0371 ft = 0.445 inch

12.51 Assuming viscosity of particle-laden flow is same as air,

$c = \sqrt{1.4(287)288} = 340$ m/s

$M_e = 50/340 = 0.147 \rightarrow \bar{f}(x_* - x_{0.147})/D = 29.2; \quad p_e/p_* = 7.44$

$Re = 50(0.2)/(1.44 \times 10^{-5}) = 6.94 \times 10^5; \quad k_s/D = 0.00025; \quad f = 0.0158$

$\bar{f}\,\Delta x/D = [\bar{f}(x_* - x_M)/D] - [\bar{f}(x_* - x_{0.147})/D]$

$\bar{f}\,\Delta x/D = 0.0158(150)/0.2 = 11.8$

$\bar{f}(x_* - x_M)/D = 29.2 + 11.8 = 41.0 \rightarrow M_1 = 0.13; \quad p_1/p_* = 8.41$

$p_1/p_e = (p_1/p_*)(p_*/p_e) = 8.41/7.45 = 1.13$

$p_1 = 1.13(100) = \underline{\underline{113 \text{ kPa}}}$

$V_1 = 0.13(340) = \underline{\underline{44.2 \text{ m/s}}}$

$T_1 = T_t/(1+0.2M_1^2) = 288/(1+0.2(0.13)^2) = 287$

$\rho_1 = (113 \times 10^3)/(287 \times 287) = \underline{\underline{1.37 \text{ kg/m}^3}}$

12.52

$c_1 = \sqrt{1.31(518)320} = 466$ m/s

$\rho_1 = 10^6/(518 \times 320) = 6.03$ kg/m^3

$M_1 = 20/466 = 0.043$

$\bar{f}(x_* - x_{0.043})/D = 407$ by Eq. (12-75)

$p_1/p_* = 25.0$ by Eq. (12-79)

$Re = 20(0.15)6.03/(1.5 \times 10^{-5}) = 1.2 \times 10^6$; $k_s/D = 0.00035$

$f = 0.0162$; $f\Delta x/D = 0.0162(3{,}000)/0.15 = 324$

$[f(x_* - x_{0.043})/D] - [\bar{f}(x_* - x_M)/D] = f\Delta x/D$

$f(x_* - x_M)/D = 407 - 324 = 83 \rightarrow M_e = 0.093$

by Eq. (12-79) $p_e/p_* = 11.5$

$p_e = (p_e/p_*)(p_*/p_1)\ p_1 = (11.5/25.0)10^6 = \underline{\underline{460\text{ kPa}}}$

12.53

$R = 4{,}127$ J/kgK; $k = 1.41$; $\nu = 1.01 \times 10^{-4}$ m^2/s

Speed of sound at entrance $= \sqrt{(1.41)(4{,}127)(288)} = 1{,}294$ m/s

$\therefore M = 200/1{,}294 = 0.154$

$\therefore kM^2 = 0.0334$; $\sqrt{k}M = 0.183$

Reynolds number $= (200)(0.1)/(1.01 \times 10^{-4}) = 1.98 \times 10^5$

From Figs 10-8 and 10-9 $f = 0.019$

At entrance $f(x_m - x_1)/D = \ln(0.0334) + (1-0.0334)/0.0334 = 25.5$

At exit $f(x_M - x_2)/D = f(x_M - x_1)/D + f(x_1 - x_2)/D = 25.5 - (0.019)(50)/0.1$

$= 25.5 - 9.5 = 16.0$

From Fig. 12-22, $kM^2 = 0.05$ or $\sqrt{k}M = 0.2236$

$p_2/p_1 = (p_m/p_1)(p_2/p_m) = 0.183/0.2236 = 0.818 \therefore p_2 = 163.6$ kPa

$\Delta p = \underline{\underline{36.4\text{ kPa}}}$

12.54

$R = 2{,}077$ J/kgK; $k = 1.66$; $\nu = 1.14 \times 10^{-4} m^2/s$

$c = \sqrt{(1.66)(2.077)(288)} = 996$ m/s

$p_2/p_1 = 100/120 = 0.833$

Iterative solution:

V_1	M_1	Re $\times 10^{-4}$	f	kM_1^2	$\frac{f(x_T - x_M)}{D}$	$\frac{f(x_T - x_e)}{D}$	kM_2^2	p_2/p_1
100	0.100	4.4	0.022	0.0166	55.1	11.1	0.0676	0.495
50	0.050	2.2	0.026	0.00415	234.5	182.5	0.0053	0.885
55	0.055	2.4	0.025	0.00502	192.9	142.9	0.006715	0.864
60	0.060	2.6	0.25	0.00598	161.1	111.1	0.008555	0.836
61	0.061	2.6	0.25	0.00618	155.8	105.8	0.00897	0.830
60.5	0.0605	2.6	0.25	0.006076	158.5	108.5	0.00875	0.833

$\therefore \quad \rho = 120 \times 10^3/(2{,}077)(288) = 0.201$ kg/m^3

$\dot{m} = (0.201)(60.6)(\pi/4)(0.05)^2 = \underline{\underline{0.0239 \text{ kg/s}}}$

CHAPTER THIRTEEN

13.1 Because $V = \sqrt{2\Delta p/(\rho C_p)}$ a 2% deviation in C_p from unity will yield a 1% error in V when the equation is applied by assuming $C_p = 1$. Thus, find Re where $C_p = 1.02$. From Fig. 13-1, Re $\approx$ 60.

Then $Vd/\nu = 60$; $V = 60\nu/d$ where $\nu = 1.45 \times 10^{-5} m^2/s$ and $d = 0.002$ m

Thus $V = 60 \times 1.45 \times 10^{-5}/0.002 = \underline{0.43 \text{ m/s}}$

13.2 From Prob. 13.1, Re ≈ 60; $V = 60\ \nu/d$ where $\nu = 10^{-6} m^2 s$

Then $V = 60 \times 10^{-6}/0.002 = \underline{0.03 \text{ m/s}}$

13.3 $\rho_{air} = 1.25\ kg/m^3$; $\Delta h_{air} = 0.0008 \times 1{,}000/1.25 = 0.64$ m

Then $V = \sqrt{2g\Delta h} = 3.54$ m/s

Check C_p: $Re = Vd/\nu = 3.54 \times 0.002/(1.41 \times 10^{-5}) = 503$

$C_p \approx 1.002$; $V = 3.54/\sqrt{c_p} = 3.54/\sqrt{1.002} = \underline{3.54 \text{ m/s}}$

13.4 $\nu \approx 1.4 \times 10^{-5} m^2/s$

Then $Re = Vd/\nu = 12 \times 0.002/(1.4 \times 10^{-5}) = 1{,}714$

From Fig. 13.1 $C_p \approx 1.00$; $\Delta p = \rho V^2/2$

where $\rho = p/RT = 98{,}000/(287 \times (273 + 10)) = 1.21\ kg/m^3$

Then $\Delta p = 9{,}810\Delta h = 1.21 \times 12^2/2$; $\Delta h = 8.88 \times 10^{-3} m = \underline{8.88 \text{ mm}}$

13.5 $\nu = 1.55 \times 10^{-5}$; $\rho = p/RT = 100{,}000/(287 \times 298) = 1.17\ kg/m^3$

$V = \sqrt{2\Delta p/\rho} = \sqrt{2 \times 5/1.17} = 2.92$ m/s

Check C_p: $Re = Vd/\nu = 2.92 \times 0.002/(1.55 \times 10^{-5}) = 377$; $C_p = 1.002$

% error $= (1 - 1/\sqrt{1.002}) \times 100 = \underline{0.1\%}$

13.6 $\rho = p/RT = 100{,}000/(410 \times 573) = 0.426\ kg/m^3$

$1.4 = \Delta p/(\rho V_0^2/2)$ or $V_0 = \sqrt{2\Delta p/(1.4\rho)}$

where $\Delta p = 0.01$ m $\times$ 9,810 Pa and $\rho = 0.426\ kg/m^3$

Then $V_0 = \sqrt{2 \times 98.1/(1.4 \times 0.426)} = \underline{18.1 \text{ m/s}}$

13.7 $\gamma_{water,20°C}$ = 9,790 N/m³; Then $\dot{w} = w/t$ = 10,000/(4 x 60) = 41.67 N/s

But γ = 9,790 N/m³ so $Q = \dot{w}/\gamma$ = 41.67/9,790 = 4.26×10^{-3} m³/s

---

13.8 $Q = \forall/t$ = 78/372 = 0.210 m³/s

Q = 0.210 m³/s/(0.02832 m³/s/cfs) = 7.42 cfs

Q = 7.42 cfs x 449 gpm/cfs = 3,332 gpm

13.9

r(m)	V(m/s)	$2\pi Vr$	area(by trapezoidal rule)
0	8.7	0	
0.01	8.6	0.54	0.0027
0.02	8.4	1.06	0.0080
0.03	8.2	1.55	0.0130
0.04	7.7	1.94	0.0175
0.05	7.2	2.26	0.0210
0.06	6.5	2.45	0.0236
0.07	5.8	2.55	0.0250
0.08	4.9	2.46	0.0250
0.09	3.8	2.15	0.0231
1.10	2.5	1.57	0.0186
0.105	1.9	1.25	0.0070
0.11	1.4	0.97	0.0056
0.115	0.7	0.51	0.0037
0.12	0	0	0.0013

Q = 0.196 m³/s

$V_{mean} = Q/A = 0.196/(0.785(0.24)^2)$

$= 4.33$ m/s

$V_{max}/V_{mean} = 8.7/4.33 = 2.0$

Laminar flow

13.10 r = 8 - y

y(in.)	r(in.)	V(ft/s)	$2\pi rV$(ft²/s)	area(ft³/s)
0.0	8.0	0	0	
0.1	7.9	72	297.8	1.24
0.2	7.8	79	322.6	2.58
0.4	7.6	88	350.2	5.61
0.6	7.4	93	360.3	5.92
1.0	7.0	100	366.5	12.11
1.5	6.5	106	360.8	15.15
2.0	6.0	110	345.6	14.72
3.0	5.0	117	306.3	27.16
4.0	4.0	122	255.5	23.41
5.0	3.0	126	197.9	18.89
6.0	2.0	129	135.1	13.88
7.0	1.0	132	69.1	8.51
8.0	0.0	135	0	2.88

Q = 152.1 ft³/s = 9,124 cfm

$V_{mean} = Q/A = 152.1/(0.785(1.33)^2) = 109$ ft/s

$V_{max}/V_{mean} = 135/109 = 1.24$; appears to be turbulent

$\rho = 14.3(144)/((53.3)(530)) = 0.0728$ lbm/ft³; $\dot{m} = 0.0728(152.1) = 11.1$ lbm/s

13.11 a) $\pi r_m^2 = (\pi/4)[(D/2)^2 - r_m^2]$

$(r_m/D)^2 = 1/16 - (r_m/D)^2(1/4)$

$5/4(r_m/D)^2 = 1/16; \quad 5(r_m/D)^2 = 1/4$

$r_m/D = \sqrt{1/20} = \underline{\underline{0.2236}}$

b) $r_c A = \int_{0.2236D}^{D/2} [r \sin(\alpha/2)/(\alpha/2)](\pi/4)2r dr = 0.9(\pi/2)(r^3/3)\Big|_{0.2236D}^{0.5D}$

$(r_c)(\pi/4)[(D/2)^2 - (0.2236D)^2] = 0.90(\pi/6)[(0.5D)^3 - (0.2236D)^3]$

$r_c/D = \underline{\underline{0.341}}$

c) $\rho = p/(RT) = 110 \times 10^3/(400 \times 573) = 0.480 \text{ kg/m}^3$

$V = \sqrt{2\Delta p/\rho_g} = \sqrt{(2)\rho_w g\Delta h/\rho_g} = \sqrt{(2)(1{,}000)(9.81)/0.48}\sqrt{\Delta h} = 202.2\sqrt{\Delta h}$

Station	Δh	V
1	0.012	7.00
2	0.011	6.71
3	0.011	6.71
4	0.009	6.07
5	0.0105	6.55

$\dot{m} = \Sigma A_{sector}\rho V_{sector} = A_T\rho(\Sigma V/5)$

$= (\pi(2)^2/4)(0.480)(6.61) = 9.96 \text{ kg/s}$

13.12

a) $\pi r_m^2 = (\pi/6)[(D/2)^2 - r_m^2]$

$7/6(r_m/D)^2 = (1/6)(1/4)$

$(r_m/D)^2 = 1/28$

$r_m/D = \underline{\underline{0.189}}$

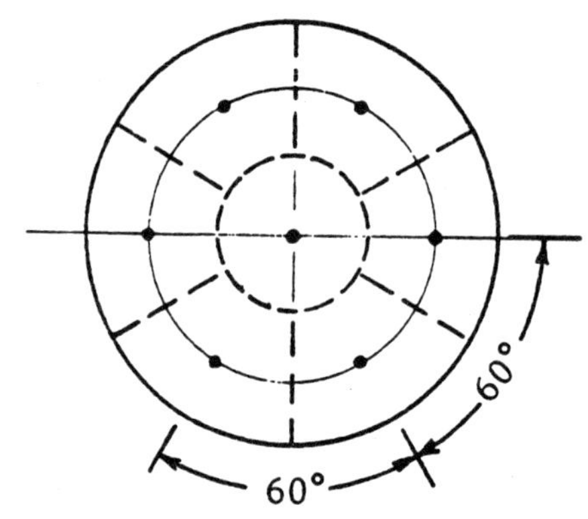

b) $r_c A = 1/6\int_{0.189D}^{0.5D} [r \sin(\alpha/2)/(\alpha/2)]2\pi r\, dr$

$(\pi r_c/6)[(D/2)^2 - (r_m)^2] = 0.955(\pi/3)(r^3/3)\Big|_{0.189D}^{0.50D}$

$r_c(0.5^2 - 0.189^2) = 0.955(6/9)[0.5^3 - 0.189^3]D$

$r_c/D = (0.955)6(0.118)/(9(0.2143)) = \underline{\underline{0.351}}$

c) $\rho = p/RT = 115 \times 10^3/((420)(250+273)) = 0.523 \text{ kg/m}^3$

$V = \sqrt{2g\rho_w\Delta h/\rho_g} = \sqrt{(2)(9.81)(1{,}000)/0.523}\sqrt{\Delta h} = 193.7\sqrt{\Delta h}$

$\dot{m} = (\pi D^2/4)\rho V_{avg} = ((\pi)(1.5)^2/4)(0.523)(17.75) = \underline{\underline{16.40}} \text{ kg/s}$

13.12 (continued)

Station	Δh (mm)	V
1	8.2	17.54
2	8.6	17.96
3	8.2	17.54
4	8.9	18.27
5	8.0	17.32
6	8.5	17.86
7	8.4	17.75

Average = 17.75

13.13 $Q = \Sigma V_i A_i$

V	A	V x A
1.32 m/s	7.6 m^2	10.0
1.54	21.7	33.4
1.68	18.0	30.2
1.69	33.0	55.8
1.71	24.0	41.0
1.75	39.0	68.2
1.80	42.0	75.6
1.91	39.0	74.5
1.87	37.2	69.6
1.75	30.8	53.9
1.56	18.4	28.7
1.02	8.0	8.2

$Q = 549.1\ m^3/s$

13.14

Fringe spacing $\Delta x = \lambda/2 \sin \phi$

$= (4{,}880 \times 10^{-10})/(2 \sin 10°)$

$= 1.41 \times 10^{-6} m$

$\Delta t = 8\mu s/4 = 2\mu s = 2 \times 10^{-6} s$ (4 intervals)

$V = \Delta x/\Delta t = (1.41 \times 10^{-6})/(2 \times 10^{-6}) = \underline{0.70\ m/s}$

13.15 Assume $V_j = \sqrt{2g \times 1.90}$ Then $C_v = V_j/V_{theor} = \sqrt{2g \times 1.90}/\sqrt{2g \times 2}$

$C_v = \sqrt{1.90/2.0} = \underline{0.975}$

$C_c = A_j/A_o = (8/10)^2 = \underline{0.640}$

$C_d = C_v C_c = 0.975 \times 0.64 = \underline{0.624}$

13.16 $C_c = A_j/A_o = (1.75/2)^2 = \underline{\underline{0.766}}$

13.17 If the angle is 90° the orifice and expected flow pattern is shown in Fig. A below.

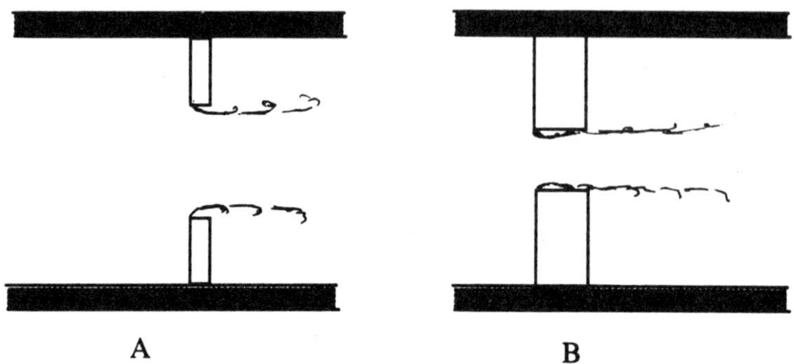

It would seem that the flow through the orifice would separate at the sharp edge just as it does for the orifice with a knife edge. Therefore, the flow pattern and flow coefficient K should be the same as with the knife edge (acute angle). However, if the orifice were very thick relative to the orifice diameter (Fig. B) then the flow could reattach to the metal of the orifice thus creating a different flow pattern and different flow coefficient K than the knife edge orifice.

13.18 Some of the possible changes that might occur are listed below:

A. Blunting (rounding) of the sharp edge might occur because of erosion or corrosion. This would probably increase the value of the flow coefficient because C_c would probably be increased.

B. Because of corrosion or erosion the face of the orifice might become rough. This would cause the flow next to the face to have less velocity

13.18 (Cont'd)

than when it was smooth. With this smaller velocity in a direction toward the axis of the orifice it would seem that there would be less momentum of the fluid to produce contraction of the jet which is formed downstream of the orifice. Therefore, as in case A, it appears that K would increase but the increase would probably be very small.

C. Some sediment might lodge in the low velocity zones next to and upstream of the face of the orifice. The flow approaching the orifice (lower part at least) would not have to change direction as abruptly as without the sediment. Therefore, the C_c would probably be increased for this condition and K would also be increased.

13.19 $d/D = 0.60$

$$Re_d = 4Q/(\pi d\nu) = 4 \times 3/(\pi \times 0.5 \times 1.22 \times 10^{-5}) = 6.3 \times 10^5$$

from Fig. 13.13: $K = 0.65$; $A = (\pi/4) \times 0.5^2 = 0.196 \text{ ft}^2$

Then $\Delta h = (Q/KA)^2/2g = (3/(0.65 \times 0.196))^2/64.4 = 8.61$ ft of water

$h = \Delta h/12.6 = 0.683 \text{ ft} = \underline{8.2 \text{ in.}}$

13.20 $Q = KA_o(2g\Delta h)^{0.5}$

$\Delta h = (1.0)(13.55 - 1) = 12.55$ ft ; $(d/D) = 0.50$

Assume $T = 60°F$, $\nu = 1.22 \times 10^{-5} \text{ ft}^2/\text{s}$

$(2g\Delta h)^{0.5}d/\nu = (2g \times 12.55)^{0.5}(0.5)/(1.22 \times 10^{-5}) = 1.17 \times 10^6$

$K = 0.625$ (from Fig. 13.13)

$Q = 0.625\ (\pi/4)(0.5^2)(64.4 \times 12.55)^{0.5} = \underline{3.49 \text{ cfs}}$

13.21 A rough pipe will have a greater maximum velocity at the center of the pipe relative to the mean velocity than would a smooth pipe. Because more flow is concentrated near the center of the rough pipe less radial flow is required as the flow passes through the orifice; therefore, there will be less contraction of the flow. Consequently the coefficient of contraction will be larger for the rough pipe. So by using the K from Fig. 13.13 for the rough pipe would probably result in an estimated discharge that is too small.

13.22 $Q = KA\sqrt{2g\Delta h}$; where $\Delta h = 5$ ft; $A = (\pi/4) \times (3/12)^2 = 0.0491$ ft^2

$\nu = 1.22 \times 10^{-5}$ ft^2/s; $d/D = 3/4$

Then $\sqrt{2g\Delta h}\, d/\nu = 3.7 \times 10^5$

$K = 0.73$ (from Fig. 13.13)

Thus $Q = 0.73 \times 0.0491\sqrt{2 \times 32.2 \times 5} = \underline{\underline{0.64 \text{ cfs}}}$

13.23 $Q = KA_o (2\Delta p/\rho)^{0.5}$; $\rho = 814$ kg/m^3 ; $\nu = 2.37 \times 10^{-6}$ m^2/s

$Re_d/K = (2\Delta p/\rho)^{0.5} (d/\nu)$

$= (2 \times 10 \times 10^3/814)^{0.5} (0.6 \times 0.02/(2.37 \times 10^{-6}))$

$= 4.96 \times 5{,}063 = 2.51 \times 10^4$

$K = 0.67$ (from Fig. 13.13)

$Q = VA_p = KA_o (2\Delta p/\rho)^{0.5}$

$V_{pipe} = K(A_o/A_p) (2\Delta p/\rho)^{0.5} = (0.67) (0.60^2) (4.96)$

$= \underline{\underline{1.20 \text{ m/s}}}$

13.24 $\Delta p_{AB} \neq \Delta p_{DE}$

Get Δh: $Q = KA\sqrt{2g\Delta h}$ where $Q = 0.10$ m^3/s; $A = (\pi/4)d^2 = (\pi/4)(0.10)^2$

$= 7.85 \times 10^{-3}$ m^2

Then $4Q/(\pi d\nu) = 4 \times 0.10/(\pi \times 0.10 \times 1.31 \times 10^{-6}) = 9.7 \times 10^5$

$K = 0.60$ (from Fig. 13.13)

Thus $\Delta h = Q^2/(K^2A^2 2g) = 0.1^2/(0.6^2 \times (7.85 \times 10^{-3})^2 \times 2 \times 9.81)$

$\Delta h = 22.97$ m of water

$p_A - p_B = \gamma\Delta h = 9{,}790 \times 22.97 = \underline{224.9 \text{ kPa}}$

$((p_D/\gamma) + z_D) - ((p_E/\gamma) + z_E) = \Delta h = 22.97$ ft

13.24 (continued)

$p_D - p_E = (22.97 - 0.30) \times 9{,}790 = \underline{\underline{221.9 \text{ kPa}}}$

Deflection on manometer = $22.97/(13.6 - 1) = \underline{\underline{1.82 \text{ m.}}}$ The deflection will be the same for both manometers.

13.25 Assume large Re

Then $Q_{15} = K_{15}A_{15}\sqrt{2g\Delta h}$; $K_{15} = 0.62$

$Q_{15} = 0.62 \times (\pi/4)(0.15)^2\sqrt{2g\Delta h}$; $K_{20} = 0.685$

$Q_{20} = 0.685 \times (\pi/4)(0.20)^2\sqrt{2g\Delta h}$

$Q_{15} = 0.01395(\pi/4)\sqrt{2g\Delta h}$

$Q_{20} = 0.0274(\pi/4)\sqrt{2g\Delta h}$

Thus the % increase is $(0.0274 - 0.01395/0.01395) \times 100 = \underline{\underline{96\%}}$

13.26 $\Delta h = (p_1/\gamma + z_1) - (p_2/\gamma + z_2)$

$\Delta h = ((60{,}000/9{,}790) + 0.3) - ((50{,}000/9{,}790) + 0)$

$\Delta h = 1.321$ m of water; $d/D = 10/50 = 0.20$

Then $Re/K = \sqrt{2 \times 9.81 \times 1.32} \times 0.10/10^{-6} = 5 \times 10^5$;

$K = 0.60$ (from Fig. 13.13)

Then $Q = 0.60 \times (\pi/4) \times (0.10)^2\sqrt{2 \times 9.81 \times 1.321} = \underline{\underline{0.0240 \text{ m}^3/\text{s}}}$

13.27

$$p_{T,1} = p_1 + \gamma\ell_1 \text{ (hydrostatic relation)}$$
$$p_{T,1} = p_2 - \gamma\ell_2 \text{ (hydrostatic relation)}$$

so

$$p_{T,2} - p_{T,2} = p_1 + \gamma\ell_1 - p_2 + \gamma\ell_2$$
$$= p_1 - p_2 + \gamma(\ell_1 + \ell_2)$$

But

$$\ell_1 + \ell_2 = z_1 - z_2$$
$$p_{T,1} - p_{T,2} = p_1 - p_2 + \gamma(z_1 - z_2)$$

or

$$\underline{\underline{p_{T,1} - p_{T,2} = (p_1 + \gamma z_1) - (p_2 + \gamma z_2)}}$$

13.28 $Re = 4Q/(\pi d\nu)$

$= 4 \times 20/(\pi \times 1 \times 1.41 \times 10^{-5}) = 1.8 \times 10^6$

Then for $d/D = 0.50$ $K = 0.625$

$Q = KA\sqrt{2g\Delta h}$ or $\Delta h = (Q/(KA))^2/2g$ where $A = \pi/4$

Then $\Delta h = (20/(0.625 \times (\pi/4)))^2/2g$

$\Delta h = 25.8$ ft; $\Delta p = \gamma\Delta h = 62.4 \times 25.8 = \underline{1{,}608 \text{ psf}}$

To get power supplied write the energy equation from the upstream reservoir water surface to downstream reservoir surface:

$p_1/\gamma + V_1^2/2g + z_1 + h_p = p_2/\gamma + V_2^2/2g + z_2 + \Sigma h_L$

$0 + 0 + 10 + h_p = 0 + 0 + 5 + \Sigma h_L$

$h_p = -5 + V^2/2g(K_e + K_E + fL/D) + h_{L,\text{ orifice}}$

$K_e = 0.5$; $K_E = 1.0$

The orifice head loss will be like that of an abrupt expansion:

$h_L = (V_j - V_{pipe})^2/(2g)$

Here, V_j is the jet velocity as the flow comes from the orifice.

$V_j = Q/A_j$ where $A_j = C_c A_0$

Assume $C_c \simeq 0.65$ Then $V_j = 20/((\pi/4) \times 1^2 \times 0.65) = 39.2$ ft/s

Also $V_p = Q/A_p = 20/\pi = 6.37$ ft/s

Then $h_{L,orifice} = (39.2 - 6.37)^2/(2g) = 16.74$ ft

Finally, $h_p = -5 + (6.37^2/(2g))(0.5 + 1.0 + (0.015 \times 300/2)) + 16.74$

$h_p = 14.10$ ft

$P = Q\gamma h_p/550$

$= 20 \times 62.4 \times 14.10/550 = \underline{\underline{32.0 \text{ horsepower}}}$

The HGL and EGL are shown below:

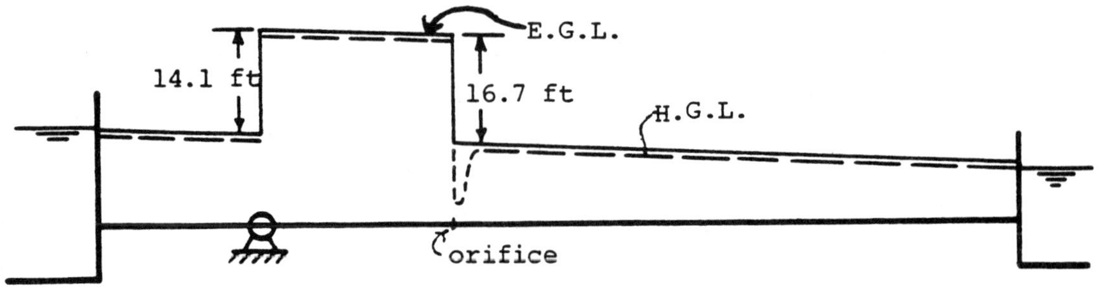

13.29 $\Delta h = 12.6 \times 1 = 12.6$ m of water

Assume $K = 0.7$; $A = Q/(K\sqrt{2g\Delta h})$

or $d^2 = (4/\pi)Q/(K\sqrt{2g\Delta h})$

$d^2 = (4/\pi) \times 0.03/(0.7\sqrt{2g \times 12.6}) = 3.47 \times 10^{-3} m^2$; $d = 5.89$ cm

$d/D = 0.39$; $Re_d = 4 \times 0.03/(\pi \times 0.0589 \times 10^{-6}) = 6.5 \times 10^5$

$K = 0.62$ so $d = \sqrt{(0.7/0.62)} \times 0.0589 = 0.0626$ m

$d = \underline{\underline{6.26 \text{ cm}}}$

13.30 Assume $T = 20°C$; $\nu = 4 \times 10^{-7} m^2/s$ (Fig. A-3); $d/D = 0.60$;

$\Delta h = \Delta p/\gamma = 35{,}000/(0.68 \times 9{,}810) = 5.25$ m

Then $\sqrt{2g\Delta h}\, d/\nu = \sqrt{2 \times 9.81 \times 5.25} \times 0.06/(4 \times 10^{-7}) = 1.52 \times 10^6$

$K = 0.650$; Then $Q = KA\sqrt{2g\Delta h} = 0.650 \times (\pi/4)(0.06)^2\sqrt{2g \times 5.25}$

$\underline{\underline{Q = 0.0187 \text{ m}^3/\text{s}}}$

13.31 Follow same procedure as for P13.29:

$\Delta h = 8$ m of water; $Q = 2 m^3/s$; $D = 1$ m; $K = 0.65$ (assume)

Then $d^2 = (4/\pi) \times 2/((0.65\sqrt{2g \times 8})) = 0.313$; $d = 0.56$ m

Try again: $d/D = 0.56$; $Re = 4Q/(\pi d\nu) \simeq 4.5 \times 10^6$; $K = 0.63$ (Fig. 13.13)

Then $d = \sqrt{0.65/0.63} \times 0.56 = 0.569$ m; $K = 0.63$ (same)

Thus $\underline{\underline{d = 56.9 \text{ cm}}}$

13.32 Assume $K = 0.65$; $T = 20°C$; $\Delta h = \Delta p/\gamma = 50{,}000/9{,}790 = 5.11$ m

Then following the procedure for P13.29:

$d^2 = (4/\pi) \times 3.0/(0.65\sqrt{2 \times 9.81 \times 5.11}) = 0.587$; $d = 0.766$ m

Check K: $Re_d = 4Q/(\pi d\nu) = 4 \times 3.0/(\pi \times 0.766 \times 10^{-6}) = 5 \times 10^6$

$d/D = 0.766/1.2 = 0.64$ Thus, $K = 0.67$ (from Fig. 13.13)

Try again: $d = \sqrt{(0.65/0.67} \times 0.766 = 0.754$

Check K: $Re_d = 5 \times 10^6$ and $d/D = 0.63$ so $K = 0.67$ (Fig. 13.13)

$d = \sqrt{(0.65/0.670)} \times 0.766 = \underline{0.754 \text{ m}}$

13.33 $p_1 + \rho V_1^2/2 = p_2 + \rho V_2^2/2$

$V_1A_1 = V_2A_2; \quad V_1 = V_2A_2/A_1$

$V_2 = \sqrt{2(p_1 - p_2)/\rho}/\sqrt{1-(A_2^2/A_1^2)}$

or $Q = (A_2/\sqrt{1-(A_2^2/A_1^2)})\sqrt{2\Delta p/\rho}$

but $A_2 = C_cA_0$ where A_0 is area of orifice

Then $Q = (C_cA_0/\sqrt{1-(A_2^2/A_1^2)})\sqrt{2\Delta p/\rho}$

or $Q = KA_0\sqrt{2\Delta p/\rho}$ where K is the flow coefficient

Assume K = 0.65; Also $A = (\pi/8) \times 0.30^2 = 0.0353\ m^2$

Then $Q = 0.65 \times 0.0353\sqrt{2 \times 80{,}000/1{,}000} = \underline{\underline{0.290\ m^3/s}}$

13.34 $Re_d = 4 \times 0.57/(\pi \times 0.30 \times 1 \times 10^{-6}) = 2.4 \times 10^6$; $d/D = 0.50$; $K = 1.02$

$\Delta h = (Q/(KA))^2/(2g) = (0.57/(1.02 \times (\pi/4) \times 0.3^2))^2/(2 \times 9.81) = 3.19\ m$

deflection $h = 3.19/12.6 = \underline{\underline{0.253\ m}}$

13.35

Assume K = 1.01; Assume T = 20°C

$Q = KA\sqrt{2g\Delta h}$ where $\Delta h = 200{,}000\ Pa/9{,}790\ N/m^3 = 20.4\ m$

Then $A = Q/(K\sqrt{2g\Delta h})$ or $\pi d^2/4 = Q/(K\sqrt{2g\Delta h})$

$d = (4Q/(\pi K\sqrt{2g\Delta h}))^{1/2}$

$d = (4 \times 10/(\pi \times 1.01\sqrt{2g \times 20.4}))^{1/2} = 0.794\ m$

Check K: $Re = 4Q/(\pi d\nu) = 1.6 \times 10^7$; $d/D = 0.4$ so $K \approx 1.0$ (from Fig. 13.13)

Try again: $d = (1.01/1.0)^{1/2} \times 0.794 = \underline{\underline{0.798\ m}}$

13.36 $\Delta h = 4$ ft and $d/D = 0.33$

$Re_d/K = (1/3)\sqrt{2 \times 32.2 \times 4}/1.22 \times 10^{-5}) = 4.4 \times 10^5$;

T = 60°F; K = 0.97 (Estimated from Fig. 13.13)

$Q = KA\sqrt{2g \times 4} = 0.97(\pi/4) \times (0.333)^2\sqrt{2g} \times 4$

$Q = \underline{\underline{1.36\ cfs}}$

13.37 The correct choice is $\underline{\underline{b)\ -10\ psi < p < 0}}$

13.38 Assume $T = 20°C$; $\nu = 10^{-6} m^2/s$

$\Delta p = 100$ kPa so $\Delta h = \Delta p/\gamma = 100{,}000/9{,}790 = 10.2$ m

Then $\sqrt{2g\Delta h}\, d/\nu = \sqrt{2 \times 9.81 \times 10.2} \times 1/10^{-6} = 1.4 \times 10^7$

Then $K \approx 1.02$ (extrapolated from Fig. 13.13)

Thus $Q = KA\sqrt{2g\Delta h} = 1.02 \times (\pi/4) \times 1^2\sqrt{2g \times 10.2}$

$Q = 11.33\ m^3/s$

13.39

Because of the streamline curvature (concave toward wall) near the pressure tap the pressure sensed at point 2 at the wall will be less than the average pressure across the section. Therefore, Q_o will be too large as determined by the formula. Thus, $K < 1$.

13.40

$\Delta p = 6.20$ psi $= 6.2 \times 144$ psf or $\Delta h = 6.20 \times 144/62.4 = 14.3$ ft

Then $\sqrt{2g\Delta h}\, d/\nu = \sqrt{2 \times 32.2 \times 14.3} \times (6/12)/(1.4 \times 10^{-5}) = 10 \times 10^5$

$K = 1.02$ Then $Q = KA\sqrt{2g\Delta h} = 1.02 \times (\pi/4) \times (6/12)^2\sqrt{2 \times 32.2 \times 14.3} = 6.08$ cfs

---

13.41

$\Delta h = 45{,}000/(0.69 \times 9{,}810) = 6.65$ m $\nu = \mu/\rho = 3 \times 10^{-4}/690 = 4.3 \times 10^{-7}$

Then $\sqrt{2 g\Delta h}\, d/\nu = \sqrt{2 \times 9.81 \times 6.65} \times 0.20/(4.3 \times 10^{-7}) = 5.3 \times 10^6$

$K = 1.02$ (from Fig. 13.13)

Then $Q = KA\sqrt{2g\Delta h} = 1.02 \times (\pi/4) \times (0.20)^2\sqrt{2 \times 9.81 \times 6.65} = 0.366\ m^3/s$

13.42

$Q = KA(2\Delta p/\rho)^{0.5}$

$Re_d/K = (2\Delta p/\rho)^{0.5}\ (d/\nu) = ((2 \times 10 \times 10^3)/(1{,}000))^{0.5}(0.02/10^{-6}) = 8.9 \times 10^4$

$K = 1.00$ (from Fig. 13-12)

$Q = (1.0)\ (\pi/4)\ (0.02^2)\ (2 \times 10 \times 10^3/10^3)^{0.5} = 0.0014\ m^3/s$

13.43 Assume $C_d = 0.98$ Then from Eq. 13.5

$$K = C_d/\sqrt{1-(A_2/A_1)^2}$$

$$= 0.98/\sqrt{1-0.75^2}$$

$$K = 1.48$$

$$A_2 = 0.00147 \text{ m}^2$$

Then $Q = (1.48)(0.00147 \text{ m}^2)(2.0 \times 9.81 \times 1)^{0.5}$

$\underline{Q = 0.00964 \text{ m}^3/\text{s}}$

13.44 $h_L = (V_1 - V_0)^2/2g$

but $V_0A_0 = V_1A_1$; $V_1 = V_0A_0/A_1 = V_0 \times (3/1)^2 = 9V_0$

Then $h_L = (9V_0 - V_0)^2/2g = \underline{\underline{64V_0^2/2g}}$

13.45 $S = nD/V$

$V = nD/S = (50)(0.01)/(0.2) = 2.5 \text{ m/s}$

$Q = VA = (2.5)(\pi/4)(0.05^2) = \underline{\underline{0.0049 \text{ m}^3/\text{s}}}$

13.46 $F_D = \text{wgt.}$

$F_D = C_DA\rho V^2/2 = mg$; $\rho V^2 = \text{const.}$

$V/V_{std.} = (\rho_{std.}/\rho)^{0.5}$ and $Q = VA$

$\therefore$ $\underline{\underline{Q/Q_{std.} = (\rho_{std.}/\rho)^{0.5}}}$ (1)

Correct by calculating ρ for the actual conditions and then use Eq. (1) to correct Q.

13.47 $\rho = 0.8 \text{ kg/m}^3$ $\rho_{standard} = 1.2 \text{ kg/m}^3$

The deflection of the rotometer is a function of the drag on the rotating element or:

$$F_D = A_p\rho V^2/2$$

$V \propto Q \propto (1/\rho)^{0.5}$ and $Q_s \propto (1/\rho_s)^{0.5}$

Thus $Q/Q_s \propto (\rho_s/\rho)^{0.5} \Rightarrow Q = Q_s \times (1.2/0.8)^{0.5}$

$Q = 3 \times (1.5)^{0.5} = 3.67\ \ell/\text{s}$

13.48 a) $t_1 = L/(c+V)$; $t_2 = L/(c-V)$

$$\Delta t = t_2 - t_1 = L[(1/(c-V)) - (1/(c+V))] = +L(2V)/(c^2 - V^2)$$

$$\therefore \quad (c^2 - V^2)\Delta t = 2LV$$

$$V^2\Delta t + 2LV - c^2\Delta t = 0$$

$$V^2 + (2LV/\Delta t) - c^2 = 0$$

Solving for V:

$$V = [(-2L/\Delta t) \pm \sqrt{(2L/\Delta t)^2 + 4c^2}]/2 = (-L/\Delta t) + \sqrt{(L/\Delta t)^2 + c^2}$$

(only reasonable root)

$$\underline{\underline{V = (L/\Delta t)[-1 + \sqrt{1 + (c\Delta t/L)^2}]}}$$

b) From above

$$\Delta t = 2LV/c^2 \text{ for } c >> V$$

$$\underline{\underline{V = c^2\Delta t/2L}}$$

c) $V = (300)^2(10 \times 10^{-3})/((2)(20)) = \underline{\underline{22.5 \text{ m/s}}}$

13.49 $Q = K\sqrt{2g}\, LH^{3/2}$ where $L = 2$m; $H = 0.13$m; $H/P = 0.43$

Then $K = 0.40 + 0.05 \times 0.43 = 0.422$

$$Q = 0.422\sqrt{2 \times 9.81} \times 2 \times (0.13)^{3/2} = \underline{\underline{0.175 \text{ m}^3/\text{s}}}$$

13.50 $Q = 0.179\sqrt{2g}\, H^{5/2}$

$$Q = 0.179\sqrt{2g}\,(0.20)^{5/2} = \underline{\underline{0.014 \text{ m}^3/\text{s}}}$$

13.51 Correct choice is $\underline{\underline{\text{c) } Q_A < Q_B}}$ because of the side contractions on A.

13.52 Correct choice is b) $(H_1/H_2) < 1$ because K is larger for smaller height of weir as shown by Eq. (13-10); therefore, less head is required for the smaller P value.

13.53

Solution procedure: Assume reasonable value of K then solve for H after which P can be solved for (P = 2.0m - H). Then iterate to get more precise answer. Assume fully ventilated weir.

$$Q = K\sqrt{2g}\, L\, H^{3/2}$$

First assume H/P = 0.60. Then K = 0.40 + 0.05 (0.60) = 0.43

Solve for H: $H = (Q/(K\sqrt{2g}\, L))^{2/3}$

$$= (4/(0.43\sqrt{(2)(9.81)}\,(3))^{2/3} = 0.788\text{m}$$

Iterate: H/P = 0.788/(2 - 0.788) = 0.65 ; K = 0.40 + .05 (.65) = 0.433

$$H = 4/(0.433\sqrt{(2)(9.81}\,(3))^{2/3} = 0.785\text{m}$$

Thus: P = 2.0 H = 2.00 - 0.785 = 1.215m

13.54 Correct choice is c).

13.55 With a head of 1 ft. on the weir $H/P = \frac{1}{2}$ thus $K_i = 0.40 + 0.05H/P = 0.425$. When H = 1in. H/P = (1/12)/2 and $K_f \approx 0.400$. As a simplification, assume K is constant at K=0.413 for the conditions of the problem

$$Q = 0.413\sqrt{2g}LH^{3/2}$$

For a period of dt the volume of water leaving the basin is equal to $A_B dH$ where A_B is the plan area of the basin. Also this volume is equal to Qdt. Equating these two volumes yields:

$$Qdt = A_B dH$$

or

$$0.413\sqrt{2g}LH^{3/2}dt = A_B dH$$

$$dt = (A_B/(0.413\sqrt{2g}LH^{3/2}))dH$$

$$dt = (A_B/0.413\sqrt{2g}L)H^{-3/2}dH$$

Integrate

$$\Delta t = (A_B/(0.413\sqrt{2g}L)\int H^{-3/2}dH$$

$$= -(A_B/(0.413\sqrt{2g}L)2H^{-1/2}|_1^{1/12}$$

$$= -(2A_B/(0.413\sqrt{2g}L)\left(\left(\frac{1}{1/12}\right)^{1/2} - 1\right)$$

$$= (2 \times 2 \times 50/(0.413 \times 8.02 \times 2))(3.46 - 1)$$

$$= 74.3\text{s}$$

13.56 Weir equation:

$$Q = K\sqrt{2g}LH^{3/2}$$

Assume $H \approx \frac{1}{2}$ft. Then $K = 0.4 + 0.05(\frac{1}{2}/3) = .41$, then

$$Q = 0.41\sqrt{64.4} \times 2H^{3/2}$$
$$Q = 6.58H^{3/2} \quad (1)$$

Energy equation written from reservoir water surface to channel water surface:

$$p_1/\gamma + V_1^2/2g + z_1 = p_2/\gamma + V_2^2/2g + z_2 + \sum h_L$$
$$0 + 0 + 100 = 0 + 0 + 3 + H + \sum h_L \quad (2)$$

where

$$\sum h_L = (V^2/2g)(K_e + fL/D + 2K_b + K_E)$$
$$= (V^2/2g)(0.5 + f(100/(1/3)) + 2 \times 0.35 + 1)$$

Assume f = 0.02 (first try)
Then $\sum h_L = 8.2V^2/2g$
Eq. (2) then becomes

$$97 = H + 8.2V^2/2g \quad (3)$$

But V=Q/A so Eq. (3) is written as

$$97 = H + 8.2Q^2/(2gA^2) \text{ where } A^2 = ((\pi/4)(1/3)^2)^2 = .00762\text{ft}^4$$
$$97 = H + 8.2Q^2/(2g \times 0.00762) \quad (4)$$
$$97 = H + 16.72Q^2$$

Solve for Q and H between Eqs. (1) and (4)

$$97 = H + 16.72Q^2$$
$$97 = H + 16.72(6.58H^{3/2})^2$$
$$H = 0.51\text{ft and } Q = 2.397\text{ft}^3/\text{s}$$

Now check Re and f

$$V = Q/A = 27.5\text{ft/s}; \text{Re} = VD/\nu = 27.5 \times (1/3)/(1.4 \times 10^{-5})$$
$$\text{Re} = 6.5 \times 10^5$$
$$f = 0.017 \text{ (Figs.10.8 and 10.9)}$$

Then Eq. (3) becomes

$$97 = H + 7.3V^2/2g$$

and Eq. (4) is $97 = H + 14.88Q^2$
Solve for H and Q again:

$$\underline{\underline{H = 0.53\text{ft}}} \text{ and } \underline{\underline{Q = 2.54\text{ft}^3/\text{s}}}$$

13.57 This involves a trial-and-error solution. First assume the head on the orifice is 1.05 m. Then

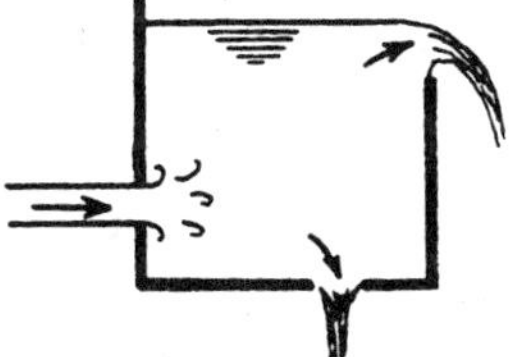

$$Q_{orifice} = KA_0\sqrt{2gh}; \quad k \approx 0.595$$

$$Q_{orifice} = 0.595 \times (\pi/4) \times (0.10)^2\sqrt{2 \times 9.81 \times 1.05} = 0.0212 \text{ m}^3/\text{s}$$

13.57 (Cont'd)

Then $Q_{weir} = K\sqrt{2g}\, LH^{3/2}$; $H_{weir} \approx (Q/(K\sqrt{2g}\, L)^{2/3}$ where $K \approx 0.405$

$H_{weir} = ((0.10 - 0.0212)/(0.405\sqrt{2 \times 9.81} \times 1))^{2/3} = 0.124$ m

Try again: $Q_{orifice} = (1.124/1.05)^{1/2} \times 0.0212 m^3/s = 0.0219\ m^3/s$

$H_{weir} = ((0.10 - 0.0219)/(0.405\sqrt{2 \times 9.81} \times 1))^{2/3} = \underline{\underline{0.124\ m}}$

H_{weir} is same as before, so iteration is complete.

Depth of water in tank is $\underline{\underline{1.124\ m}}$

13.58 With a sharp edged weir the flow will break free of the sharp edge and a definite (repeatable) flow pattern will be established. That assumes that the water surfaces both above and below the nappe are under atmospheric pressure. However, if the top of the weir were not sharp (rounded) then the lower part of the flow may follow the rounded portion of the weir plate a slight distance downstream of the front face of the plate before breaking free of the plate. This would probably lessen the degree of contraction of the flow. With less contraction the flow coefficient would undoubtedly be larger than given by Eq. (13.10).

If the weir is not ventilated below the Nappe such as might be the case for a weir that extends the full width of a rectangular channel (as shown in Fig. 13.18) then as the water plunges into the downstream pool air bubbles would be entrained in the flow and some of the air from under the Nappe would be carried downstream. Therefore, as the air under the Nappe becomes evacuated a pressure less than atmospheric would be established in that region. This would draw the Nappe downward and cause higher velocities to occur near the weir crest. Therefore, greater flow would occur than indicated by use of Eqs. 13.9 and 13.10.

13.59 $Q = K(2g)^{0.5}LH^{3/2}$

$Q = (0.40 + (1/2)(0.05))(64.4)^{0.5} \times 10 \times 1$

$= \underline{34.1 \text{ cfs}}$

13.60 First determine the discharge through the venturi meter.

$$Q = KA_2\sqrt{2\Delta p/\rho}$$
$$= 1 \times (12/144)\sqrt{2 \times 10 \times 144/1.94}$$
$$= 3.21 \text{ft}^3/\text{s}$$

Now determine the head on the weir by the weir equation.

$$Q = 0.179\sqrt{2g}H^{5/2}$$
$$3.21 = 0.179\sqrt{64.4}H^{5/2}$$
$$\underline{\underline{H = 1.38 \text{ft} = 16.5 \text{in.}}}$$

13.61 $Q_{weir} = K\ (2g)^{0.5}LH^{3/2}$

$K = 0.40 + 0.05\ (1/2) = 0.425$

$Q_{out} = 0.425\ (8.025)\ (2)\ (1) = 6.821 \text{ cfs}$

$Q_{in} = V_AA_A + V_BA_B$

$= 4(\pi/4)\ (1^2) + 4(\pi/4)\ (0.5^2)$

$= \pi(1.25) = 3.927 \text{ cfs}$

$Q_{in} << Q_{out}$; therefore, water level is falling

13.62 $Q = (0.40 + .05(1/2))\sqrt{2g}\ (3H_R)H_R^{1.5}$ (1)

$Q = 0.179\ \sqrt{2g}\ H_T^{2.5}$ (2)

Equate Eqs. (1) and (2)

$(0.425)\sqrt{2g}\ (3)\ H_R^{2.5} = 0.179\ \sqrt{2g}\ H_T^{2.5}$

$(H_R/H_T)^{2.5} = 0.179/(3 \times 0.425)$

$\underline{H_R/H_T = 0.456}$

13.63 The water level in the first reservoir will start to rise as soon as the flow is increased. It will continue to rise until the outflow over the rectangular weir is equal to the inflow to the reservoir. The same process will occur in the second reservoir until the outflow over the triangular weir is equal to the inflow to the first reservoir.

Determine the increase in head on the rectangular weir with an increase in discharge of 50%.

Initial conditions: $H_R/P = 0.5$ so $K = 0.4 + .05 \times .5 = 0.425$

Then $Q_{Ri} = 0.425LH_{Ri}^{3/2}$ (1)

Assume $K_f = K_i = 0.425$ (first try)

Then $Q_{Rf} = 0.425LH_{Rf}^{3/2}$ (where $Q_{Rf} = 1.5Q_i$) (2)

Divide Eq. (2) by Eq. (1)

$$Q_{Rf}/Q_{Ri} = (0.425L/0.425L)(H_{Rf}/H_{Ri})^{3/2}$$

$$H_{Rf}/H_{Ri} = (1.5)^{2/3} = 1.31$$

Check K_i : $K=0.40 + .05 \times .5 \times 1.31 = .433$

Recalculate H_{Rf}/H_{Ri} .

$$H_{Rf}/H_{Ri} = ((0.425/0.433) \times 1.5)^{2/3} = 1.29$$

The final head on the rectangular weir will be 29% greater than the initial head. Now determine the increase in head on the triangular weir with a 50% increase in discharge.

$$Q_{Tf}/Q_{Ti} = (H_{Tf}/H_{Ti})^{5/2}$$

$$\text{or } H_{Tf}/H_{Ti} = (Q_{Tf}/Q_{Ti})$$

$$= (1.5)^{2/5}$$

$$= 1.18$$

The head on the triangular weir will be 18% greater with the 50% increase in discharge.

13.64 $Q = K\sqrt{2g}\, LH^{3/2}$ where L = 3m, Q = 6 m³/s

Assume $K \approx 0.41$ then $H = (Q/(0.41\sqrt{2g} \times 3))^{2/3}$

$H = (6/(0.41 \times \sqrt{2 \times 9.81} \times 3))^{2/3} = 1.10$ m

Then $P \approx 2.0 - 1.10 = 0.90$ m $H/P \approx 1.22$

Try again: $K = 0.40 + 1.22 \times 0.05 = 0.461$

$H = (6/(0.461 \times \sqrt{2 \times 9.81} \times 3))^{2/3} = 0.986$ m

So height of weir $P = 2.0 - 0.986 = \underline{\underline{1.01\text{ m}}}$; $H/P = 0.976$

Try again: $K = 0.40 + 0.976 \times 0.05 = 0.449$

$H = (6/(0.449 \times \sqrt{2 \times 9.81} \times 3))^{2/3} = 1.00$ m

$P = 2.00 - 1.00 = \underline{\underline{1.00\text{ m}}}$

13.65 $Q = 0.179\sqrt{2g}\, H^{5/2}$ where H = 1.5 ft

Then $Q = 0.179\sqrt{2g} \times (1.5)^{5/2} = \underline{\underline{3.96\text{ ft}^3/\text{sec}}}$

13.66 $Q = (8/15)\, C_d\, (2g)^{0.5} \tan(\theta/2)\, H^{5/2}$

$Q = (8/15)\,(0.60)\,(64.4)^{0.5} \tan(22.5°)\, H^{5/2}$

$Q = 1.064\, H^{5/2}$

$H = (Q/1.064)^{2/5} = (10/(60 \times 1.064))^{2/5} = \underline{\underline{0.476\text{ ft}}}$

13.67 $k_s/D = 0.001$ (from Fig. 10.9) ; assume $f = 0.02$

Write energy Eq. from well water surface to tank water surface.

$p_1/\gamma + V_1^2 2g + z_1 + h_p = p_2/\gamma + V_2^2/2g + z_2 + \Sigma h_L$

$0 + 0 + 0 + h_p = 0 + 0 + (2+h) + (V^2/2g)(k_e + (fL/D) + K_E)$

where

h = depth of water in tank

$20 = (2+h) + (V^2/2g)(0.5 + (0.02 \times 2.5/0.05) + 1)$

$18 = h + 0.127\, V^2$

$V = ((18-h)/0.127)^{0.5}$

13.67 (continued)

$Q = VA = ((18-h)/0.127)^{0.5}(\pi/4)(0.05)^2 =$
$0.00551\ (18-h)^{0.5}$ (1)

Triangular Weir:

$Q = 0.179\sqrt{2g}\,H^{2.5}$ where $H = h - 1$

Then

$Q = 0.179\sqrt{2g}\,(h-1)^{2.5} = 0.793\ (h-1)^{2.5}$ (2)

Equate (1) and (2)

$0.00551\ (18-h)^{0.5} = 0.793\ (h-1)^{2.5}$

$0.00695\ (18-h)^{0.5} = (h-1)^{2.5}$

Solve for h: $\underline{h = 1.24\text{ m}}$

Also, upon checking Re we find our assumed f is OK.

13.68 $p_t/p_1 = (1 + (k-1/2)M^2)^{k/(k-1)}$

$= (1 + 0.2M^2)^{3.5}$ for air

$(140/100) = (1 + 0.2M^2)^{3.5}$; $M = \underline{0.710}$

$T_t/T = 1 + 0.2\ M^2$

$T = 300/1.10 = 273$

$c = \sqrt{(1.4)(287)(273)} = \underline{331\text{ m/s}}$

$V = Mc = (0.71)(331) = \underline{235\text{ m/s}}$

13.69 The purpose of the algebraic manipulation is to express p_1/p_{t_2} as a function of M_1 only.

For convenience, express the group of variables below as

$F = 1 + ((k-1)/2)M^2$ and $G = kM^2 - ((k-1)/2)$

$p_1/p_{t_2} = (p_1/p_{t_1})(p_{t_1}/p_{t_2}) = (p_1/p_{t_1})(p_1/p_2)(F_1/F_2)^{k/k-1}$

From Eq. 12-38, $p_1/p_2 = (1+kM_2^2)/(1+kM_1^2)$

So $p_1/p_{t_2} = (p_1/p_{t_1})((1+kM_2^2)/(1+kM_1^2))(F_1/F_2)^{k/k-1}$

13.69 (continued)

From Eq. 12-40, we have

$(M_1/M_2) = ((1+kM_1^2)/(1+kM_2^2))(F_2/F_1)^{1/2}$

Thus, we can write

$(p_1/p_{t_2}) = (p_1/p_{t_1})(M_2/M_1)(F_1/F_2)^{k+1/(2(k-1))}$

But, from Eq. 12-41

$M_2 = (F_1/G_1)^{1/2}$ Also, $p_1/p_{t_1} = 1/(F_1^{k/k-1})$

So, $p_1/p_{t_2} = 1/(F_1^{k/k-1})(F_1^{1/2}/G_1^{1/2})(1/M_1)(F_1/F_2)^{k+1/(2(k-1))}$

$= (G_1^{-1/2}/M_1)F_2^{-(k+1)/2(k-1)}$

However, $F_2 = 1 + ((k-1)/2)M_2^2 = 1 + ((k-1)/2)(F_1/G_2) = ((k+1)/2)M_1)^2/G_1$

Substituting for F_2 in expression for p_1/p_{t_2} gives

$p_1/p_{t_2} = (1/M_1)(G_1^{1/k-1})/((k+1)/2M_1)^{k+1/k-1}$

Multiplying numerator and denominator by $(2/k+1)^{1/k-1}$ gives

$p_1/p_{t_2} = ((2kM_1^2/(k+1)-(k-1)/(k+1))^{1/(k-1)}/((\frac{k+1}{2})M_1^2)^{k/k-1}$

13.70 Using the Rayleigh pitot tube formula

$54/200 = (1.1667\ M_1^2 - 0.1667)^{2.5}/(1.893M_1^2)$

and solving for M_1 gives $M_1 = \underline{1.57}$

$\therefore\ T_1 = 350/(1+0.2(1.57)^2) = 234\ K$

$c_1 = \sqrt{(1.4)(287)(234)} = 307\ m/s$

$V_1 = 1.57 \times 307 = \underline{\underline{482\ m/s}}$

13.71

$P_1 = 120\ kPa$ $P_2 = 80\ kPa$ $K = 1.66$ $D_2/D_1 = 0.5$

$T_1 = 17°C$ $R = 2{,}077\ J/kgK$

$\rho_1 = 120 \times 10^3/(2{,}077 \times 290) = 0.199\ kg/m^3$; $p_1/\rho_1 = 6.03 \times 10^5$

Using Eq. 13-16,

$V_2 = ((5)(6.03 \times 10^5)(1-0.666^{0.4})/(1-(0.666^{1.2} \times 0.54)))^{1/2} = 686\ m/s$

13.71 (continued)

$\rho_2 = (p_2/p_1)^{1/k}\rho_1 = (0.666)^{0.6}\rho_1 = 0.784\ \rho_1 = 0.156\ kg/m^3$

$\dot{m} = (0.005)^2(0.785)(0.156)(686) = \underline{0.0021\ kg/s}$

13.72 $p_1 = 150$ kPa $p_2 = 110$ kPa $T = 300$ K $R = 518$ J/kgK $k = 1.31$

$\rho_1 = 150 \times 10^3/(518 \times 300) = 0.965\ kg/m^3$ $d = 0.8$ cm $d/D = 0.5$

$2g\Delta h = 2\Delta p/\rho_1 = (2(40x10^3))/0.965 = 8.29 \times 10^4$; $\nu = 1.6 \times 10^{-5} N{\cdot}s/m^2$

$Re/K = ((0.008)/(1.6 \times 10^{-5}))\sqrt{8.29 \times 10^4} = 1.43 \times 10^5$

From Fig. 13.13:

$K = 0.62$

$V = 1 - ((1/1.31)(1-(110/150))(0.41 + 0.35(0.4)^4)) = 0.915$

$\dot{m} = (0.62)(0.915)(0.785)(0.008)^2\sqrt{(2)(0.965)(40\times10^3)} - \underline{0.00792\ kg/s}$

13.73

Expansion factor: $Y = 1 - \{(1/k)(1 - (p_2/p_1))(0.41 + 0.35(A_0/A_1)^2)\}$

$A_0/A_1 = (1/2)^2 = 0.25$; $A_0 = 7.85 \times 10^{-5}\ m^2$

$Y = 1 - \{(1/1.4)\ (1 - (100/150)\ (0.41 + 0.35\ (.25)^2)\}$

$= 0.897$

$\dot{m} = YA_0K\ (2\rho_1\ (p_1 - p_2))^{0.5}$

$Re_d/K = (2\Delta p/\rho)^{0.5}\ d/\nu = (2 \times 50 \times 10^3/1.8)^{0.5}\ (.01/(1.8 \times 10^{-5}))$

$= 236 \times 556 = 1.31 \times 10^5$

$K = 0.63$ (From Fig. 13.13)

$\dot{m} = (0.897)(7.85 \times 10^{-5})(0.63)(2 \times 1.8 \times 50 \times 10^3)^{0.5} = \underline{1.88 \times 10^{-2} kg/s}$

13.74 $d/D = 0.50$

$d = 0.5 \times 0.02\ m = 0.01\ m$

For hydrogen @ T = 15°C = 288K:

$k = 1.41$, $R = 4127$ J/kgK

$\rho = 0.0851\ kg/m^3$ (from Table A.2)

13.74 (Cont'd)

$$A_0 = (\pi/4)(0.01)^2 = 7.85 \times 10^{-5} m^2$$

$$\dot{m} = YA_0K(2\rho_1\Delta p)$$

$$\dot{m} = (1)(7.85 \times 10^{-5})(0.62)(2(0.0851)(1,000))^{0.5}$$

$$\underline{\dot{m} = 6.35 \times 10^{-4} \text{ kg/s}}$$

13.75

$$\text{Hole area} = (\pi/4)(0.25)^2 = 0.049/in^2 = 3.41 \times 10^{-4} \text{ ft}^2$$

For natural gas ; $k = 1.31$, $R = 3098$ ft-lbf/slug ^oR

$$\dot{m} = (p_t A_* / \sqrt{RT})(k^{0.5})(2/(k+1))^{(k+1/(2(k-1))}$$

$$p_t = (50 + 14) = 64 \text{ psia} = 9216 \text{ psfa}$$

$$T = (460 + 70) = 530 \ ^oR$$

$$\dot{m} = ((9216 \times 3.41 \times 10^{-4})/\sqrt{3098 \times 530})(1.31)^{0.5} (2/2.31)^{3.726} = \underline{\underline{0.00164 \text{ slug/s}}}$$

$$= \underline{\underline{0.0528 \text{ lbm/s}}}$$

13.76 Some of the physical effects that might occur are:

A. Abrasion might cause the weir crest to be rounded and this would undoubtedly produce greater flow than indicated by Eqs. 13.9 and 13.10 (see the answer to problem 13.58).

B. If solid objects such as floating sticks come down the canal and hit the weir they may dent the weir plate. Such dents would be slanted in the downstream direction and may even cause that part of the weir crest to be lower than the original crest. In either case these effects should cause the flow to be contracted less than before thus increasing the flow coefficient.

13.76 (Cont'd)

C. Another physical effect that might occur in an irrigation canal is that sediment might collect upstream of the weir plate. Such sediment accumulation would force flow away from the bottom before reaching the weir plate. Therefore, with this condition less flow will be deflected upward by the weir plate and less contraction of the flow would occur. With less contraction the flow coefficient would be increased. For all of the physical effects noted above flow would be increased for a given head on the weir.

13.77 A jet to be studied can be produced by placing an orifice in the side of a rectangular tank as shown below.

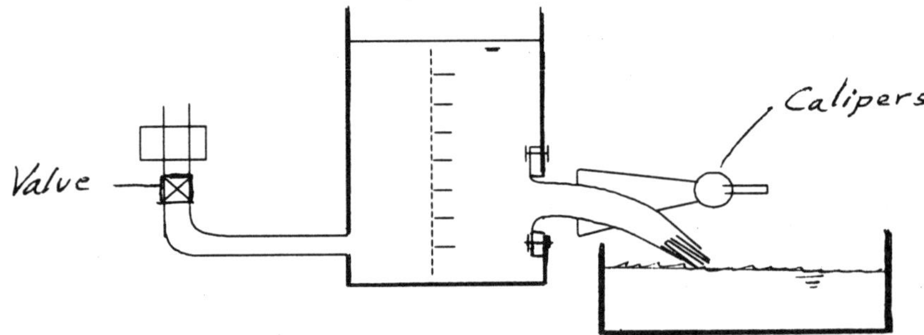

The plate orifice could be machined from a brass plate so that the upstream edge of the orifice would be sharp. The diameter of the orifice could be measured by inside calipers and a micrometer. The contracted jet could be measured by outside calipers and micrometer. Thus the coefficient of contraction could be computed as $C_c = (d_j/d)^2$. However, there may be more than desired error in measuring the water jet diameter by means of a

13.77 (Cont'd)

caliper. Another way to estimate d_j is to solve for it from A_j where A_j is obtained from $A_j = Q/V_j$. Then $d_j = (4A_j/\pi)^{1/2}$. The discharge, Q, could be measured by means of an accurate flow meter or by a weight measurement of the flow over a given time interval. The velocity at the vena contract could be fairly accurately determined by means of the Bernoulli equation. Measure the head on the orifice and compute Vj from $V_j = \sqrt{2gh}$ where h is the head on the orifice. Because the flow leading up to the vena contract is converging it will be virtually irrotational; therefore, Bernoulli's equation will be valid.

Another design decision that must be made is how to dispose of the discharge from the orifice. This could be collected into a tank and then discharged into the lab reservoir through one of the grated openings.

13.78 First, decisions have to be made regarding the physical setup. This should include:

A. How to connect the 2 in. pipe to water source.

B. Providing means of discharging flow back into the lab reservoir. Probably have pipe discharging directly into reservoir through one of the grated openings.

C. Locating control valves in the system.

D. Deciding a length of 2" pipe on which measurements will be made. It is desirable to have enough length of pipe to yield a measurable amount of head loss.

To measure the head loss, one can tap into the pipe at several points along the pipe (six or eight points should be sufficient). The differential pressure between the upstream tap and downstream tap can first be measured.

Then measure the differential pressure between the next tap and the downstream tap, etc. until the pressure difference between the downstream tap and all others has completed. From all these measurements the slope of the hydraulic grade line could be computed.

The discharge could be measured by weighing a sample of the flow for a period of time and then computing the volume rate of flow. Or the discharge could be measured by an electromagnetic flow meter if one is installed in the supply pipe.

The diameter of the pipe should be measured by inside calipers and micrometer. Even though one may have purchased 2 inch pipe, the nominal diameter is usually not the actual diameter. With this diameter one can calculate the cross-sectional area of the pipe. Then the mean velocity can be computed for each run: $V=Q/A$.

Then for a given run, the resistance coefficient, f, can be computed with Eq. (10.22). Other things that should be considered in the design:

A. Make sure the pressure taps are far enough downstream of the control valve or any other pipe fitting so that uniform flow is established in the section of pipe where measurements are taken.

B. The differential pressure measurements could be made by either differential pressure transducers or a differential manometer.

13.78 (Cont'd)

C. Appropriate valving and manifolding could be designed in the system so that only one pressure transducer or manometer is needed for all pressure measurements.

D. The water temperature should be taken so that the specific weight of the water can be found.

E. The design should include means of purging the tubing and manifolds associated with the pressure differential measurements so that air bubbles can be eliminated from the measuring system. Air bubbles often produce erroneous readings.

13.79 Most of the design setup for this equipment will be the same as for Prob. (13.78) except that the valve to be tested would be placed about midway along the two inch pipe. Pressure taps should be included both upstream and downstream of the valve so that hydraulic grade lines can be established both upstream and downstream of the valve (see Fig. 10.15). Then as shown in Fig. (10.15) the head loss due to the valve can be evaluated. The velocity used to evaluate K_V is the mean velocity in the 2 in. pipe so that could be evaluated in the same manner as given in the solution for Prob. (13.78).

CHAPTER FOURTEEN

14.1 $C_T = F_T/\rho D^4 n^2 = 0.048$

Thus, $F_T = 0.048\rho D^4 n^2 = 0.048 \times 1.05 \times 3^4 \times (1,400/60)^2 = \underline{\underline{2,223\ N}}$

14.2 $V_0/nD = (80,000/3,600)/((1,400/60) \times 3) = 0.317$

$C_T = 0.020$ (from Fig. 14-3)

Then $F_T = 0.020 \times \rho D^4 n^2 = 0.020 \times 1.05 \times 3^4 \times (1,400/60)^2 = \underline{\underline{926\ N}}$

$C_p = 0.011$ (from Fig. 14-3)

Then $P = 0.011\rho D^5 n^3 = 0.011 \times 1.05 \times 3^5 \times (1,400/60)^3 = \underline{\underline{35.7\ kW}}$

14.3 $D = 8$ ft; $n = 1,000/60 = 16.67$ rev/sec; $V_0 = 30$ mph $= 44$ fps

$V_0/nD = 44/(16.67 \times 8) = 0.33$; from Fig. 14-3 $C_T = 0.0182$; $C_p = 0.011$

$F_T = 0.0182 \times 0.0024 \times 8^4 \times 16.67^2 = \underline{\underline{49.7\ lbf}}$

Power $= 0.011 \times 0.0024 \times 8^5 \times 16.67^3 = 4,005$ ft-lb/sec $= \underline{\underline{7.3\ hp}}$

If $V_0 = 0$, $C_T = 0.0475$

$F_T = 0.0475 \times 0.0024 \times 8^4 \times 16.67^2 = \underline{\underline{130\ lbf}}$

14.4 $D = 6$ ft; $V_0 = 30$ mph $= 44$ fps

From Fig. 14-3, at maximum efficiency, $V_0/(nD) = 0.285$

$n = 44/(0.285 \times 6) = 25.73$ rps; $N = \underline{\underline{1,544\ rpm}}$

14.5 At maximum efficiency $C_T = 0.023$ and $C_p = 0.012$

$F_T = 0.023 \times 0.0024 \times 6^4 \times 25.73^2 = \underline{\underline{47.4\ lbf}}$

Power $= 0.012 \times 0.0024 \times 6^5 \times 25.73^3 = 3,815$ ft-lb/sec $= \underline{\underline{6.94\ hp}}$

14.6 $P = C_p \rho n^3 D^5$ $\qquad \rho = p/RT = 60 \times 10^3/((287)(273)) = 0.766\ kg/m^3$

$T = C_T \rho n^2 D^4$

$T = \text{Drag} = \text{Lift}/30 = (1,000)(9.81)/(30) = 327\ N$

$327 = (0.025)(0.766)(3,000/60)^2 D^4$

14.6 (continued)

$D = 1.62$ m

$L = \text{wgt.} = C_L\ (1/2)\rho V_0^2 S$

$L/(C_L S) = (\rho V_0^2/2) = (1{,}000)\ (9.81)/((0.40)\ (10)) = 2{,}452$

$V_0^2 = (2{,}452)\ (2)/(0.766) = 6{,}402$; $V_0 = 80$ m/s

14.7 $V_{tip} = 0.9c = 0.9 \times 335 = 301.5$ m/s

$V_{tip} = \omega r = n(2\pi)r;\ \ n = 301.5/(2\pi r) = 301.5/(\pi D)$

for D = 2m,	n = 48.0 rev/sec;	N = 2,879 rpm
3m	32.0 rev/sec	1,919 rpm
4m	24.0 rev/sec	1,440 rpm

14.8 At maximum efficiency $V_0/(nD) = 0.285$

Then $n = V_0/(0.285D) = (50{,}000/3{,}600)/(0.285 \times 2) = 24.4$ rev/sec

$N = 24.4 \times 60 = 1{,}462$ rpm

14.9 At maximum efficiency $C_T = 0.023$ and $C_P = 0.012$

$F_T = 0.023 \times 1.1 \times 2^4 \times (24.36)^2 = 240$ N

$P = 0.012 \times 1.1 \times 2^5 \times (24.36)^3 = 6.11$ kW

14.10 $C_T = 0.048$

Thus $F_T = 0.048\rho D^4 n^2 = 0.048 \times 1.1 \times 2^4 \times (1{,}000/60)^2 = 235$ N

$a = F/M = 235/300 = 0.782$ m/s^2

14.11 $D = 40$ cm; $n = 1{,}000/60 = 16.67$ rev/sec; $\Delta h = 3$ m

$C_H = \Delta hg/D^2n^2 = 3 \times 9.81/((0.4)^2 \times (16.67)^2) = 0.662$

from Fig. 14-6, $C_Q = Q/(nD^3) = 0.625$

Then $Q = 0.625 \times 16.67 \times (0.4)^3 = 0.667$ m^3/s

14.12 $n = 690/60 = 11.5$ rev/sec; $D = 71.2$ cm; $\Delta h = 10$ m

$C_H = \Delta hg/(n^2D^2) = 10 \times 9.81/((0.712)^2(11.5)^2) = 1.46$

from Fig. 14-6, $C_Q = 0.40$ and $C_P = 0.76$

$Q = C_Q nD^3 = 0.40 \times 11.5 \times 0.712^3 = \underline{\underline{1.66\ m^3/s}}$

$\text{Power} = C_p \rho D^5 n^3 = 0.76 \times 1{,}000 \times 0.712^5 \times 11.5^3 = \underline{\underline{211\ kW}}$

14.13

$D = 35.6$ cm; $n = 11.5$ r/s

Writing the energy equation from the reservoir surface to the center of the pipe at the outlet,

$$p_1/\gamma + V_1^2/(2g) + z_1 + h_p = p_2/\gamma + V_2^2/(2g) + z_2 + \Sigma h_L$$

$$h_p = 21.5 - 20 + [Q^2/(A^2 2g)](1 + fL/D + k_e + k_b)$$

$L = 64$ m assume $f = 0.014$ $r_b/D = 1$

from Table 10-2, $k_b = 0.35$ $k_e = 0.1$

(1) $h_p = 1.5 + [Q^2((0.014(64)/0.356) + 0.35 + 0.1 + 1)]/[2(9.81)(\pi/4)^2(0.356)^4] = 1.5 + 20.42Q^2$

$C_Q = Q/(nD^3) = Q/[(11.5)(0.356)^3] = 1.93Q$

(2) $h_p = C_H n^2 D^2/g = C_H(11.5)^2(0.356)^2/9.81 = 1.71C_H$

Q(m/s)	C_Q	C_H	h_p(1) (m)	h_p(2) (m)
0.10	0.193	2.05	1.70	3.50
0.15	0.289	1.70	1.96	2.91
0.20	0.385	1.55	2.32	2.65
0.25	0.482	1.25	2.78	2.13
0.30	0.578	0.95	3.34	1.62
0.35	0.675	0.55	4.00	0.94

Then plotting the system curve and the pump curve, we obtain the operating condition:

$Q = 0.22\ m^3/s$; Power = $\underline{\text{6.5 kW (from Fig. 14.7)}}$

14.13 (continued)

Graph for solution of Problems 14.13 & 14.14:

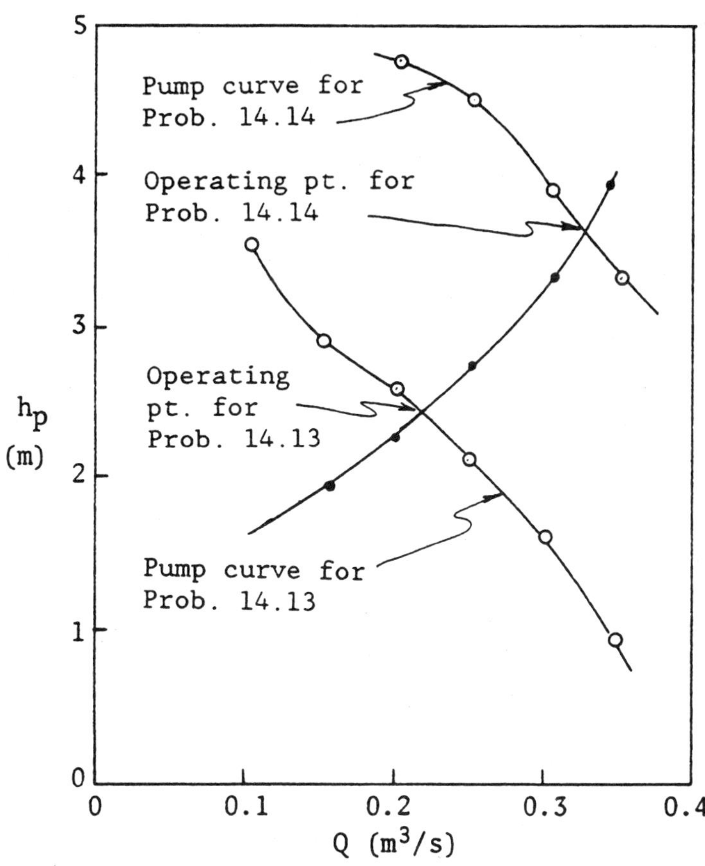

14.14 The system curve will be the same as in Prob. 14.13.

$C_Q = Q/[nD^3] = Q/[15(0.356)^3] = 1.48Q$

$h_p = C_H n^2 D^2/g = C_H (15)^2 (0.356)^2/9.81 = 2.91C_H$

Q	C_Q	C_H	h_p
0.20	0.296	1.65	4.79
0.25	0.370	1.55	4.51
0.30	0.444	1.35	3.92
0.35	0.518	1.15	3.34

Plotting the pump curve with the system curve gives the operating condition:

$Q = \underline{0.32\ m^3/s}$; $C_Q = 1.48(0.32) = 0.474$

Then from Fig. 14-6, $C_p = 0.70$

Power $= C_p n^3 D^5 \rho = 0.70(15)^3(0.356)^5 1,000 = \underline{13.5\ kW}$

14.15 At maximum efficiency, from Fig. 14-6,

$C_Q = 0.64$; $C_P = 0.60$; and $C_H = 0.75$

$D = 1.5$ ft; $n = 1{,}100/60 = 18.33$ rev/sec

$Q = C_Q nD^3 = 0.64 \times 18.33 \times 1.5^3 = \underline{\underline{39.6 \text{ cfs}}}$

$\Delta h = C_H n^2 D^2/g = 0.75 \times 18.33^2 \times 1.5^2/32.2 = \underline{\underline{17.6 \text{ ft}}}$

$\text{Power} = C_P \rho D^5 n^3 = 0.60 \times 1.94 \times 1.5^5 \times 18.33^3 = 54{,}437 \text{ ft-lb/sec} = \underline{\underline{99.0 \text{ hp}}}$

14.16 At maximum efficiency, from Fig. 14-6,

$C_Q = 0.64$; $C_P = 0.60$: $C_H = 0.75$; $D = 0.50$ m; $n = 45$ rps

$Q = C_Q nD^3 = 0.64 \times 45 \times 0.5^3 = \underline{\underline{3.60 \text{ m}^3\text{/s}}}$

$\Delta h = C_H n^2 D^2/g = 0.75 \times 45^2 \times 0.5^2/9.81 = \underline{\underline{38.7 \text{ m}}}$

$\text{Power} = C_P \rho D^5 n^3 = 0.60 \times 1{,}000 \times 0.5^5 \times 45^3 = \underline{\underline{1.709 \text{ MW}}}$

14.17 $D = 14/12 = 1.167$ ft; $n = 900/60 = 15$ r/s

$\Delta h = C_H n^2 D^2/g = C_H (15)^2 (1.167)^2/32.2 = 9.52\, C_H$ ft

$Q = C_Q nD^3 = C_Q 15(1.167)^3 = 23.8\, C_Q$ cfs

C_Q	C_H	Q(cfs)	Δh(ft)
0.0	2.9	0.0	27.60
0.1	2.55	2.38	24.27
0.2	2.0	4.76	19.04
0.3	1.7	7.15	16.18
0.4	1.5	9.53	14.28
0.5	1.2	11.91	11.42
0.6	0.85	14.29	8.09

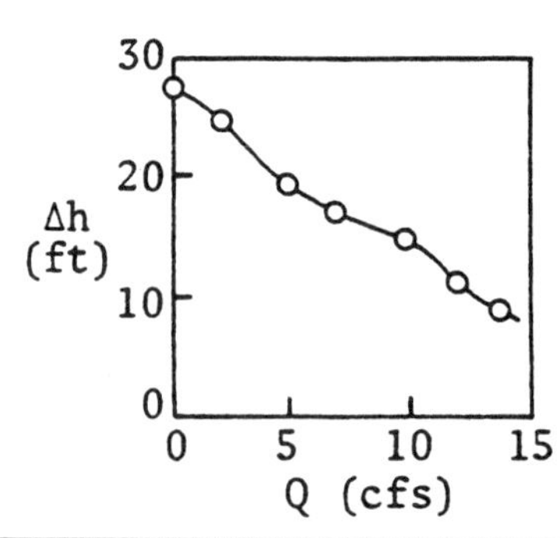

14.18 $D = 60$ cm $= 0.60$ m; $N = 690$ rpm $= 11.5$ rev/sec

Then $\Delta h = C_H D^2 n^2/g = 4.853\, C_H$

$Q = C_Q nD^3 = 2.484\, C_Q$

C_Q	C_H	$Q(\text{m}^3\text{/s})$	h(m)
0.0	2.90	0.0	14.1
0.1	2.55	0.248	12.4
0.2	2.00	0.497	9.7
0.3	1.70	0.745	8.3
0.4	1.50	0.994	7.3
0.5	1.20	1.242	5.8
0.6	0.85	1.490	4.2

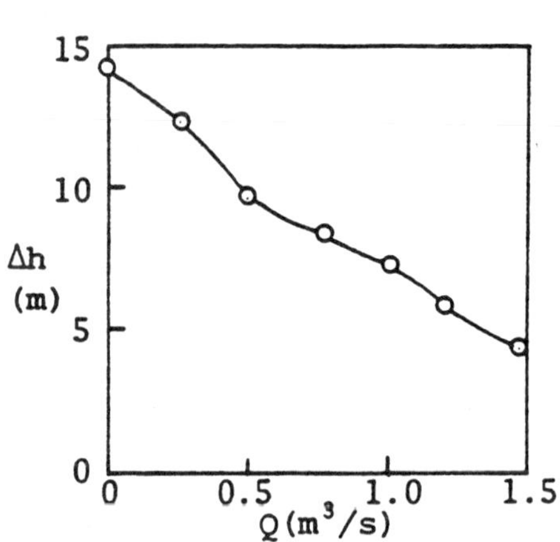

14.19 $D = 0.371 \times 2 = 0.742$ m; $n = 2{,}133.5/(2 \times 60) = 17.77$ rps

from Fig. 14-10, at peak efficiency $C_Q = 0.121$, $C_H = 5.15$

$\Delta h = C_H n^2 D^2/g = 5.15(17.77)^2(0.742)^2/9.81 = \underline{\underline{91.3 \text{ m}}}$

$Q = C_Q nD^3 = 0.121(17.77)(0.742)^3 = \underline{\underline{0.878 \text{ m}^3/\text{s}}}$

14.20 $Q = VA = (60)(\pi/4)(1)^2 = 47.1 \text{ m}^3/\text{s}$

$C_Q = Q/(nD^3) = (47.1)/((1{,}800/60)(2)^3 = 0.196$

$C_P = 0.90$ (From Fig. 14-6)

$P = C_P \rho D^5 n^3 = (0.90)(1.2)(2)^5(30)^3) = 933$ kW

Power needed to drive the fan = $\underline{\underline{933 \text{ kW}}}$

14.21 $D = 0.36$ m; $L = 610$ m; $\Delta z = 450 - 366 = 84$ m

Assume $\Delta h = 90$ m[$>\Delta z$], then from Fig. 14-9, $Q = 0.24 \text{ m}^3/\text{s}$

$V = Q/A = 0.24/[(\pi/4)(0.36)^2] = 2.36$ m/s; $k_s/D = 0.00012$

Assuming $T = 20°\text{C}$, $Re = VD/\nu = 2.36(0.36)/10^{-6} = 8.5 \times 10^5$

from Fig. 10-8, $f = 0.014$

$h_f = (0.014(610)/0.36)((2.36)^2/(2 \times 9.81)) = 6.73$ m; $h \approx 84 + 6.7 = 90.7$ m

from Fig. 14-9, $Q = 0.23\text{m}^3/\text{s}$; $V = 0.23/((\pi/4)(0.36)^2) = 2.26$ m/s

$h_f = [0.014(610)/0.36](2.26)^2/(2 \times 9.81) = 6.18$ m

so $\Delta h = 84 + 6.2 = 90.2$ m and from Fig. 14-9 $\underline{\underline{Q = 0.225 \text{ m}^3/\text{s}}}$

14.22 $D = 0.371$ m $= 1.217$ ft; $n = 1{,}500/60 = 25$ rps

$\Delta h = C_H n^2 D^2/g$ so $C_H = 150(32.2)/[(25)^2(1.217)^2] = 5.22$

from Fig. 14-10 $C_Q = 0.120$

Then $Q = C_Q nD^3 = 0.120(25)(1.217)^3 = \underline{\underline{5.4 \text{ cfs}}}$

14.23 $C_H = \Delta H g/D^2 n^2$ Since $\overline{C_H}$ will be the same for the maximum head condition, then $\Delta H \propto n^2$ or $H_{1,500} = H_{1,000} \times (1{,}500/1{,}000)^2$

$H_{1,500} = 102 \times 2.25 = \underline{\underline{229.5 \text{ ft}}}$

14.24 $H \propto n^2$ so $H_{30}/H_{35.6} = (30/35.6)^2$

or $H_{30} = 104 \times (30/35.6)^2 = \underline{73.8 \text{ m}}$

14.25 $D = 0.40$ m $n = 25$ rps

$C_H = \Delta hg/(n^2D^2) = 50(9.81)/[(25)^2(0.40)^2] = 4.91$

from Fig. 14-10 $C_Q = 0.136$

then $Q = C_Q nD^3 = 0.136(25)(0.40)^3 = \underline{0.218 \text{ m}^3/\text{s}}$

14.26 $D = 20$ cm; $N = 5{,}000$ rpm $= 83.33$ rps; $\rho = 814$ kg/m^3

from Fig. 14-10: At maximum efficiency $C_Q = 0.125$; $C_H = 5.15$; $C_p = 0.69$

$Q = C_Q nD^3 = (0.125)(83.33)(0.20)^3 = \underline{0.0833 \text{ m}^3/\text{sec}}$

$\Delta h = C_H D^2 n^2/g = (5.15)(0.20)^2(83.33)^2/(9.81) = \underline{145.8 \text{ m}}$

$P = C_p \rho D^5 n^3 = (0.69)(814)(0.20)^5(83.33)^3 = \underline{104.0 \text{ kW}}$

14.27 $D = 1.52$ m; $n = 500/60 = 8.33$ rps

$Q = C_Q nD^3 = C_Q(8.33)(1.52)^3 = 29.27\, C_Q$ m^3/s

$\Delta h = C_H n^2 D^2/g = C_H(8.33)^2(1.52)^2/9.81 = 16.36\, C_H$ m

C_Q	Q	C_H	Δh
0.0	0.0	5.80	94.9
0.04	1.17	5.80	94.9
0.08	2.34	5.75	94.1
0.10	2.93	5.60	91.6
0.12	3.51	5.25	85.9
0.14	4.10	4.80	78.5
0.16	4.68	4.00	65.4

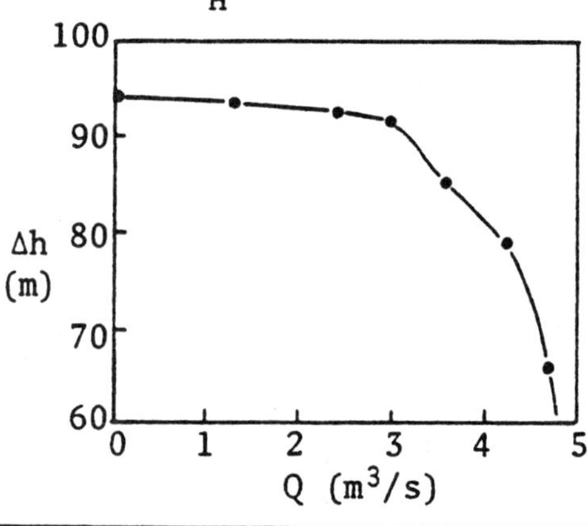

14.28 $N_{ss} = NQ^{1/2}/(\text{NPSH})^{3/4}$

$N = 690$ rpm

NPSH $\simeq$ 14.7 psi x 2.31 ft/psi - $h_{vap.press.} \simeq 33$ ft

$Q = 0.22$ m^3/s x 264.2 gallons/s x 60 s/min = 3487 gpm

$N_{ss} = 690 \times (3487)^{1/2}/(33)^{3/4} = \underline{2960}$

$\underline{N_{ss}}$ **is much below 8500; therefore, it is in a safe operating range.**

14.29 $N = 1{,}500$ rpm so $n = 25$ rps; $Q = 10$ cfs; $h = 25$ ft

$$n_s = n\sqrt{Q}/[g^{3/4}h^{3/4}] = (25)(10)^{1/2}/[(32.2)^{3/4}(25)^{3/4}] = 0.62$$

Then from Fig. 14-14, $n_s > 0.60$, so use an axial flow pump.

14.30 $n = 25$ rps; $Q = 0.30\ m^3/sec$; $h = 8$ meters

$$n_s = n\sqrt{Q}/[g^{3/4}h^{3/4}] = (25(0.3)^{1/2}/[(9.81)^{3/4}(8)^{3/4}] = 0.52$$

Then from Fig. 14-14, $n_s < 0.60$ so use a mixed flow pump.

14.31 $N = 1{,}100$ rpm $= 18.33$ rps; $Q = 0.4\ m^3/sec$; $h = 70$ meters

$$n_s = n\sqrt{Q}/[g^{3/4}h^{3/4}] = (18.33)(0.4)^{1/2}/[(9.81)^{3/4}(70)^{3/4}]$$

$$= (18.33)(0.63)/[(5.54)(24.2)] = 0.086$$

Then from Fig. 14-14, $n_s < 0.23$ so use a radial flow pump.

14.32 $N = 1{,}100$ rpm $= 18.33$ rps; $Q = 10$ cfs; $h = 600$ ft

$$n_s = n\sqrt{Q}/[g^{3/4}h^{3/4}] = (18.33)(10)^{1/2}/[(32.2)^{3/4}(600)^{3/4}]$$

$$= (18.33)(3.16)/[(13.5)(121)] = 0.035$$

Then from Fig. 14-14, $n_s < 0.20$, so use a radial flow pump.

14.33 The safe operating N_{ss} is 8500, or

$$8500 = NQ^{1/2}/(NPSH)^{3/4}$$

The suction head is given as 1.5m = 4.92 ft. Then assuming that the atmospheric pressure is 14.7 psia, we compute NPSH as

$$NPSH = 14.7 \text{ psi} \times 2.31 \text{ ft/psi} + 4.92 \text{ ft} - h_{vap.press.} = 38.4 \text{ ft}$$

$$Q = 0.40\ m^3/s \times 264.2 \text{ gallons}/m^3 \times 60 \text{ s/min} = 6341 \text{ gpm}$$

Then $N = 8500 \times (NPSH)^{3/4}/Q^{1/2}$

$$= 8500 \times 15.4/79.6$$

$N = $ 1644 rpm

14.34

$n_s = n\sqrt{Q}/(g^{3/4}h^{3/4})$; $n = 10$ rps; $Q = 1.0\ m^3/s$; $h = 3 + (1.5+fL/D)V^2/(2g)$;

$V = 1.27$ m/s Assume $f = 0.01$, so

$h = 3 + (1.5 + 0.01 \times 20/1)(1.27)^2/(2 \times 9.81) = 3.14$ m

Then $n_s = 10 \times \sqrt{1}/(9.81 \times 3.14)^{3/4} = 0.76$

From Fig. 14-14, use axial flow pump.

14.35 Max. air speed = 30 m/s; Area = $0.36\ m^2$; $n = 2{,}000/60 = 33.3$ rps;

$Q = 30.0 \times 0.36 = 10.8\ m^3/s$; $\rho = 1.2\ kg/m^3$ at 20°C

From Fig. 14-6, at maximum efficiency, $C_Q = 0.63$ and $C_p = 0.60$

$C_Q = Q/nD^3$, so $D^3 = Q/(nC_Q) = 10.8/(33.3 \times 0.63) = 0.61\ m^3$

$D = 0.80$ m

$C_p = P/(\rho n^3D^5)$, so $P = C_p\rho n^3D^5 = 0.6(1.2)(33.3)^3(0.848)^5 = 11.7$ kW

14.36 Volume = $10^5 m^3$; time for discharge = 15 min = 900 sec

$N = 600$ rpm = 10 rps; $\rho = 1.22\ kg/m^3$ at 60°F

$Q = (10^5 m^3)/(900\ sec) = 111.1\ m^3/sec$

From Fig. 14-6, at maximum efficiency, $C_Q = 0.63$; $C_p = 0.60$

For two blowers operating in parallel, the discharge per blower will be 1/2, or $Q = 55.55\ m^3/sec$, then $D^3 = Q/nC_Q = (55.55)/[10 \times 0.63] = 8.815$

$D = 2.066$ meters

$P = C_p\rho D^5n^3 = (0.6)(1.22)(2.066)^5(10)^3 = 27.6$ kW

14.37

$\dot{m} = 1$ kg/s; $p_1 = 100$ kPa; $p_2 = 150$ kPa; $T_1 = 27°C$; $e = 0.65$; $k = 1.26$;

$R = 518$

$$P_{th} = (k/(k-1))Qp_1[(p_2/p_1)^{(k-1)/k} - 1] = (k\dot{m}/(k-1))RT_1[(p_2/p_1)^{(k-1)/k} - 1]$$

$$= (1.26/0.26)(1)518(300)[(1.5)^{0.26/1.26} - 1] = 65.6 \text{ kW}$$

$P_{ref} = P_{th}/e = 65.6/0.65 = \underline{\underline{101 \text{ kW}}}$

14.38

$P_{th} = 10$ kW x 0.6 = 6 kW

$$P_{th} = (k/(k-1))Qp_1[(p_2/p_1)^{(k-1)/k} - 1] = (1.3/0.3)Q \times 9 \times 10^4[(140/90)^{0.3/1.3} - 1]$$

$= 4.18 \times 10^4 Q$; $Q = 6.0/41.8 = \underline{\underline{0.143 \text{ m}^3/\text{s}}}$

14.39 $P_{th} = p_1Q_1 \ln(p_2/p_1) = \dot{m}RT_1 \ln(p_2/p_1) = 1 \times 287 \times 288 \ln 4 = 114.6$ kW

$P_{ref} = 114.6/0.5 = \underline{\underline{229 \text{ kW}}}$

14.40

Assume T = 10°C

Writing the energy equation from reservoir to turbine jet,

$$p_1/\gamma + V_1^2/2g + z_1 = p_2/\gamma + V_2^2/2g + z_2 + \Sigma h_L$$

$$0 + 0 + 650 = 0 + V_{jet}^2/2g + 0 + (fL/D)(V_{pipe}^2/2g)$$

but from continuity, $V_{pipe}A_{pipe} = V_{jet}A_{jet}$

$$V_{pipe} = V_{jet}(A_{jet}/A_{pipe}) = V_{jet}(0.16) = 0.026\ V_{jet}$$

so, $(V_{jet}^2/2g)(1 + (fL/D)0.026^2) = 650$

$$V_{jet} = [(2 \times 9.81 \times 650)/(1 + (0.016(10{,}000)/1)0.026^2)]^{1/2} = 107.3 \text{ m/s}$$

Power $= Q\gamma V_{jet}^2 e = 107.3(\pi/4)(0.16)^2 9{,}810(107.3)^2 0.83/(2 \times 9.81) = \underline{\underline{10.31 \text{ MW}}}$

$V_{bucket} = (1/2)V_{jet} = 53.7$ m/s $= (D/2)\omega$; $D = 53.7 \times 2/(360 \times (\pi/30)) = \underline{\underline{2.85 \text{ m}}}$

14.41

Referencing velocities to the bucket

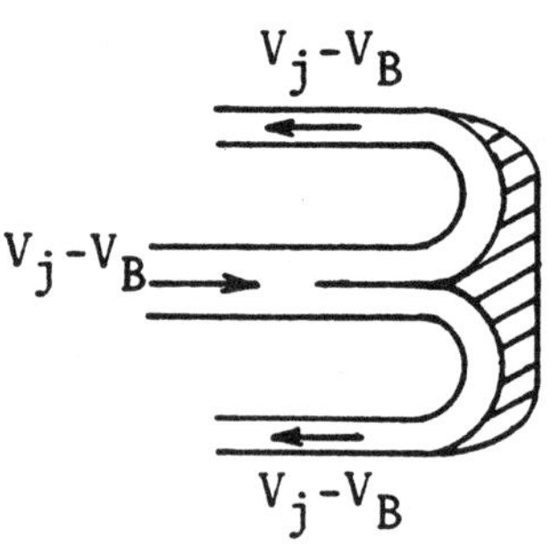

$\Sigma F_{\text{bucket on jet}} = \rho Q[-(V_j-V_B)-(V_j-V_B)]$

Then, $\Sigma F_{\text{on bucket}} = \rho V_jA_j2(V_j-V_B)$

assuming the combination of buckets to be intercepting flow at the rate of V_jA_j.

Then, Power $= FV_B = 2\rho A_j[V_j^2V_B-V_jV_B^2]$

For maximum power production, d Power/$dV_B = 0$

so, $0 = 2\rho A(V_j^2 - V_j2V_B)$

$0 = V_j - 2V_B$ or $V_B = 1/2\ V_j$

14.42 Consider the power developed from the force on a single bucket. Referencing velocities to the bucket gives

$\Sigma F_{\text{on water}} = \rho Q_{\text{rel. to bucket}}(-(1/2)V_j-(1/2)V_j)$

Then $F_{\text{on bucket}} = \rho(V_j-V_B)A_j(V_j)$

but $V_j-V_B = 1/2\ V_j$ so $F_{\text{on bucket}} = 1/2\ \rho AV_j^2$

Then Power $= FV_B = (1/2)\rho Q\ V_j^3/2$

This power is 1/2 that given by Eq. (14-20). The extra power comes from the operation of more than a single bucket at a time so that the wheel as a whole turns the full discharge; whereas, a single bucket intercepts flow at a rate of $1/2\ V_jA_j$.

14.43

$V_{r_1} = Q/(2\pi r_1B) = 126/(2\pi \times 5 \times 1) = 4.01$ m/s

$\omega = 60 \times 2\pi/60 = 2\pi$ rad./s

a) $\alpha_1 = \text{arc cot}((r_1\omega/V_{r_1}) + \cot\beta_1) = \text{arc cot}((5 \times 2\pi/4.01) + 0.577) = \underline{\underline{6.78°}}$

$\alpha_2 = \text{arc tan}(V_{r_2}/(\omega r_2)) = \text{arc tan}((4.01 \times 5/3)/(3 \times 2\pi)) = \text{arc tan } 0.355$

$= 15.5°$

b) $V_1 = V_{r_1}/\sin\alpha_1 = 4.01/0.118 = 39.97$ m/s; $V_2 = V_{r_2}/\sin\alpha_2 = 20.0$ m/s

$P = \rho Q\omega(r_1V_1\cos\alpha_1 - r_2V_2\cos\alpha_2)$

$P = 998 \times 126 \times 2\pi(5 \times 39.97 \times \cos 6.78° - 3 \times 20.0 \times \cos 15.5°) = \underline{\underline{111.1\text{ MW}}}$

c) Increase $\underline{\underline{\beta_2}}$

14.44

$V_{r_1} = 4/(2\pi \times 1.5 \times 0.3) = 1.415$ m/s; $V_{r_2} = 4/(2\pi \times 1.2 \times 0.3) = 1.768$ m/s;

$\omega = (60/60)2\pi = 2\pi \ s^{-1}$

$\alpha_1 = \text{arc cot } ((r_1\omega/V_{r_1}) + \cot \beta_1) = \text{arc cot}((1.5(2\pi)/1.415) + \cot 85°)$

$\alpha_1 = \text{arc cot}(6.66 + 0.0875) = \underline{8°25'}$

$V_{tan_1} = r_1\omega + V_{r_1} \cot \beta_1 = 1.5(2\pi) + 1.415(0.0875) = 9.549$

$V_{tan_2} = r_2\omega + V_{r_2} \cot \beta_2 = 1.2(2\pi) + 1.768(-3.732) = 0.940$

$T = \rho Q(r_1 V_{tan_1} - r_2 V_{tan_2}) = 1{,}000(4)((1.5 \times 9.549) - 1.2 \times 0.940) = \underline{\underline{52{,}780 \text{ N-m}}}$

Power $= T\omega = 52{,}780 \times 2\pi = \underline{331.6 \text{ kW}}$

14.45 $\omega = 120/60 \times 2\pi = 4\pi \ s^{-1}$; $V_{r_1} = 113/(2\pi(2.5)0.9) = 7.99$ m/s

$\alpha_1 = \text{arc cot}((r_1\omega/V_{r_1}) + \cot \beta_1) = \text{arc cot } ((2.5(4\pi)/7.99) + \cot 45°)$

$= \text{arc cot } (3.93 + 1) = \underline{\underline{11°28'}}$

14.46 To get power write the energy equation:

$p_1/\gamma + V_1^2/2g + z_1 = p_2/\gamma + V_2^2/2g + z_2 + \Sigma h_L + h_t$

$0 + 0 + 3{,}000 = 0 + 0 + 2{,}600 + \Sigma h_L + h_t$

$\Sigma h_L = (V^2/2g)\ (f(L/D) + K_E + K_e + 2K_b)$

Assume $k_e = 0.50$; $K_E = 1.0$; $k_b = 0.2$; $k_s/D = 0.00016$

$V = Q/A = 8/((\pi/4)\ (1)^2) = 10.19$ ft/s ;

$Re = VD/\nu = (10.19)\ (1)/(1.2 \times 10^{-5}) = 8.5 \times 10^5$

$f = 0.0145$

$\Sigma h_L = ((10.19)^2/(64.4))[(0.0145)(1{,}000/1) + 1.0 + 0.5 + 2 \times 0.2]$

$\Sigma h_L = 1.612\ (16.4) = 26.44$ ft

$h_t = 3{,}000 - 2{,}600 - 26.44 = 373.6$ ft

Power to turbine $= \gamma Q\ h_t/550 = (8)\ (62.4)\ (373.6)/550$

$= 339$ horsepower

14.46 (Cont'd)

Power output from turbine = 339 x **Eff.** = 339 x 0.8 = $\underline{\underline{271 \text{ hp}}}$

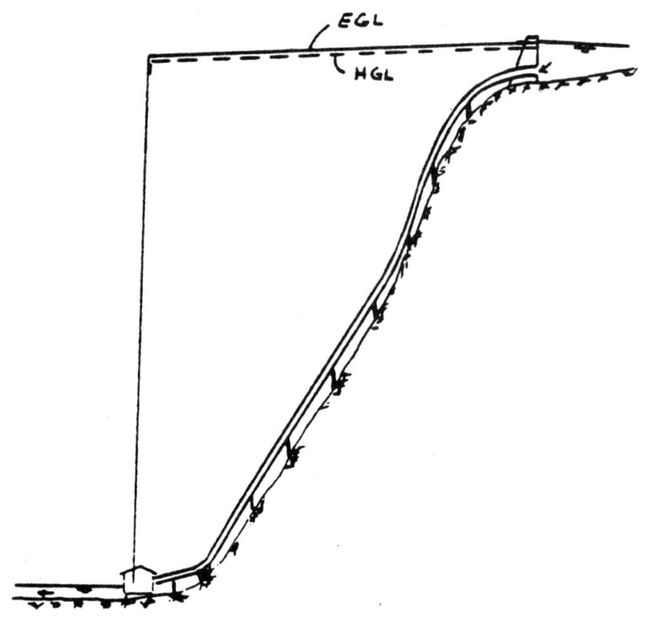

14.47

$P_{max} = (16/54)\rho U^3 A = (16/54) \times 1.2 \times (50{,}000/3{,}600)^3 \pi \times 2^2/4 = \underline{\underline{2.99 \text{ kW}}}$

14.48

$P_{max} = (16/54)\rho U^3 A$

so $A_{min.} = P \times (54/16)/\rho U^3 = 500 \times (54/16)/(1.2 \times (20{,}000/3{,}600)^3) = \underline{\underline{8.20 \text{ m}^2}}$

14.49 **Power** = $(16/27)\ (\rho AV^3/2)$

$= (16/27)\ (0.07/32.2)\ (\pi/4)\ (10)^2\ (44)^3/2 = 4{,}307$ ft-lbf/s

0.80 x **Power** = $\gamma Q h_p = (0.80)\ (4{,}307) = 3{,}446$ ft-lbf/s

$Q = (3{,}446)/((62.4)\ (10)) = 5.52$ cfs = 331 cfm

$= \underline{\underline{2{,}476 \text{ gpm}}}$

14.50 Assume that this system will be used on a daily basis; therefore, some safety should be included in the design. That is, include more than one pump so that if one malfunctions there will be at least another one or two to satisfy the demand. Also, periodic maintenance may be required; therefore, when one pump is down there should be another one or two to provide service. The degree of required safety would depend on the service. For this problem, assume that three pumps will be used to supply the maximum discharge of 1 m^3/s. Then each pump should be designed to supply a flow of water of 0.333 m^3/s (5278 gpm). Also assume, for the first cut at the design, that the head loss from reservoir to pump will be no greater than 1 meter and that each pump itself will be situated in a pump chamber at an elevation 1 m below the water surface of the reservoir. Thus, the NPSH will be approximately equal to the atmospheric pressure head, or 34 ft.

Assume that the suction specific speed will be limited to a value of 8500:

$$N_{ss} = 8500 = NQ^{1/2}/(NPSH)^{3/4}$$

$$\text{or } NQ^{1/2} = 8500 \times (34)^{3/4} \quad (1)$$

$$= 119,681$$

Assume that 60 cycle A.C. motors will be used to drive the pumps and that these will be synchronous speed motors. Common synchronous speeds in rpm are: 1200, 1800, 3600; however, the normal speed will be about 97% of synchronous speed*. Therefore, assume we have speed choices of 1160 rpm, 1750 rpm and 3500 rpm. Then from Eq.(1) we have the following maximum discharges for the different speeds of operation:

N(rpm)	Q(gpm)	Q(m /s)
1160	10,645	0.672
1750	1169	0.295
3500	1169	0.074

Based upon the value of discharge given above, it is seen that a speed of 1160 rpm is the choice to make if we use 3 pumps. The pumps should be completely free of cavitation.

Next, calculate the impeller diameter needed. From Fig. 14.10 for maximum efficiency $C_Q \approx 0.12$ and $C_H \approx 5.2$ or

$$0.12 = Q/nD^3 \qquad (2)$$

and $$5.2 = \Delta H/(D^2n^2/g) \qquad (3)$$

Then for N = 1160 rpm (n = 19.33 rps) and Q = 0.333 m^3/s we can solve for D from Eq. (2).

$$D^3 = Q/(0.12\ n)$$

$$= 0.333/(0.12 \times 19.33)$$

$$= 0.144$$

or $$D = 0.524\ m$$

Now with a D of 0.524 m the head produced will be

$$\Delta H = 5.2\ D^2n^2/g \quad \text{(from Eq. (3)}$$

$$= 5.2\ (0.524)^2\ (19.33)^2/(9.81)$$

$$= 54.4m$$

*Given in Conference Proceedings on"Pumping Station Design", Bozeman, MT,1981.

With a head of 54.4 m determine the diameter of pipe required to produce a discharge of 1 m^3/s.

From the solution to Prob. 10.100 (as an approximation to this problem), we have

$$h_p = 50m + (V^2/2g)(2.28 + fL/D)m$$

Assume $f = 0.012$

$L = 400$ m

so $h_p = 50\ m + (V^2/2g)(2.28 + 4.8/D)m$

$$54m = 50 + (V^2/2g)(2.28 + 4.8/D) \qquad (4)$$

Equation (4) may be solved for D by an iteration process: Assume D, then solve for V and then see if Eq. (4) is satisfied, etc. The iteration was done for D of 60 cm, 70 cm and 80 cm and it was found that the closest match came with D = 70 cm. Now compute the required power for an assumed efficiency of 92%.

$$P = Q\gamma h_p/\text{eff.}$$

$$= 0.333 \times 9{,}810 \times 54/0.92$$

$$P = 192\ kW = 257\ hp$$

In summary, D = 70 cm, N = 1160 rpm,

Q per pump = 0.333 m^3/s, P = 192 kW

The above calculations yield a solution to the problem. That is, a pump and piping system has been chosen that will produce the desired discharge. However, a truly valid design should include the economics of the problem. For example, the first cost of the pipe and equipment should be expressed in terms of cost per year based upon the expected life of the equipment. Then the annual cost of power should be included in the total cost. When this is done, the size of pipe becomes important (smaller size yields higher annual cost of power). Also, pump manufacturers have a multiple number of pump designs to choose from which is different than for this problem. We had only one basic design although considerable variation was available with different diameters and speed.

The design could also include details about how the piping for the pumps would be configured. Normally this would include 3 separate pipes coming from the reservoir, each going to a pump, and then the discharge pipes would all feed into the larger pipe that delivers water to the elevated tank. Also, there should be gate valves on each side of a pump so it could be isolated for maintenance purposes, etc. Check valves would also be included in the system to prevent back flow through the pumps in event of a power outage.

CHAPTER FIFTEEN

15.1 Check Froude number:

$Fr = V/\sqrt{gy} = 18/\sqrt{32.2 \times 0.333} = 5.5$

The Froude number is greater than 1 so the flow is supercritical.

$E = y + V^2/g$

$E = 0.333 + 18^2/(2 \times 32.2) = 5.36$ ft

Solving for the alternate depth for an E of 5.36 yields

$y_{alt} = 5.34$ ft

15.2 $V = Q/A = 900/(16 \times 3) = 18.75$ ft/s

$F = V/\sqrt{gy} = 18.75/\sqrt{32.2 \times 3} = 1.91$ (supercritical)

15.3 $V = Q/A$; $A = Q/V = 350/3 = 116.7$ ft^2; $d = 116.7/15 = 7.78$ ft

$F = V/\sqrt{gy} = 3/\sqrt{32.2 \times 7.78} = 0.19$ (subcritical)

15.4 $F = V/\sqrt{gy} = (Q/A)/\sqrt{gy} = (Q/(By))/\sqrt{gy}$

$F = Q/(B\sqrt{g}\, y^{3/2})$

$F_{0.3} = 10/(3\sqrt{9.81} \times 0.3^{3/2}) = 6.49$ (supercritical)

$F_{1.0} = 10/(3\sqrt{9.81} \times 1.0^{3/2}) = 1.06$ (supercritical)

$F_{2.0} = 10/(3\sqrt{9.81} \times 2.0^{3/2}) = 0.38$ (subcritical)

$y_c = (q^2/g)^{1/3} = ((Q/B)^2/9.81)^{1/3} = 1.04$ m

15.5 $E_{30} = y_1 + q^2/(2gy_1^2) = 0.30 + (12/3)^2/(2 \times 9.81 \times 0.3^2) = 9.36$ m

Then $y_2 + q^2/(2gy_2^2) = 9.36$ m where $q = (12/3)$ m^2/s

Solving: $y_{alt} = 9.35$ m

15.6 $V_c^2/g = y_c$; $y_c = 1.12$ ft

15.7 $Q = (1/n)AR^{2/3}S^{1/2}$

$9 = (1/0.014) \times 4\bar{y}(4\bar{y}/(B+2y))^{2/3} \times (0.005)^{1/2}$

Solving for y gives: $y = 0.693$ m and $V = Q/(By) = 3.25$ m/s

Then $F = V/\sqrt{gy} =$ 1.24 (supercritical)

15.8 First determine V:

$V = Q/A = 12m^3/s/((3\times 1m^2) + 1^2m^2) = 3.00$ m/s

$D = A/T = 4m^2/5m = 0.80$ m

Then $Fr = V/(gD)^{0.5} = 3.00/(9.81 \times 0.80)^{0.50} = 1.07$

$Fr > 1$ ∴ The flow is supercritical.

15.9 $V/\sqrt{gD} = 1$ for critical flow condition

or $(V/\sqrt{D}) = \sqrt{g}$ for critical flow condition

$V = Q/A = 20/(3y + y^2)$

$D = A/T = (3y + y^2)/(3 + 2y)$

$(20/(3y+y^2))/((3y+y^2)/(3+2y))^{0.5} = \sqrt{9.81} = 3.132$

Solving for y yields: $y_{cr} = 1.4$m

15.10

y (m)	E (m)
0.25	7.59
0.30	5.40
0.40	3.27
0.50	2.33
0.60	1.87
0.70	1.64
0.80	1.52
0.90	1.47
1.00	1.46
1.10	1.48
1.40	1.63
2.00	2.11
4.00	4.03
7.00	7.01

$E = y + q^2/(2gy^2)$ (for a rectangular channel)

For this problem $q = Q/B = 18/6 = 3m^2/s$

So $E = y + 3^2/(2gy^2)$

or $E = y + 0.4587/y^2$

E vs. y is shown in table at left.

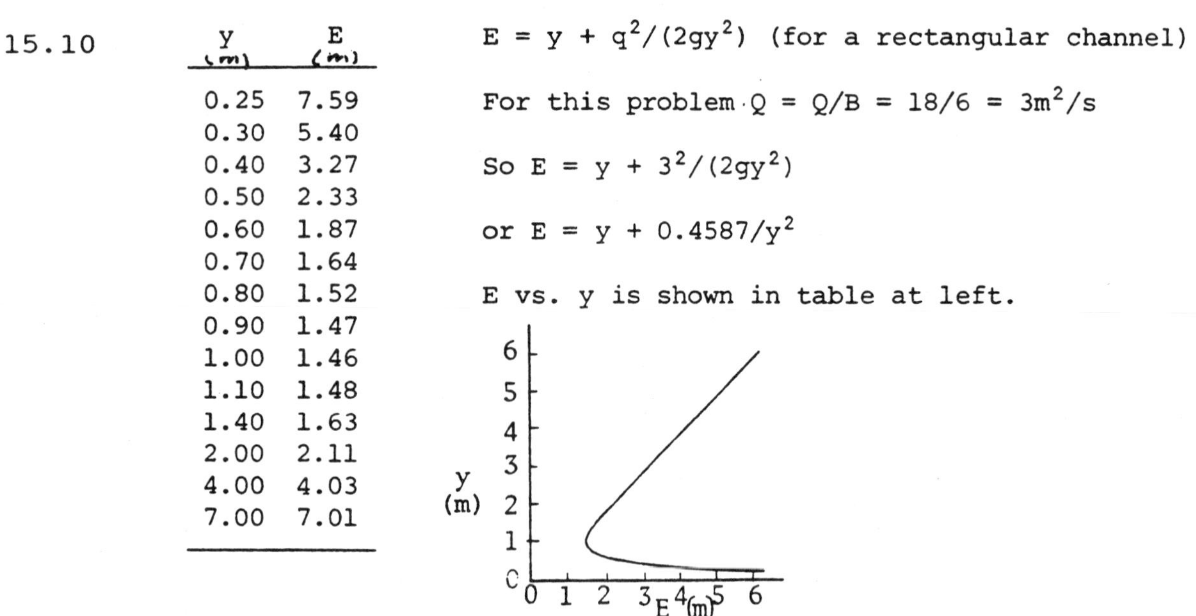

The alternate depth to $y = 0.30$ is $y = 5.38$ m

15.10 (continued)

Sequent depth: $y_2 = (\bar{y}_1/2)(\sqrt{1 + 8F_1^2} - 1)$

$F_1 = V/\sqrt{gy_1} = (3/0.3)/\sqrt{9.81 \times 0.30} = 5.83$

Then $y_2 = (0.3/2)(\sqrt{1 + 8 \times 5.83^2} -1) = \underline{\underline{2.33 \text{ m}}}$

15.11 $d_{brink} \approx 0.71\ y_c = 0.71(q^2/g)^{1/3}$

Then for $d_{brink} = 0.40$ m $\quad q = (0.40 \times g^{1/3}/0.71)^{3/2}$

$q = 1.324\ m^2/s$

Then $\underline{Q = 3q = \underline{3.97\ m^3/s}}$

15.12 Solution like that for Prob. 15.11:

$q = (1.20 \times (32.2)^{1/3}/0.71)^{3/2}$

$q = 12.47\ m^2/s$

Then $Q = 15 \times 12.47 = \underline{\underline{187 \text{ cfs}}}$

15.13 $y_{brink} = 0.71\ y_c$ where $y_c = \sqrt[3]{q^2/g}$; $q = 400/15 = 26.67\ ft^2/s$

$y_{brink} = 0.71\ \sqrt[3]{26.67^2/32.2} = \underline{\underline{1.99 \text{ ft}}}$

15.14 H = 1.5 ft , P = 2 ft , L = 10 ft

$H/(H+P) = (1.5/3.5) = 0.43 \Rightarrow C = 0.89$ (Fig. 15.7)

$Q = 0.385\ CL\sqrt{2g}\ H^{1.5}$

$Q = 0.385\ (0.89)\ (10)\ \sqrt{64.4}\ (1.5)^{1.5}$

$\underline{\underline{Q = 50.5 \text{ cfs}}}$

15.15 H = 0.50 m , L = 6 m , P = 2 m

$H/(H+P) = (0.5/(2.5)) = 0.20 \Rightarrow C = 0.860$ (Fig. 15.7)

$Q = 0.385\ (0.860)\ (6)\ \sqrt{2 \times 9.81}\ (0.50)^{1.5}$

$\underline{\underline{Q = 3.11\ m^3/s}}$

15.16 $L = 10$ m $Q = 25\ m^3/s$ $C \approx 0.85$ (Fig. 15.7)

$Q = 0.385\ C\ L\ \sqrt{2g}\ H^{3/2}$

$25 = 0.385\ (0.85)\ (10)\ \sqrt{2 \times 9.81}\ H^{3/2}$

$(H)^{3/2} = 1.725$; $H = 1.438$,

Water surface elevation = 101.44 m

15.17 $L = 50$ ft $Q = 1600$ cfs $C \approx 0.85$ (Fig. 15.7)

$Q = 0.385\ C\ L\ \sqrt{2g}\ H^{3/2}$

$1600 = 0.385\ (0.85)\ (50)\ \sqrt{64.4}\ H^{3/2}$

$H = 5.30$ ft

Water surface elevation = 305.30 ft

15.18 $V_1 = 3$ m/s So $E_1 = y_1 + V_1^2/2g = 3 + 3^2/(2 \times 9.81) = 3.46$ m

$F_1 = V_1/\sqrt{gy_1} = 3/\sqrt{9.81 \times 3} = 0.55$ (subcritical)

Then $E_2 = E_1 - \Delta z_{step} = 3.46 - 0.30 = 3.16$ m

$y_2 + q^2/(2gy_2^2) = 3.16$ m

$y_2 + 9^2/(2gy_2^2) = 3.16$

$y_2 + 4.13/y_2^2 = 3.16$

Solving for y_2 yields $y_2 = 2.49$ m

Then $\Delta y = y_2 - y_1 = 2.49 - 3.00 = -0.51$ m

W.S. drops 0.21 m

For a downward step $E_2 = E_1 + \Delta z_{step} = 3.46 + 0.3 = 3.76$ m

$y_2 + 4.13/y_2^2 = 3.76$ Solving: $y_2 = 3.40$ m

Then $\Delta y = y_2 - y_1 = 3.40 - 3 = 0.40$ m

W.S. elevation change = +0.10 m

Max. upward step before altering upstream conditions:

$y_c = y_2 = \sqrt[3]{q^2/g} = \sqrt[3]{9^2/9.81} = 2.02$

$E_1 = \Delta z_{step} + E_2$ where $E_2 = 1.5\ y_c = 1.5 \times 2.02 = 3.03$ m

Max. $z_{step} = E_1 - E_2 = 3.46 - 3.03 = 0.43$ m

15.19

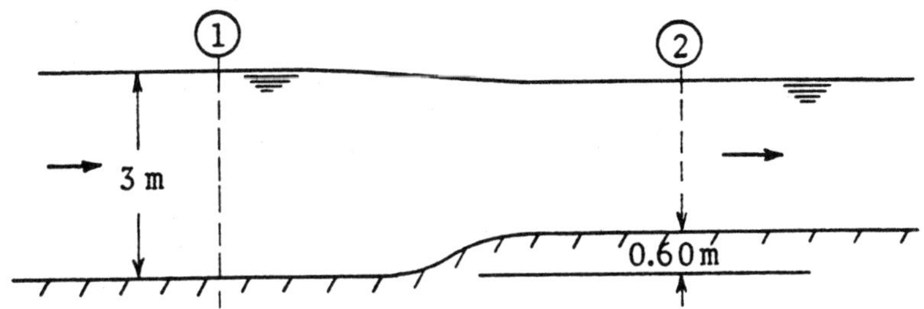

$E_2 = E_1 - 0.60$; $V_1 = 2$ m/s; $F_1 = V_1/\sqrt{gy_1} = 2/\sqrt{9.81 \times 3} = 0.369$

Then $E_2 = 3 + (2^2/(2 \times 9.81)) - 0.60 = 2.60$ m

Solve for y_2: $y_2 + q^2/(2gy_2^2) = 2.60$ where $q = 2 \times 3 = 6 m^3/s/m$

Then $y_2 + 6^2/(2 \times 9.81 \times y_2^2) = 2.60$

$y_2 + 1.83/y_2^2 = 2.60$

Solving, one gets $y_2 = 2.24$ m; $\Delta y = y_2 - y_1 = 2.24 - 3.00 = \underline{\underline{-0.76\text{m}}}$

Water surface drops $\underline{\underline{0.16 \text{ m}}}$

For downward step of 15 cm we have

$E_2 = (3 + (2^2/(2 \times 9.81)) + 0.15 = 3.35$ m

$y_2 + 6^2/(2 \times 9.81 \times y_2^2) = 3.35$

$y_2 + 1.83/y_2^2 = 3.35$

Solving: $y_2 = 3.17$ m or $y_2 - y_1 = 3.17 - 3.00 = \underline{\underline{+0.17 \text{ m}}}$

Water surface rises $\underline{\underline{0.02 \text{ m}}}$

The maximum upstep possible before affecting upstream water surface levels is for $y_2 = y_c$

$y_c = \sqrt[3]{q^2/g} = 1.54$ m

Then $E_1 = \Delta z_{step} + E_{2,crit}$

$\Delta z_{step} = E_1 - E_{2,crit} = 3.20 - (y_c + V_c^2/2g) = 3.20 - 1.5 \times 1.54$

$z_{step} = \underline{\underline{+0.89 \text{ m}}}$

15.20 $y_c = (q^2/g)^{1/3} = (6^2/9.81)^{0.333} = 1.542$ m

y_c is depth allowed over the hump for the given conditions.

$E_1 = E_2$; $V_1 = q/y_1 = 6/3 = 2$ m/s ; $V_2 = 6/1.542 = 3.891$ m/s

$V_1^2/2g + y_1 = V_2^2/2g + y_2 + \Delta z$

$2^2/2g + 3 = (3.891^2/(2 \times 9.81)) + 1.542 + \Delta z$

$\Delta z = 3.204 - 0.772 - 1.542 = \underline{\underline{0.89 \text{ m}}}$

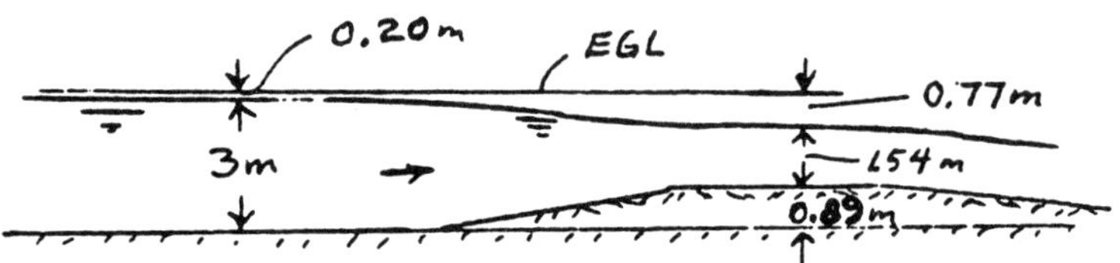

15.21 $F_1 = V_1/\sqrt{gy_1} = 3/\sqrt{9.81 \times 3} = 0.55$ (subcrit)

$E_1 = E_2 = y_1 + V_1^2/2g = 3 + 3^2/\ 2 \times 9.81 = 3.46$ m

$q_2 = Q/B_2 = 27/2.6 = 10.4\ m^3/s/m$

Then $y_2 + q^2/(2gy_2^2) = y_2 + (10.4)^2/(2 \times 9.81 \times y_2^2) = 3.46$

$y_2 + 5.50/y_2^2 = 3.46$

Solving: $y_2 = 2.71$ m

$\Delta z_{\text{water surface}} = \Delta y = y_2 - y_1 = 2.71 - 3.00 = \underline{\underline{0.29 \text{ m}}}$

Max. contraction without altering the upstream depth will occur with $y_2 = y_c$

$E_2 = 1.5\ y_c = 3.46;\quad y_c = 2.31$ m

Then $V_c^2/2g = y_c/2 = 2.31/2$ or $V_c = 4.76$ m/s

$Q_1 = Q_2 = 27 = B_2 y_c V_c;\quad B_2 = 27/(2.31 \times 4.76) = 2.46$ m

The width for max. contraction = $\underline{\underline{2.46 \text{ m}}}$

15.22 Write the energy equation from a section in the channel upstream of the ship to a section where the ship is located.

$E_1 = E_2$

$V_1^2/2g + y_1 = V_2^2/2g + y_2$

$A_1 = 35 \times 200 = 7{,}000\ m^2$; $V_1 = 5 \times 0.515 = 2.575\ m/s$

$2.575^2/(2 \times 9.81) + 35 = (Q/A_2)^2/(2 \times 9.81) + y_2$ (1)

where $Q = V_1 A_1 = 2.575 \times 7{,}000\ m^3/s$ (2)

$A_2 = 200\ m \times y_2 - 29 \times 63$ (3)

Substituting Eq's (2) and (3) into Eq. (1) and solving for y_2

yields $y_2 = 34.70$ m

Therefore, the ship squat is $y_1 - y_2 = 35.0 - 34.7 = \underline{\underline{0.30\ m}}$

15.23 Apply the momentum equation for a unit width

$\Sigma F_x = \Sigma V_x \rho \underline{V} \cdot \underline{A}$

$\gamma y_1^2/2 - \gamma y_2^2/2 - 2{,}000 = -\rho V_1^2 y_1 + \rho V_2^2 y_2$

let $V_1 = q/y_1$ and $V_2 = q/y_2$ and divide by γ

$y_1^2/2 - y_2^2/2 - 200/\gamma = (-q_1^2 y_1/(g y_1^2) + q_2^2 y_2/(g y_2^2)$

$1/2 - y_2^2/2 - 3.205 = (+(20)^2/32.2)(-1 + 1/y_2)$

Solving for y_2 yields: $y_2 = \underline{\underline{1.43\ ft}}$

15.24

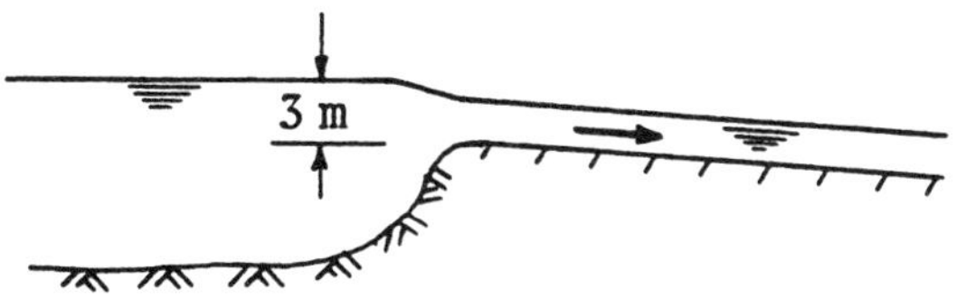

Assume negligible velocity in the reservoir and negligible energy loss. Then the channel entrance will act like a broad crested weir.

Thus $Q = 0.545\sqrt{g}\, LH^{3/2}$ where L = 4 m and H = 3 m

Then $Q = 0.545\sqrt{9.81} \times 4 \times 3^{3/2} = \underline{\underline{35.5\ m^3/s}}$

15.25 $V = \sqrt{gy} = \sqrt{32.2 \times 1.00} = \underline{5.67 \text{ ft/s}}$

15.26 $V = \sqrt{gy}$; $2.0 = \sqrt{9.81}\sqrt{y}$

$y = 2^2/9.81 = \underline{0.408 \text{ m}}$

15.27 As the waves travel into shallower water their speed is decreased ($V = \sqrt{gy}$); therefore, the wave lags that in deeper water. Thus, the wave crests tend to become parallel to the shoreline.

15.28 Let the upstream section (where y = 3 ft) be section 1 and the downstream section (y = 2 ft) be section 2.

Then $V_1 = 18/3 = 6$ ft/s and $V_2 = 18/2 = 9$ ft/s

$$y_1 + V_1^2/2g + z_1 = y_2 + V_2^2/2g + z_2 + h_L$$

$$3 + 6^2/(2 \times 32.2) + 2 = 2 + 9^2/(2 \times 32.2) + h_L$$

$$h_L = \underline{2.30 \text{ ft}}$$

$P = Q\gamma h_L/550$

$= 18 \times 62.4 \times 2.3/550 = \underline{4.70 \text{ horsepower}}$

Determine the force of ramp by writing the momentum equation between section 1 and 2. Let F_x be the force of the ramp on the water and assume x positive in the direction of flow. Then

$$\Sigma F_x = \rho q(V_{2x} - V_{1x})$$

$$\gamma y_1^2/2 - \gamma y_2^2/2 + F_x = 1.94 \times 18(9-6)$$

$$(62.4/2)(3^2-2^2) + F_x = 104.8$$

$$F_x = -51.2 \text{ lbf}$$

The ramp exerts a force of 51.2 lbs opposite to the direction of flow.

15.29 $y_0 + q^2/(2gy_0^2) = y_1 + q^2/(2gy_1^2)$; $q = 2 \text{ m}^3/\text{s/m}$; $y_0 = 5$ m

$5 + 2^2/(2(9.81)5^2) = y_1 + 2^2/(2(9.81)y_1^2) \rightarrow y_1 = 0.206$ m

$F_1 = q/\sqrt{gy_1^3} = 2/\sqrt{9.81(0.206)^3} = 6.829$

$y_2 = (y_1/2)(\sqrt{1 + 8F_1^2} - 1) = (0.206/2)(\sqrt{1 + 8(6.829^2)} - 1) = \underline{1.89\text{m}}$

15.30 $F_1 = q/\sqrt{gy^3} = 2.0/\sqrt{32.2(10/12)^3} = 4.63 > 1$

$\therefore$ Jump can form.

$y_2 = (y_1/2)(\sqrt{1+8F_1^2} - 1) = (10/12)(\sqrt{1+8(4.63)^2} - 1 = \underline{5.06 \text{ ft}}$

15.31 First develop the expression for y_1 and $V_{theor.}$. Write the energy equation from the upstream pool level to y_1.

$V_0^2/2g + z_0 = V_1^2/2g + z_1$; assume V_0 is negligible

$0 + 100 = V^2_{theor.}/2g + y_1$ (1)

But $V_{theor.} = V_{act.}/0.95$ (2)

and $V_{act.} = q/y_1$ (3)

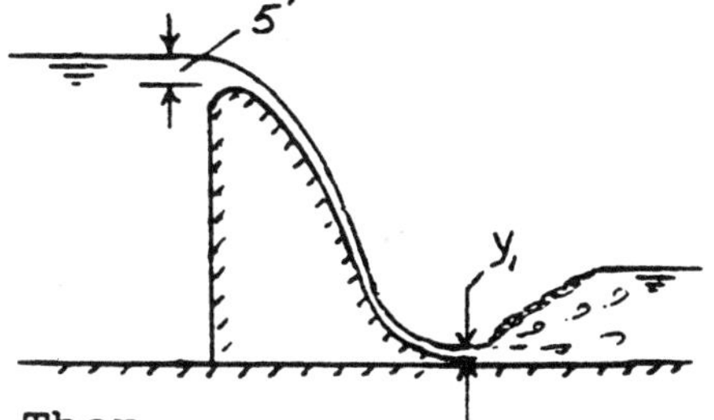

Consider a unit width of spillway. Then

$q = Q/L = K\sqrt{2g}\ H^{1.5}$

$= 0.5\sqrt{2g}\ (5^{1.5})$

$q = 44.86$ cfs/ft (4)

Solving Eqs. (1), (2), (3), and (4) yields

$y_1 = 0.59$ ft and $V_{act.} = 76.03$ ft/sec

$Fr_1 = V/\sqrt{gy_1} = 76.03/\sqrt{(32.2)(0.59)} = 17.44$

Now solve for the depth of flow on the apron:

$y_2 = (y_1/2)((1 + 8\ Fr_1^2)^{0.5} - 1)$

$= (0.59/2)((1 + 8\ (17.44^2))^{0.5} - 1) = \underline{14.3 \text{ ft}}$

15.32 $y_2 = (y_1/2)\ ((1 + 8Fr_1^2)^{0.5} - 1)$

where $Fr_1^2 = V_1^2/(gy_1) = q^2/(gy_1^3)$

Then $y_2 = (y_1/2)\ ((1 + 8q^2/(gy_1^3))^{0.5} - 1)$

$y_2 - y_1 = (y_1/2)\ [((1 + 8q^2/(gy_1^3))^{0.5} - 1) - 2]$

However $y_2 - y_1 = 14.0$ ft (given) ; $q = 65$ ft^2/s

Therefore 14.0 ft $= (y_1/2)\ [((1 + 8 \times 65^2/(gy_1^3))^{0.5} - 1) - 2]$

$\underline{y_1 = 1.08 \text{ ft}}$

15.33 $F_1 = V/\sqrt{gy} = 5/\sqrt{9.81(0.4)} = 2.52$

$y_2 = (0.4/2)(\sqrt{1+8(2.52)^2}-1) = \underline{\underline{1.24\text{ m}}}$

15.34 Check Fr upstream to see if it is really supercritical flow

$Fr = V/(gD)^{0.5}$

$D = A/T = (By + y^2)/(B + 2y)$

$D_{y=0.4} = (5 \times 0.4 + 0.4^2)/(5 + 2 \times 0.4) = 0.372\text{ m}$

Then $Fr_1 = 10\text{ m/s}/((9.81\text{ m/s}^2)(0.372))^{0.5}$

$Fr_1 = 5.23$ (Flow is supercritical; jump will form)

Momentum Equation (Eq. 15.23):

$\bar{p}_1 A_1 + \rho Q V_1 = \bar{p}_2 A_2 + \rho Q V_2$ (1)

Evaluate $\bar{p}_1$ by considering the hydrostatic forces on the trapezoidal section divided into rectangular plus triangular areas as shown below:

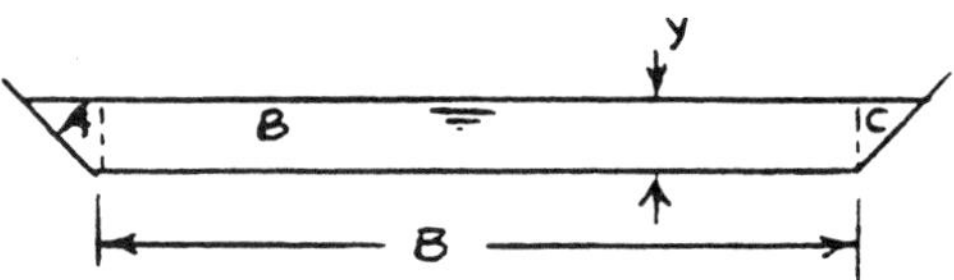

Then $\bar{p}_1 A_1 = \bar{p}_A A_A + \bar{p}_B A_B + \bar{p}_C A_C$

$= (\gamma y_1/3)(y_1^2/2) + (\gamma y_1/2)By_1 + (\gamma y_1/3)(y_1^2/2)$

$= \gamma(y_1^3/6) + \gamma B(y_1^2/2) + \gamma(y_1^3/6)$

$= \gamma(y_1^3/3) + \gamma B(y_1^2/2)$

$\bar{p}_1 A_1 = \gamma((y_1^3/3) + B(y_1^2/2))$

Also $\rho Q V_1 = \rho Q Q/A_1 = \rho Q^2/A_1$

Equation (1) is then written as

$\gamma((y_1^3/3)+(B(y_1^2/2))+\rho Q^2/A_1 = \gamma((y_2^3/3)+B(y_2^2/2))+\rho Q^2/(By_2+y_2^2)$

$\gamma = 9{,}810\text{ N/m}^2$, $B = 5$ m, $y_1 = 40$ cm $= 0.40$ m,
$Q = V_1 A_1 = 21.6\text{ m}^3/\text{s}$

$A_1 = 5 \times 0.4 + 0.4^2 = 2.16\text{ m}^2$

solving for y_2 yields: $\underline{\underline{y_2 = 2.45\text{ m}}}$

15.35 $y_2/y_1 = 13/0.5 = 0.5((1 + 8Fr_1^2)^{0.5} - 1)$ (15.26)

Solving for Fr_1 yields $Fr_1 = 19.089$

Then $Fr_1 = V/(gy_1)^{0.5}$

$V = Fr_1(gy_1)^{0.5} = 19.089\ (32.2 \times 0.5)^{0.5}$

$= 76.59$ ft/s

Finally $q = V_1y_1 = (76.59 \text{ ft/s})(0.50 \text{ ft}) = \underline{38.3 \text{ ft}^2/\text{s}}$

15.36 $y_c = (q^2/g)^{1/3}$ $q = 500/20 = 25$ cfs/ft

$y_c = (25^2/32.2)^{1/3} = 2.687$ ft

Solve for $y_{n,1}$:

$Q = (1.49/n)AR^{2/3}S_0^{1/2}$; $A = 20y$; $R = 20y/(20 + 2y)$

$500 = (1.49/0.015)\ (0.01)^{1/2}\ (20y)^{5/3}/(20 + 2y)^{2/3}$

$(20y)^{5/3}/(20 + 2y)^{2/3} = 50.34$

Solving yields $y_{n,1} = 1.86$ ft

Thus one concludes that normal depth in reach 1 would be supercritical, in reach 2 subcritical and in reach 3 critical.

If reach 2 is long then the flow would be near normal depth in reach 2. Thus, the flow would probably go from supercritical flow in reach 1 to subcritical in reach 2. In going from sub to supercritical a hydraulic jump would form.

Determine jump height and location:

$y_2 = (y_1/2)\ ((1 + 8\ Fr_1^2)^{0.5} - 1)$

$Fr_1 = V_1/(gy_1)^{0.5} = (25/1.86)/(32.2 \times 1.86)^{0.5} = 1.737$

$y_2 = (1.86/2)\ ((1 + 8 \times 1.737^2)^{0.5} - 1) = \underline{3.73 \text{ ft}}$

Because y_2 is less than the normal depth in reach 2 the jump will probably occur in reach 1. The water surface profile could occur as shown below.

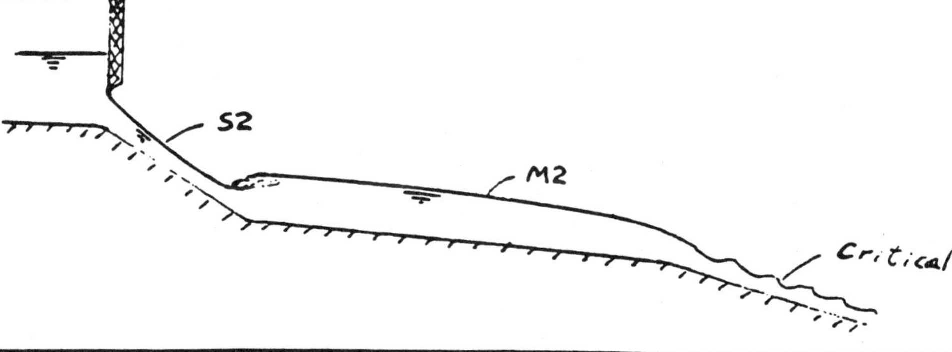

15.37 $Fr_1 = V_1/\sqrt{gy_1} = 10/\sqrt{9.81 \times 0.10} = 10.1$ (supercritical)

$$V_2 = q/y_2 = (0.10 \text{ m})(10 \text{ m/s})/(1.1 \text{ m}) = 0.91 \text{ m/s}$$

$$Fr_2 = V_2/(gy_2)^{0.5} = 0.91/(9.81 \times 1.1)^{0.5} = 0.277$$

A hydraulic jump will form because flow goes from supercritical to subcritical. Determine y_1 for a y_2 of 1.1 m:

$$y_1 = (y_2/2)((1 + 8\,Fr_2^2)^{0.5} - 1)$$

$$= (1.1/2)((1 + 8 \times .277^2)^{0.5} - 1)$$

$$= 0.14 \text{ m}$$

Therefore the jump will start at about the 29 m distance downstream of the sluice gate. Profile and energy grade line:

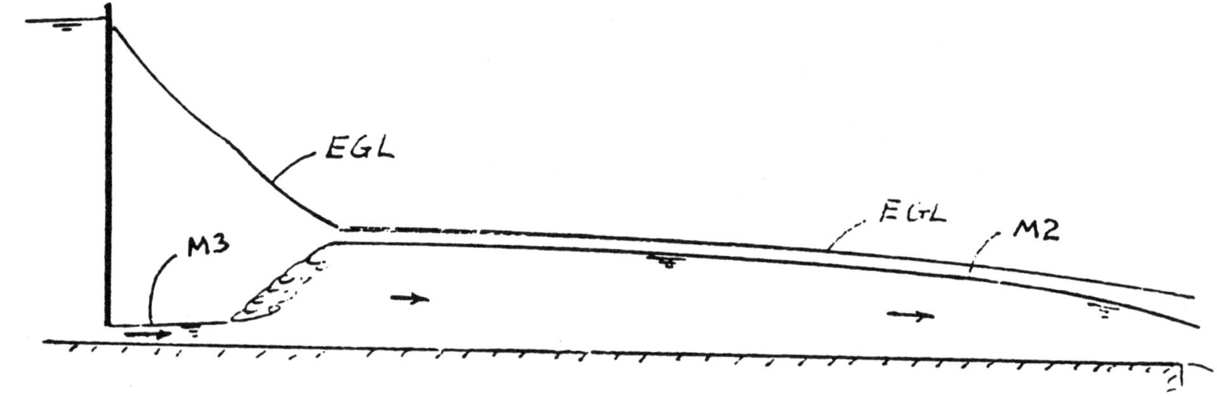

15.38 $Re_x = (10 \times 0.5)/(10^{-6}) = 5 \times 10^6$

$$c_f = (0.058)/(Re_x^{0.2}) \quad \text{(from Chapt. 9 for } Re_x < 10^7\text{)}$$

$$= (0.058)/(21.9) = 0.0026$$

$$\tau_0 = c_f \rho V^2/2 = (0.0026)(998)(10^2/2)$$

$$= 130 \text{ N/m}^2$$

Therefore; the correct choice is d) $\tau > 40 \text{ N/m}^2$

15.39 Assume negligible energy loss for flow under the sluice gate. Write the Bernoulli equation from a section upstream of the sluice gate to a section immediately downstream of the sluice gate.

$$y_0 + V_0^2/2g = y_1 + V_1^2/2g$$

$$65 + \text{neglig.} = 1 + V_1^2/2g$$

$$V_1 = \sqrt{64 \times 64.4} = 64.2 \text{ ft/s}$$

$$F_1 = V_1/\sqrt{gy_1} = 64.2/\sqrt{32.2 \times 1} = 11.3$$

15.39 (continued)

Now solve for the depth after the jump:

$$y_2 = (y_1/2)(\sqrt{1 + 8F_1^2} - 1)$$

$$= (1/2)(\sqrt{1 + 8 \times 11.3^2} - 1) = 15.5 \text{ ft}$$

$$h_L = (y_2 - y_1)^3/(4y_1y_2)$$

$$= (14.51)^3/(4 \times 1 \times 15.51) = \underline{\underline{49.2 \text{ ft}}}$$

$$P = Q\gamma h_L/550$$

$$= (64.2 \times 1 \times 5) \times 62.4 \times 49.2/550 = \underline{\underline{1{,}793 \text{ horsepower}}}$$

15.40 For this experiment it is necessary to first produce supercritical flow in the flume and then force this flow to become subcritical. The supercritical flow could be produced by means of a sluice gate as shown in Prob. 15.39 and the jump could be forced by means of another sluice gate farther down the flume. Therefore, one needs to include in the design an upstream chamber that will include a sluice gate from which the high velocity flow will be discharged.

The relevant equation for the hydraulic jump is Eq. (15.28). Therefore, to verify this equation y_1, y_2 and V_1 will have to be measured or deduced by some other means. A fairly accurate measurement of y_2 can be made by means of a point gage or piezometer. The depth y_1 could also be measured in the same way; however, the degree of accuracy of this measurement will be less than for y_2 because y_1 is much smaller than y_2. Perhaps a more accurate measure of y_1 would be to get an accurate reading of the gate opening of the sluice gate and apply a coefficient of contraction to that

15.40 (Cont'd)

reading to get y_1. The C_C for a sluice gate could be obtained from the literature.

The velocity, V_1, which will be needed to compute F_{r1} can probably be best calculated by Bernoulli's equation knowing the depth of flow in the chamber upstream of the sluice gate. Therefore, a measurement of that depth must be made.

Note that for use of V_1 and y_1 just downstream of the sluice gate, the hydraulic jump will have to start very close to the sluice gate because the depth will increase downstream due to the channel resistance. The jump location may be changed by operation of the downstream sluice gate.

Other things that could or should be considered in the design:

A. Choose maximum design discharge. This will be no more than 5 cfs (see Prob. 13.77).

B. Choose reasonable size of chamber upstream of sluice gate. A 10 ft depth would be ample for a good experiment.

C. Choose width, height and length of flume.

D. work out details of sluice gates and their controls.

15.41

$V = (1/n)R^{2/3}S_0^{1/2}$ where $n = 0.015$ (assume)

$R = A/P = (0.4 \times 10)/(2 \times 0.4 + 10) = 0.370$ m

Then $V = (1/0.015)(0.370)^{2/3} \times (0.04)^{1/2} = 6.87$ m/s

Then $F_1 = V/\sqrt{gy_1} = 6.87/\sqrt{9.81 \times 0.40} = 3.47$ (supercritical)

Then $y_2 = (y_1/2)(\sqrt{1+8 \times F_1^2} - 1) = (0.40/2)(\sqrt{1+8 \times (3.47)^2} - 1) = \underline{\underline{1.77 \text{ m}}}$

15.42

Assume the shear stress will be the average of τ_{0_1}, uniform approaching the jump, and τ_{0_2}, uniform flow leaving the jump.

$\tau_0 = f\rho V^2/8$ (10-21)

where $f = f(Re, k_s/4R)$

$R_{e_1} = V_1(4R_1)/\nu$

From sol. to P15.41:

$R_{e_1} = 6.87 \times (4 \times 0.37)/10^{-6}$

$R_{e_1} = 10^7$

assume $k_s = 3 \times 10^{-3}$m

$k_s/4R_1 = 3 \times 10^{-3}/(4 \times 0.37)$

$k_s/4R_1 = 2 \times 10^{-3}$

From Fig. 10-8, $f_1 = 0.024$

Then

$\tau_{0_1} = 0.024 \times 1{,}000 \times (6.87)^2/8$

$\tau_{0_1} = 142$ N/m^2

$R_{e_2} = V_2 \times (4R_2)/\nu$

$V_2 = V_1 \times 0.4/1.77 = 1.55$ m/s

$R_2 = A/P = (1.77 \times 10)/(2 \times 1.77 + 10) = 1.31$ m

$R_{e_2} = 1.55 \times (4 \times 1.31)/10^{-6}$

$R_{e_2} = 8 \times 10^6$

$k_s/4R_2 = 3 \times 10^{-3}/(4 \times 1.31)$

$k_s/4R_2 = 6 \times 10^{-4}$

$f_2 = 0.018$

$\tau_{0_2} = 0.018 \times 1{,}000 \times (1.55)^2/8$

$\tau_{0_2} = 5.4$ N/m^2

$$\tau_{avg} = (142 + 5.4)/2 = 74 \text{ N/m}^2$$

Then $F_s = \tau_{avg}A_s = \tau_{avg}PL$

where $L \approx y_2$, $P \approx B + (y_1+y_2)$

Then $F_s \approx 74(10+(0.40+1.77))(6 \times 1.77) = 9{,}560$ N

$F_H = (\gamma/2)(y_2^2-y_1^2)B = (9{,}810/2)((1.77)^2-(0.40)^2) \times 10 = 145{,}820$ N

Thus, $F_s/F_H = 9{,}560/145{,}820 = \underline{0.066}$

Note: The above estimate probably gives an excessive amount of wgt. to τ_{0_1} because τ_0 will not be linearly distributed. A better estimate might be to assume a linear distribution of velocity with an average f and integrate $\tau_0 dA$ from one end to the other.

15.43

$q = 0.40 \times 10 = 4.0\ m^3/s/m$

Then $y_c = \sqrt[3]{q^2/g} = \sqrt[3]{(4.0)^2/9.81} = 1.18$ m

Then we have $y < y_n < y_c$; therefore, the water surface profile will be an S3.

Shear stress: Assume a boundary layer develops similar to a flat plate downstream of the plane of the sluice gate.

Then $Re_x \approx V \times 0.5/\nu$

$Re_x = 10 \times 0.5/10^{-6} = 5 \times 10^6$

$c_f = 0.058/Re_x^{1/5} = 0.00265$ (from Ch. 9)

Then $\tau_0 = c_f \rho V_0^2/2 = 0.00265 \times 998 \times 10^2/2 = \underline{132\ N/m^2}$

15.44 $y_n = 2$ ft $y_c = (q^2/g)^{1/3} = (10^2/32.2)^{1/3} = 1.46$ ft.

$y > y_n > y_c$ $\therefore$ profile is M1 (see Fig. 15-16)

Choice c) is the correct choice.

15.45 The correct choice is d) A2.

15.46

$q = 5/3$ $F_1 = q/\sqrt{gy^3} = (5/3)/\sqrt{9.81(0.3)^3} = 3.24 > 1$ (supercritical)

Flow over weir, $Q = (0.40+0.05\ H/P)L\sqrt{2g}\ H^{3/2}$

$5 = (0.40+0.05\ H/1.6)\times 3\sqrt{2(9.81)}H^{3/2}$

Solving by iteration gives H = 0.917 m

Depth upstream of weir = 0.917 + 1.6 = 2.52 m

$F_2 = (5/3)/\sqrt{9.81(2.52)^3} = 0.133 < 1$ (subcritical)

$\therefore$ A hydraulic jump forms. $y_2 = (0.3/2)(\sqrt{1+8(3.24)^2} - 1) = 1.23$ m

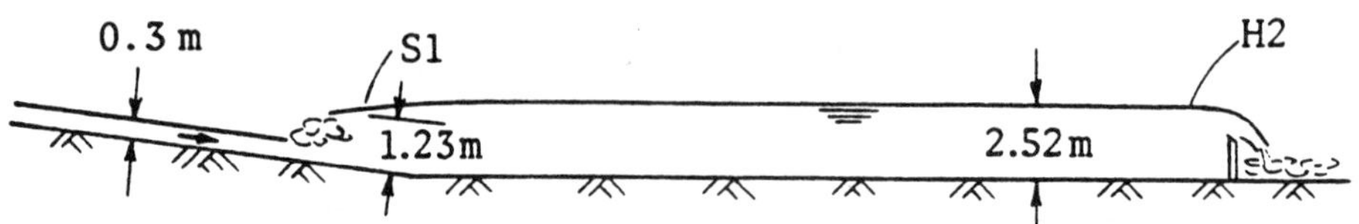

15.47 The profile might be an M profile or an S profile depending upon whether the slope is mild or steep. However, if it is a steep slope the flow would be uniform right to the brink. Check to see if M or S slope:

$$Q = (1.49/n)\ A\ R^{0.667}\ S^{0.5}$$

$$AR^{2/3} = Q/((1.49/n)(S^{0.5}));\ \text{assume } n = 0.012$$

$$= 120/((1.49/0.012)(0.0001)^{0.5})$$

$$(by)(by/(10+2y))^{.667} = 96.6$$

With b=10 ft we can solve for y: y = 5.2 ft

Then V = Q/A = 120/52 = 2.31 ft/sec

$$F = V/\sqrt{gy} = 2.31/(\sqrt{32.2 \times 5.2}) = 0.18\ \text{(subcritical)}$$

Therefore, the water surface profile will be an <u>M2</u>.

15.48 First determine the depth upstream of the weir.

$$Q = K\sqrt{2g}\ L\ H^{3/2} \qquad (13.9)$$

where K = 0.40 + 0.05 H/P. By trial and error (first assume K then solve for H, etc.) solve for H yielding H = 2.06 ft.

Then the velocity upstream of the weir will be

$$V = Q/A = 108/(4.06 \times 10) = 2.66\ \text{ft/sec}$$

$$Fr = V/\sqrt{gy} = 2.66/(32.2 \times (4.06))^{0.5} = 0.23\ \text{(subcritical)}$$

The Froude number just downstream of the sluice gate will next be determined:

$$V = Q/A = 108/(10 \times 0.40) = 27\ \text{ft/sec}$$

$$Fr = V/\sqrt{gy} = 27/\sqrt{32.2 \times 0.40} = 7.52\ \text{(supercritical)}$$

Because the flow is supercritical just downstream of the sluice gate and subcritical upstream of the weir a jump will form someplace between these two sections.

15.48 (continued)

Now determine the approximate location of the jump. Let y_2 = depth downstream of the jump and assume it is approximately equal to the depth upstream of the weir ($y \approx$ 4.06 ft). By trial and error (utilizing Eq. (15.25)) it can be easily shown that a depth of 0.40 ft is required to produce the given y_2. Thus the jump will start immediately downstream of the sluice gate and it will be approximately 25 ft long. Actually, because of the channel resistance y_2 will be somewhat greater than y_2 = 4.06 ft; therefore, the jump may be submerged against the sluice gate and the water surface profile will probably appear as shown below.

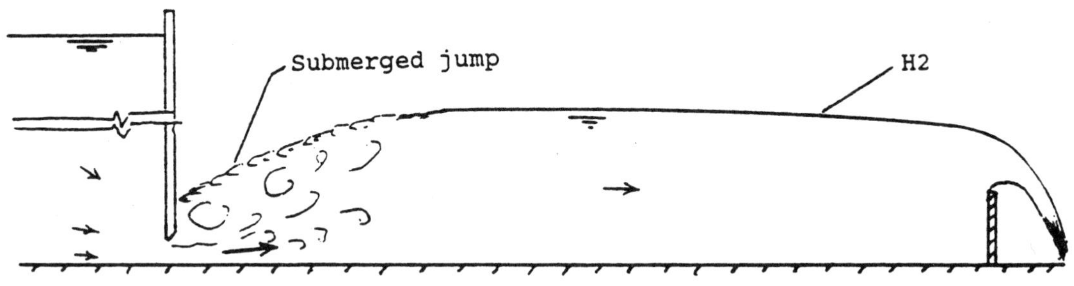

15.49 $y_c = \sqrt[3]{q^2/g} = \sqrt[3]{20^2/32.2}$ = 2.32 ft. Thus the slopes in parts 1 and 3 are steep.

If part 2 is very long, then a depth greater than critical will be forced in part 2 (the part with adverse slope). In that case a hydraulic jump will be formed and it may occur on part 2 or it may occur on part 1. These two possibilities are both shown below. The other possibility is for no jump to form on the adverse part. Also see this below:

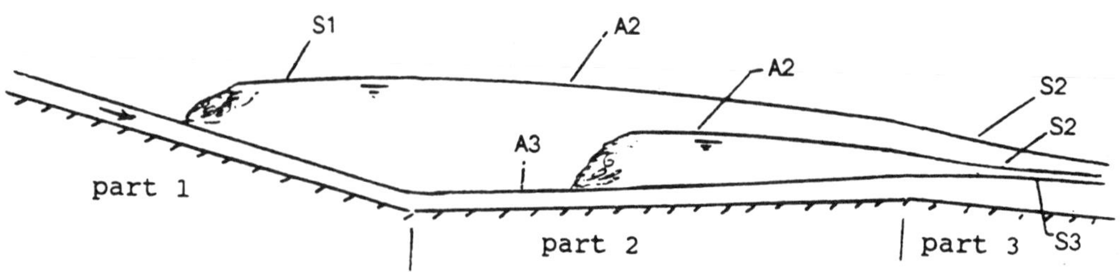

15.50 $F_1 = q/\sqrt{gy^3} = 3/\sqrt{9.81(0.2)^3} = 10.71$

$F_2 = 3/\sqrt{9.81(0.6)^3} = 2.06$

∴ Continuous H-3 profile

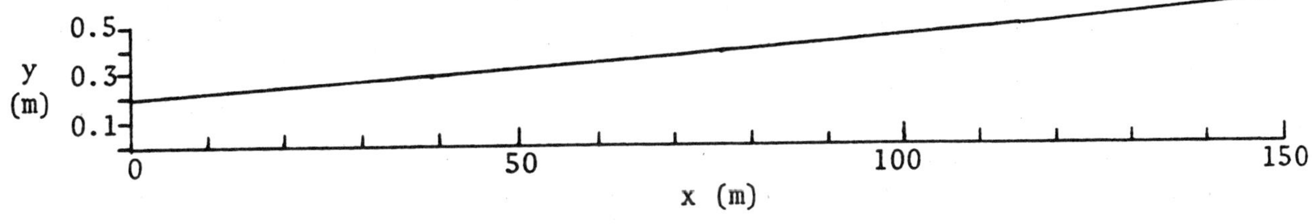

y	$\bar{y}$	V	$\bar{V}$	E	ΔE	S_f	Δx	x
0.2		15		11.6678				0
	0.25		12.5		6.2710	0.1593	39.4	
0.3		10		5.3968				39.4
	0.35		8.75		2.1298	0.0557	38.2	
0.4		7.5		3.2670				77.6
	0.45		6.75		0.9321	0.0258	36.1	
0.5		6.0		2.3349				113.7
	0.55		5.5		0.4607	0.0140	32.9	
0.6		5.0		1.8742				146.6

15.51

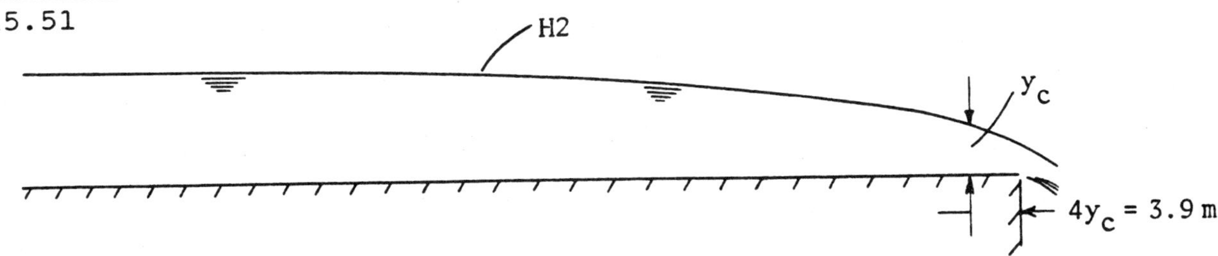

$q = Q/B = 12/4 = 3\ m^3/s/m$

$y_c = \sqrt[3]{q^2/g} = 0.972$ m (This depth occurs near brink.)

Carry out a step solution for the profile upstream from the brink.

$Re \approx V \times 4R/\nu \approx 3 \times 1/10^{-6} \approx 3 \times 10^6$; $k_s/4R \approx 0.3 \times 10^{-3}/4 \approx 0.000075$;

$f \approx 0.010$

See page 339 for soultion table.

Check Δy from 1.2 to 1.1

$V_{avg} = Q/((A_1+A_2)/2) = 12/(4.8+4.4)/2 = 2.61$ ft/s

$V_m^2 = 6.805$

15.51 (continued)

$R_{1.2} = 1.2 \times 4/(6.4) = 0.75$

$R_{1.1} = 1.1 \times 4/(6.2) = 0.7097$ $\quad R_m = 0.729$ m

$h_f/L = S_f = fV^2/(8gR_m) = (0.01 \times 6.805)/(8 \times 9.81 \times 0.729) = 1.189 \times 10^{-3}$

$\Delta h_f = S_f \times 33 = 0.0392$ m

$y_1 + V_1^2/2g = y_2 + V_2^2/2g + \Delta h_f$

$y_1 - y_2 = 0.379 - 0.319 + 0.0392 = 0.0996$ m OK

15.52 Upstream of jump the profile will be an H3.

Downstream of jump the profile will be an H2.

The baffle blocks will cause the depth upstream of A to increase; therefore, the jump will move towards the sluice gate.

15.53

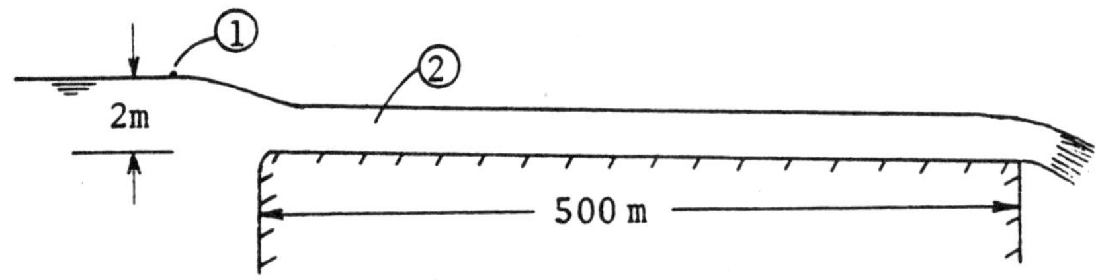

The channel is steep; therefore, critical depth will occur just inside the channel entrance. Then write the energy equation from the reservoir, (1), to the entrance section (2).

$y_1 + V_1^2/2g = y_2 + V_2^2/2g$; Assume $V_1 = 0$

Then $2 = y_2 + V_2^2/2g = y_c + 0.5\, y_c$

Solving for y_c: $y_c = 2/1.5 = 1.33$ m

Get $V_c = V_2$: $V_c^2/g = y_c = 1.33$ or $V_c = 3.62$ m/s

Then $Q = V_c A_2 = 3.62 \times 1.33 \times 4 = \underline{\underline{19.2 \text{ m}^3/\text{s}}}$

Solution Table for Problem 15.51

Section number upstream of y_c	Depth y,m	Velocity at section V,m/s	Mean Velocity in reach $(V_1+V_0)/2$	V^2	Hydraulic Radius R=A/P,m	Mean Hydraulic Radius $R_m=(R_1+R_2)/2$	$S_f=fV^2_{mean}/8gR_{mean}$	$\Delta x = ((y_2+V_2^2/2g)-(y_1+V_1/2g))/S_f$	Distance upstream from brink x,m
1(at $y=y_c$)	0.972	3.086			0.654				3.9 m
			3.073	9.443		0.656	1.834×10^{-3}	0.1 m	4.0 m
2	0.98	3.060			0.658				
			3.045	9.272		0.660	1.790×10^{-3}	0.4 m	4.4 m
3	0.99	3.030			0.662				
			2.986	8.916		0.669	1.698×10^{-3}	1.7 m	6.1 m
4	1.02	2.941			0.675				
			2.886	8.327		0.684	1.551×10^{-3}	4.7 m	10.9 m
5	1.06	2.830			0.693				
			2.779	7.721		0.701	1.403×10^{-3}	7.7 m	18.6 m
6	1.10	2.727			0.710				
			2.613	6.828		0.730	1.192×10^{-3}	33.2 m	51.8 m
7	1.20	2.500			0.750				
			2.404	5.779		0.769	9.576×10^{-4}	55.3 m	107.1 m
8	1.30	2.308			0.788				
			2.225	4.951		0.806	7.83×10^{-4}	80.0 m	187.1 m
9	1.40	2.143			0.824				
			2.0715	4.291		0.841	6.501×10^{-4}	107.4 m	294.5 m
10	1.50	2.00			0.857				

The depth 300 m upstream is approximately <u>1.51 m</u>

15.54 a) Assume uniform flow is established in the channel except near the downstream end. Then if the energy equation is written from the reservoir to a section near the upstream end of the channel, we have:

$2.5 \approx V_n^2/2g + y_n$ (1)

Also, $V_n = (1/n)R^{2/3}S^{1/2}$ or $V_n^2/2g = (1/n^2)R^{4/3}S/2g$ (2)

where $R = A/P = 3.5y_n/(2y_n + 3.5)$ (3)

Then combining Eqs. (1), (2) and (3) we have

$2.5 = ((1/n^2)((3.5y_n/(2y_n + 3.5))^{4/3}S/2g) + y_n$ (4)

Assuming n = 0.012 and solving Eq. (4) for y_n yields:

$y_n = 2.16$ m; Also solving (2) yields $V_n = 2.58$ m/s

Then Q = VA = 2.58 x 3.5 x 2.16 = $\underline{\underline{19.5\ m^3/s}}$

b) With only a 100 m-long channel, uniform flow will not become established in the channel; therefore, a trial-and-error type of solution is required. Critical depth will occur just upstream of the brink, so assume a value of y_c, then calculate Q and calculate the water surface profile back to the reservoir. Repeat the process for different values of y_c until a match between the reservoir water surface elevation and the computed profile is achieved.

15.55 $q = 10\ m^3/s/m$ $y_c = \sqrt[3]{q^2/g} = \sqrt[3]{10^2/9.81} = 2.17$ m

y	$\bar{y}$	V	$\bar{V}$	E	ΔE	$S_f \times 10^4$	Δx	x	elev.
52.17		0.1917		52.170				0	52.17
	51.08		0.1958		2.168	0.00287	-5,429		
50		0.20		50.002				-5,430	52.17
	45		0.2222		9.999	0.00419	-25,024		
40		0.25		40.003				-30,450	52.18
	35		0.2857		9.997	0.00892	-25,048		
30		0.333		30.006				-55,550	52.22
	25		0.400		9.993	0.02447	-25,146		
20		0.50		20.013				-80,650	52.26
	15		0.6667		9.962	0.11326	-25,631		
10		1.00		10.051				-106,280	52.51
	9		1.1111		1.971	0.5244	-5,671		
8		1.25		8.080				-111,950	52.78
	7		1.4286		1.938	1.1145	-6,716		
6		1.667		6.142				-118,670	53.47

15.56 First, one has to determine whether the uniform flow in the channel is super or subcritical. Determine y_n and then see if for this y_n the Froude number is greater or less than unity.

$Q = (1.49/n)\ AR^{2/3} S^{1/2}$; Assume n = 0.015

$12 = (1.49/0.015) \times y \times y^{2/3} \times (0.04)^{1/2}$

$y_n = 0.739$ ft and $V = Q/y_n = 16.23$ ft/s

$F = V/\sqrt{gy_n} = 3.33$ Therefore, uniform flow in the channel is supercritical and one can surmise that a hydraulic jump will occur upstream of the weir. One can check this by determining what the sequent depth is. If it is less than the weir height plus head on the weir then the jump will occur.

Get sequent depth:

$$y_2 = (y_1/2)(\sqrt{1 + 8F_1^2} - 1)$$

$$= (0.739/2)(\sqrt{1 + 8 \times 3.33^2} - 1)$$

$$y_2 = \underline{\underline{3.13 \text{ ft}}}$$

Get head on weir:

$$Q = K\sqrt{2g}\ LH^{3/2} \qquad \text{Assume } K = 0.42$$

$$12 = 0.42\sqrt{64.4} \times 1 \times H^{3/2}$$

$H = 2.33$ ft; $H/P = 2.33/3 = 0.78$ so $K = 0.40 + 0.05 \times 0.78$

Better estimate for H: $H = 2.26$ ft $= 0.44$

Then depth just upstream of weir = 3 + 2.26 = $\underline{\underline{5.56 \text{ ft.}}}$
Therefore, it is proved that a jump will occur.
A rough estimate for the distance to where the jump will occur may be found by applying Eq.(15.35) with a single step computation. A more accurate calculation would include several steps.

The single-step calculation is given below:

$$\Delta x = ((y_1 - y_2) + (V_1^2 - V_2^2))/2g/(S_f - S_0)$$

where $y_1 = 3.13$ ft; $V_1 = q/y_1 = 12/3.13 = 3.83$ ft/s; $V_1^2 = 14.67\ \text{ft}^2/\text{s}^2$

$y_2 = 5.56$ ft; $V_2 = 2.16$ ft/s $V_2^2 = 4.67\ \text{ft}^2/\text{s}^2$

$S_f = fV_{avg}^2/(8gR_{avg})$; $V_{avg} = 3.00$ ft/s; $R_{avg} = 4.34$ ft

Assume $k_s = 0.001$ ft; $k_s/4R = 0.00034$

$$Re = V \times 4R/\nu = ((3.83 + 2.16)/2) \times 4 \times 4.34/(1.22 \times 10^{-5}) = 4.3 \times 10^6$$

15.56 (continued)

Then $f = 0.015$ and $S_f = 0.015 \times 3.0^2/(8 \times 32.2 \times 4.34) = 0.000121$

$\Delta x = ((3.13 - 5.56) + (14.67 - 4.67)/(64.4))/(0.000121 - 0.04) = \underline{\underline{57.0\ ft}}$

Thus, the water surface profile is shown below:

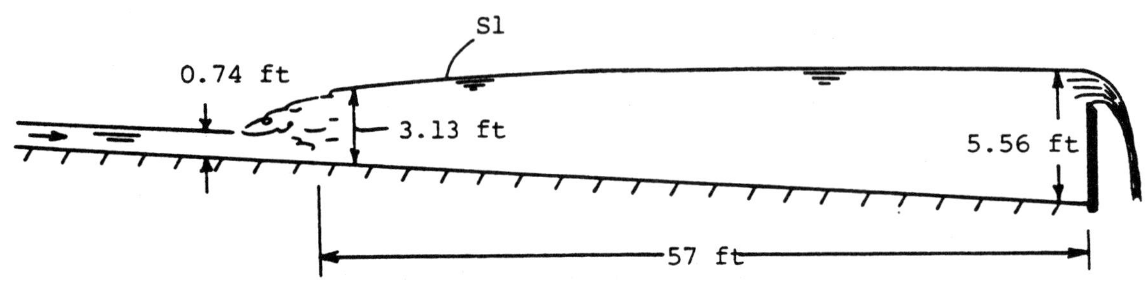

CHAPTER SIXTEEN

16.1

$$y_{i+1} = y_i + y_i'\Delta x + y_i''\frac{\Delta x^2}{2!} + y_i'''\frac{\Delta x^3}{3!} \cdots \quad (1)$$

$$y_{i+2} = y_i + 2y_i'\Delta x + y_i''\frac{4\Delta x^2}{2!} + y_i'''\frac{8\Delta x^3}{3!} \quad (2)$$

Multiply Eq.(1) by 4 and subtract Eq.(2) gives

$$y_i' = \frac{-3y_i + 4y_{i+1} - y_{i+2}}{2\Delta x} + O(\Delta x^2)$$

16.2

$$r^2\frac{dv}{dr}\bigg|_{i+\frac{1}{2}} = r^2_{i+\frac{1}{2}}\frac{(v_{i+1} - v_i)}{\Delta r}$$

$$r^2\frac{dv}{dr}\bigg|_{i-\frac{1}{2}} = r^2_{i-\frac{1}{2}}\frac{(v_i - v_{i-1})}{\Delta r}$$

$$\frac{d}{dr}(r^2\frac{dv}{dr}) = \frac{r^2\frac{dv}{dr}\Big|_{i+\frac{1}{2}} - r^2\frac{dv}{dr}\Big|_{i-\frac{1}{2}}}{\Delta r}$$

$$\frac{d}{dr}(r^2\frac{dv}{dr}) - r\frac{dv}{dr} - v = 0$$

$$r^2_{i+\frac{1}{2}}\frac{(v_{i+1} - v_i)}{\Delta r^2} - r^2_{i-\frac{1}{2}}\frac{(v_i - v_{i-1})}{\Delta r^2} - r_i\frac{(v_{i+1} - v_{i-1})}{2\Delta r} - v_i = 0$$

$$v_{i+1}\left(\frac{r^2_{i+\frac{1}{2}}}{\Delta r^2} - \frac{r_i}{2\Delta r}\right) - v_i\left(\frac{r^2_{i+\frac{1}{2}}}{\Delta r^2} + \frac{r^2_{i-\frac{1}{2}}}{\Delta r^2} + 1\right) + v_{i-1}\left(\frac{r^2_{i-\frac{1}{2}}}{\Delta r^2} + \frac{r_i}{2\Delta r}\right) = 0$$

16.3

$$\mu_{i+\frac{1}{2}}u_{i+1} - (\mu_{i+\frac{1}{2}} + \mu_{i-\frac{1}{2}})u_i + \mu_{i-\frac{1}{2}}u_{i-1} = 0$$

$$T = 313.2 + 60\ y/h$$

$$\mu = 6.24 \times 10^{-9}\exp(5900/T)$$

VELOCITY DISTRIBUTION BETWEEN PLATES *****

I	U(I)
1	0.00000
2	0.04114
3	0.12151
4	0.27133
5	0.53914
6	1.00000

16.4 $$\psi_{i+1} = \psi_i + a\Delta x\psi'_i + a^2\frac{\Delta x^2}{2!}\psi''_i \ldots \quad (1)$$

$$\psi_{i-1} = \psi_i - \Delta x\psi'_i + \frac{\Delta x^2}{2}\psi''_i \quad (2)$$

Subtract Eq.(2) from Eq.(1) to obtain

$$\psi_{i+1} - \psi_{i-1} = (a+1)\Delta x\psi'_i + \frac{\Delta x^2}{2!}(a^2-1) \ldots$$

$$\psi'_i = \frac{\psi_{i+1}-\psi_{i-1}}{(a+1)\Delta x} + O(\Delta x)$$

$$\left.\frac{d\psi}{dx}\right|_{i+\frac{1}{2}} = \frac{\psi_{i+1}-\psi_i}{a\Delta x} \qquad \left.\frac{d\psi}{dx}\right|_{i-\frac{1}{2}} = \frac{\psi_i-\psi_{i-1}}{\Delta x}$$

$$\frac{d^2\psi}{dx^2} = \left(\frac{\psi_{i+1}-\psi_i}{a\Delta x} - \frac{\psi_i-\psi_{i-1}}{\Delta x}\right)/\frac{(a+1)\Delta x}{2}$$

$$\frac{d^2\psi}{dx^2} = \left(\frac{2}{a(a+1)}\right)\frac{\psi_{i+1}-\psi_i(1+a)+a\psi_{i-1}}{\Delta x^2}$$

16.5 a) $$\frac{d}{dr}\left(\mu r^3\frac{d\omega}{dr}\right) = 0$$

or

$$\frac{d}{dr}\left(r^3\frac{d\omega}{dr}\right) = 0$$

$$\frac{r^3_{i+\frac{1}{2}}\left.\frac{d\omega}{dr}\right|_{i+\frac{1}{2}} - r^3_{i-\frac{1}{2}}\left.\frac{d\omega}{dr}\right|_{i-\frac{1}{2}}}{\Delta r} = 0$$

$$\frac{r^3_{i+\frac{1}{2}}(\omega_{i+1}-\omega_i) - r^3_{i-\frac{1}{2}}(\omega_i-\omega_{i-1})}{\Delta r^2} = 0$$

$$r^3_{i+\frac{1}{2}}\omega_{i+1} - (r^3_{i+\frac{1}{2}} + r^3_{i-\frac{1}{2}})\omega_i + r^3_{i-\frac{1}{2}}\omega_{i-1} = 0$$

Analytic solution:

$$\mu r^3\frac{d\omega}{dr} = \text{const}$$

$$d\omega = \frac{C_1\,dr}{r^3}$$

$$\omega = -\frac{C_1}{2r^2} + C_2$$

16.5 (continued)

$\omega = 10$ at $r = 0.5$

$\omega = 0$ at $r = 1.0$

$$v = \frac{10}{3}\left(\frac{1}{r} - r\right)$$

VELOCITY DISTRIBUTION BETWEEN PLATES

I	U(I)
1	5.00000
2	3.56608
3	2.44014
4	1.50892
5	0.70846
6	0.00000

Analytic result

I	U(I)
1	5.0
2	3.5555
3	2.4286
4	1.5000
5	0.7037
6	0.0

b)

VELOCITY DISTRIBUTION BETWEEN PLATES *****

I	U(I)
1	9.50000
2	7.55940
3	5.63973
4	3.74035
5	1.86063
6	0.00000

Linear profile

I	U(I)
1	9.5
2	7.6
3	5.7
4	3.8
5	1.9
6	0.0

c) $T = 313.2 + 60\,\dfrac{\ell n(r)}{\ell n(0.5)}$ (r in cm)

Finite difference equation

$$(\mu r^3)_{i+\frac{1}{2}}\omega_{i+1} - [(\mu r^3)_{i+\frac{1}{2}} + (\mu r^3)_{i-\frac{1}{2}}]\omega_i + (\mu r^3)_{i-\frac{1}{2}}\omega_{i-1} = 0$$

$$T = 2\pi\mu_1 r_1^3\left(\frac{\omega_2 - \omega_1}{\Delta r}\right)$$

VELOCITY DISTRIBUTION BETWEEN PLATES *****

I	U(I)
1	5.00000
2	1.94530
3	0.79862
4	0.32211
5	0.10623
6	0.00000

TORQUE ON INNER CYLINDER = 0.348E-05 N-M/CM OF LENGTH

16.6 a) $\frac{d}{dy}(\mu\frac{du}{dy}) = -\frac{dp_z}{ds}$

$$\frac{d^2u}{dy^2} = -\frac{1}{\mu}\frac{dp_z}{ds}$$

$$\frac{d^2\bar{u}}{d\bar{y}^2} = -1 \qquad \text{when } \bar{u} = \frac{u\mu}{h^2}\left(\frac{dp_z}{ds}\right)^{-1} = \frac{u}{\beta}\,,\ \bar{y} = \frac{y}{h}$$

Analytic solution $\bar{u} = \bar{y}(1-\bar{y})/2$

Finite difference equation

$$\bar{u}_{i-1} - 2\bar{u}_i + \bar{u}_{i+1} = -\Delta\bar{y}^2$$

B = h**2/u * dP/dS

U/B DISTRIBUTION BETWEEN PLATES *****

I	U/B
1	0.00000
2	0.08000
3	0.12000
4	0.12000
5	0.08000
6	0.00000

Analytic solution

I	U/B
1	0.0
2	0.08
3	0.12
4	0.12
5	0.08
6	0.0

b) $\frac{d}{d\bar{y}}(\bar{\mu}\frac{d\bar{u}}{d\bar{y}}) = -1$ when $\bar{u} = \frac{\mu_o u}{h^2}\left(\frac{dp_z}{ds}\right)^{-1}$

$$T = 303 + 60\,\bar{y}$$

$$\bar{\mu} = \exp\left(\frac{1770}{T}\right)/\exp\left(\frac{1770}{303}\right)$$

$$\bar{\mu}_{i-\frac{1}{2}}\bar{u}_{i-1} - (\bar{\mu}_{i-\frac{1}{2}}+\bar{\mu}_{i+\frac{1}{2}})\bar{u}_i + \bar{\mu}_{i+\frac{1}{2}}\bar{u}_{i+1} = -\Delta\bar{y}^2$$

VELOCITY DISTRIBUTION BETWEEN PLATES ***

I	U(I)
1	0.00000
2	0.10642
3	0.18275
4	0.20815
5	0.15726
6	0.00000

MEAN VELOCITY = 0.1091 m/s

DISCHARGE = 0.1091 cubic meters per meter of width

16.7 Continuity eqn. $\frac{d}{dt} \int \rho dv + \rho Q = 0$

$$V = \frac{\pi (h \tan 15)^2 h}{3}$$

$$\therefore \rho \frac{d}{dt}(\frac{\pi}{3} h^3 \tan^2 15) + \rho Q = 0$$

$$\frac{dh}{dt} = - \frac{Q(t)}{\pi h^2 \tan^2 15}$$

Explicit: $h_{n+1} = h_n - \frac{Q_o}{\pi} \frac{[1+\cos(\omega t_n)]\Delta t}{h_n^2 \tan^2 15}$

Implicit: $h_{n+1} + \frac{\Delta t}{\pi} Q_o \frac{[1+\cos(\omega t_{n+1})]}{h_{n+1}^2 \tan^2 15} = h_n$

16.8

$$\frac{\partial^2 \psi}{\partial x^2} = \frac{2}{a(a+1)}[\frac{\psi_E + a\psi_N - (a+1)\psi_p}{\Delta x^2}]$$

$$\frac{\partial^2 \psi}{\partial y^2} = \frac{2}{b(b+1)}[\frac{\psi_N + b\psi_S - (b+1)\psi_p}{\Delta y^2}]$$

$$\frac{\partial^2 \psi}{\partial x^2} + \frac{\partial^2 \psi}{\partial y^2} = 0$$

$$\psi_p = \frac{ab\Delta x^2 \Delta y^2}{b\Delta y^2 + a\Delta x^2}[\frac{\psi_E + a\psi_W}{a(a+1)\Delta x^2} + \frac{\psi_N + b\psi_S}{b(b+1)\Delta y^2}]$$

16.9 $u = 0.5 + 0.5y$

$$\nabla^2 \psi = -\Omega_z = \frac{\partial v}{\partial x} - \frac{\partial u}{\partial y} = 0.5$$

Finite difference equation

$$\psi_{i,j-1} - 4\psi_{i,j} + \psi_{i,j+1} = -\psi_{i-1,j} - \psi_{i+1,j} + 0.5\Delta \ell^2$$

$$\psi = \psi_{CL} + \int u dy = 0.5y + 0.25y^2$$

16.9 (continued)

***STREAM FUNCTION DISTRIBUTION IN DUCT WITH A CONTRACTION RATIO OF TWO AND A CONSTANT VORTICITY OF .5**

NIT=137 ERROR= 0.99E-02

J=	1	3	5	7	9	11	13	15	17	19	21
1	0.000	0.052	0.110	0.172	0.240	0.312	0.390	0.472	0.560	0.652	0.750
4	0.000	0.054	0.113	0.177	0.246	0.318	0.396	0.477	0.564	0.654	0.750
7	0.000	0.056	0.117	0.183	0.252	0.326	0.403	0.483	0.568	0.657	0.750
10	0.000	0.059	0.123	0.190	0.261	0.335	0.412	0.492	0.574	0.660	0.750
13	0.000	0.063	0.130	0.201	0.274	0.350	0.427	0.505	0.584	0.666	0.750
16	0.000	0.069	0.141	0.217	0.294	0.372	0.449	0.526	0.601	0.675	0.750
19	0.000	0.076	0.156	0.239	0.322	0.405	0.485	0.562	0.633	0.696	0.750
22	0.000	0.087	0.177	0.270	0.364	0.455	0.542	0.621	0.691	0.750	0.000
25	0.000	0.100	0.204	0.311	0.420	0.527	0.625	0.712	0.000	0.000	0.000
28	0.000	0.113	0.233	0.359	0.492	0.630	0.750	0.000	0.000	0.000	0.000
31	0.000	0.125	0.258	0.400	0.558	0.750	0.000	0.000	0.000	0.000	0.000
34	0.000	0.133	0.273	0.421	0.580	0.750	0.000	0.000	0.000	0.000	0.000
37	0.000	0.137	0.280	0.430	0.587	0.750	0.000	0.000	0.000	0.000	0.000
40	0.000	0.139	0.283	0.433	0.589	0.750	0.000	0.000	0.000	0.000	0.000
43	0.000	0.140	0.284	0.434	0.590	0.750	0.000	0.000	0.000	0.000	0.000
46	0.000	0.140	0.285	0.435	0.590	0.750	0.000	0.000	0.000	0.000	0.000
49	0.000	0.140	0.285	0.435	0.590	0.750	0.000	0.000	0.000	0.000	0.000

16.10

$$\psi = \frac{1}{x^2+y^2}$$

$$u = \frac{\partial\psi}{\partial y} = \frac{-2y}{(x^2+y^2)^2} \quad \frac{\partial u}{\partial y} = \frac{-2}{(x^2+y^2)^2} + \frac{8y^2}{(x^2+y^2)^3}$$

$$v = -\frac{\partial\psi}{\partial x} = \frac{2x}{(x^2+y^2)^2} \quad \frac{\partial v}{\partial x} = \frac{2}{(x^2+y^2)^2} - \frac{8x^2}{(x^2+y^2)^3}$$

$$\frac{\partial v}{\partial x} - \frac{\partial u}{\partial y} = \frac{4}{(x^2+y^2)^2} - \frac{8(x^2+y^2)}{(x^2+y^2)^3} \neq 0$$

Flow is rotational.

16.11

$$u = \frac{\partial\psi}{\partial y} = x$$

$$v = -\frac{\partial\psi}{\partial x} = -y$$

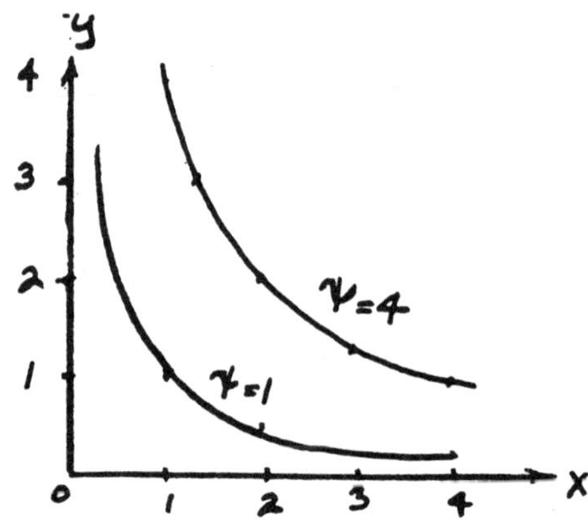

16.12 $u = U, \quad v = 0$

$$\frac{\partial \psi}{\partial x} = -v = 0 \qquad \therefore \psi = f(y)$$

$$\frac{d\psi}{dy} = U, \quad \psi = Uy + \text{const}$$

16.13 $y = ax^3 + bx^2 + cx + d$

$y(-2) = 2 \quad y'(-2) = -1, \quad y(2) = 0 \quad y'(2) = 0$

$y = 0.125x^2 - 0.5x + 0.5$

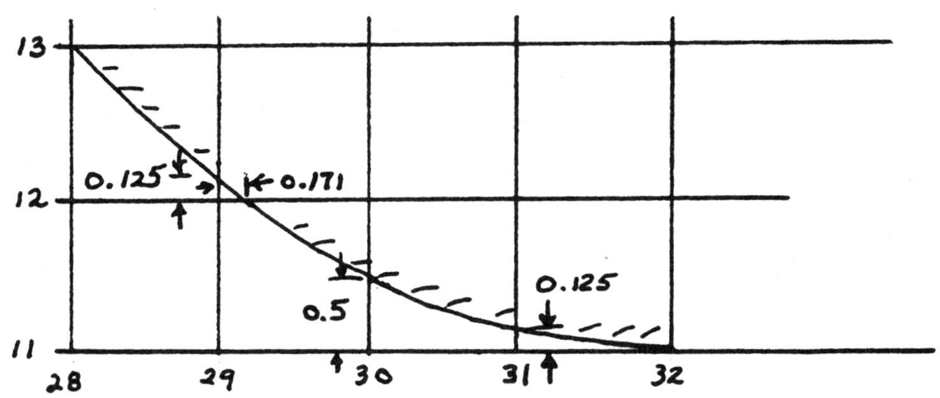

General finite difference equation:

$$C_n\psi_{i,j+1} + C_s\psi_{i,j-1} + C_e\psi_{i+1,j} + C_w\psi_{i-1,j} = (C_n+C_s+C_e+C_w)\psi_{i,j}$$

at point (29,12), a(east) = 0.171 b(north) = 0.125

CN(29,12) = 14.2222 CE(29,12) = 9.9880
CS(29,12) = 1.7778 CW(29,12) = 1.7079

at point (30,11) b(north) = 0.5

CN(30,11) = 2.6667
CS(30,11) = 1.3333

at point (31,11) b(north) = 0.125

CN(31,11) = 14.2222
CS(31,11) = 1.7778

16.13 (continued)

*STREAM FUNCTION DISTRIBUTION IN DUCT WITH A CONTRACTION RATIO OF TWO

NIT=109 ERROR= 0.98E-02

J=	1	3	5	7	9	11	13	15	17	19	21
1	0.000	0.100	0.200	0.300	0.400	0.500	0.600	0.700	0.800	0.900	1.000
4	0.000	0.102	0.203	0.304	0.405	0.506	0.605	0.705	0.803	0.902	1.000
7	0.000	0.104	0.207	0.310	0.412	0.512	0.612	0.710	0.808	0.904	1.000
10	0.000	0.107	0.213	0.317	0.421	0.522	0.622	0.719	0.814	0.907	1.000
13	0.000	0.111	0.221	0.329	0.435	0.538	0.637	0.733	0.825	0.913	1.000
16	0.000	0.117	0.233	0.346	0.456	0.562	0.662	0.756	0.843	0.924	1.000
19	0.000	0.125	0.250	0.371	0.488	0.599	0.701	0.794	0.876	0.945	1.000
22	0.000	0.137	0.273	0.406	0.534	0.654	0.763	0.858	0.938	1.000	0.000
25	0.000	0.152	0.303	0.452	0.598	0.735	0.856	0.957	0.000	0.000	0.000
28	0.000	0.168	0.337	0.507	0.679	0.850	1.000	0.000	0.000	0.000	0.000
31	0.000	0.182	0.366	0.556	0.758	1.000	0.000	0.000	0.000	0.000	0.000
34	0.000	0.191	0.385	0.583	0.788	1.000	0.000	0.000	0.000	0.000	0.000
37	0.000	0.196	0.394	0.593	0.796	1.000	0.000	0.000	0.000	0.000	0.000
40	0.000	0.198	0.398	0.597	0.798	1.000	0.000	0.000	0.000	0.000	0.000
43	0.000	0.199	0.399	0.599	0.799	1.000	0.000	0.000	0.000	0.000	0.000
46	0.000	0.200	0.400	0.600	0.800	1.000	0.000	0.000	0.000	0.000	0.000
49	0.000	0.200	0.400	0.600	0.800	1.000	0.000	0.000	0.000	0.000	0.000

	PRESSURE COEFFICIENTS ALONG WALL
1	0.0116
4	0.0360
7	0.0816
10	0.1487
13	0.2556
16	0.4355
19	0.7778
22	0.4408
25	-0.2897
28	-2.2121
31	-7.9182
34	-3.6240
37	-3.1967
40	-3.0720
43	-3.0279
46	-3.0119
49	-3.0073

16.14 $\dfrac{\partial\psi}{\partial y} = U \cos\alpha; \quad \psi = U y \cos\alpha + f(x)$

$$\frac{\partial\psi}{\partial x} = -v = -U \sin\alpha = f'(x)$$

$$f(x) = -Ux \sin\alpha + \text{const}$$

$$\psi = U(y \cos\alpha - x \sin\alpha) + \text{const}$$

16.15 $\alpha = Uy(1 - \frac{a^2}{x^2+y^2})$

$$u = \frac{\partial\psi}{\partial y} = U(1 - \frac{a^2}{x^2+y^2}) + \frac{2Uy^2a^2}{(x^2+y^2)^2}$$

$$u(0,a) = 2U$$

$$v = -\frac{\partial\psi}{\partial x} = \frac{2Uyxa^2}{(x^2+y^2)^2}$$

$$v(0,a) = 0$$

$$C_p = 1 - \frac{u^2+v^2}{U^2} = -3$$

16.16 $\nabla\psi\cdot\nabla\phi = \frac{\partial\psi}{\partial x}\frac{\partial\phi}{\partial x} + \frac{\partial\psi}{\partial y}\frac{\partial\phi}{\partial y}$

$$= -v\,u + u\,v = 0$$

$\therefore$ Mutually orthogonal

16.17 $u = +\frac{\partial\psi}{\partial y} \quad v = -\frac{\partial\psi}{\partial x}, \quad U = 1$

$$C_p = 1 - (u^2+v^2)$$

Use expression for 2nd-order accuracy for velocities.

$$u_{wall} = \frac{\partial\psi}{\partial y} = (3\psi_{i,wall} - 4\psi_{i,wall-1} + \psi_{i,wall-2})/2\Delta y$$

$$u_{axis} = (4\psi_{i,2} - \psi_{i,3})/2\Delta y$$

	PRESSURE COEFFICIENT ALONG AXIS	**PRESSURE COEFFICIENT ALONG WALL**
1	**-0.0142**	**0.0064**
4	**-0.0461**	**0.0206**
7	**-0.1139**	**0.0508**
10	**-0.2296**	**0.1030**
13	**-0.4362**	**0.1981**
16	**-0.8100**	**0.3768**
19	**-1.4888**	**0.7516**
22	**-2.7076**	**0.2712**
25	**-4.7775**	**-1.2016**
28	**-7.7943**	**-5.9644**
31	**-11.0266**	**-24.9116**
34	**-13.2832**	**-16.3812**
37	**-14.3670**	**-15.3606**
40	**-14.7841**	**-15.1081**
43	**-14.9283**	**-15.0341**
46	**-14.9751**	**-15.0116**
49	**-14.9872**	**-15.0060**

16.18 $$\frac{d}{dt}\int \rho d\forall + \Sigma\, \rho \vec{V}\cdot\vec{A} = 0$$

Incompressible fluid,

$$\vec{V}\cdot\vec{A} = 0$$

$$2\pi U_r r\Delta z\Big|_{r=r_o} + \frac{\partial}{\partial r}(2\pi U_r r\Delta z)\Delta r - 2\pi U_r r\Delta z\Big|_{r=r_o}$$

$$+ 2\pi r\Delta r\, U_z\Big|_{z=z_o} + \frac{\partial}{\partial z}(2\pi r\Delta r U_z)\Delta z - 2\pi U_z r\Delta r\Big|_{z=z_o} = 0$$

$$\therefore \frac{1}{r}\frac{\partial}{\partial r}(rU_r) + \frac{\partial U_z}{\partial z} = 0$$

16.19

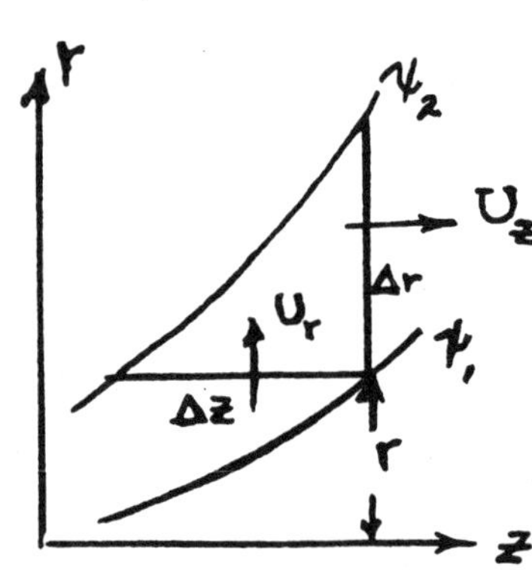

$$U_r 2\pi r\Delta z = \psi(r,z-\Delta z) - \psi(r,z)$$

$$U_r = -\frac{1}{2\pi r}\,\frac{\psi(r,z)-\psi(r,z-\Delta z)}{\Delta z}$$

In limit: $U_r = -\frac{1}{r}\frac{\partial\psi}{\partial z}$ (without 2π)

Similarly: $U_z = \frac{1}{r}\frac{\partial\psi}{\partial r}$

$$rU_r = -\frac{\partial\psi}{\partial z} \quad rU_z = \frac{\partial\psi}{\partial r}$$

$$\frac{\partial}{\partial r}(rU_r) + \frac{\partial}{\partial z}(rU_z) = -\frac{\partial^2\psi}{\partial r\partial z} + \frac{\partial^2\psi}{\partial r\partial z} = 0$$

16.20 $$\frac{\partial^2\psi}{\partial z^2} + r\frac{\partial}{\partial r}\left(\frac{1}{r}\frac{\partial\psi}{\partial r}\right) = 0$$

$$\frac{\partial^2\psi}{\partial z^2} = \frac{\psi_{i+1}-2\psi_i+\psi_{i-1}}{\Delta \ell^2}$$

$$r\frac{\partial}{\partial r}\left(\frac{1}{r}\frac{\partial\psi}{\partial r}\right) = r_j\left[\frac{1}{r_{j+\frac{1}{2}}}\frac{(\psi_{j+1}-\psi_j)}{\Delta\ell^2} - \frac{1}{r_{j-\frac{1}{2}}}\frac{(\psi_j-\psi_{j-1})}{\Delta\ell^2}\right]$$

Finite-difference equation:

$$C_e\psi_{i+1,j} + C_w\psi_{i-1,j} + C_n\psi_{i,j+1} + C_s\psi_{i,j-1} = (C_n+C_e+C_w+C_s)\psi_{i,j}$$

16.20 (continued)

In general

$C_e = C_w = 1, \quad C_n = r_j/r_{j+\frac{1}{2}}, \quad C_s = r_j/r_{j-\frac{1}{2}}$

At point (20,20)

$C_n = 1.524r_j/(r_j + 0.357\Delta r) \qquad C_s = 1.143r_j/(r_j - 0.5\Delta r)$

At point (30,11)

$C_n = 6.4r_j/(r_j + 0.125\Delta r) \qquad C_s = 1.6r_j/(r_j - 0.5\Delta r)$

$$\text{On axis} \quad u = \left.\frac{1}{r}\frac{\partial\psi}{\partial r}\right|_{r\to 0} = \frac{\lim_{r\to 0}\frac{\partial}{\partial r}\left(\frac{\partial\psi}{\partial r}\right)}{\lim_{r\to 0} 1} = \left.\frac{\partial^2\psi}{\partial r^2}\right|_{r=0} = \frac{2\,\psi_{i,2}}{\Delta r^2}$$

Boundary conditions
$\psi = r^2/2$ at i=1

$\psi = 1/2$ at wall

$\psi_{i,3} = 4\psi_{i,2}$ along j=3

*STREAM FUNCTION DISTRIBUTION IN DUCT WITH A CONTRACTION RATIO OF TWO

NIT= 75 ERROR= 0.98E-02

J=	1	3	5	7	9	11	13	15	17	19	21
1	0.000	0.005	0.020	0.045	0.080	0.125	0.180	0.245	0.320	0.405	0.500
4	0.000	0.005	0.020	0.046	0.081	0.127	0.182	0.247	0.322	0.406	0.500
7	0.000	0.005	0.021	0.047	0.083	0.130	0.185	0.250	0.324	0.407	0.500
10	0.000	0.006	0.022	0.049	0.087	0.134	0.190	0.255	0.328	0.410	0.500
13	0.000	0.006	0.024	0.053	0.093	0.142	0.199	0.265	0.336	0.414	0.500
16	0.000	0.007	0.027	0.059	0.102	0.155	0.216	0.282	0.351	0.424	0.500
19	0.000	0.008	0.031	0.069	0.119	0.178	0.244	0.313	0.381	0.445	0.500
22	0.000	0.010	0.038	0.084	0.145	0.216	0.292	0.368	0.439	0.500	0.000
25	0.000	0.012	0.048	0.106	0.184	0.276	0.370	0.459	0.000	0.000	0.000
28	0.000	0.015	0.059	0.135	0.240	0.372	0.500	0.000	0.000	0.000	0.000
31	0.000	0.017	0.070	0.161	0.296	0.500	0.000	0.000	0.000	0.000	0.000
34	0.000	0.019	0.076	0.174	0.314	0.500	0.000	0.000	0.000	0.000	0.000
37	0.000	0.020	0.079	0.178	0.318	0.500	0.000	0.000	0.000	0.000	0.000
40	0.000	0.020	0.080	0.179	0.319	0.500	0.000	0.000	0.000	0.000	0.000
43	0.000	0.020	0.080	0.180	0.320	0.500	0.000	0.000	0.000	0.000	0.000
46	0.000	0.020	0.080	0.180	0.320	0.500	0.000	0.000	0.000	0.000	0.000
49	0.000	0.020	0.080	0.180	0.320	0.500	0.000	0.000	0.000	0.000	0.000

16.20 (continued)

	PRESSURE COEFFICIENT ALONG AXIS	PRESSURE COEFFICIENT ALONG WALL
1	-0.0142	0.0064
4	-0.0461	0.0206
7	-0.1139	0.0508
10	-0.2296	0.1030
13	-0.4362	0.1981
16	-0.8100	0.3768
19	-1.4888	0.7516
22	-2.7076	0.2712
25	-4.7775	-1.2016
28	-7.7943	-5.9644
31	-11.0266	-24.9116
34	-13.2832	-16.3812
37	-14.3670	-15.3606
40	-14.7841	-15.1081
43	-14.9283	-15.0341
46	-14.9751	-15.0116
49	-14.9872	-15.0060

16.21

$$F = p_1A_1 - p_2A_2 + \frac{(p_1+p_2)}{2}(A_2-A_1)$$

$$= p_1(A_1 + A_2/2 - A_1/2) - p_2(A_2 - A_2/2 + A_1/2)$$

$$= (p_1-p_2)\frac{(A_1+A_2)}{2}$$

16.22

$$u_i = u_i^\circ + \frac{A_i}{\dot{m}}(\Delta p_{i-1} - \Delta p_i)$$

$$\rho_i = \rho_i^\circ + \frac{\partial\rho}{\partial p}\Delta p_i = \rho_i^\circ + \frac{1}{k}\frac{\rho_i}{p_i}\Delta p_i$$

$$\rho_i u_i A_i = \rho_i^\circ u_i^\circ A_i + \rho_i^\circ \frac{A_i^2}{\dot{m}}\Delta p_{i-1} - \left(\frac{\rho_i^\circ A_i^2}{\dot{m}} - \frac{u_i^\circ \rho_i^\circ A_i}{kp_i}\right)\Delta p_i$$

Continuity equation

$$\rho_i^\circ u_i^\circ A_i - \rho_{i+1}^\circ A_{i+1} u_{i+1}^\circ = -\frac{\rho_i^\circ A_i^2}{\dot{m}}\Delta p_{i-1} + \left(\frac{\rho_i^\circ A_i^2}{\dot{m}} - \frac{u_i^\circ \rho_i^\circ A_i}{kp_i} + \frac{\rho_{i+1}^\circ A_{i+1}^2}{\dot{m}}\right)\Delta p_i$$

$$- \left(\frac{\rho_{i+1}^\circ A_{i+1}^2}{\dot{m}} - \frac{u_{i+1}^\circ \rho_{i+1}^\circ A_{i+1}}{kp_{i+1}}\right)\Delta p_{i+1}$$

16.22 (continued)

***FLOW RATE, VELOCITY AND PRESSURE DISTRIBUTION FOR
FLOW THROUGH AN EXTENDED LENGTH VENTURI APPROACH SECTION
WITH A PRESSURE DIFFERENCE OF 4.00E 01 KPA

*ITERATION NO=123 FLOW RATE= 1.852E-02 KG/S
REYNOLDS NUMBER= 2.86E 06 RESIDUAL= 2.17E-06

I	DUCT DIA M	VELOCITY M/S	PRESSURE KPA
1	0.0200	49.61	100.0
11	0.0191	54.37	99.7
21	0.0181	60.94	99.3
31	0.0171	68.83	98.6
41	0.0160	78.45	97.8
51	0.0150	90.38	96.6
61	0.0140	105.57	94.9
71	0.0129	125.57	92.3
81	0.0119	153.44	87.9
91	0.0109	197.02	79.8
101	0.0100	285.87	60.0

Analytic Solution:

$$\frac{p_2}{p_1} = 0.6 \quad \frac{D_2}{D_1} = 0.5 \quad \rho_1 = 1.189 \text{ kg/m}^3 \quad p_1 = 10^5 \text{Pa} \quad A_2 = (0.01)^2(0.785)$$

From Eqn. (13-16)

$$\dot{m} = 7.85 \times 10^{-5}(0.6)^{0.7143}\left[\frac{7 \cdot 10 \cdot 1.189(1-0.6^{0.286})}{1-(0.6^{1.428}(0.5)^4}\right]^{\frac{1}{2}}$$

$$= \underline{\underline{0.0186}} \text{ kg/s}$$

16.23 $u_i > 0$

$$u\frac{\partial u}{\partial x} = \frac{u_i - u_{i-1}}{\Delta x}\left(\frac{u_i + u_i}{2}\right) + \frac{u_{i+1} - u_i}{\Delta x} \cdot \left(\frac{u_i - u_i}{2}\right)^{\to 0} = u_i \frac{(u_i - u_{i-1})}{\Delta x}$$

$u_i < 0$

$$u\frac{\partial u}{\partial x} = \frac{u_i - u_{i-1}}{\Delta x}\left(\frac{|u_i| - |u_i|}{2}\right)^{\to 0} - |u_i| \frac{u_{i+1} - u_i}{\Delta x}$$

$$= u_i \frac{(u_{i+1} - u_i)}{\Delta x}$$

16.24 Momentum equation:

$$\dot{m}u_{i+1} = \dot{m}u_i + A_{i+1}(p_i - p_{i+1}) - 8\eta u_{i+1}\pi\Delta x - \frac{4}{3}\tau_o \pi D_{i+1}\Delta x$$

Define $p_i^\circ = \frac{16}{3}\tau_o \sum_{i=N-1}^{1} \frac{\Delta x}{D_i}$

$$\dot{m}u_{i+1} = \dot{m}u_i + A_{i+1}(\bar{p}_i - \bar{p}_{i+1}) - 8\eta u_{i+1}\pi\Delta x$$

where $\bar{p} = p_i - p_i^\circ$

Thus, $u_{i+1} = \left[\dot{m}u_i + \frac{A_{i+1}}{\dot{m}}(\bar{p}_i - \bar{p}_{i+1})\right] / (\dot{m} + 8\,\eta\pi\Delta x)$

Initial guess for flow rate based on equilibration of pressure drop and shear force.

```
****FLOW OF BINGHAM PLASTIC THROUGH EXTENDED
    LENGTH VENTURI FOR

    PRESSURE DROP=  0.4724E 05PA AND
    MASS FLOW =   0.2020E-05KG/SEC
```

I	AREA(M**2)	VEL(M/S)	PRESS(PA)
1	0.3140E-03	0.4724E 05	0.4288E-05
11	0.2871E-03	0.4378E 05	0.4690E-05
21	0.2570E-03	0.4013E 05	0.5240E-05
31	0.2285E-03	0.3627E 05	0.5893E-05
41	0.2017E-03	0.3216E 05	0.6675E-05
51	0.1766E-03	0.2778E 05	0.7624E-05
61	0.1532E-03	0.2308E 05	0.8791E-05
71	0.1314E-03	0.1802E 05	0.1025E-04
81	0.1113E-03	0.1254E 05	0.1210E-04
91	0.9286E-04	0.6564E 04	0.1450E-04
101	0.7850E-04	0.5960E-06	0.1715E-04

16.24 (continued)

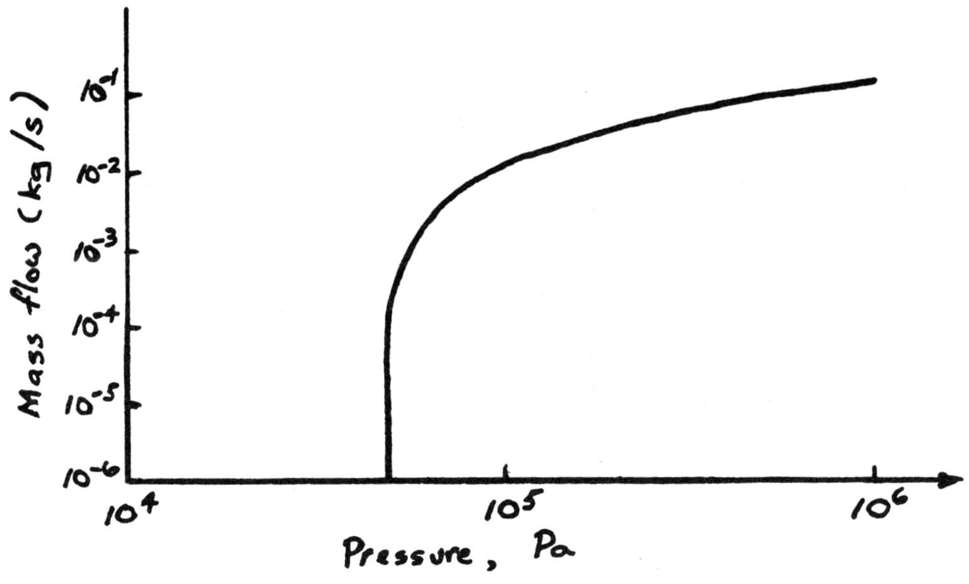

16.25

$$\frac{u(\Omega_{i+1,j} - \Omega_{i-1,j})}{2\Delta\ell} + \frac{v(\Omega_{i,j+1} - \Omega_{i,j-1})}{2\Delta\ell}$$

$$= \nu\left(\frac{\Omega_{i+1,j} + \Omega_{i-1,j} - 2\Omega_{i,j}}{\Delta\ell^2}\right) + \nu\left(\frac{\Omega_{i,j+1} + \Omega_{i,j-1} - 2\Omega_{i,j}}{\Delta\ell^2}\right)$$

$$\left(1 - \frac{u\Delta\ell}{2\nu}\right)\Omega_{i+1,j} + \left(1 + \frac{u\Delta\ell}{2\nu}\right)\Omega_{i-1,j} + \left(1 + \frac{v\Delta\ell}{2\nu}\right)\Omega_{i,j-1} + \left(1 - \frac{v\Delta\ell}{2\nu}\right)\Omega_{i,j+1} = 4\Omega_{i,j}$$

All positive if $\left|\frac{u\Delta\ell}{\nu}\right|$ and $\left|\frac{v\Delta\ell}{\nu}\right| < 2.$

16.26

At screen location

$$\dot{m}u_{i+1} = \dot{m}u_i + A_{i+1}(p_i - p_{i+1}) - \frac{\rho u_{i+1}^2}{2}(c_f P_{i+1}\Delta x + k_L)$$

At all other locations $k_L = 0$.

16.26 (continued)

***FLOW RATE, VELOCITY AND PRESSURE DISTRIBUTION FOR
FLOW THROUGH AN EXTENDED LENGTH VENTURI APPROACH SECTION
WITH A PRESSURE DIFFERENCE OF 1.00E 01 KPA

*ITERATION NO=114 FLOW RATE= 3.392E-01 KG/S
REYNOLDS NUMBER= 4.32E 04 RESIDUAL= 3.63E-05

I	DUCT DIA M	VELOCITY M/S	PRESSURE KPA
1	0.0200	1.08	10.0
11	0.0191	1.18	9.9
21	0.0181	1.32	9.7
31	0.0171	1.48	9.5
41	0.0160	1.68	9.1
51	0.0150	1.92	7.8
61	0.0140	2.21	7.3
71	0.0129	2.58	6.4
81	0.0119	3.05	5.0
91	0.0109	3.65	2.9
101	0.0100	4.32	0.0

16.27 $$\Omega = \frac{\partial v}{\partial x} - \frac{\partial u}{\partial y}$$

at wall $v = 0$, $u = \frac{\partial \psi}{\partial y}$

$$\Omega_{wall} = -\frac{\partial^2 \psi}{\partial y^2}$$

$$\psi_{wall+2} = \psi_{wall} + \frac{\partial \psi}{\partial y} 2\Delta y + \frac{\partial^2 \psi}{\partial y^2} \frac{4\Delta y^2}{2}$$

$$\frac{\partial \psi}{\partial y} = u = 0$$

$$\Omega_{wall} = (\psi_{wall} - \psi_{wall+2})/2\Delta y^2$$

16.28 $$\frac{\partial h}{\partial t} + \frac{\partial A}{\partial x} = 0$$

$$\frac{\partial A}{\partial t} + \frac{\partial B}{\partial x} = 0 \qquad B = Av + \frac{gh^2}{2}$$

$$\frac{\partial^2 h}{\partial t^2} = -\frac{\partial^2 A}{\partial x \partial t}$$

$$\frac{\partial^2 A}{\partial x \partial t} = -\frac{\partial^2 B}{\partial x^2}$$

16.28 (continued)

$$\frac{\partial^2 h}{\partial t^2} = \frac{\partial^2 B}{\partial x^2} = \frac{\partial^2}{\partial x^2}(Av) + gh\frac{\partial^2 h}{\partial x^2} + g\left(\frac{\partial h}{\partial x}\right)^2$$

$$\frac{\partial^2 h}{\partial t^2} = gh\frac{\partial^2 h}{\partial x^2}$$

16.29

$$c = \sqrt{gh} = \sqrt{9.81 \times 5} = 7 \text{ m/s}$$

$$T = 4.50/7 = 28.6 \text{ sec}$$

$$\omega_{Nat} = \frac{2\pi}{28.6} = 0.22/\text{sec}$$

```
***WATER SURFACE PROFILE OF BASIN OPENING ONTO
RESERVOIR WITH SINUSOIDALLY VARYING WATER SURFACE LEVEL
I=          1       21       41       61       81      101

TIME=   20.0
H(I)     5.29     5.27     5.13     4.87     4.65     4.52
V(I)     0.00     0.34     0.78     1.34     1.79     2.04

TIME=   40.0
H(I)     7.48     7.41     7.15     6.45     5.34     5.29
V(I)     0.00    -0.34    -0.78    -1.65    -3.04    -3.00

TIME=   60.1
H(I)     2.45     2.46     2.70     4.06     4.72     5.30
V(I)     0.00     0.50     0.59    -1.21    -1.55    -1.62

TIME=   80.1
H(I)     6.14     6.00     5.48     4.90     4.58     4.53
V(I)     0.00     1.61     2.77     3.72     4.30     4.42
```

16.30

$$\frac{\partial h}{\partial t} + \frac{\partial}{\partial x}(Vh) = 0 \qquad \text{continuity}$$

$$h\frac{\partial V}{\partial t} + V\frac{\partial h}{\partial t} + V\frac{\partial}{\partial x}(hV) + hV\frac{\partial V}{\partial x} + gh\frac{\partial h}{\partial x} = -c_f\frac{V|V|}{2}\left(\frac{P}{W}\right) \qquad \text{momentum}$$

Subtracting continuity and dividing by h:

$$\frac{\partial V}{\partial t} + V\frac{\partial V}{\partial x} = -g\frac{\partial h}{\partial x} - c_f\frac{V|V|}{2}\left(\frac{P}{Wh}\right)$$

16.31 Continuity $\frac{d}{dt} \int \rho h dx + \rho hV|_2 - \rho hV|_1 = 0$

$$\frac{\partial h}{\partial t} + \frac{\partial}{\partial x}(hv) = 0$$

Momentum $\frac{d}{dt} \int \rho hv dx + \rho hV^2|_2 - \rho hV^2|_1 = (\frac{\gamma h^2}{2}|_1 - \frac{\gamma h^2}{2}|_2) \cos \alpha$

$$+ \rho gh\Delta x \sin \alpha - c_f \frac{|V|V}{2} \rho P\Delta x$$

Assuming $\cos \alpha \simeq 1$ and $\sin \alpha = S_o$,

$$\frac{\partial}{\partial t}(hV) + \frac{\partial}{\partial x}(hV^2 + \frac{gh^2}{2}) = ghS_o - c_f \frac{|V|V}{2} P$$

16.32 Initial condition on h:

$$h = 2.5 + \frac{2.5x}{L}$$

Include gravity effect in source term:

$$C_i = -gh_i S_o + (c_f/2)| V_i | V_i P_i$$

*****WATER SURFACE PROFILE OF BASIN OPENING ONTO RESERVOIR WITH SINUSOIDALLY VARYING WATER SURFACE LEVEL**

I=	1	21	41	61	81	101
TIME=	20.0					
H(I)	3.92	4.35	4.65	4.80	5.12	5.45
V(I)	0.00	0.35	0.70	1.13	1.22	1.22
TIME=	40.0					
H(I)	3.34	3.76	4.06	4.29	4.47	4.62
V(I)	0.00	-0.07	-0.02	0.06	0.16	0.23
TIME=	60.1					
H(I)	2.86	3.18	3.62	4.05	4.46	4.86
V(I)	0.00	-0.91	-1.25	-1.39	-1.44	-1.44
TIME=	80.1					
H(I)	2.78	3.33	3.91	4.52	5.05	5.49
V(I)	0.00	0.06	0.15	0.32	0.41	0.41

16.33

$$R = \frac{A}{P} = \frac{(5)(10)}{2x5+10} = \frac{50}{20} = 2.5 \text{ m}$$

$$C = \frac{2.5^{1/6}}{0.04} = 29.1$$

$$C = (8g/f)^{1/2}, \quad f = \frac{8g}{C^2} = \frac{8 \cdot 9.81}{29.1^2} = .0926$$

$$c_f = 0.023$$

***WATER SURFACE PROFILE OF BASIN OPENING ONTO RESERVOIR WITH SINUSOIDALLY VARYING WATER SURFACE LEVEL

I=	1	21	41	61	81	101
TIME= 20.0						
H(I)	5.99	5.96	5.85	5.74	5.60	5.45
V(I)	0.00	-0.01	0.06	0.12	0.18	0.22
TIME= 40.0						
H(I)	4.49	4.49	4.49	4.49	4.54	4.62
V(I)	0.00	-0.01	-0.01	-0.03	0.03	0.10
TIME= 60.1						
H(I)	4.64	4.64	4.66	4.69	4.76	4.86
V(I)	0.00	-0.09	-0.18	-0.25	-0.35	-0.47
TIME= 80.1						
H(I)	5.81	5.78	5.73	5.65	5.57	5.49
V(I)	0.00	0.19	0.28	0.36	0.41	0.44

TRANSPARENCY MASTERS

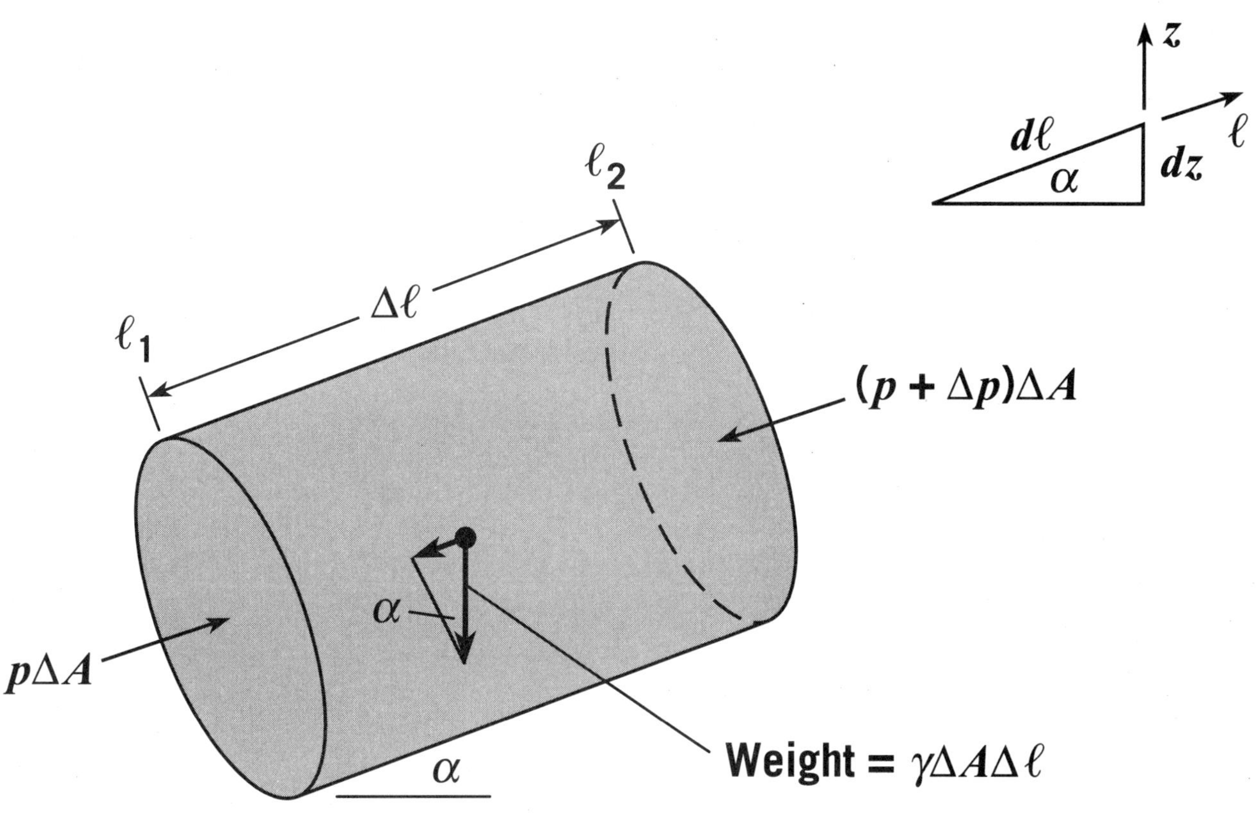

 Roberson/FLUID 6E

FIGURE 3.4 1

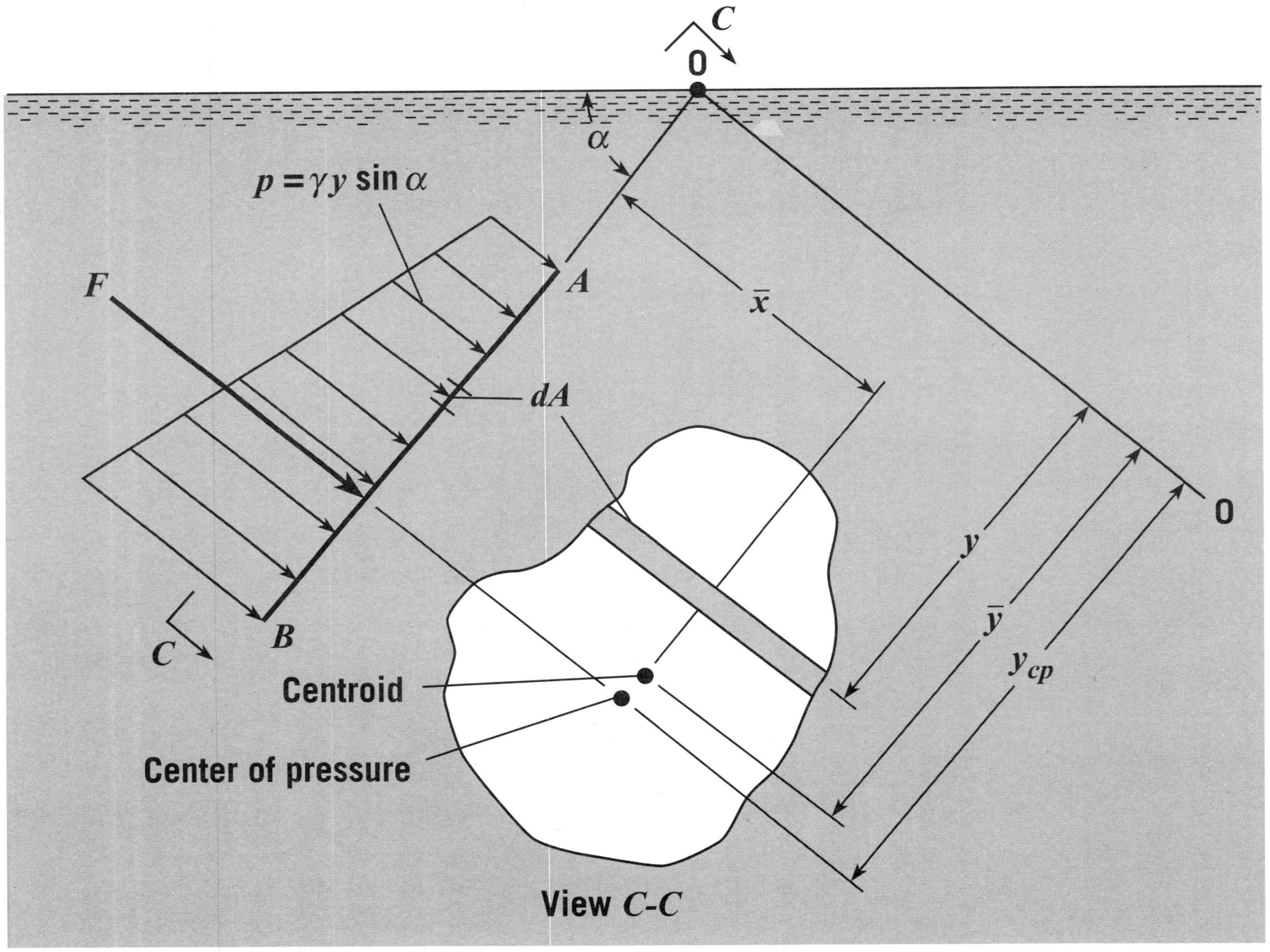

C
0
α
$p = \gamma y \sin \alpha$
F
A
$\bar{x}$
dA
0
y
$\bar{y}$
y_{cp}
B
C
Centroid
Center of pressure
View C-C

FIGURE 3.10 2

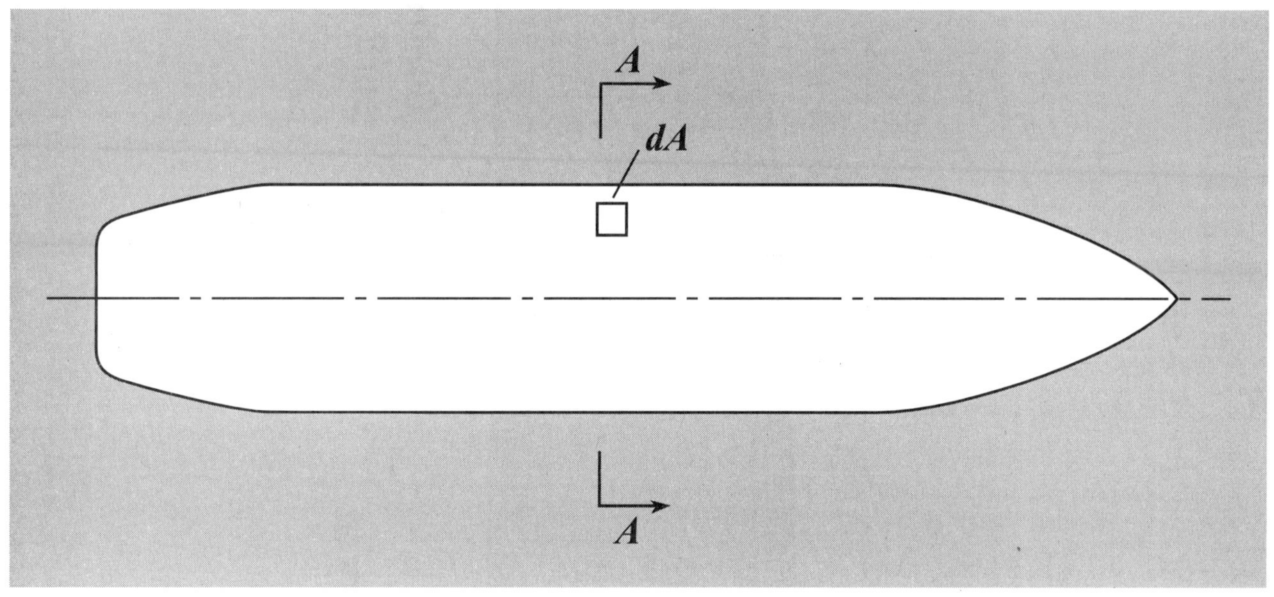

(*a*)

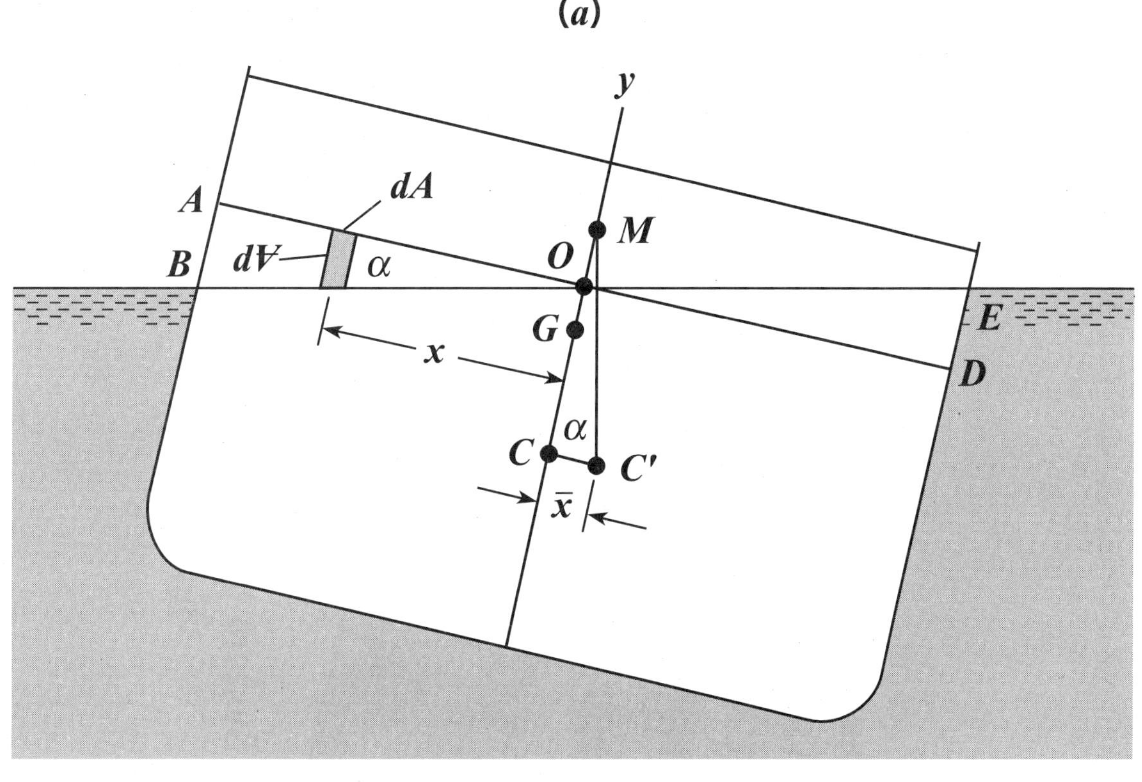

(*b*)

FIGURE 3.17 3

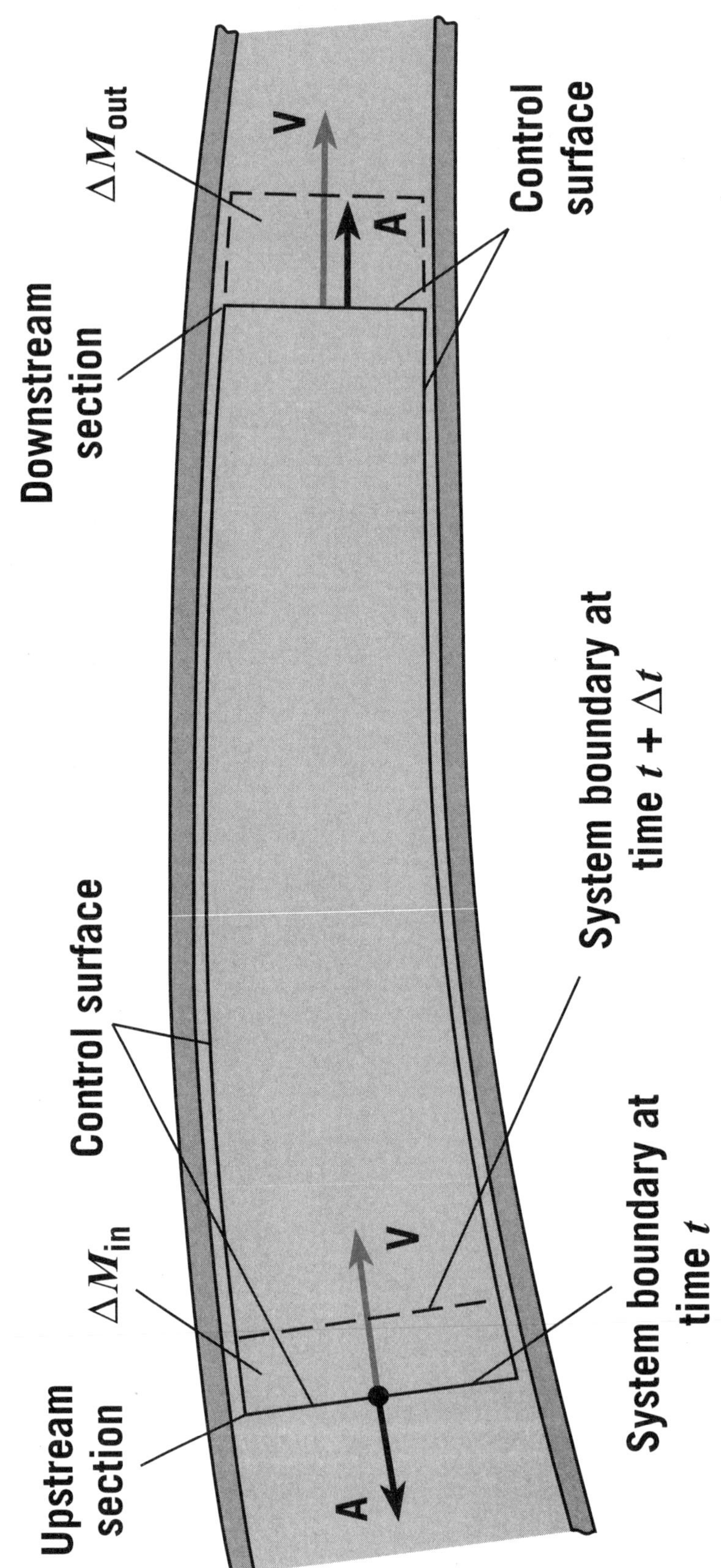

Downstream section
ΔM_{out}
V
A
Control surface
System boundary at time $t + \Delta t$
Control surface
ΔM_{in}
V
Upstream section
A
System boundary at time t

FIGURE 4.13

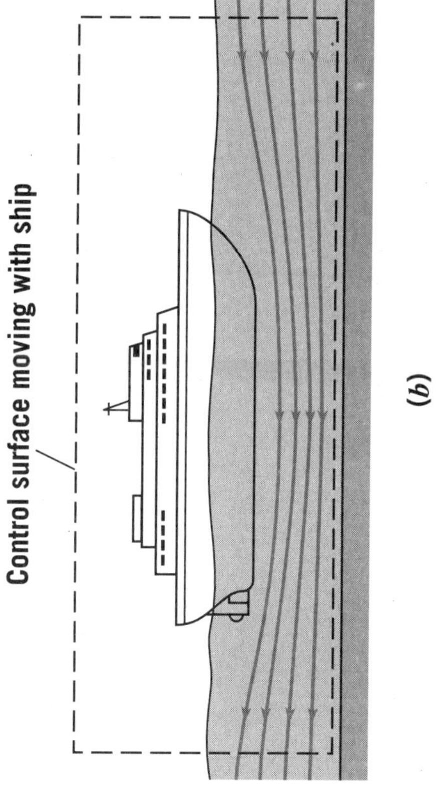

(b)

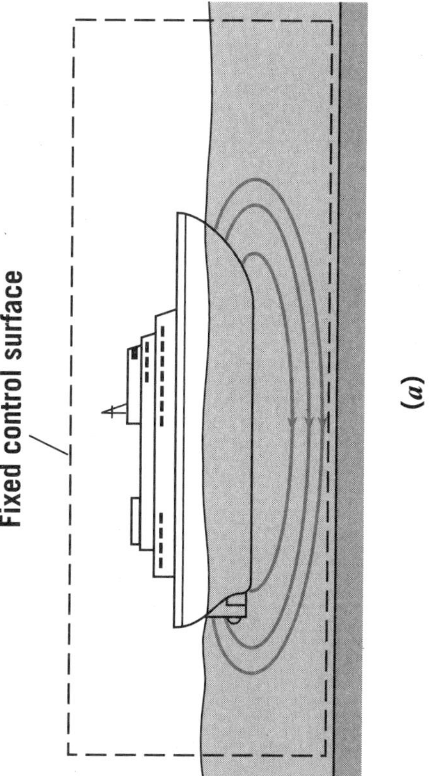

(a)

FIGURE 4.14

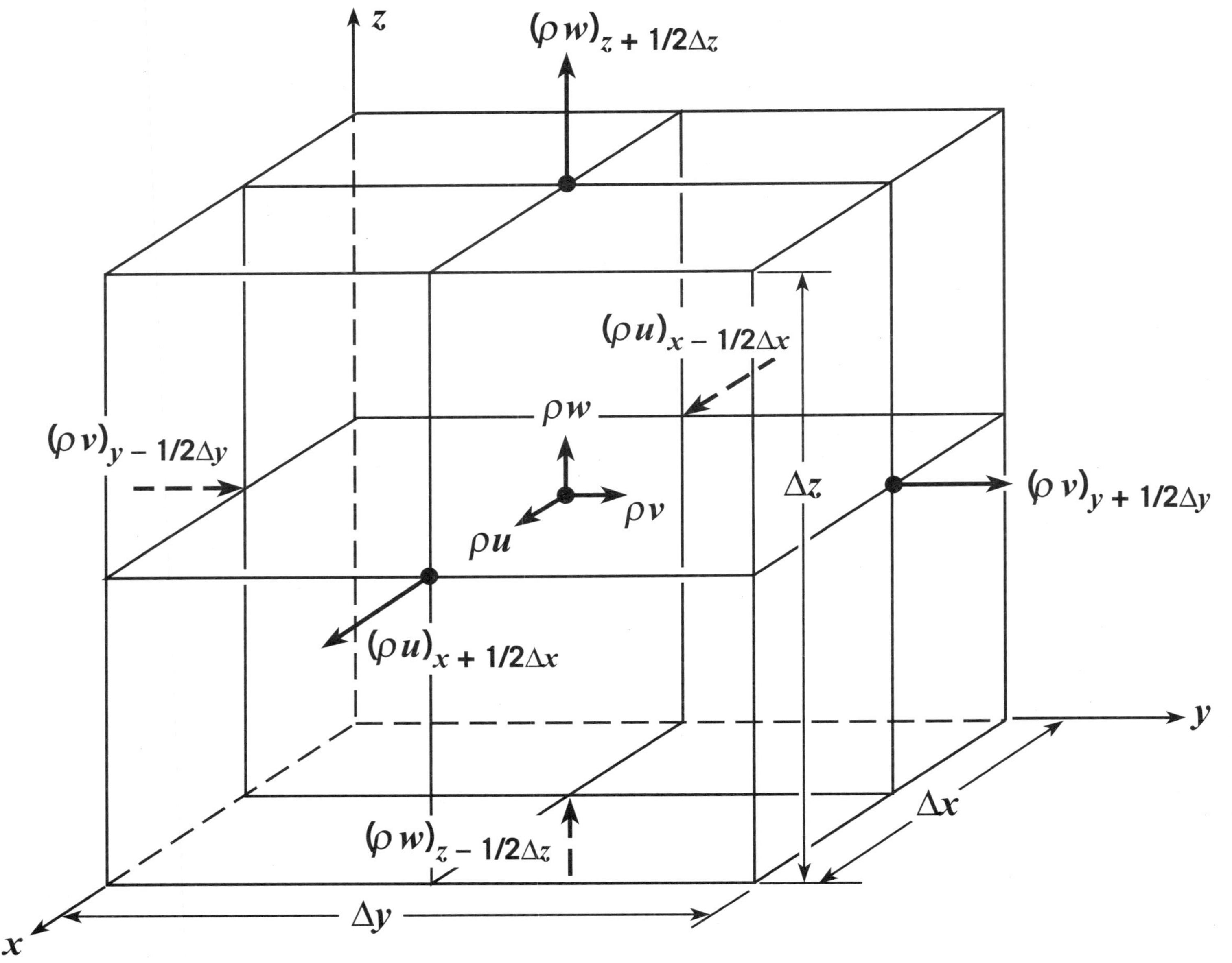

z
$(\rho w)_{z + 1/2\Delta z}$
$(\rho u)_{x - 1/2\Delta x}$
ρw
$(\rho v)_{y - 1/2\Delta y}$
Δz
$(\rho v)_{y + 1/2\Delta y}$
ρv
ρu
$(\rho u)_{x + 1/2\Delta x}$
y
Δx
$(\rho w)_{z - 1/2\Delta z}$
Δy
x

FIGURE 4.16

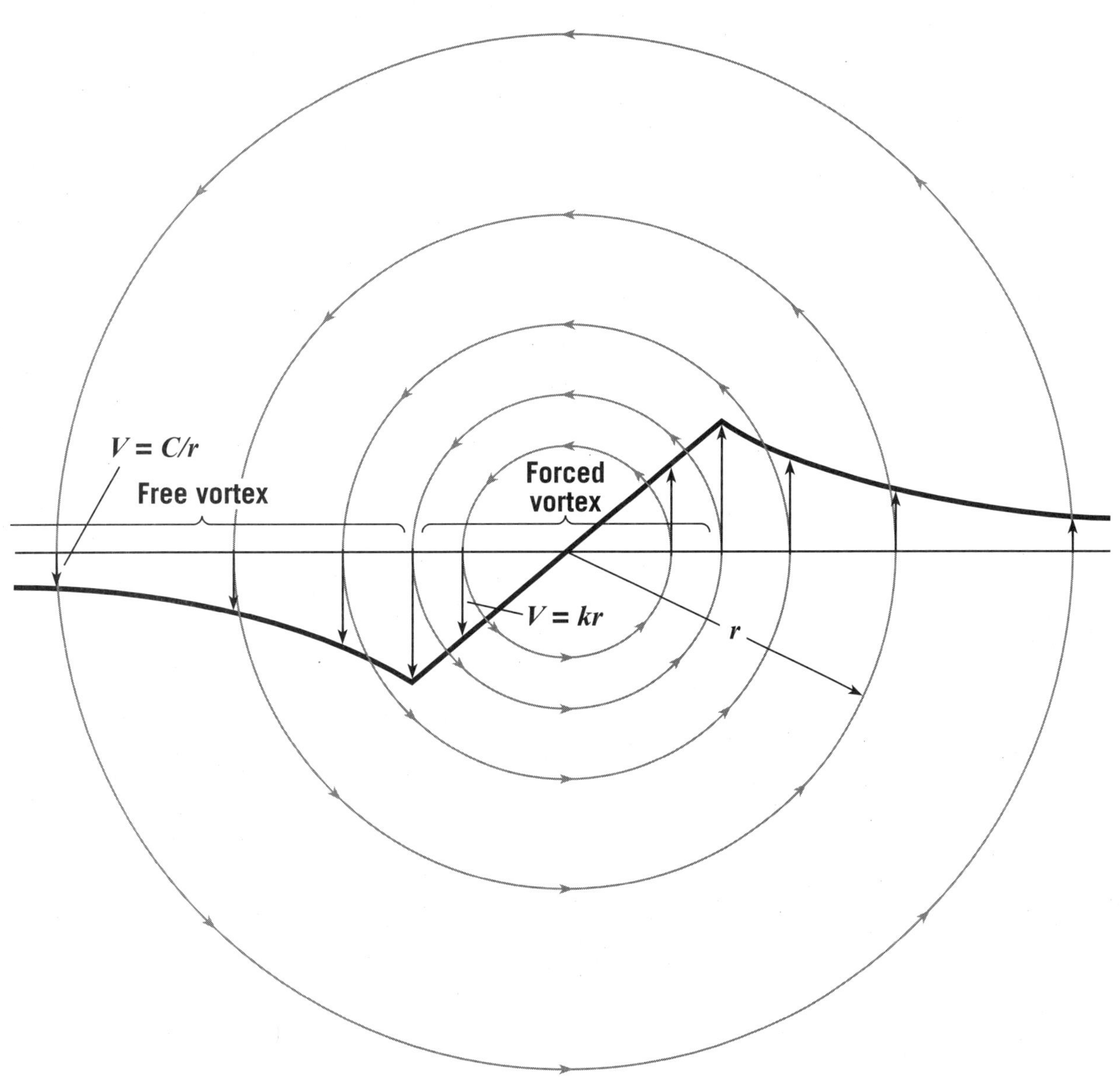

FIGURE 4.20

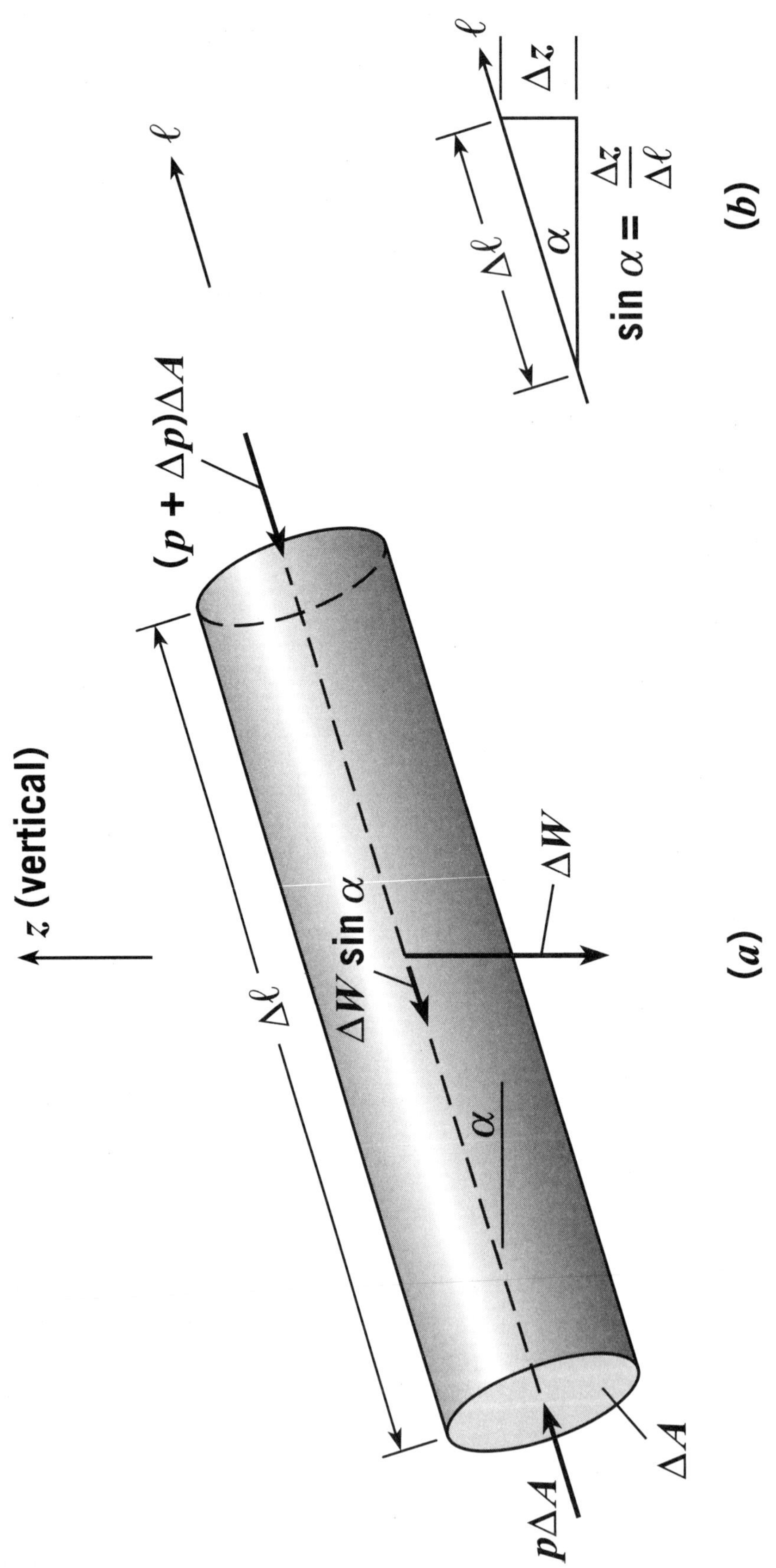

FIGURE 5.2

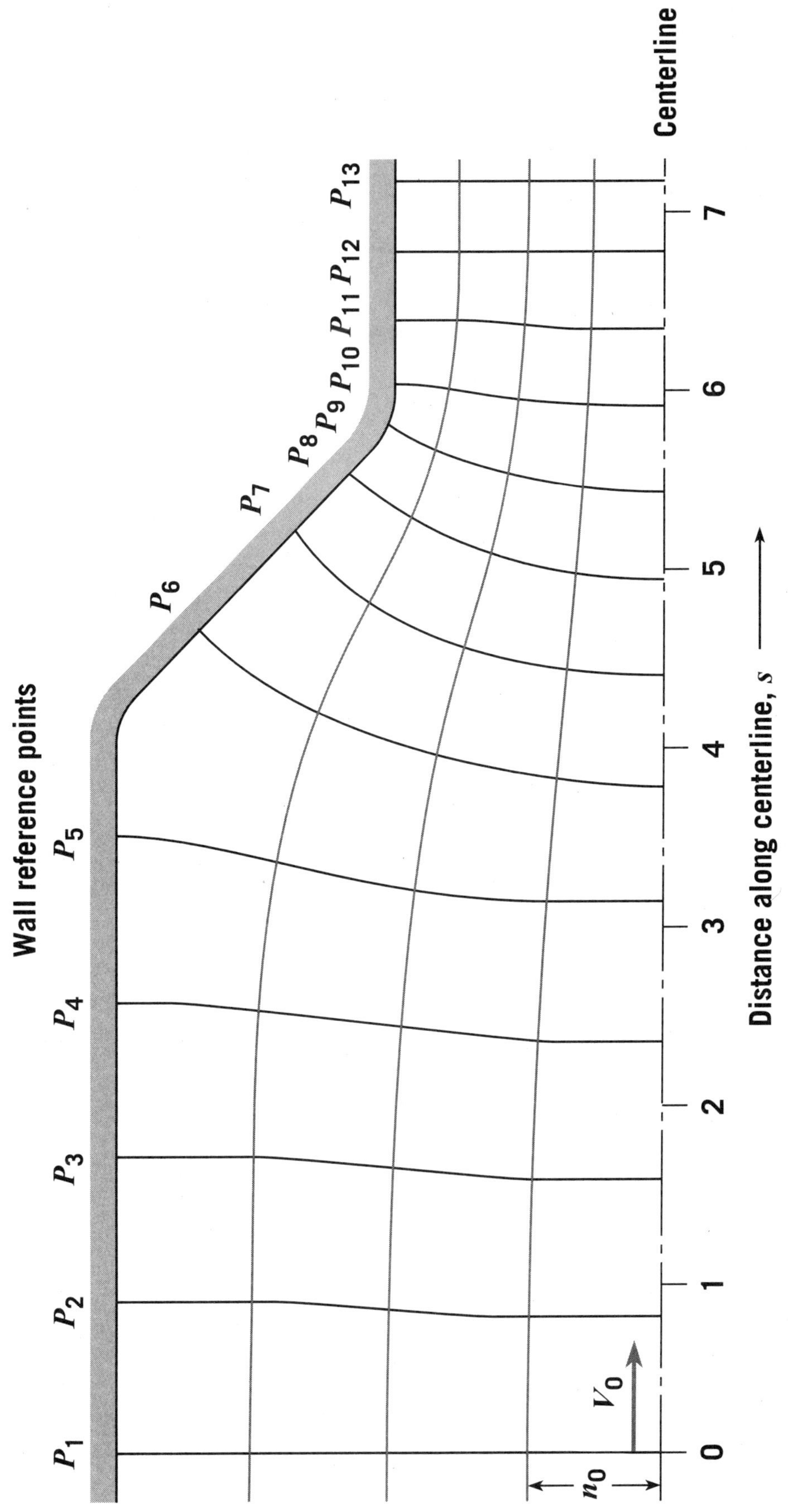
Wall reference points
P_1
P_2
P_3
P_4
P_5
P_6
P_7
P_8
P_9
P_{10}
P_{11}
P_{12}
P_{13}
Centerline
V_0
n_0
0
1
2
3
4
5
6
7
Distance along centerline, s

FIGURE 5.7

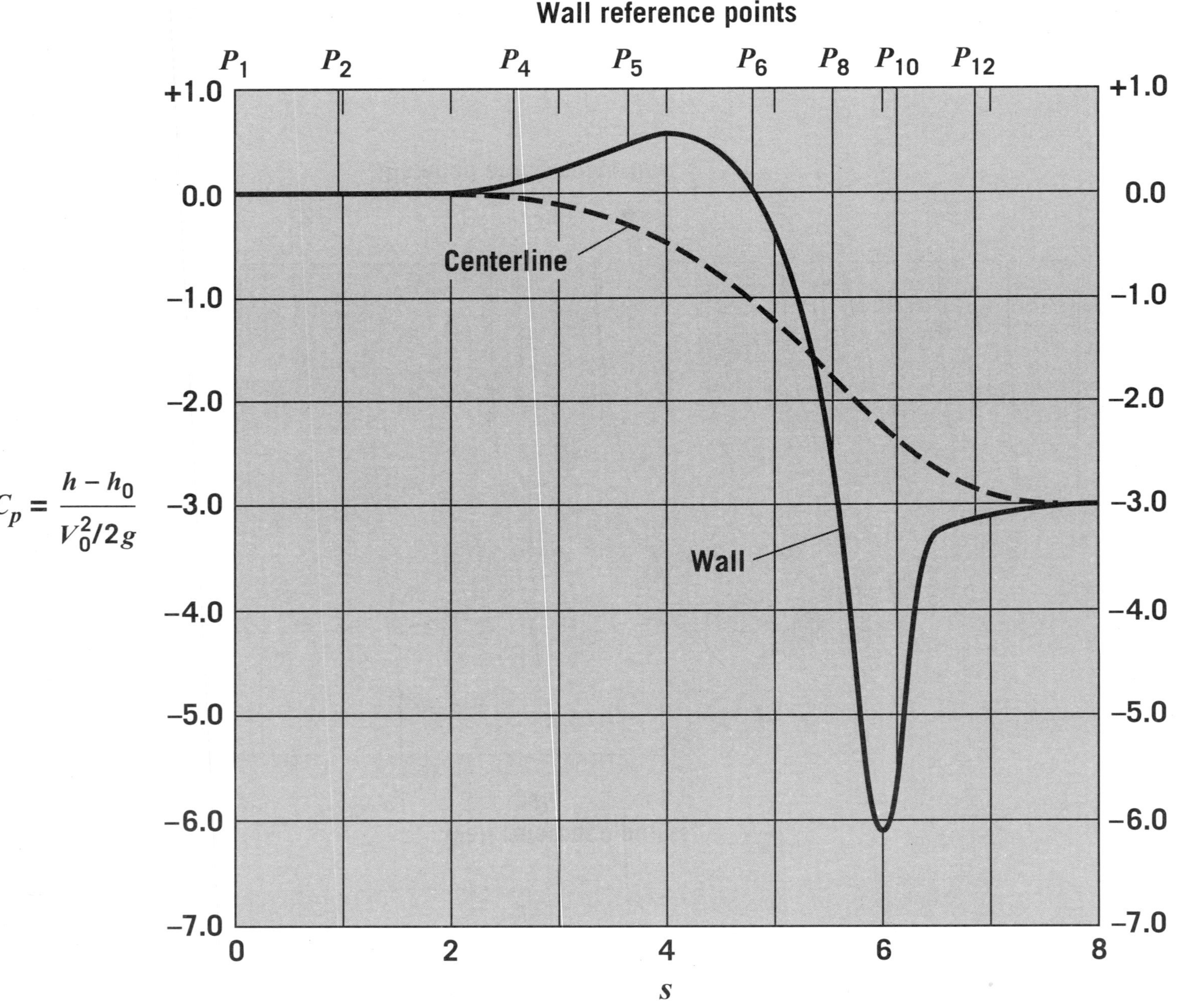
Wall reference points
P_1
P_2
P_4
P_5
P_6
P_8
P_{10}
P_{12}
+1.0
0.0
−1.0
−2.0
−3.0
−4.0
−5.0
−6.0
−7.0
0
2
4
6
8
s
Centerline
Wall
$C_p = \frac{h - h_0}{V_0^2/2g}$

FIGURE 5.8

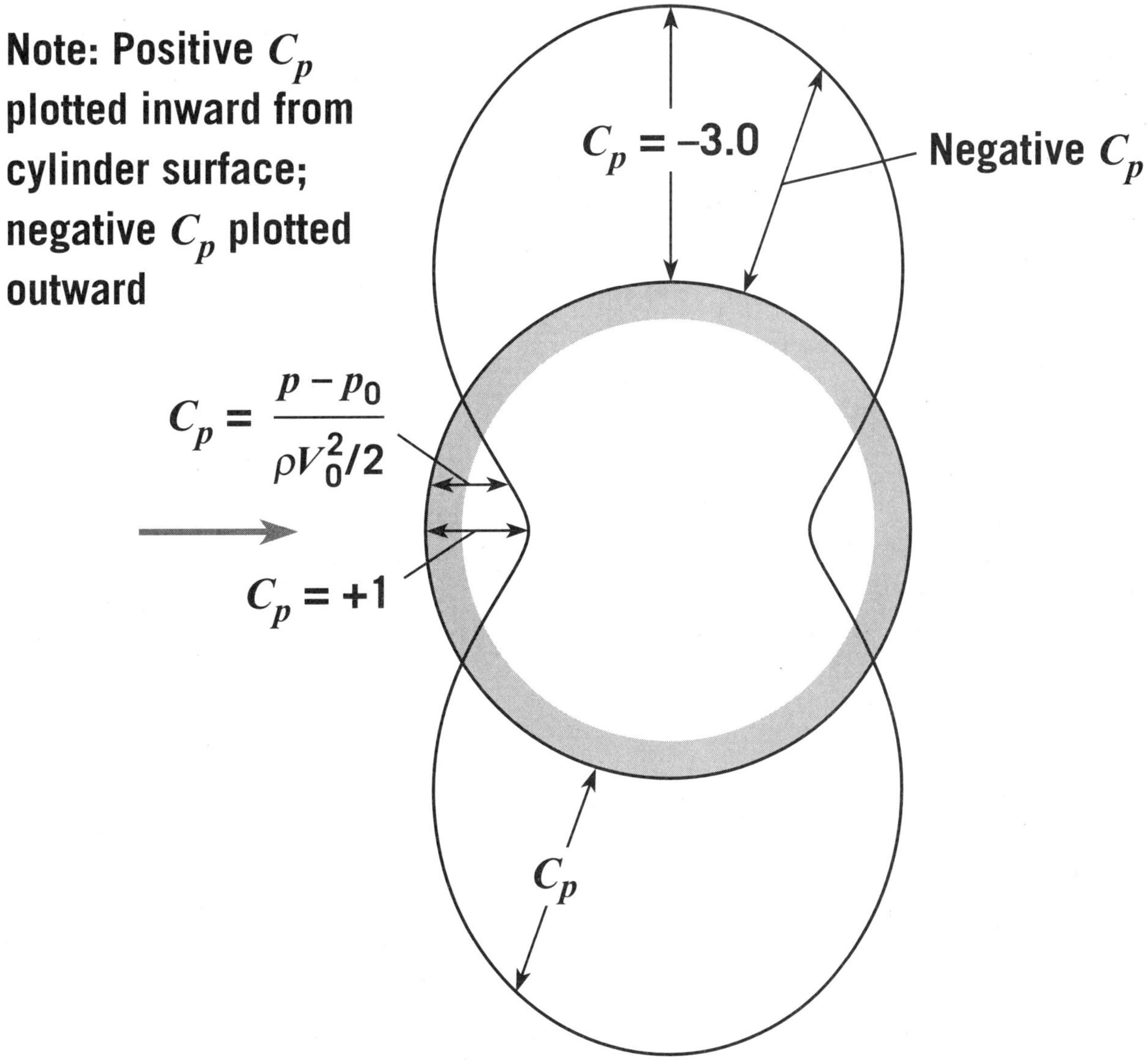

Note: Positive C_p plotted inward from cylinder surface; negative C_p plotted outward

$C_p = -3.0$

Negative C_p

$$C_p = \frac{p - p_0}{\rho V_0^2/2}$$

$C_p = +1$

C_p

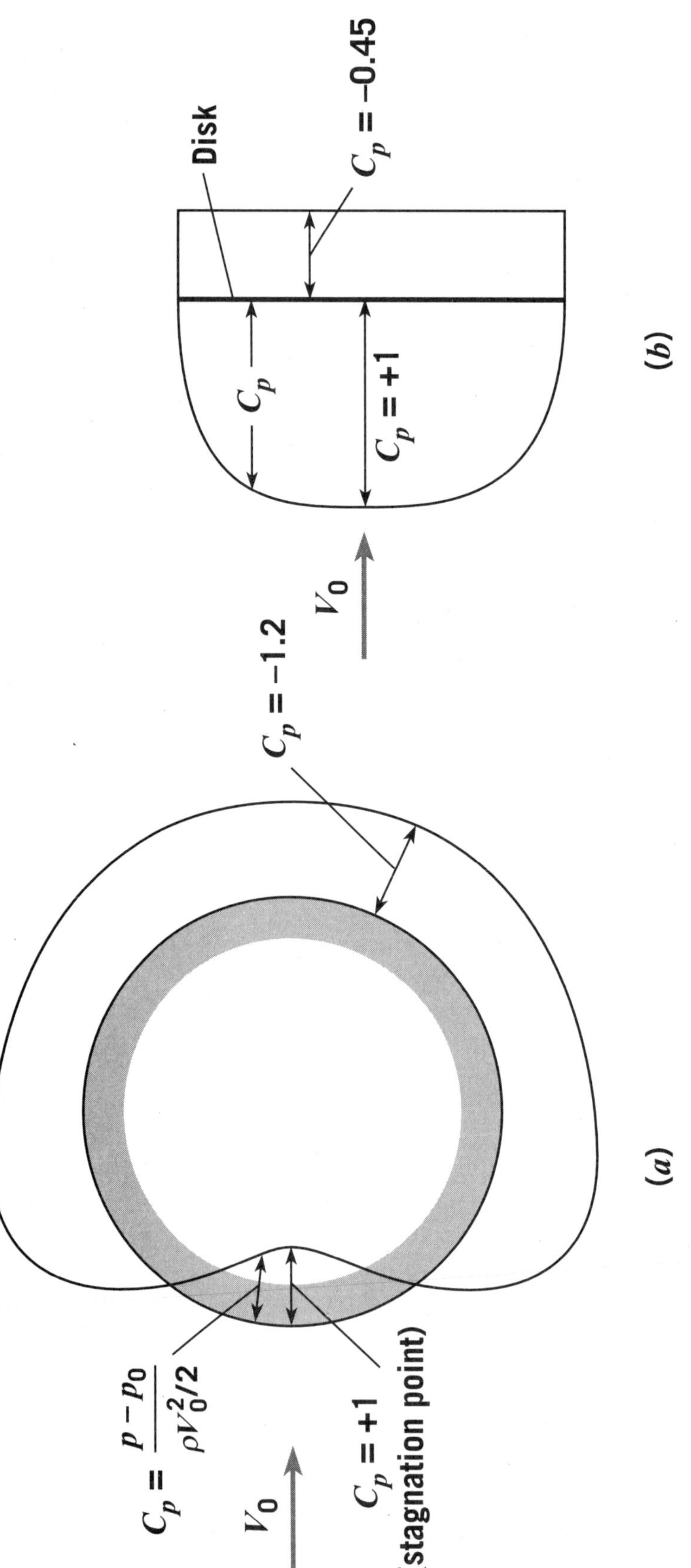

Disk
$C_p = -0.45$
C_p
$C_p = +1$
(b)
V_0
$C_p = -1.2$
(a)
$C_p = \frac{p - p_0}{\rho V_0^2/2}$
V_0
$C_p = +1$
(stagnation point)

FIGURE 5.13

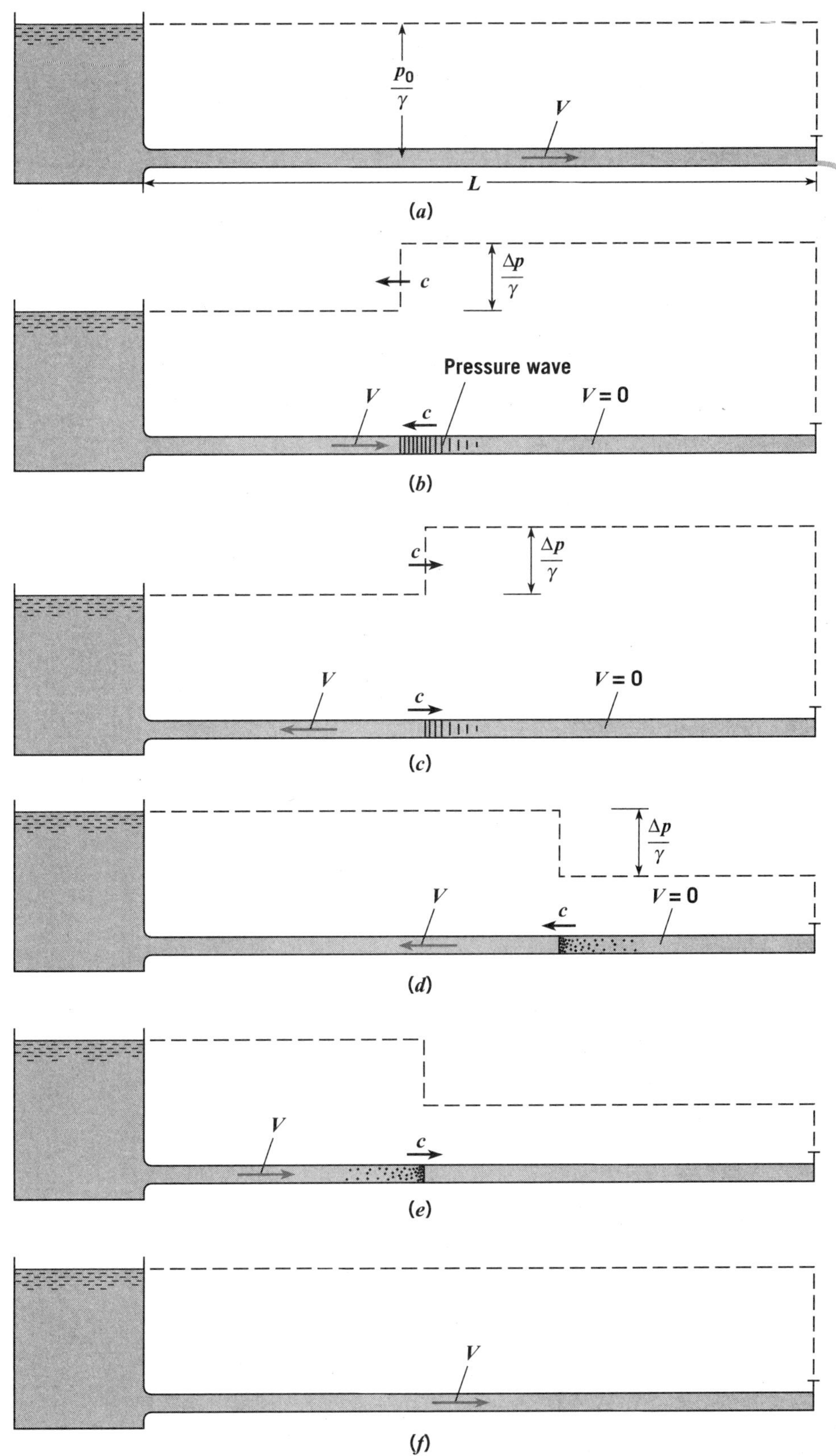

 Roberson/FLUID 6E

FIGURE 6.9

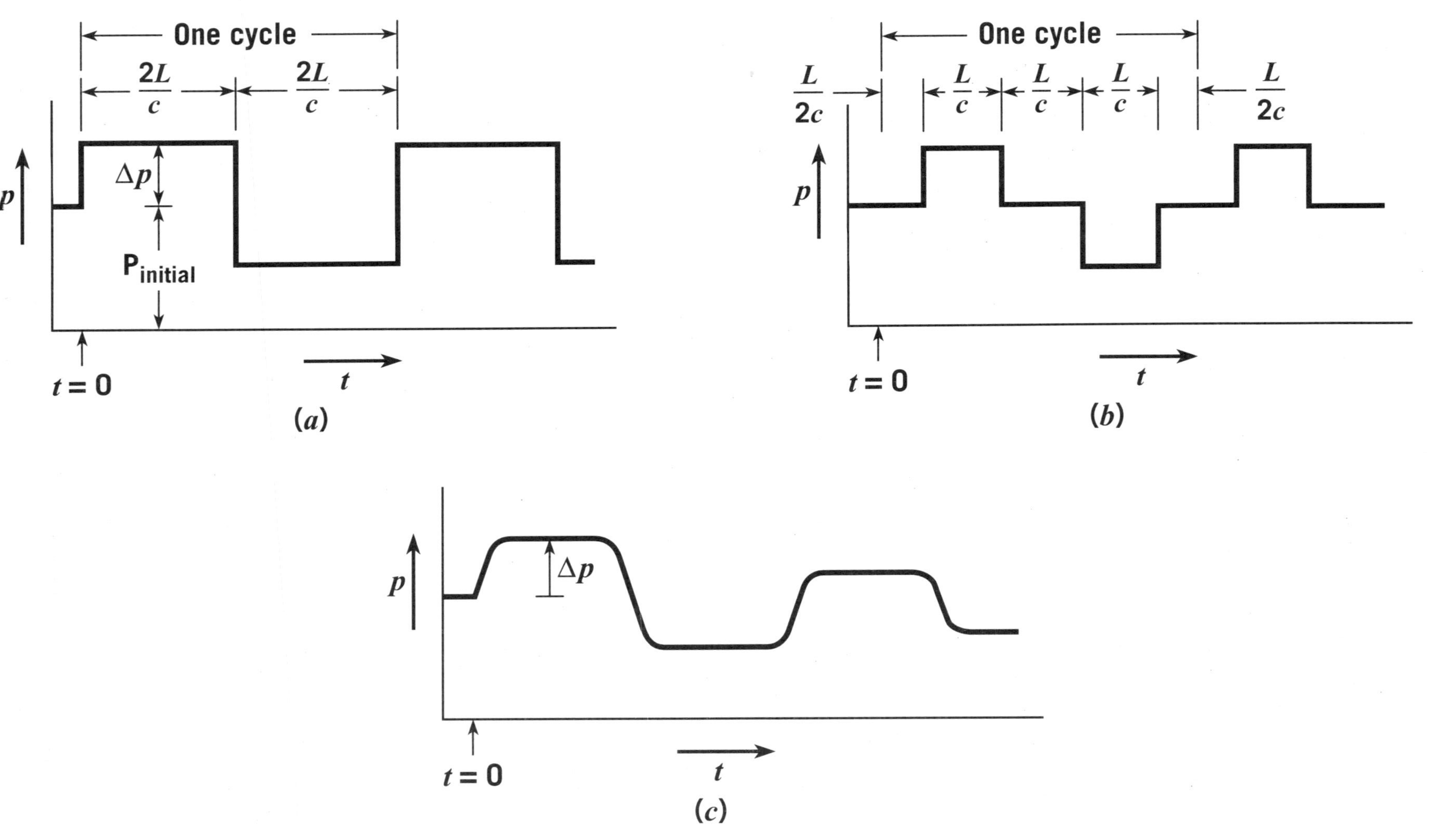
One cycle
2L/c
2L/c
p
Δp
P initial
t = 0
t
(a)
One cycle
L/2c
L/c
L/c
L/c
L/2c
p
t = 0
t
(b)
p
Δp
t = 0
t
(c)

FIGURE 6.10

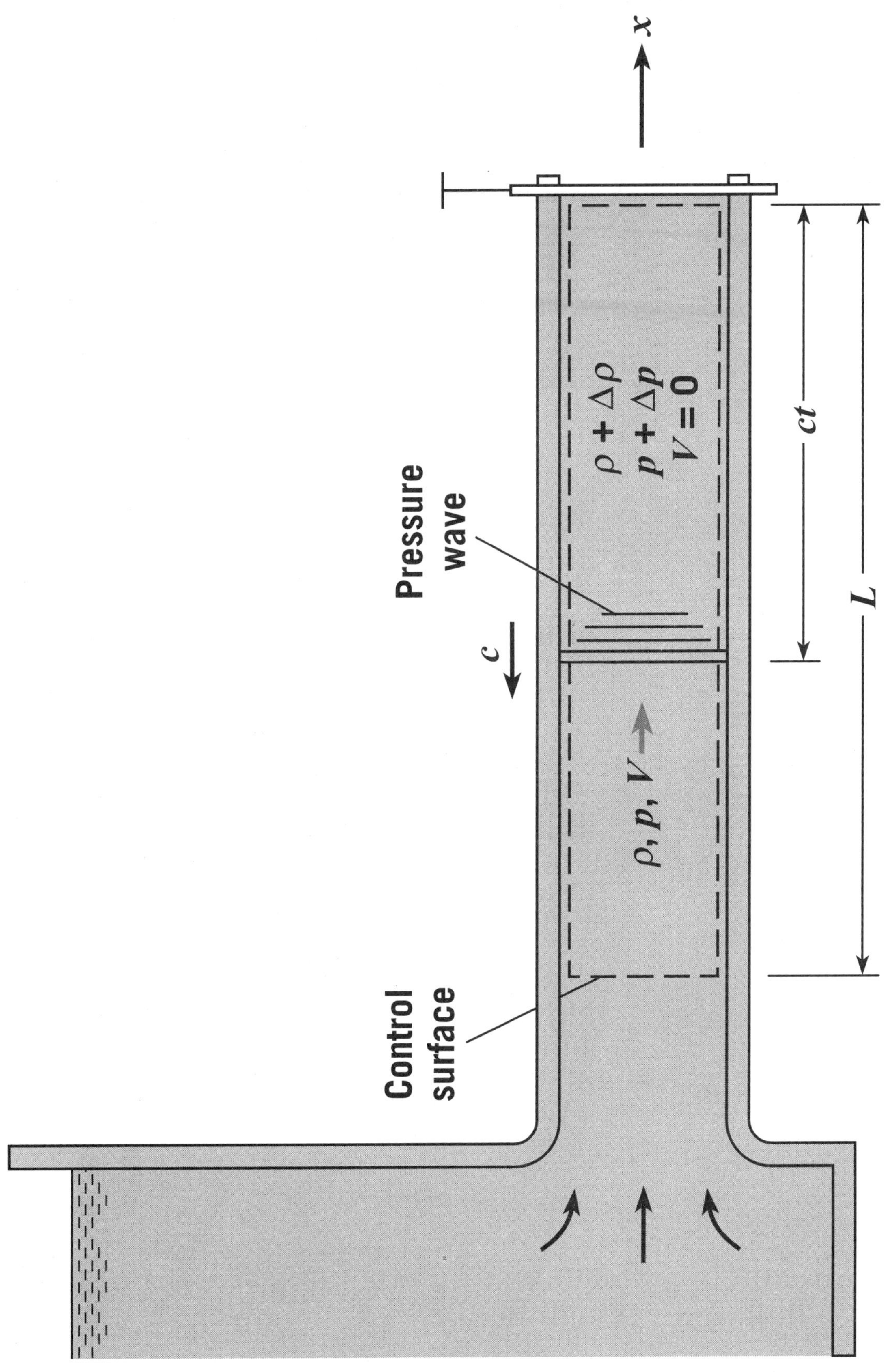

x
Pressure wave
c
ρ + Δρ
p + Δp
V = 0
ρ, p, V
ct
L
Control surface

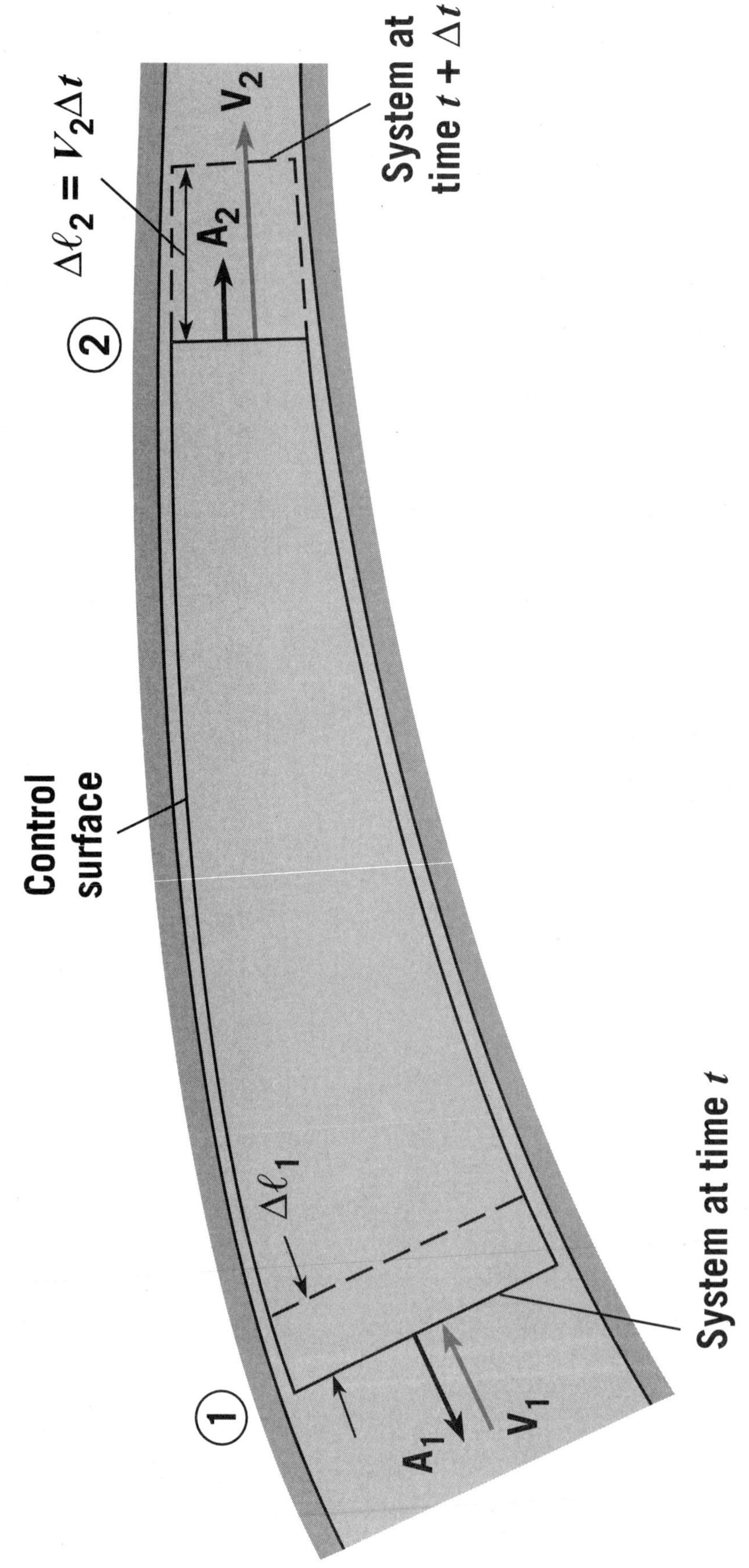

Control surface
System at time t
System at time t + Δt
Δℓ2 = V2Δt
1
2
A1
V1
A2
V2
Δℓ1

FIGURE 7.1

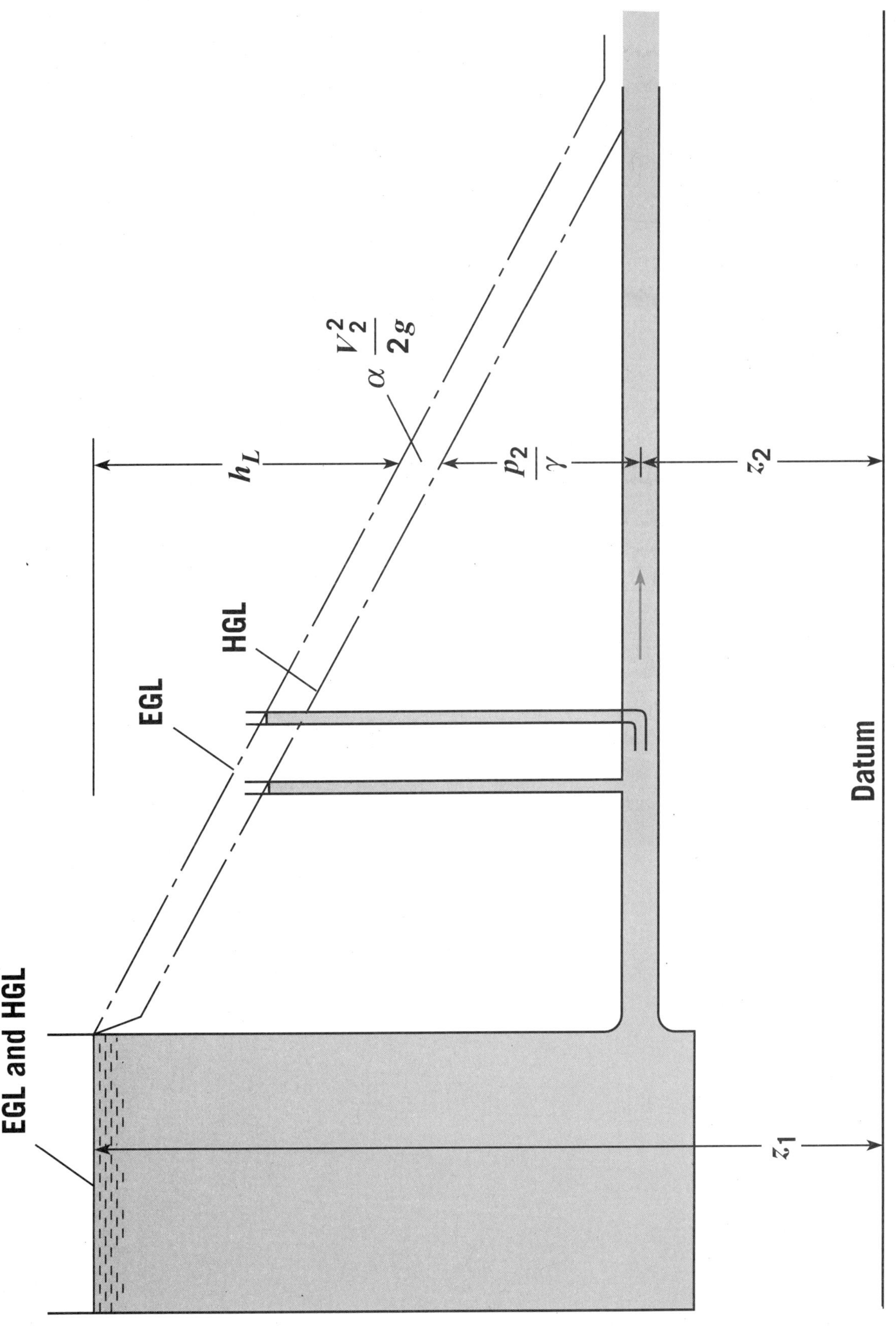

 Roberson/FLUID 6E

FIGURE 7.4

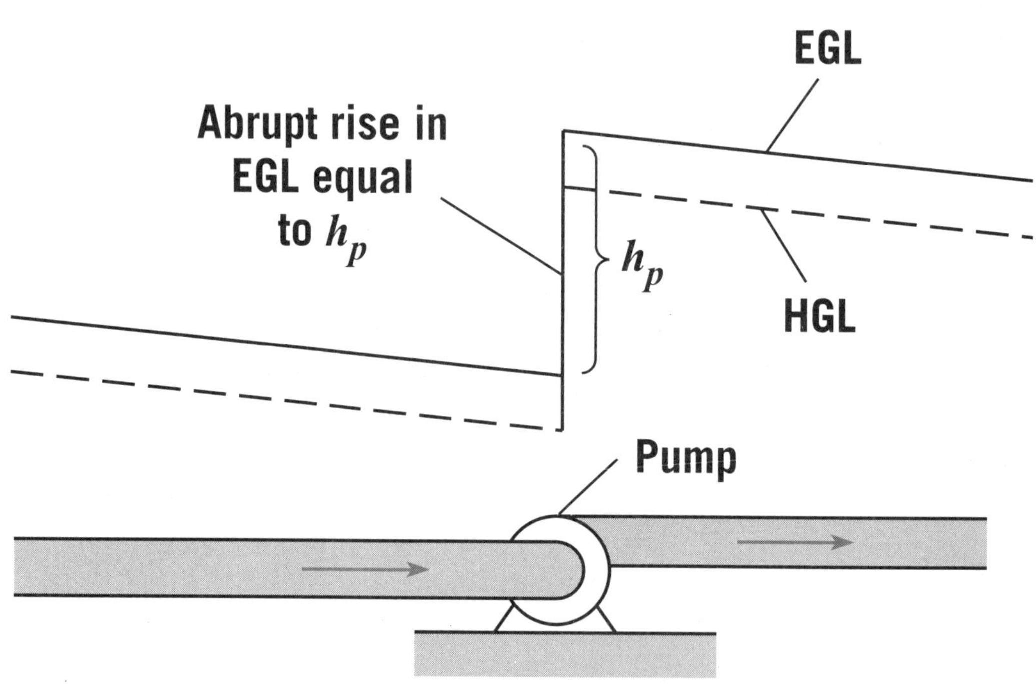

FIGURE 7.5

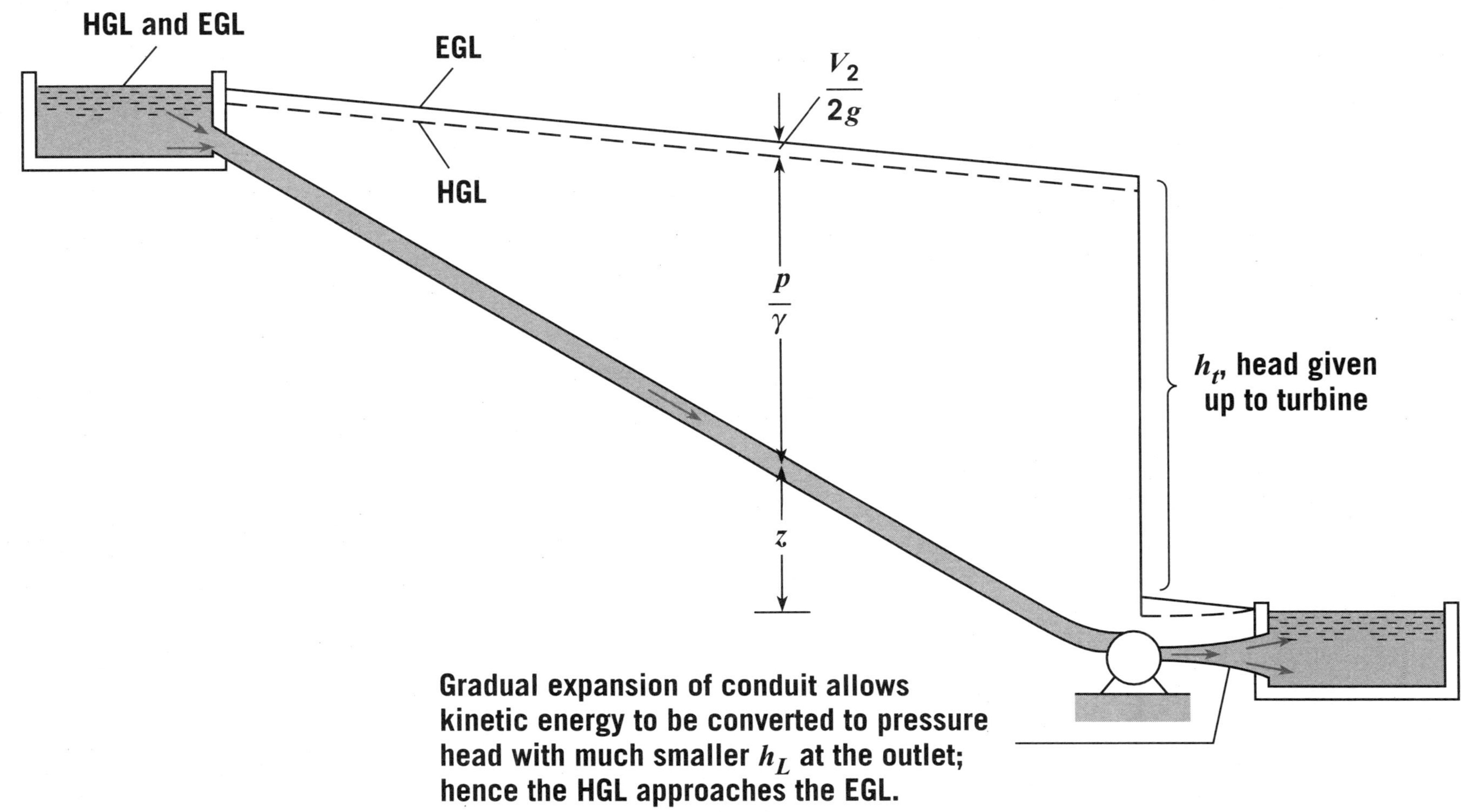

HGL and EGL
EGL
HGL
$\frac{V_2}{2g}$
$\frac{p}{\gamma}$
z
h_t, head given up to turbine
Gradual expansion of conduit allows kinetic energy to be converted to pressure head with much smaller h_L at the outlet; hence the HGL approaches the EGL.

 Roberson/FLUID 6E

FIGURE 7.6

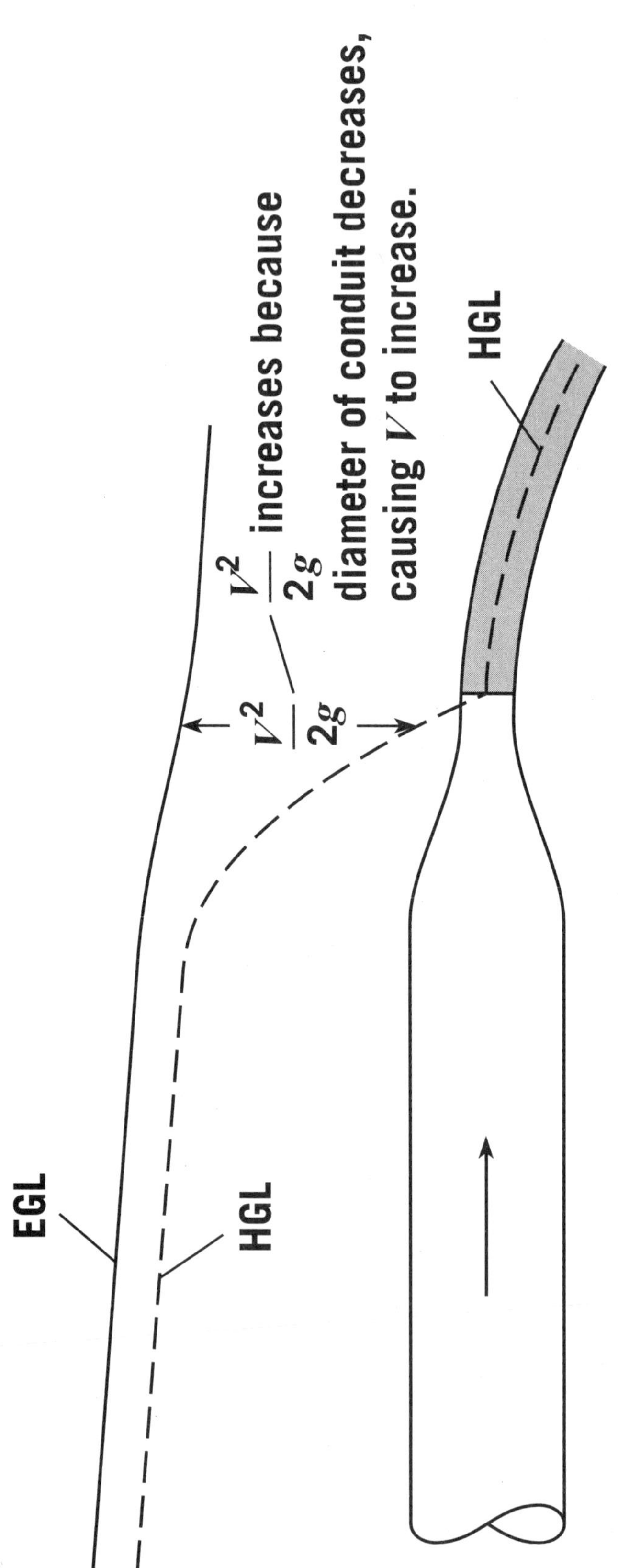

$\frac{V^2}{2g}$ increases because diameter of conduit decreases, causing V to increase.
$\frac{V^2}{2g}$
HGL
EGL
HGL

 Roberson/FLUID 6E

FIGURE 7.7

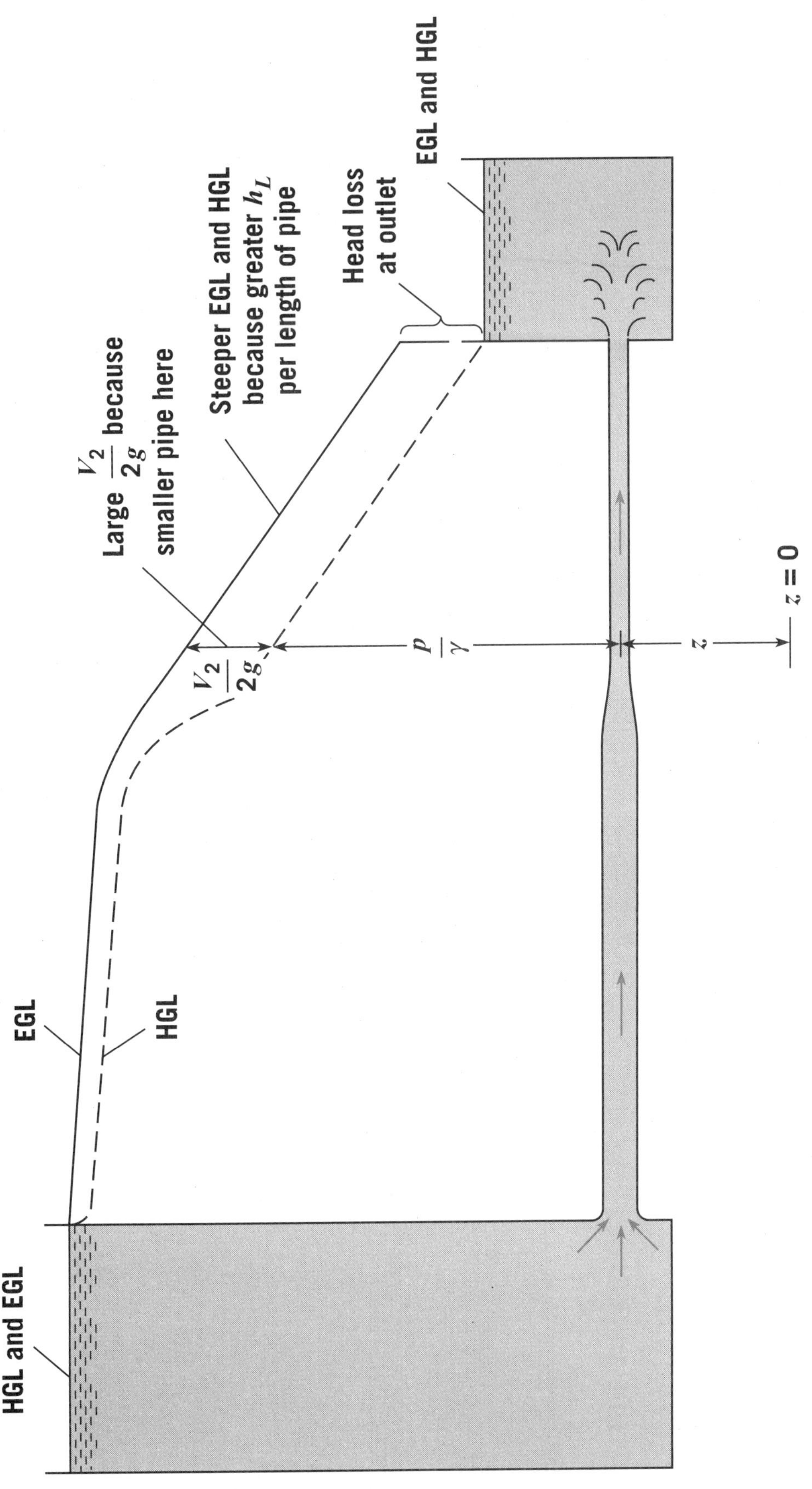

HGL and EGL
EGL
HGL
Large $\frac{V_2}{2g}$ because smaller pipe here
Steeper EGL and HGL because greater h_L per length of pipe
Head loss at outlet
EGL and HGL
$\frac{V_2}{2g}$
$\frac{p}{\gamma}$
z
$z = 0$

FIGURE 7.8

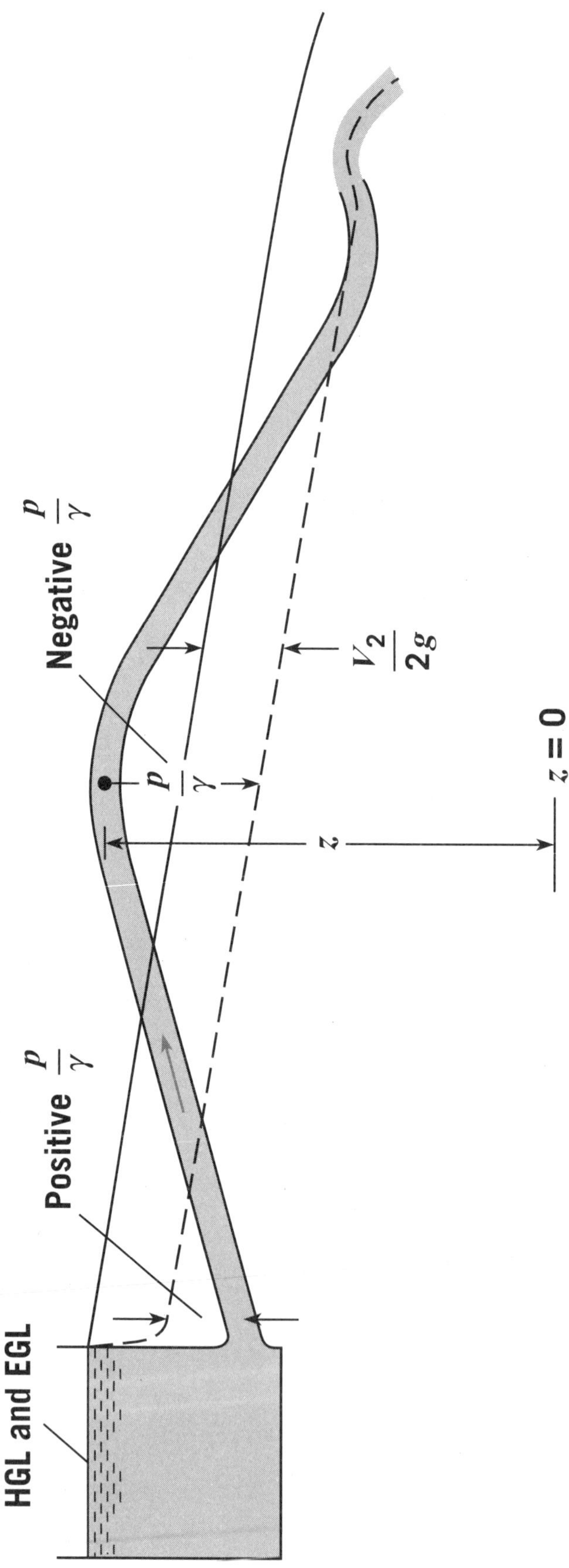
HGL and EGL
Positive $\frac{p}{\gamma}$
Negative $\frac{p}{\gamma}$
$\frac{p}{\gamma}$
$\frac{V^2}{2g}$
z
$z = 0$

FIGURE 7.9

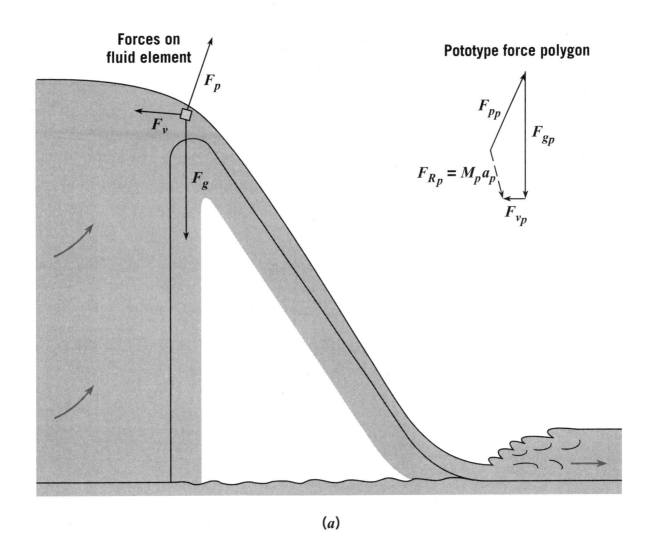

Forces on fluid element

F_p

F_v

F_g

Pototype force polygon

F_{p_p}

F_{g_p}

$F_{R_p} = M_p a_p$

F_{v_p}

(*a*)

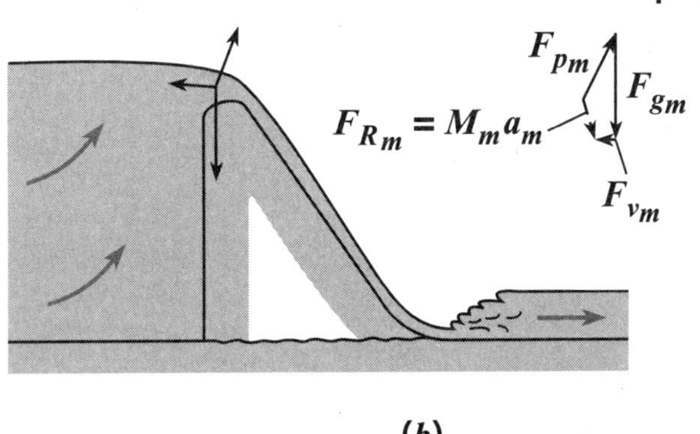

(*b*)

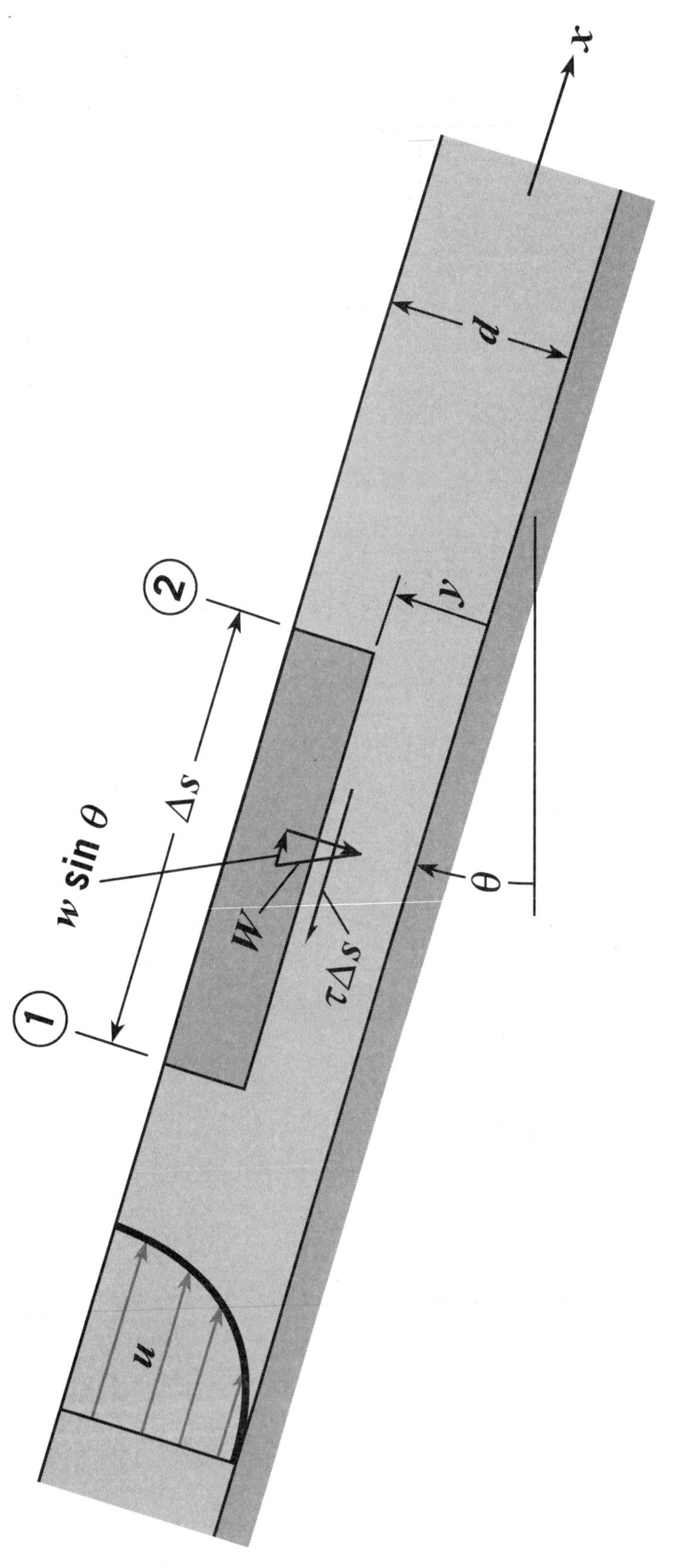

FIGURE 9.2

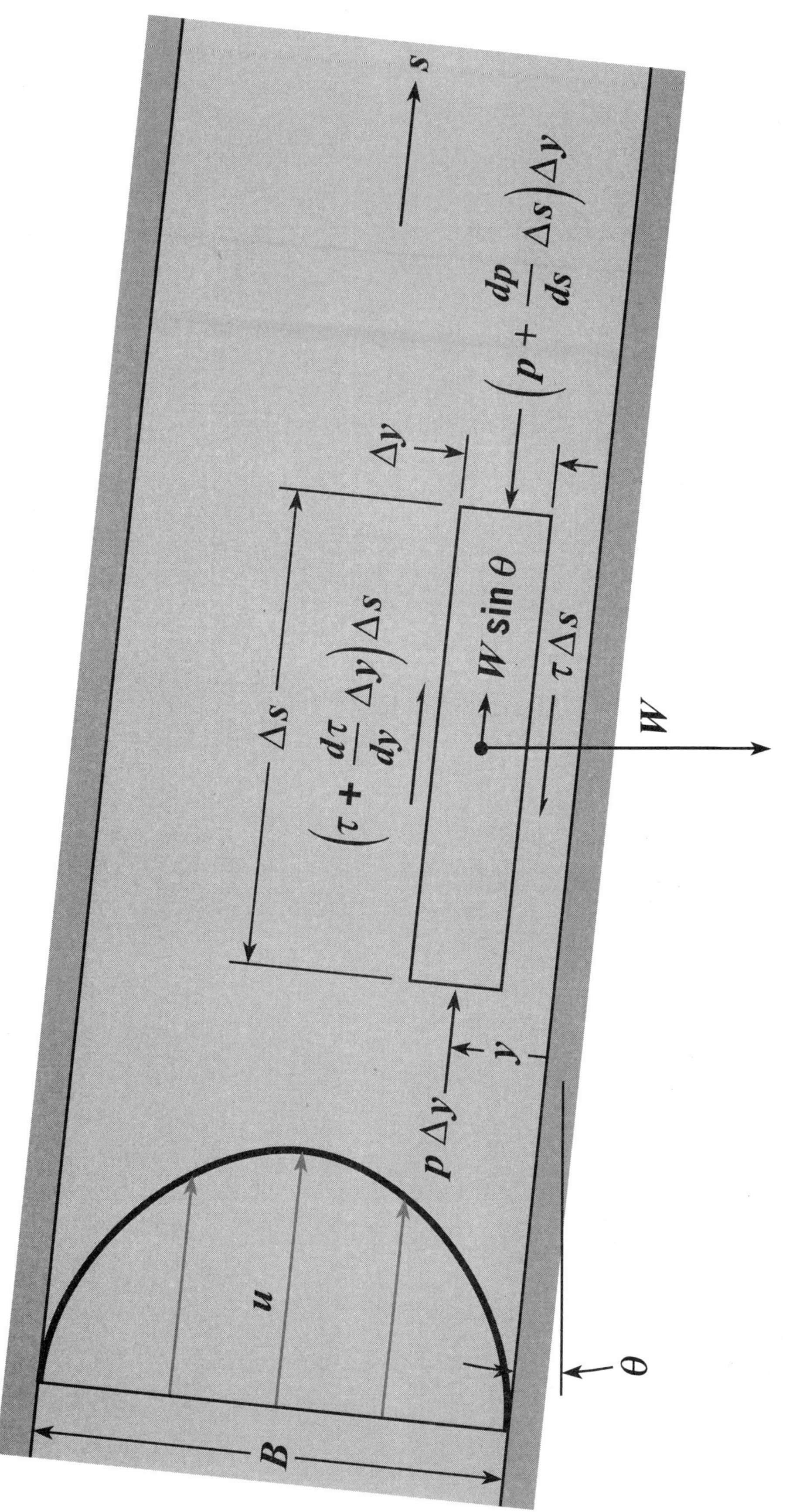

s
$\left(p + \frac{dp}{ds}\Delta s\right)\Delta y$
Δy
Δs
$\left(\tau + \frac{d\tau}{dy}\Delta y\right)\Delta s$
$W \sin\theta$
$\tau \Delta s$
W
$p\,\Delta y$
y
u
θ
B

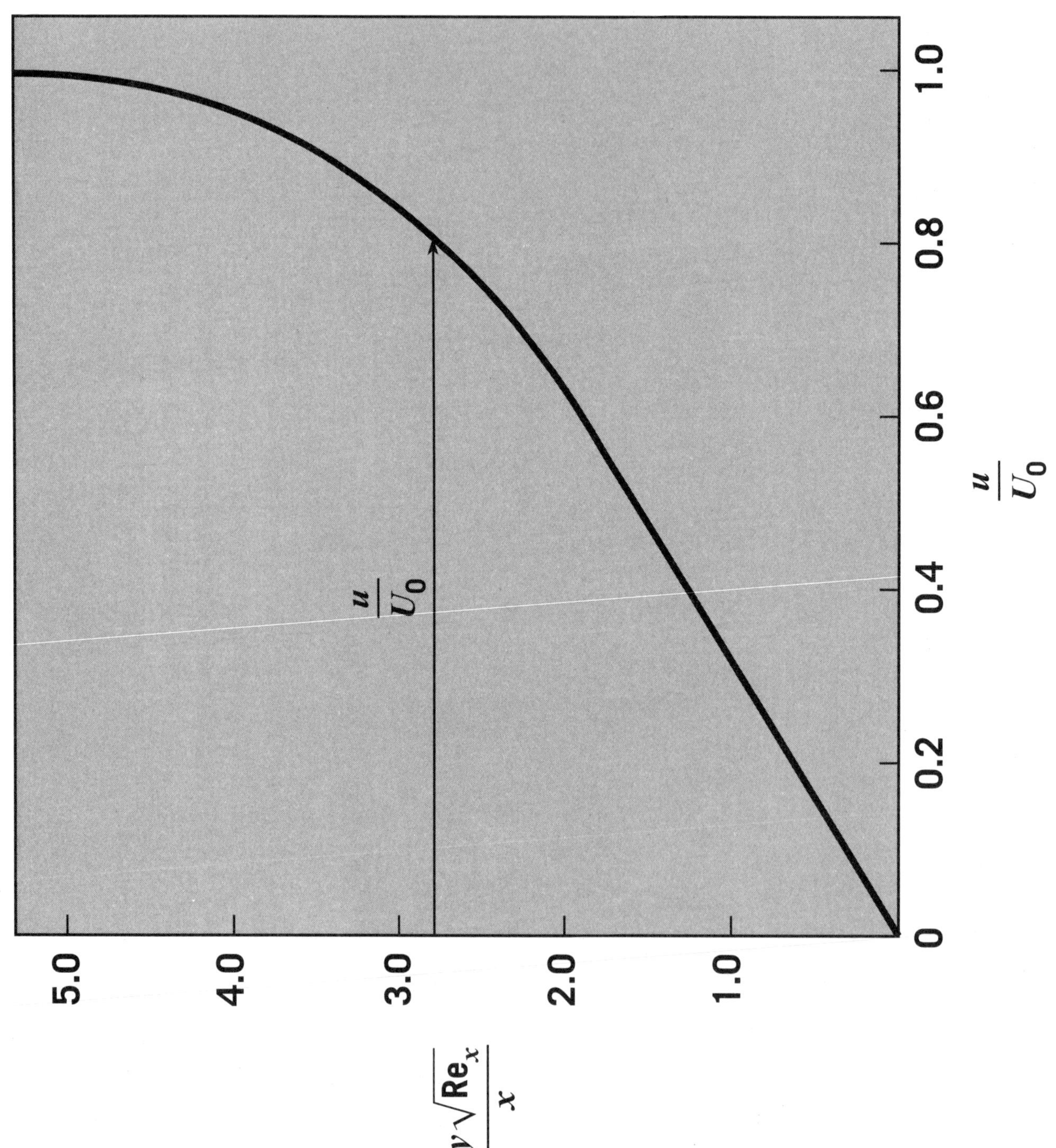
5.0
4.0
3.0
2.0
1.0
0
0.2
0.4
0.6
0.8
1.0
$\frac{u}{U_0}$
$\frac{u}{U_0}$
$\frac{y\sqrt{\mathrm{Re}_x}}{x}$

FIGURE 9.5

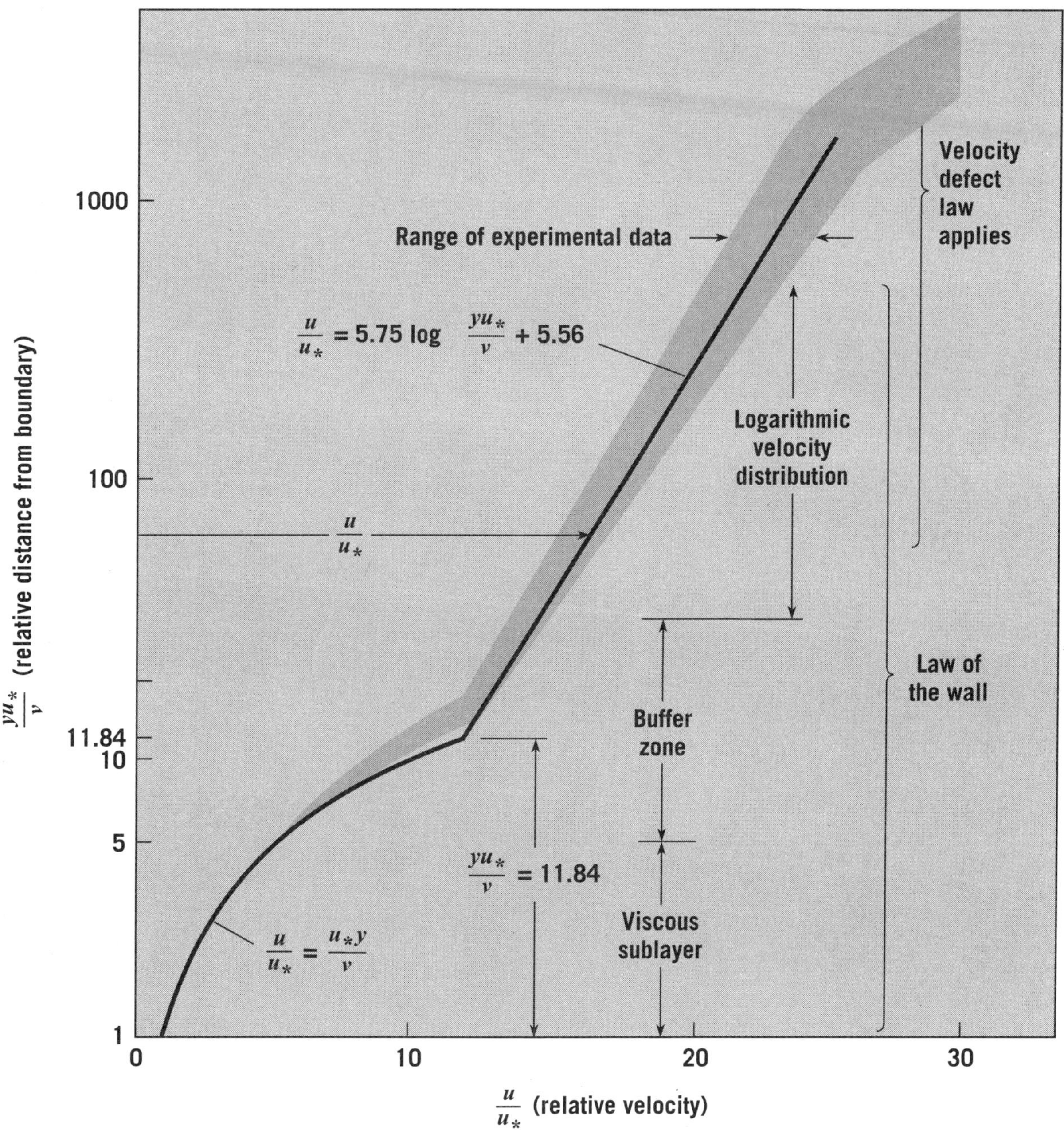
1000
100
11.84
10
5
1
0
10
20
30
$\frac{yu_*}{\nu}$ (relative distance from boundary)
$\frac{u}{u_*}$ (relative velocity)
Range of experimental data
$\frac{u}{u_*} = 5.75 \log \frac{yu_*}{\nu} + 5.56$
$\frac{u}{u_*}$
$\frac{u}{u_*} = \frac{u_* y}{\nu}$
$\frac{yu_*}{\nu} = 11.84$
Velocity defect law applies
Logarithmic velocity distribution
Law of the wall
Buffer zone
Viscous sublayer

FIGURE 9.9

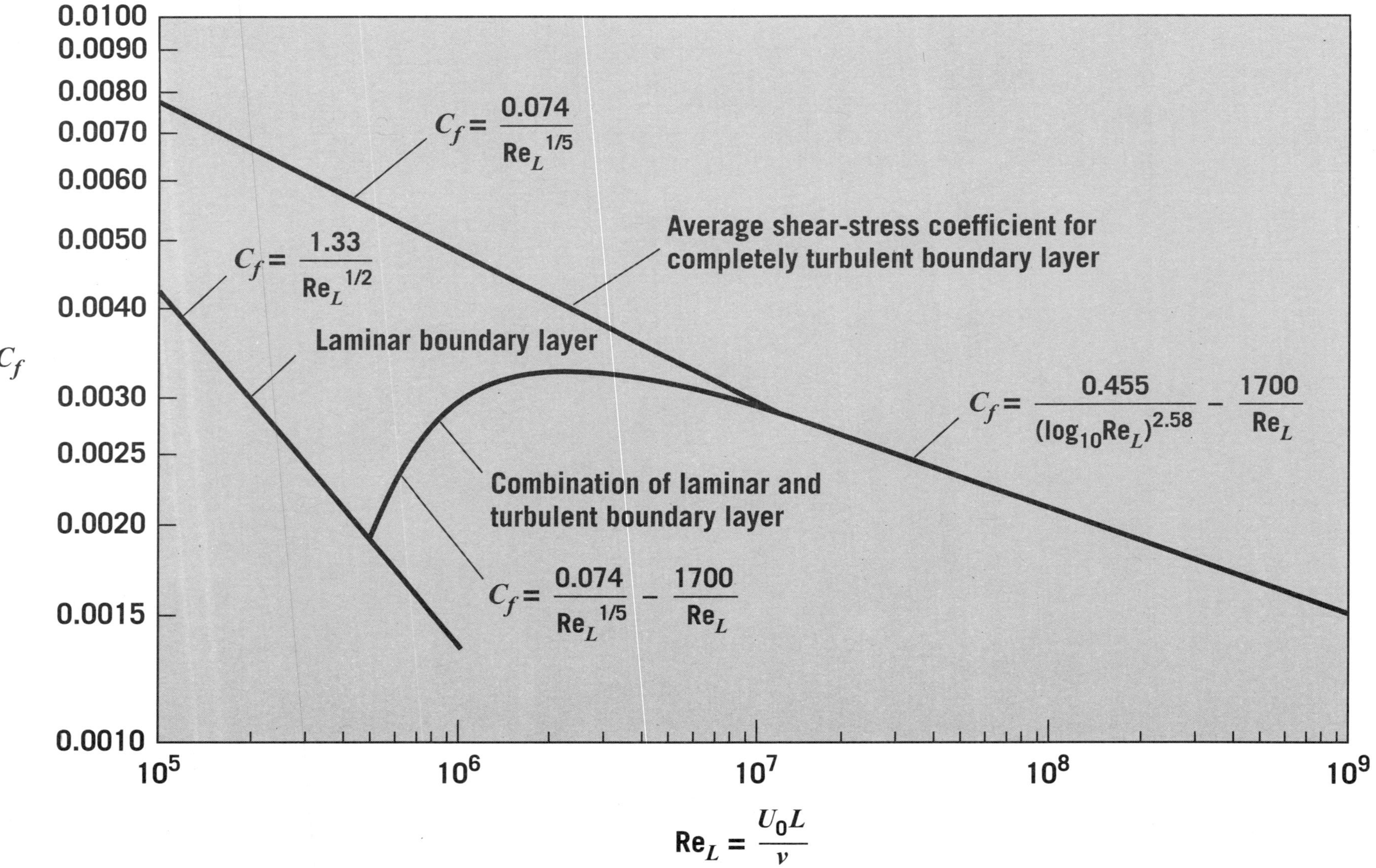

$C_f = \frac{0.074}{Re_L^{1/5}}$
Average shear-stress coefficient for completely turbulent boundary layer
$C_f = \frac{1.33}{Re_L^{1/2}}$
Laminar boundary layer
$C_f = \frac{0.455}{(\log_{10} Re_L)^{2.58}} - \frac{1700}{Re_L}$
Combination of laminar and turbulent boundary layer
$C_f = \frac{0.074}{Re_L^{1/5}} - \frac{1700}{Re_L}$
C_f
0.0100
0.0090
0.0080
0.0070
0.0060
0.0050
0.0040
0.0030
0.0025
0.0020
0.0015
0.0010
10^5
10^6
10^7
10^8
10^9
$Re_L = \frac{U_0 L}{\nu}$

FIGURE 9.13

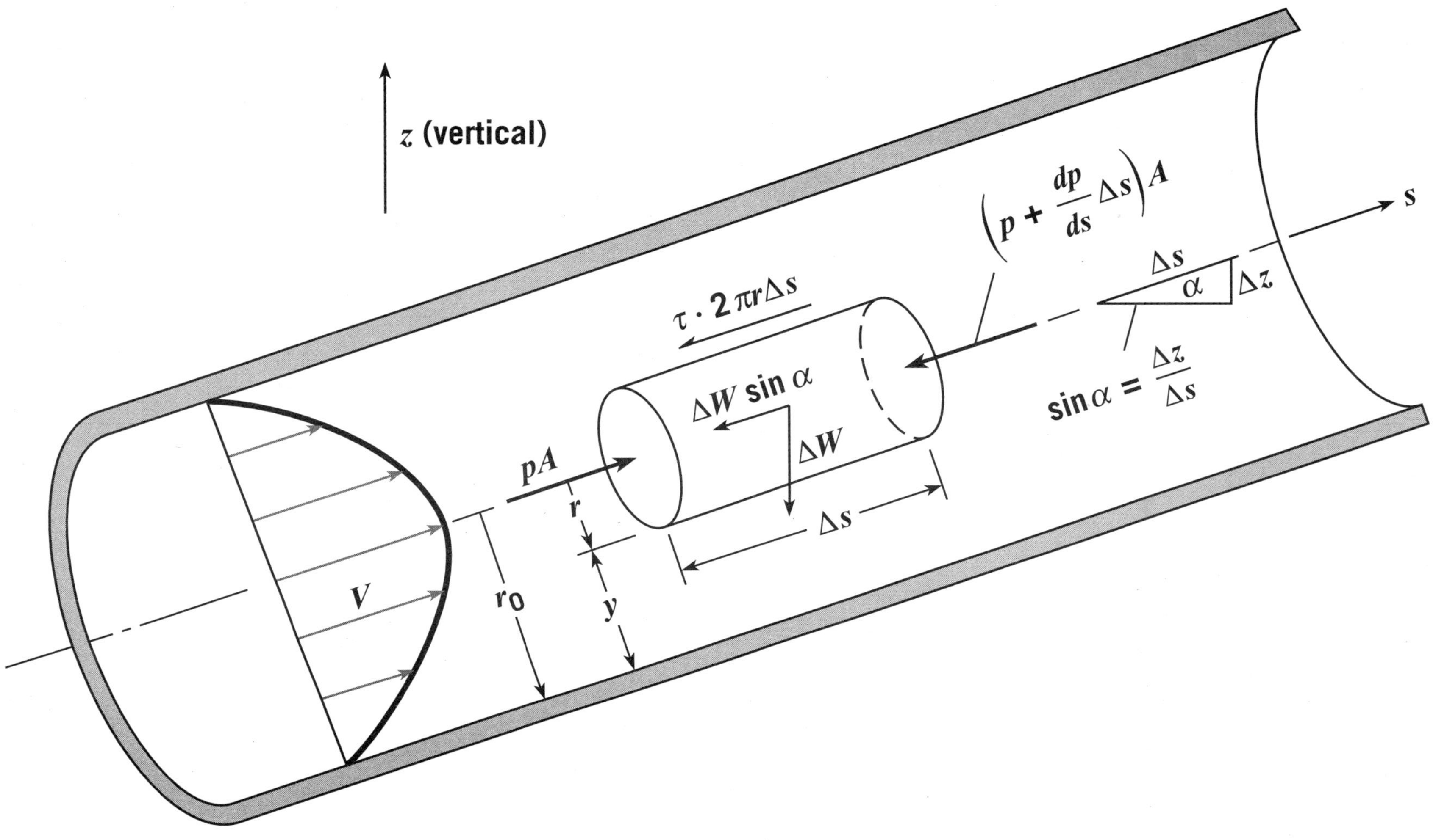

z (vertical)
$\left(p + \frac{dp}{ds}\Delta s\right)A$
s
$\tau \cdot 2\pi r \Delta s$
Δs
α
Δz
$\Delta W \sin\alpha$
$\sin\alpha = \frac{\Delta z}{\Delta s}$
pA
ΔW
r
Δs
V
r_0
y

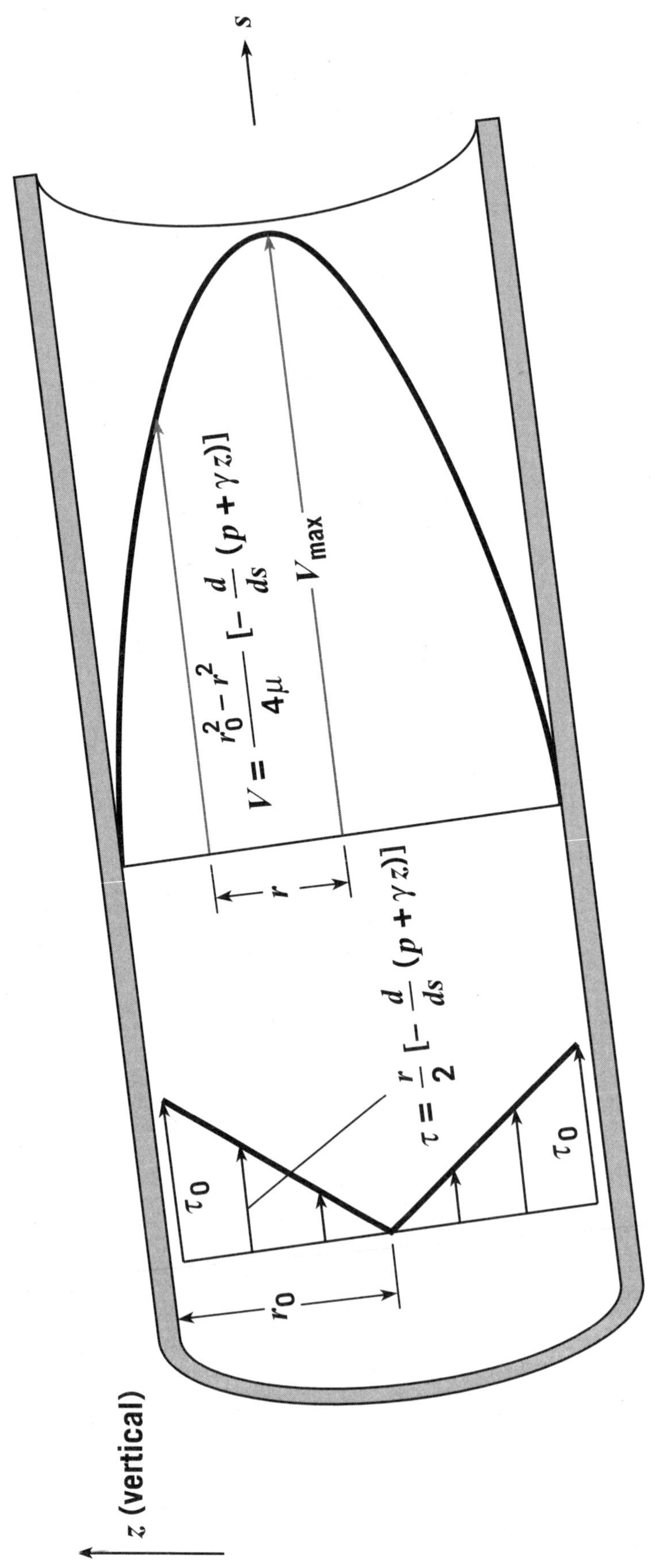

 Roberson/FLUID 6E

FIGURE 10.2

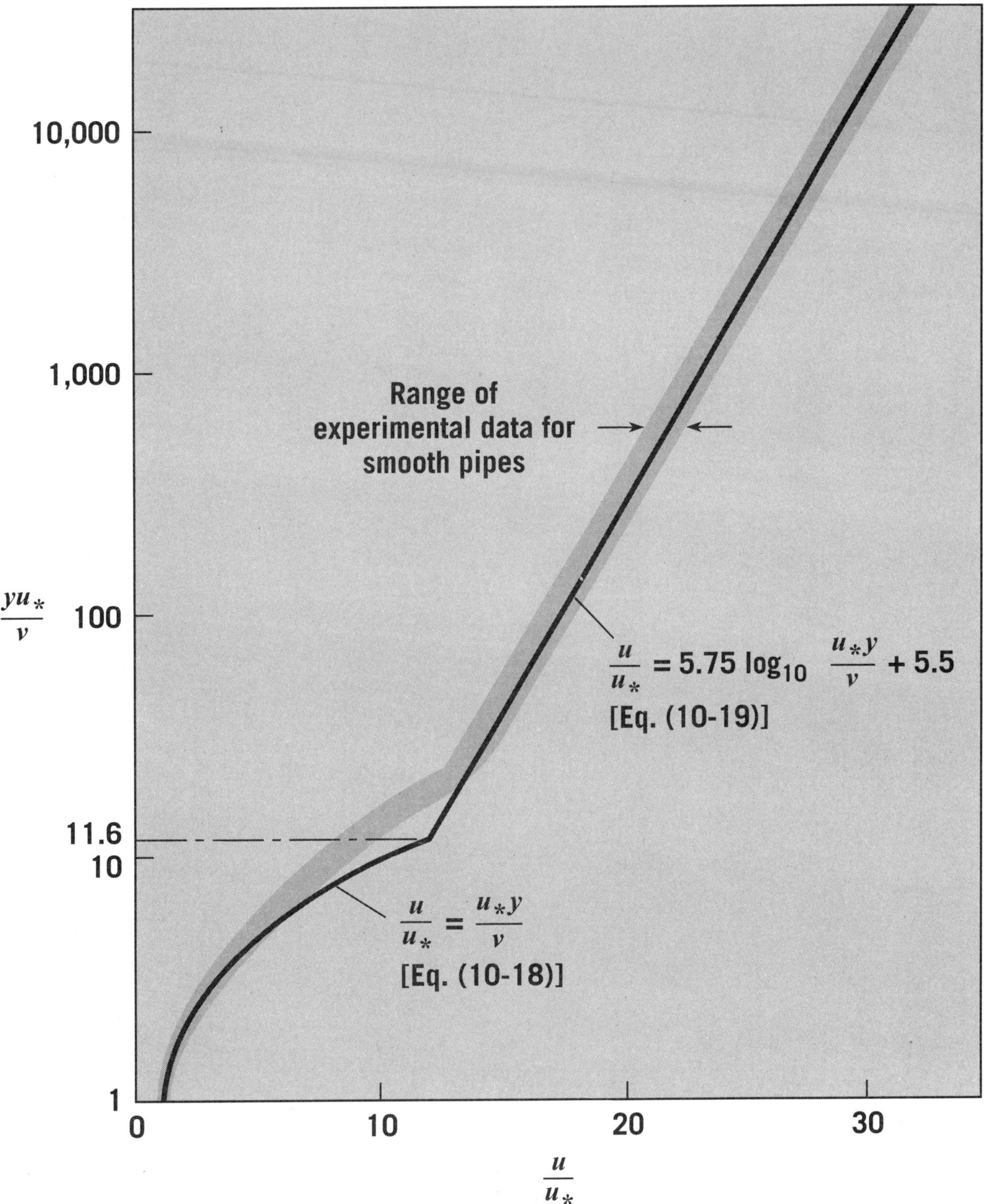

 Roberson/FLUID 6E

FIGURE 10.5

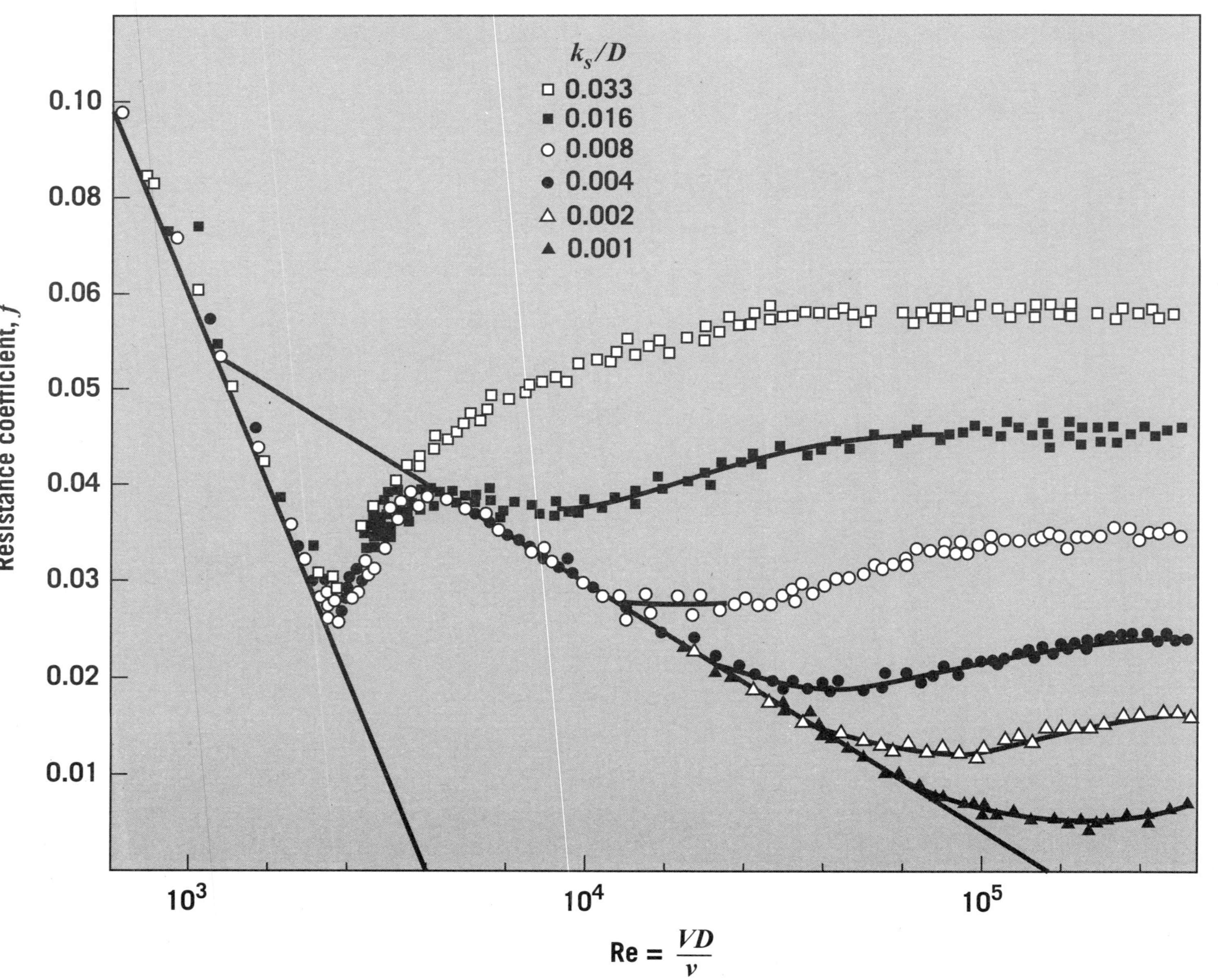

Resistance coefficient, f
0.10
0.08
0.06
0.05
0.04
0.03
0.02
0.01
10^3
10^4
10^5
$\text{Re} = \frac{VD}{\nu}$
k_s/D
□ 0.033
■ 0.016
○ 0.008
● 0.004
△ 0.002
▲ 0.001

FIGURE 10.7

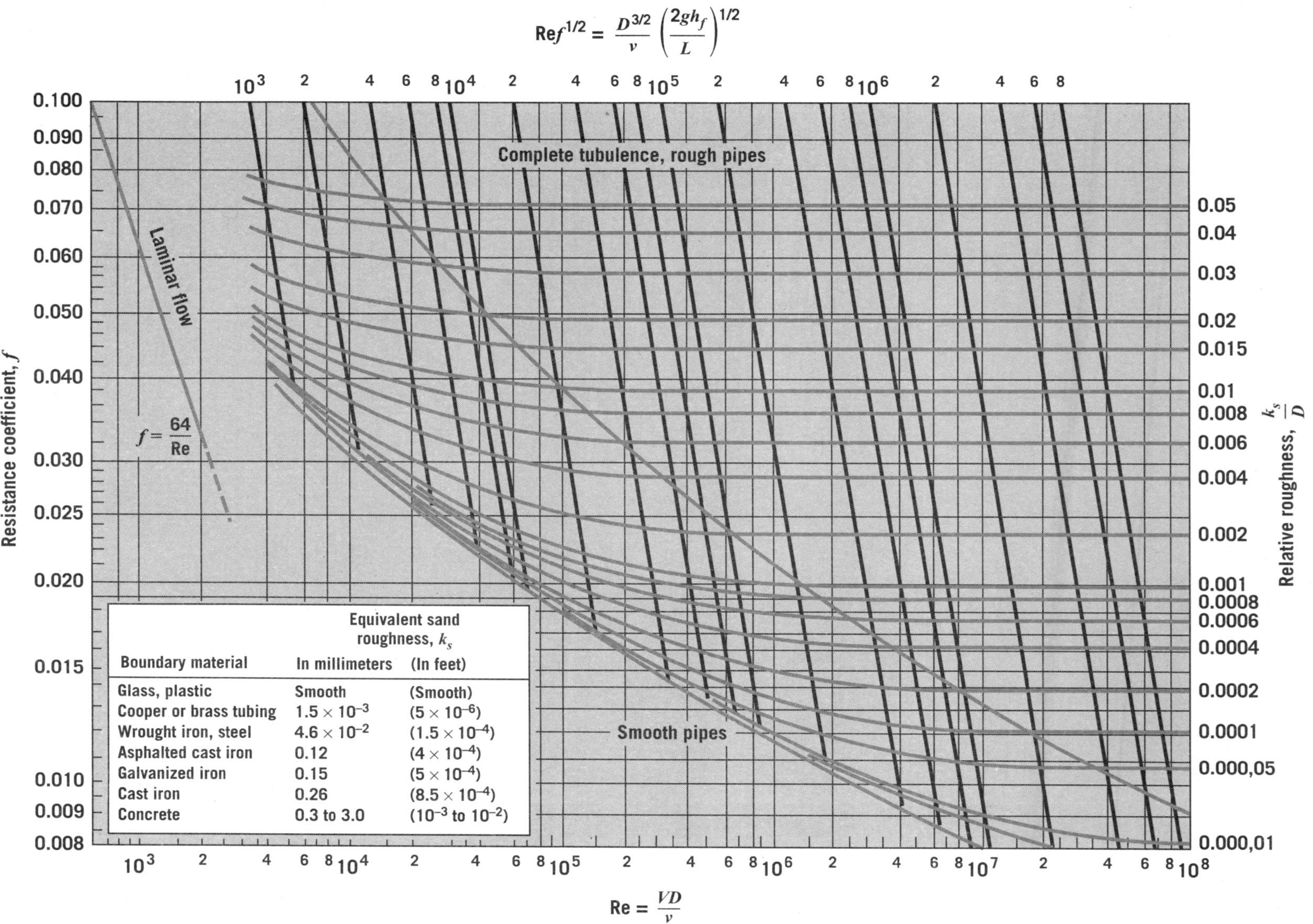

Boundary material	Equivalent sand roughness, k_s In millimeters	(In feet)
Glass, plastic	Smooth	(Smooth)
Cooper or brass tubing	1.5×10^{-3}	(5×10^{-6})
Wrought iron, steel	4.6×10^{-2}	(1.5×10^{-4})
Asphalted cast iron	0.12	(4×10^{-4})
Galvanized iron	0.15	(5×10^{-4})
Cast iron	0.26	(8.5×10^{-4})
Concrete	0.3 to 3.0	(10^{-3} to 10^{-2})

FIGURE 10.8

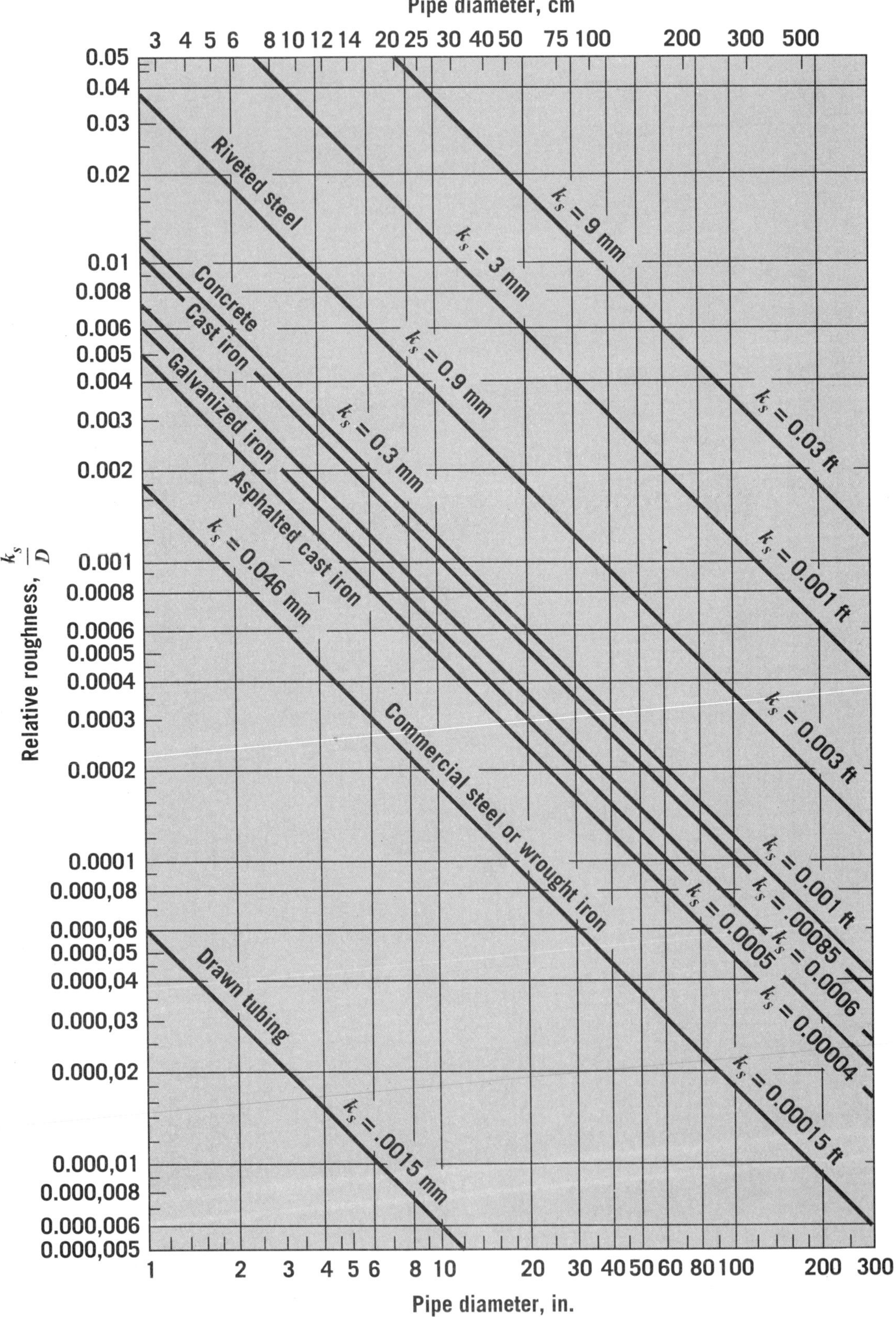

 Roberson/FLUID 6E

FIGURE 10.9 34

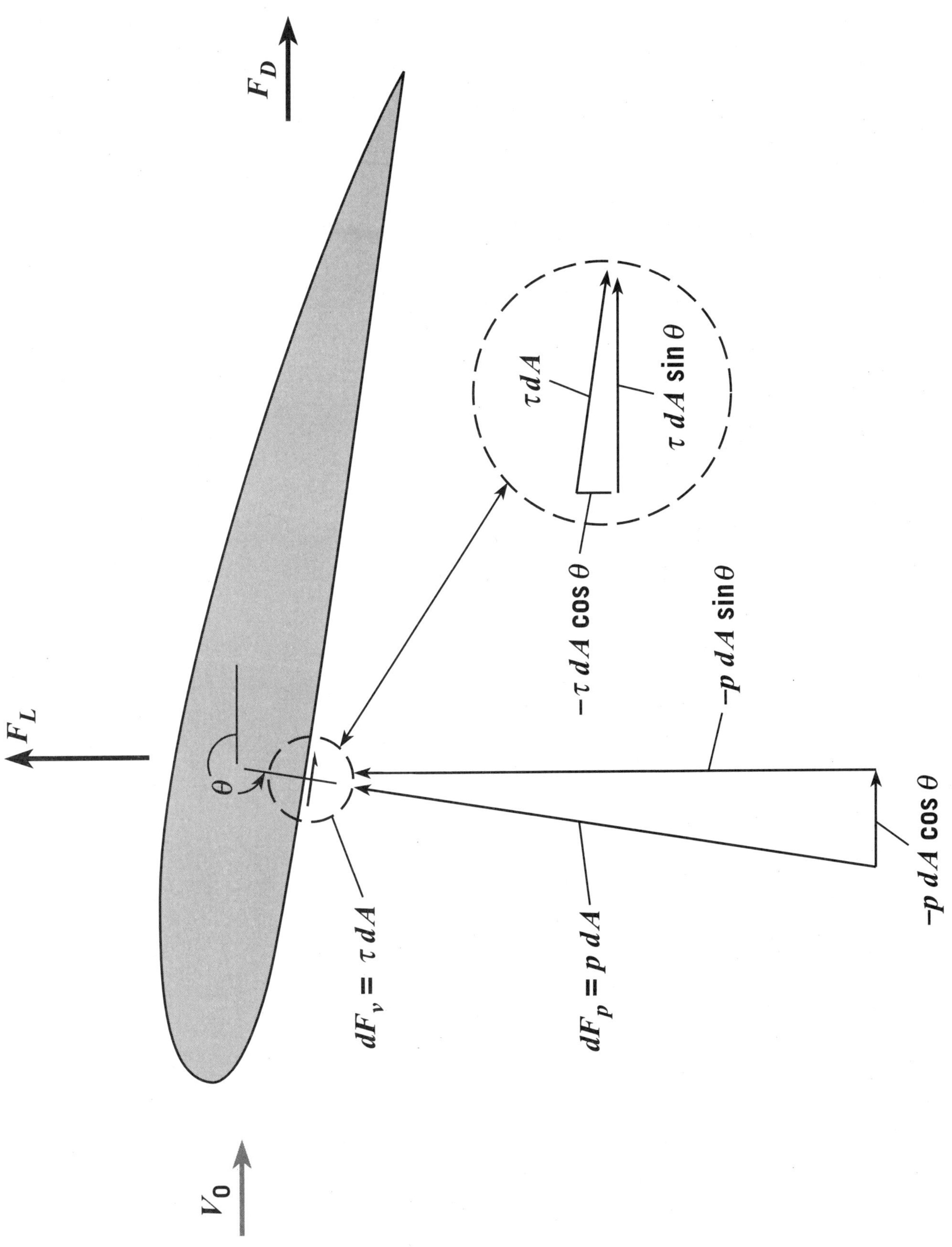

F_D
F_L
θ
$\tau\, dA$
$\tau\, dA \sin\theta$
$-\tau\, dA \cos\theta$
$-p\, dA \sin\theta$
$-p\, dA \cos\theta$
$dF_v = \tau\, dA$
$dF_p = p\, dA$
V_0

FIGURE 11.2

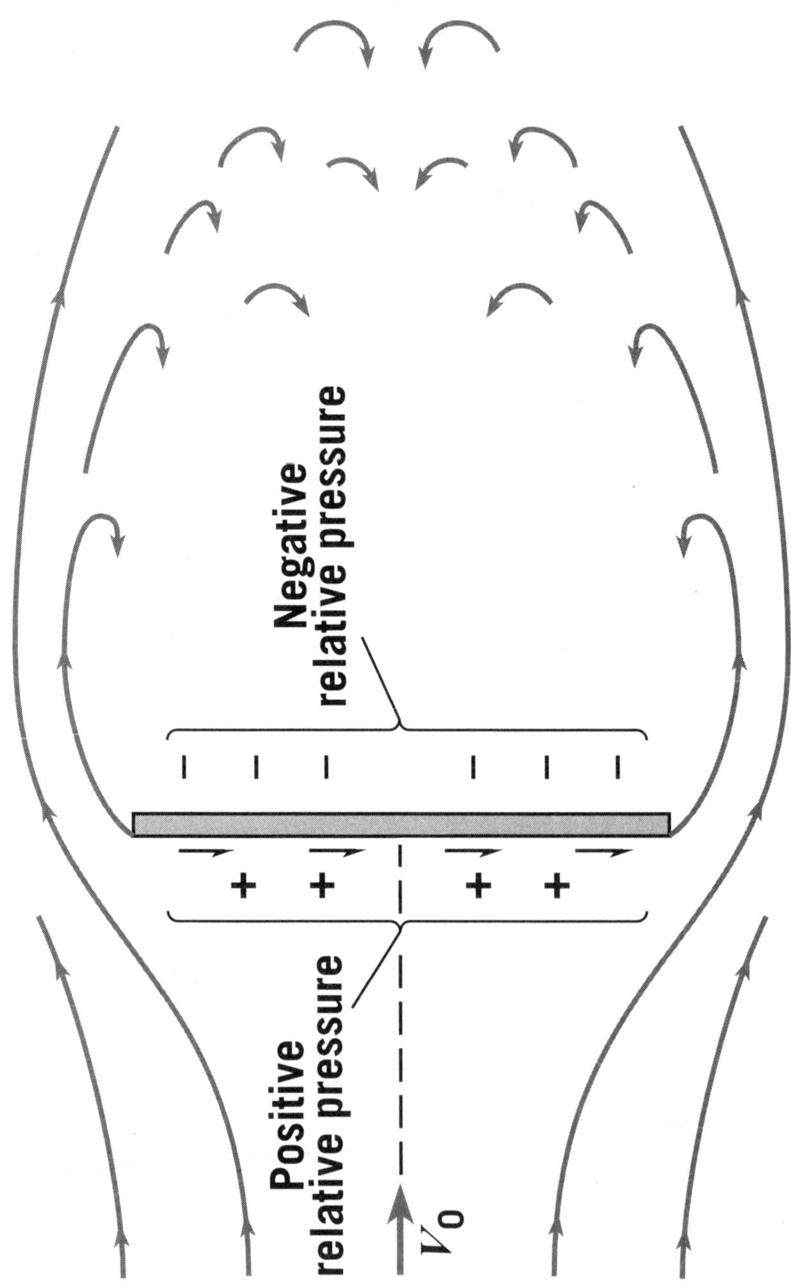

Negative relative pressure
Positive relative pressure
V_0

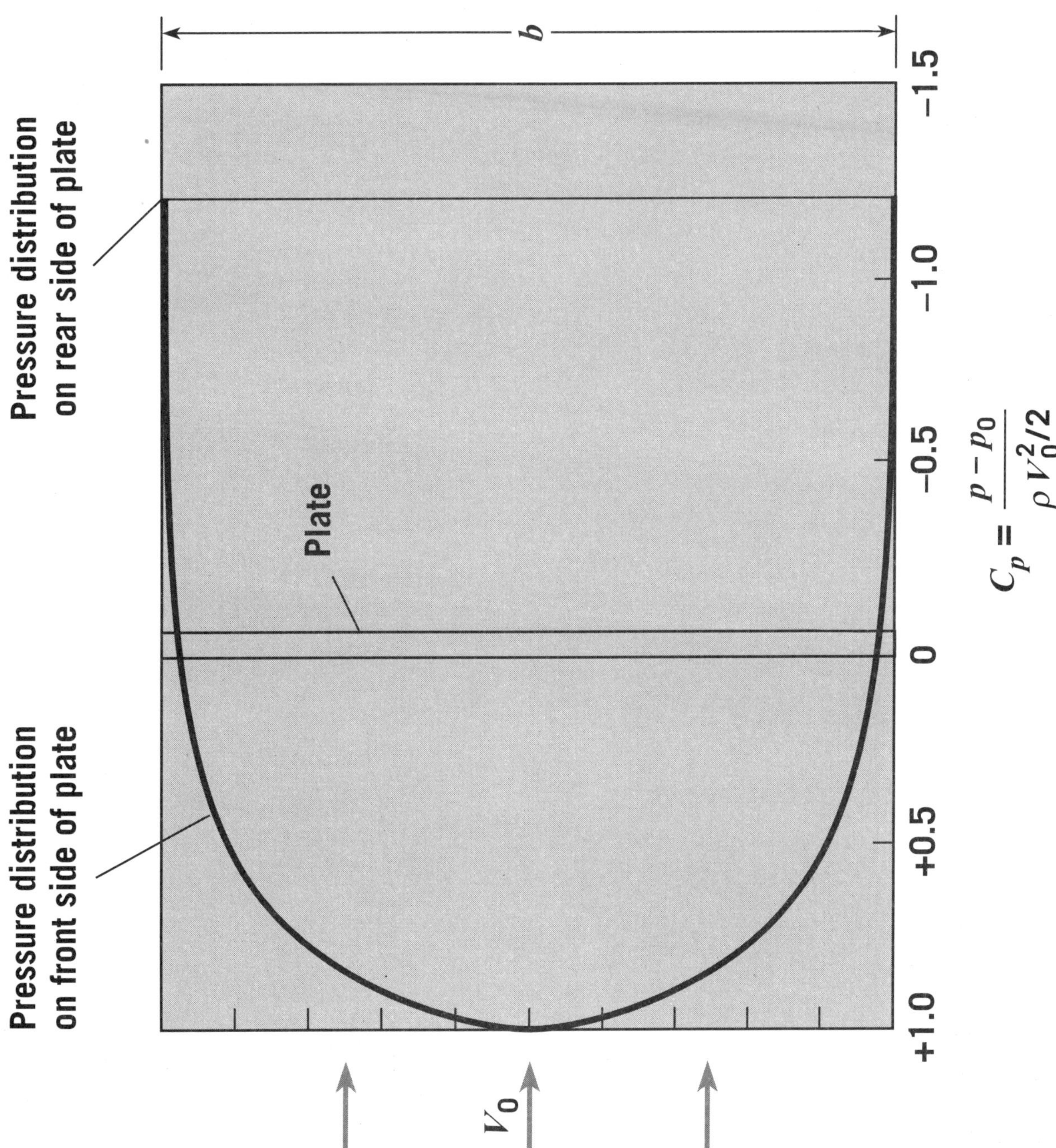

Pressure distribution on rear side of plate
Pressure distribution on front side of plate
Plate
b
V_0
+1.0
+0.5
0
−0.5
−1.0
−1.5
$C_p = \frac{p - p_0}{\rho V_0^2/2}$

FIGURE 11.4

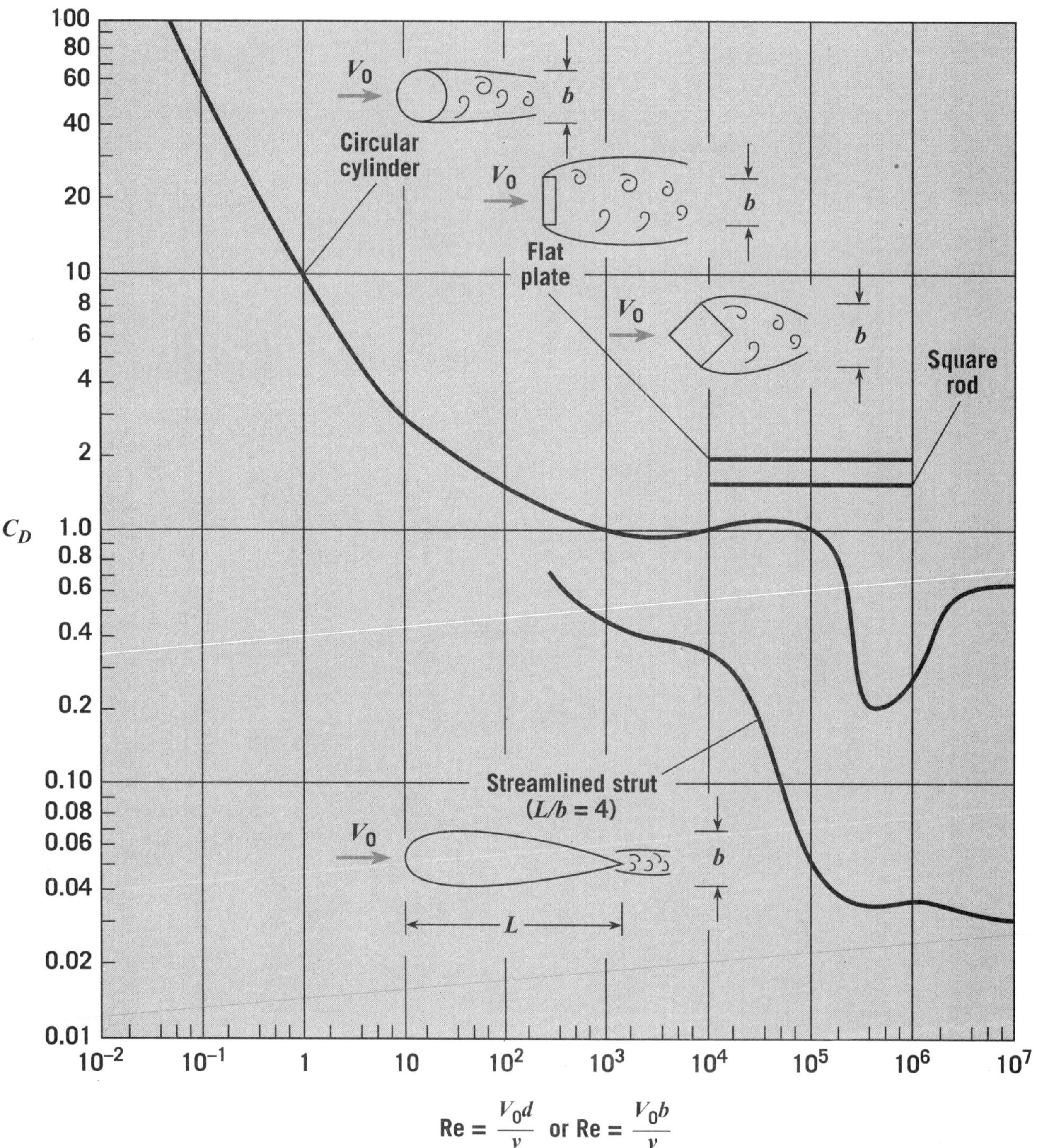

100
80
60
40
20
10
8
6
4
2
C_D
1.0
0.8
0.6
0.4
0.2
0.10
0.08
0.06
0.04
0.02
0.01
V_0
b
Circular cylinder
V_0
b
Flat plate
V_0
b
Square rod
Streamlined strut ($L/b = 4$)
V_0
b
L
10^{-2}
10^{-1}
1
10
10^2
10^3
10^4
10^5
10^6
10^7
$\text{Re} = \frac{V_0 d}{\nu}$ or $\text{Re} = \frac{V_0 b}{\nu}$

FIGURE 11.5

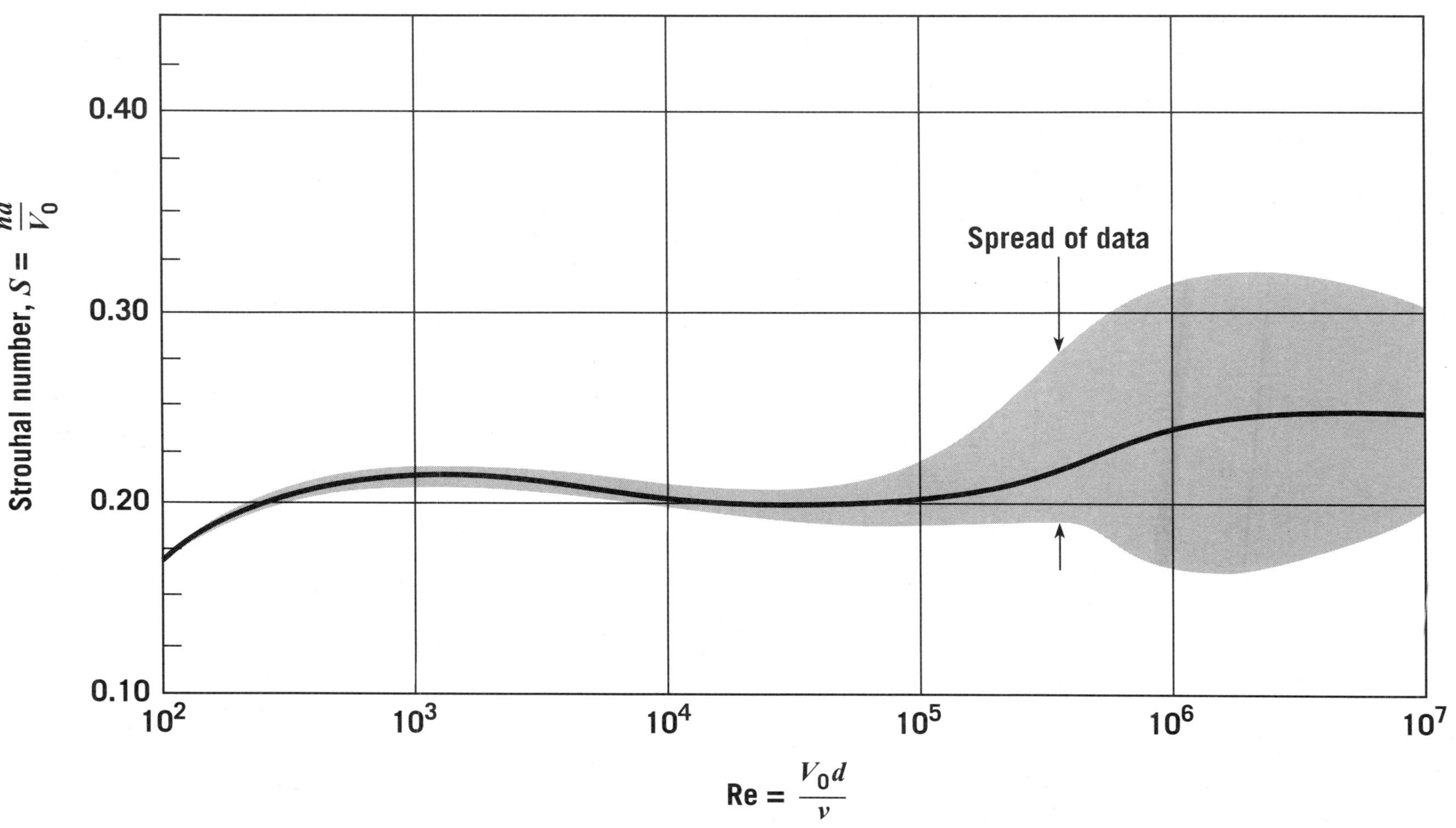

Spread of data
Strouhal number, $S = \frac{nd}{V_0}$
0.40
0.30
0.20
0.10
10^2
10^3
10^4
10^5
10^6
10^7
$\mathrm{Re} = \frac{V_0 d}{\nu}$

FIGURE 11.10

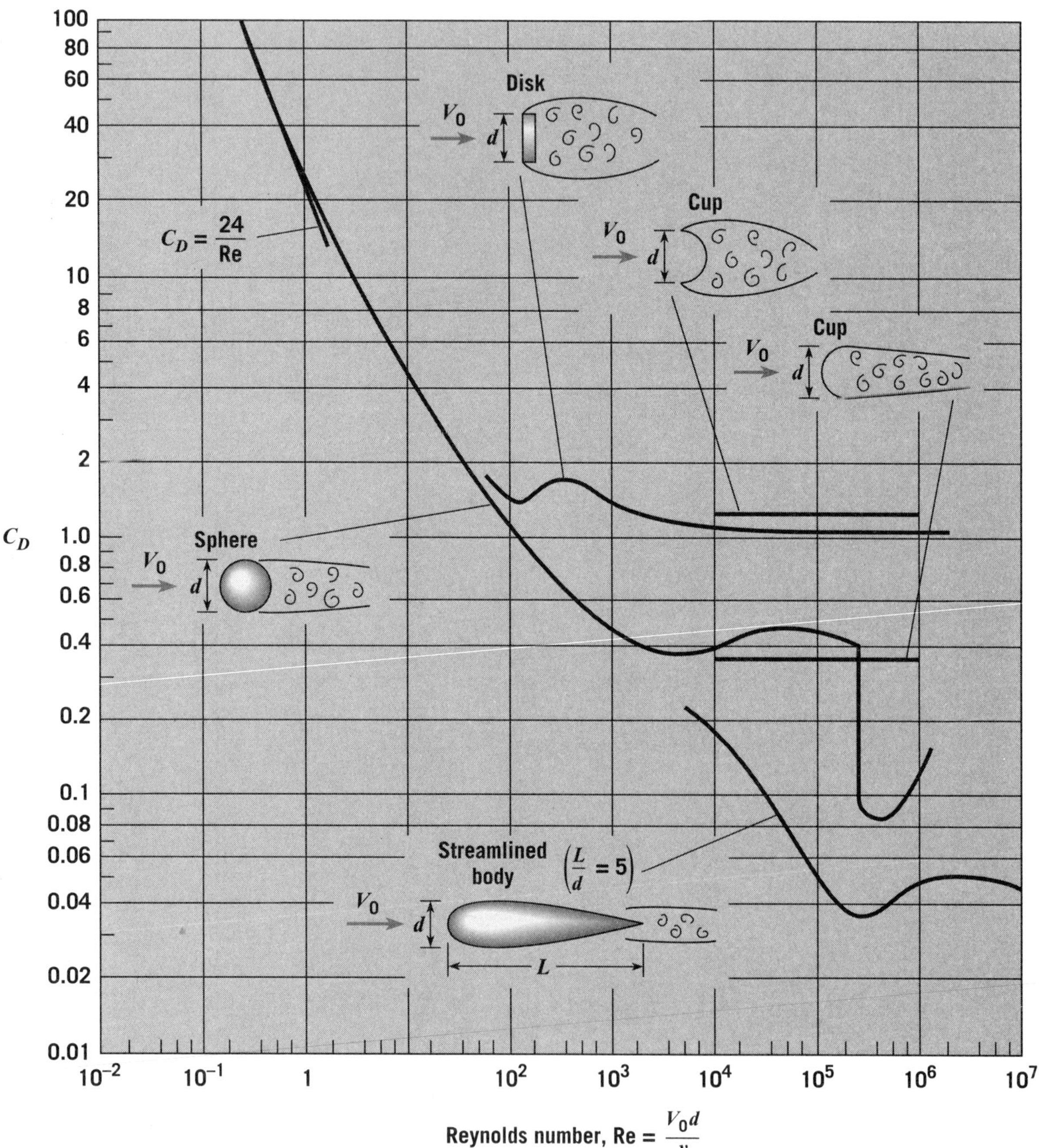
Disk
Cup
Cup
Sphere
Streamlined body
$\left(\frac{L}{d} = 5\right)$
V_0
d
L
$C_D = \frac{24}{\mathrm{Re}}$
C_D
Reynolds number, $\mathrm{Re} = \frac{V_0 d}{\nu}$
100
80
60
40
20
10
8
6
4
2
1.0
0.8
0.6
0.4
0.2
0.1
0.08
0.06
0.04
0.02
0.01
10^{-2}
10^{-1}
1
10^2
10^3
10^4
10^5
10^6
10^7

FIGURE 11.11

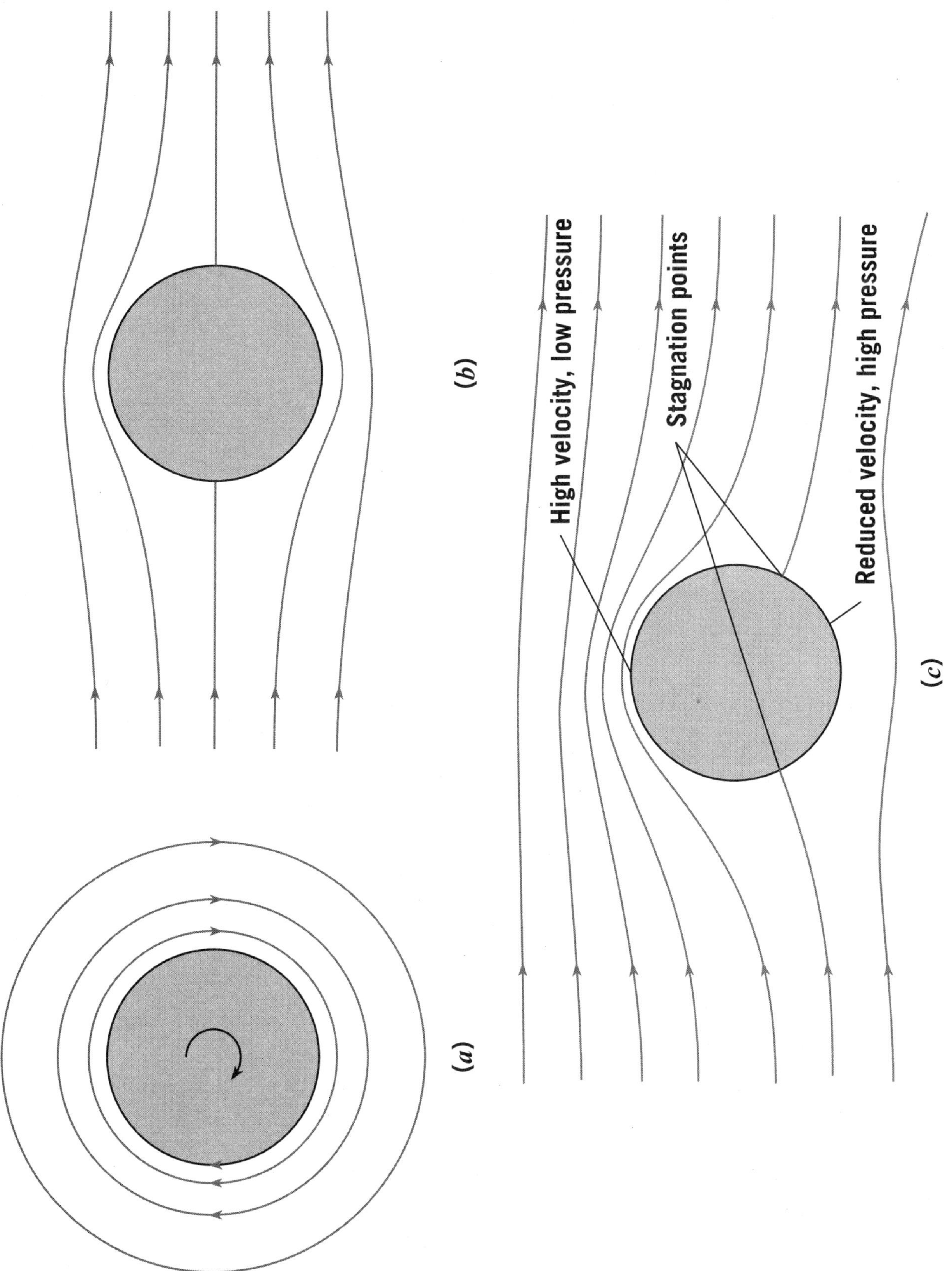
(a)
(b)
High velocity, low pressure
Stagnation points
Reduced velocity, high pressure
(c)

FIGURE 11.15

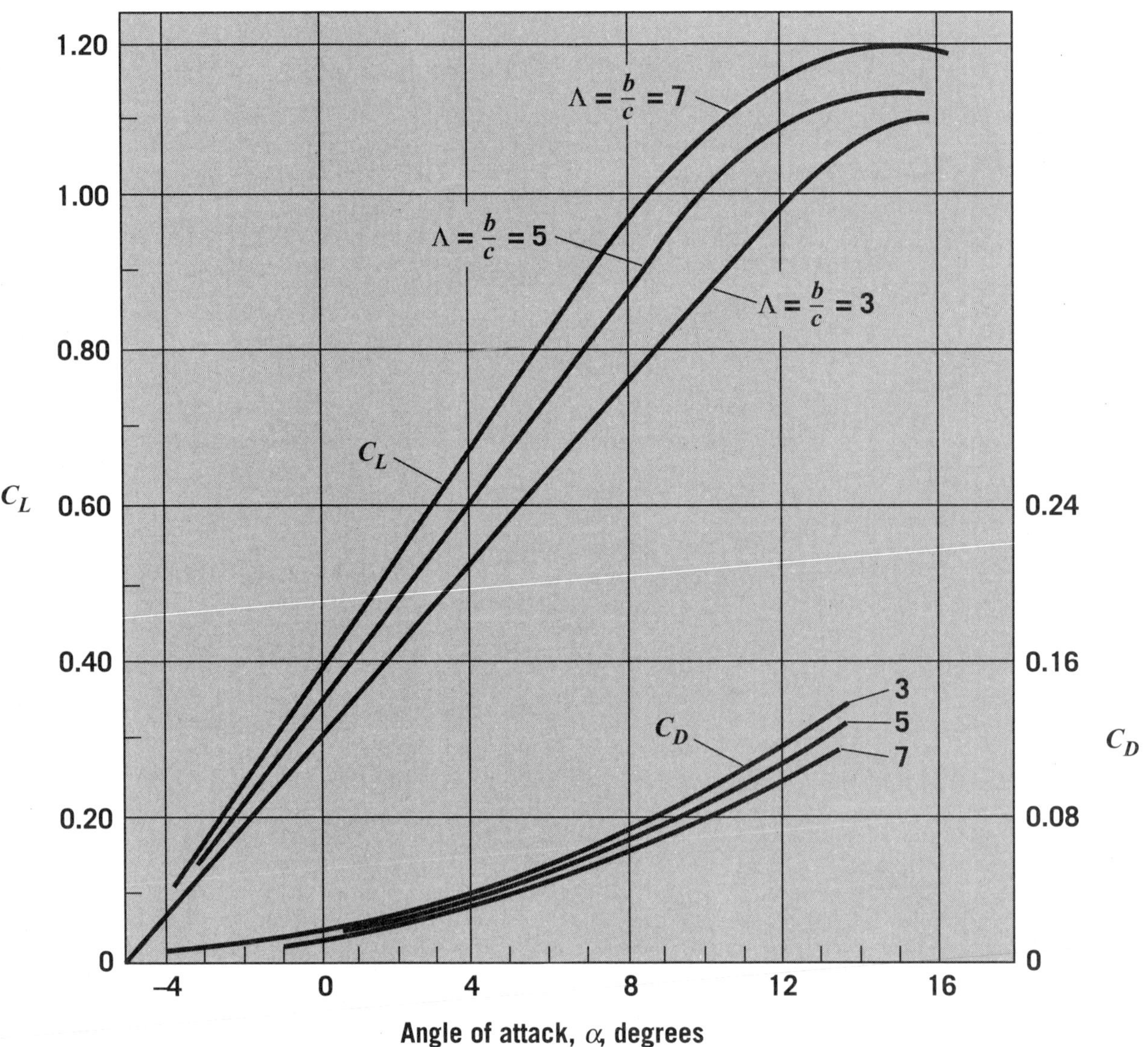

FIGURE 11.23 42

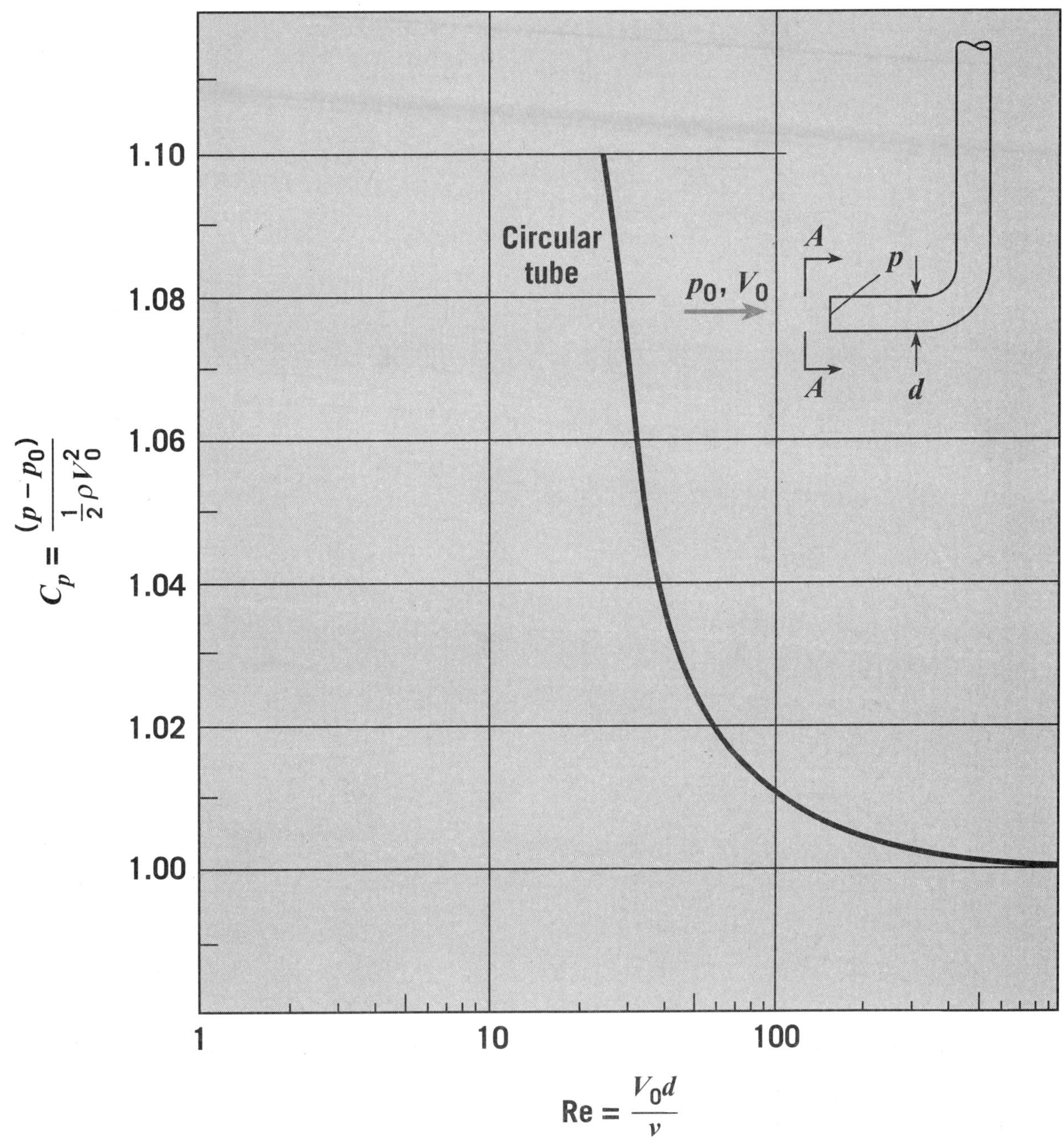

 Roberson/FLUID 6E

FIGURE 13.1

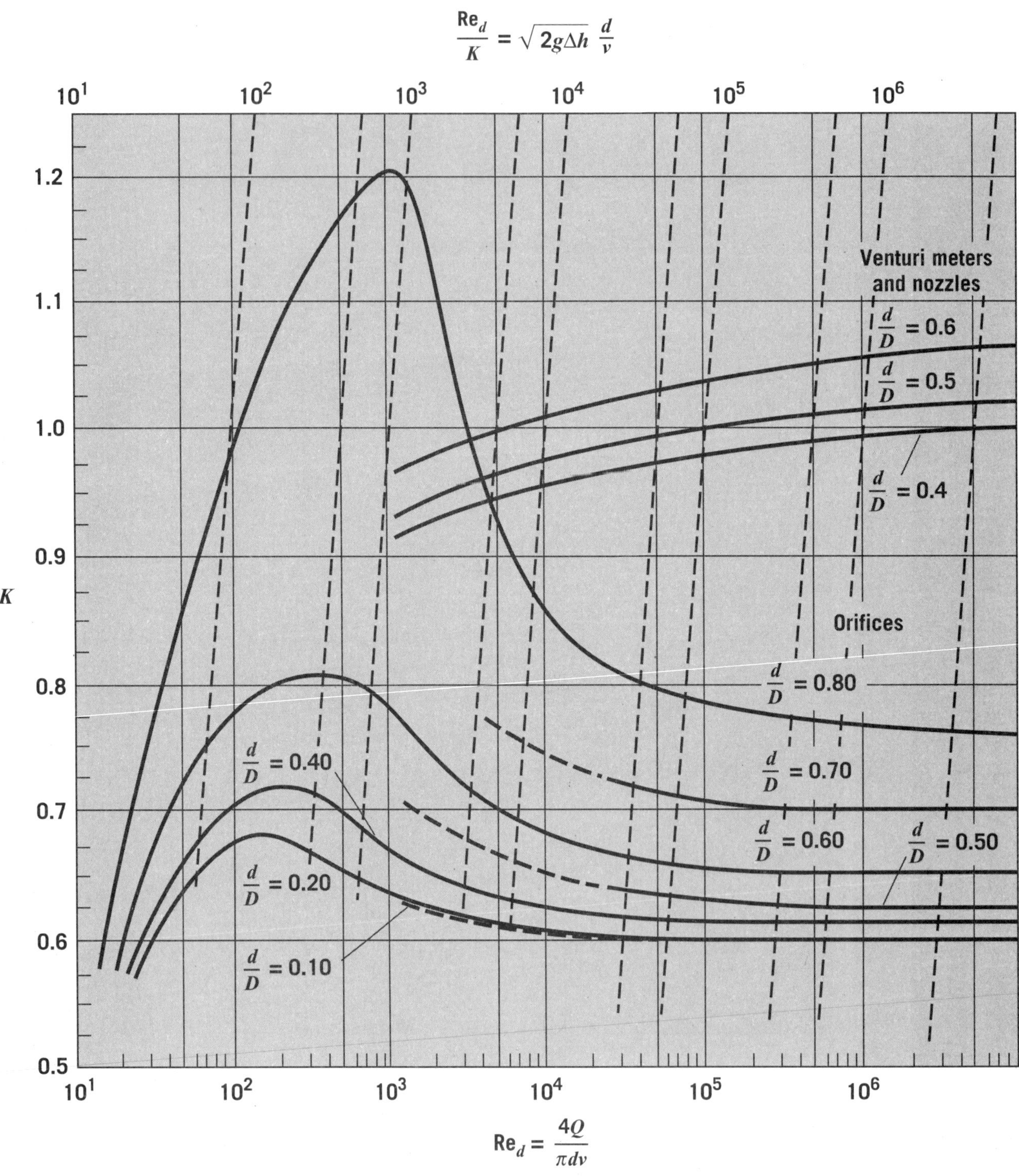

$\frac{Re_d}{K} = \sqrt{2g\Delta h}\ \frac{d}{\nu}$
10^1
10^2
10^3
10^4
10^5
10^6
1.2
1.1
1.0
0.9
0.8
0.7
0.6
0.5
K
Venturi meters and nozzles
$\frac{d}{D} = 0.6$
$\frac{d}{D} = 0.5$
$\frac{d}{D} = 0.4$
Orifices
$\frac{d}{D} = 0.80$
$\frac{d}{D} = 0.70$
$\frac{d}{D} = 0.60$
$\frac{d}{D} = 0.50$
$\frac{d}{D} = 0.40$
$\frac{d}{D} = 0.20$
$\frac{d}{D} = 0.10$
$Re_d = \frac{4Q}{\pi d \nu}$

FIGURE 13.13

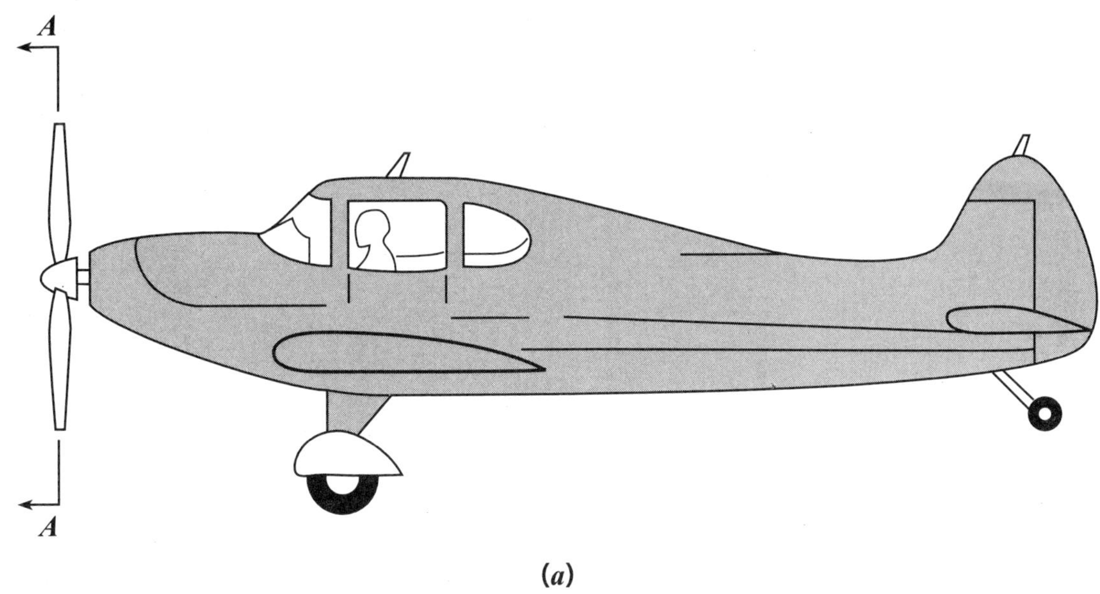

(a)

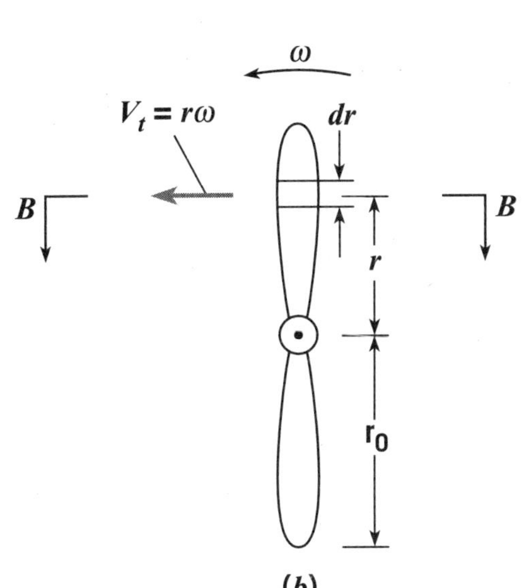

(b)

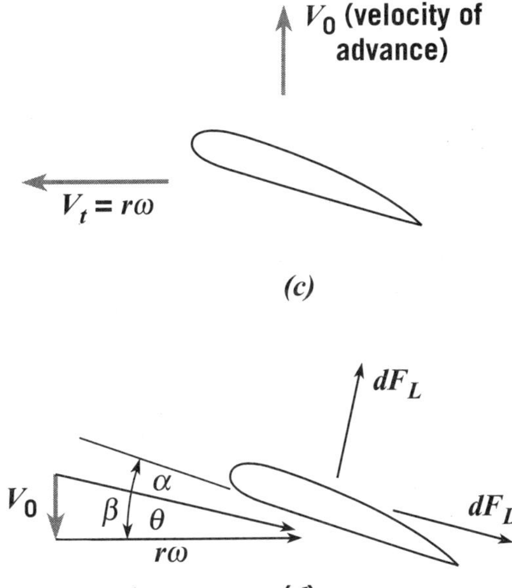

(c)

(d)

FIGURE 14.1

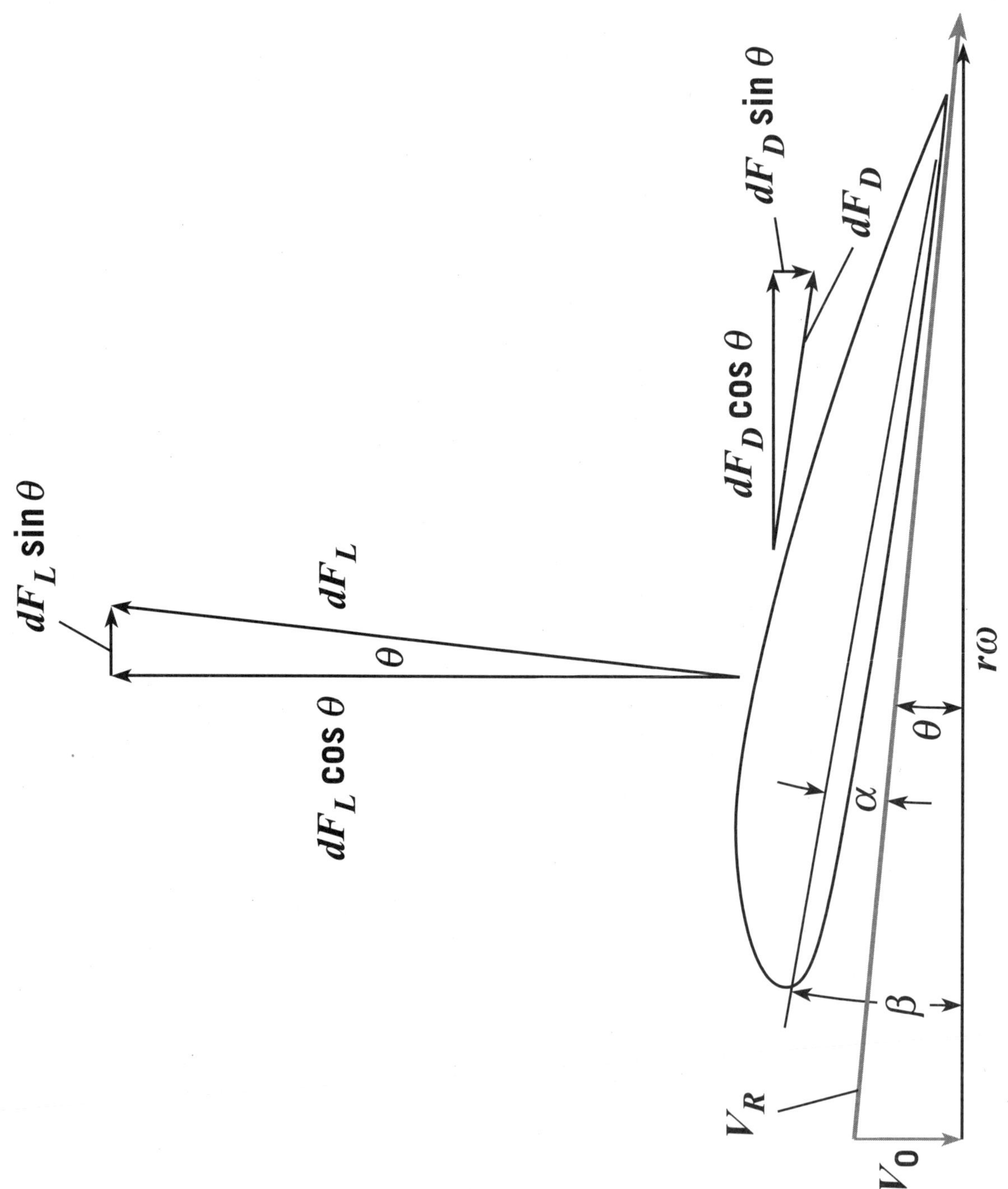

 Roberson/FLUID 6E

FIGURE 14.2

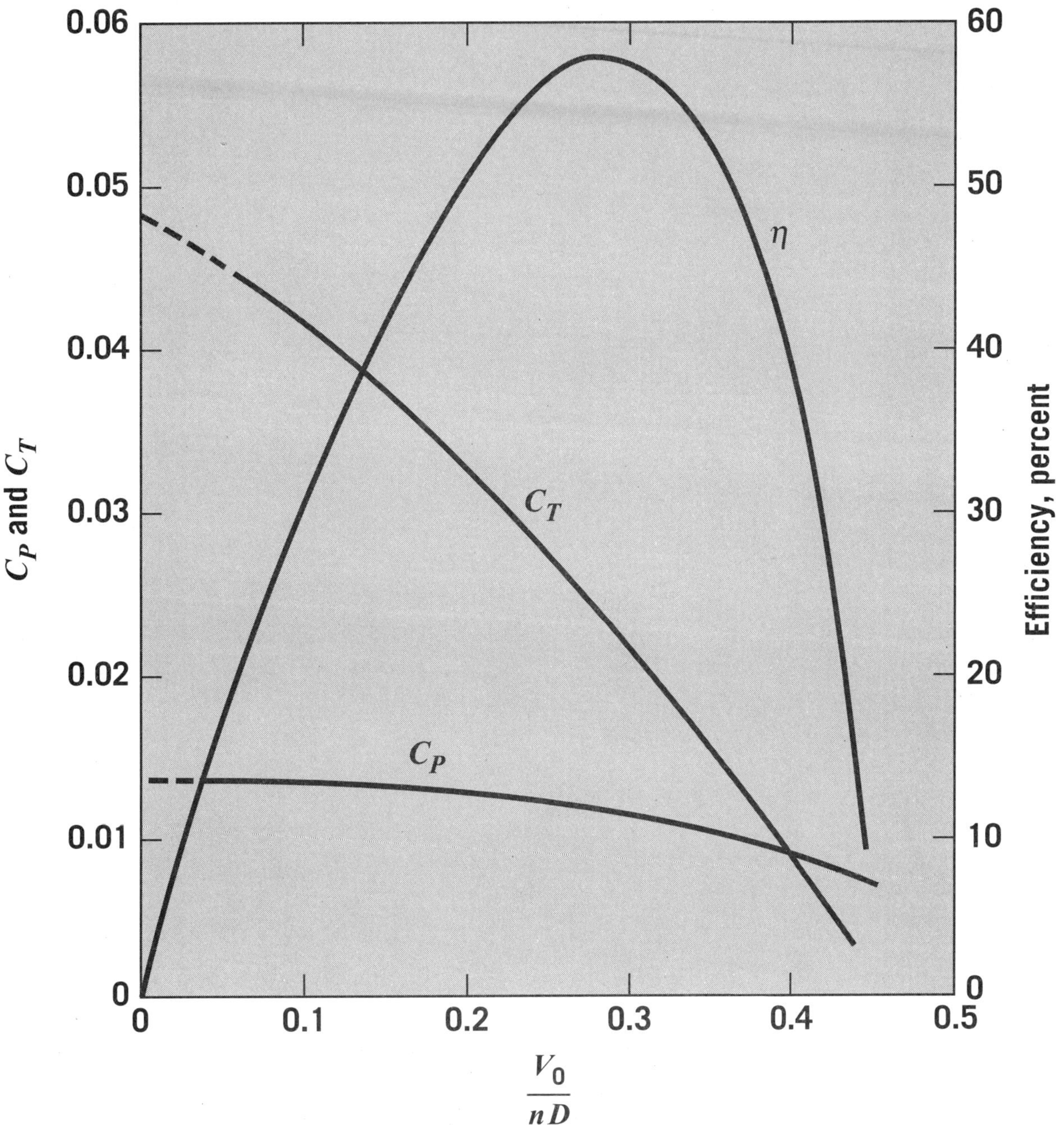
0.06
0.05
0.04
0.03
0.02
0.01
0
0
0.1
0.2
0.3
0.4
0.5
60
50
40
30
20
10
0
C_P and C_T
Efficiency, percent
η
C_T
C_P
$\frac{V_0}{nD}$

FIGURE 14.3

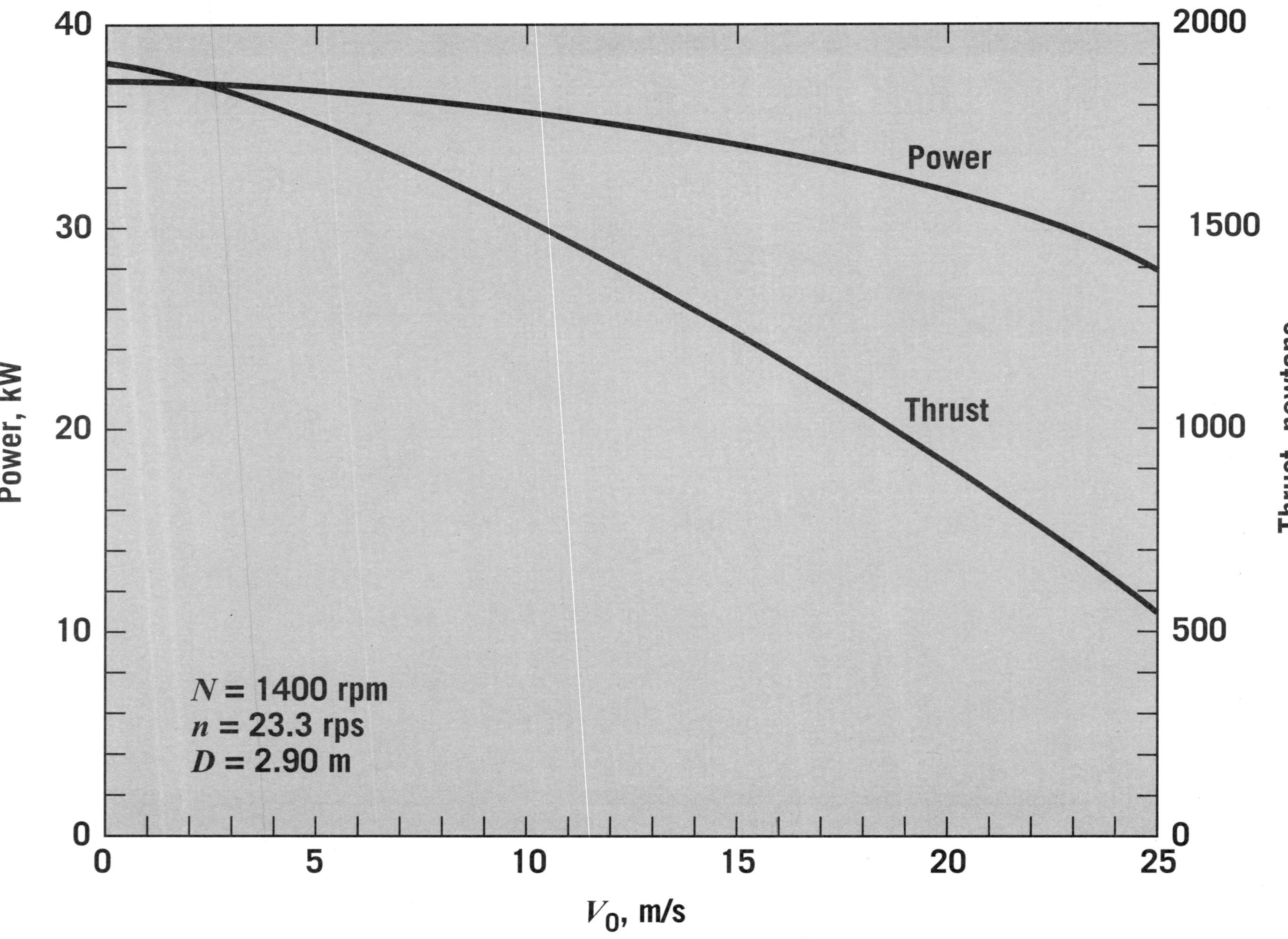
Power
Thrust
N = 1400 rpm
n = 23.3 rps
D = 2.90 m
Power, kW
40
30
20
10
0
Thrust, newtons
2000
1500
1000
500
0
0
5
10
15
20
25
V_0, m/s

FIGURE 14.4

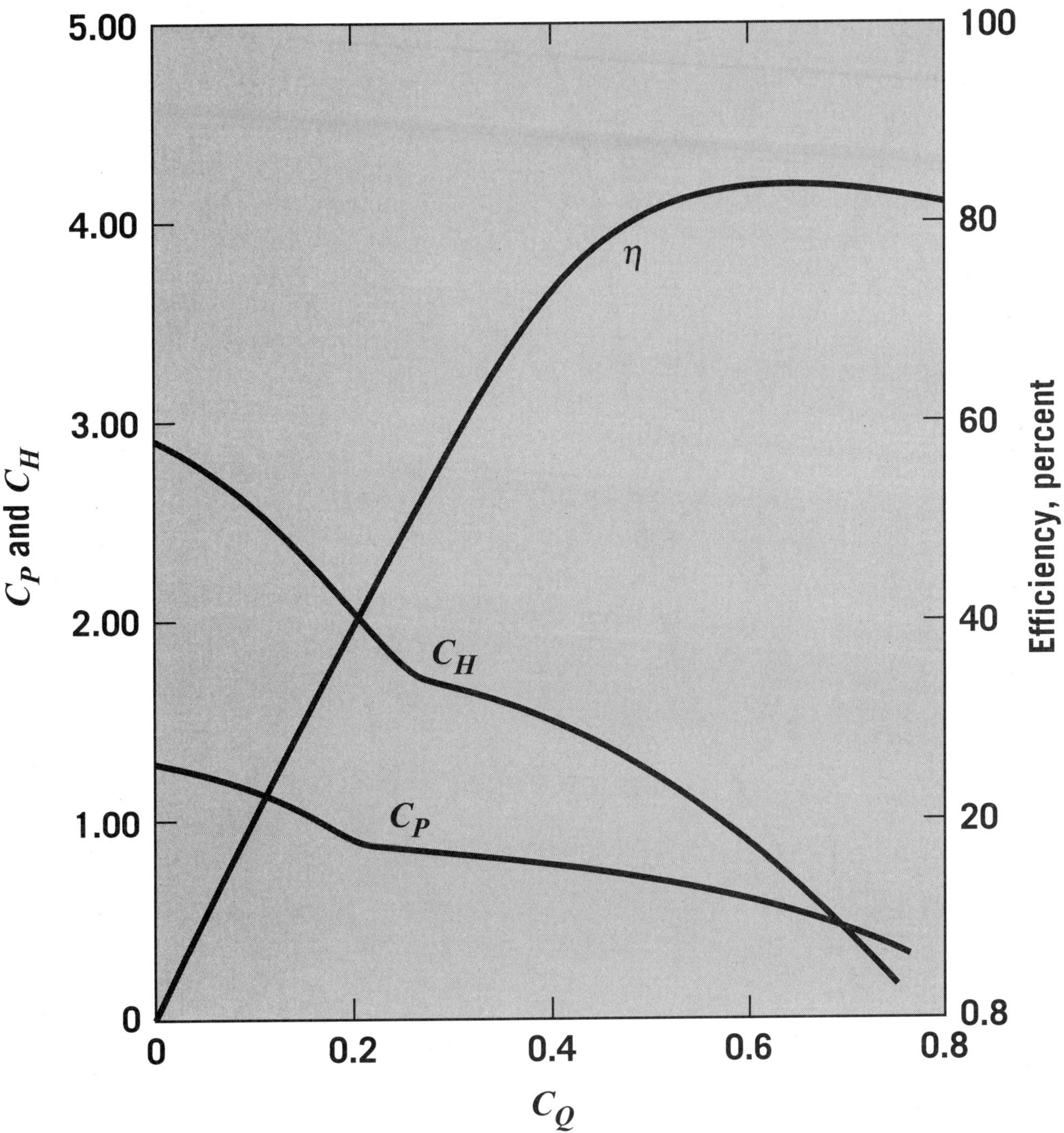

 Roberson/FLUID 6E

FIGURE 14.6

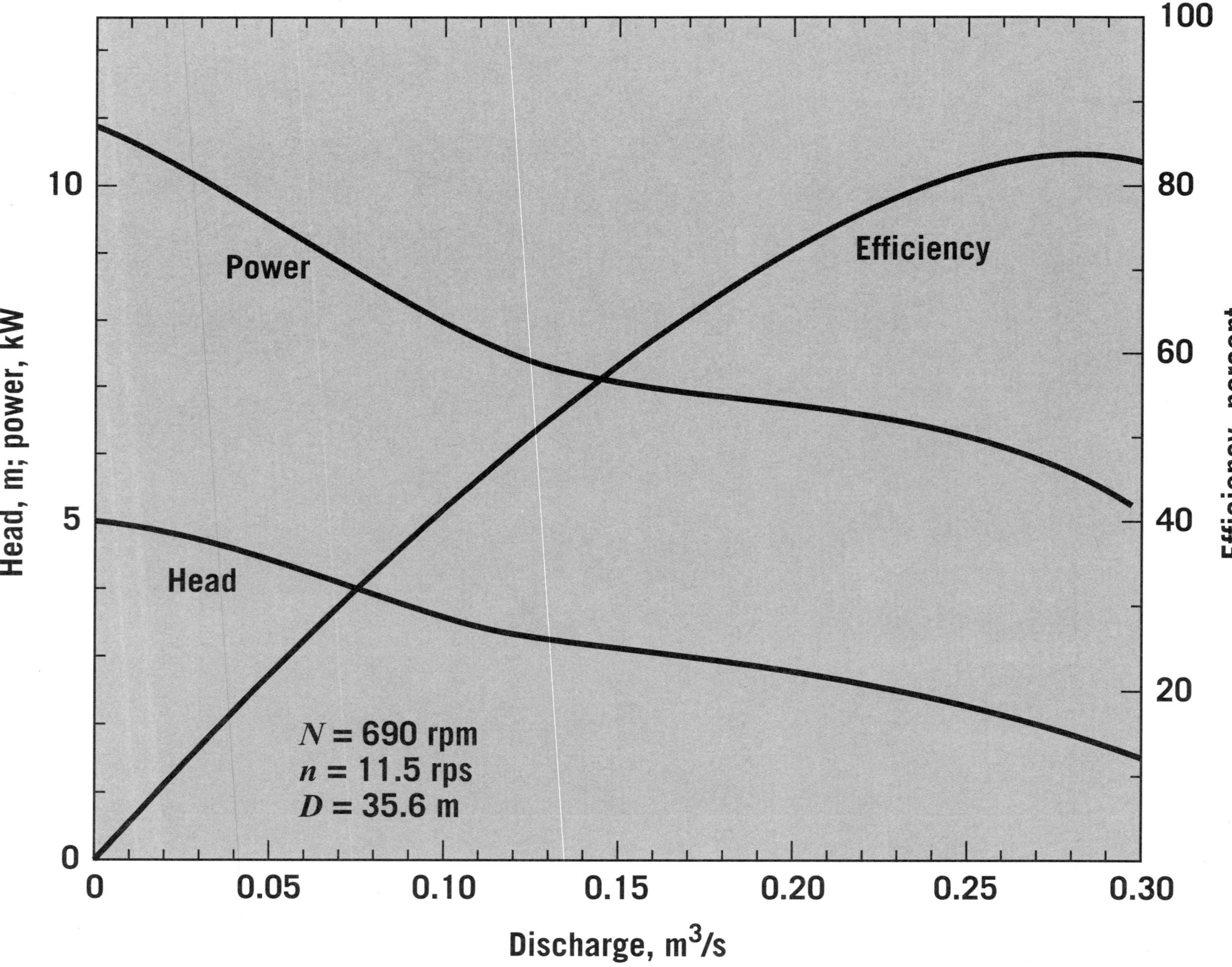
Power
Efficiency
Head
N = 690 rpm
n = 11.5 rps
D = 35.6 m
Head, m; power, kW
10
5
0
Efficiency, percent
100
80
60
40
20
0
0.05
0.10
0.15
0.20
0.25
0.30
Discharge, m3/s

FIGURE 14.7

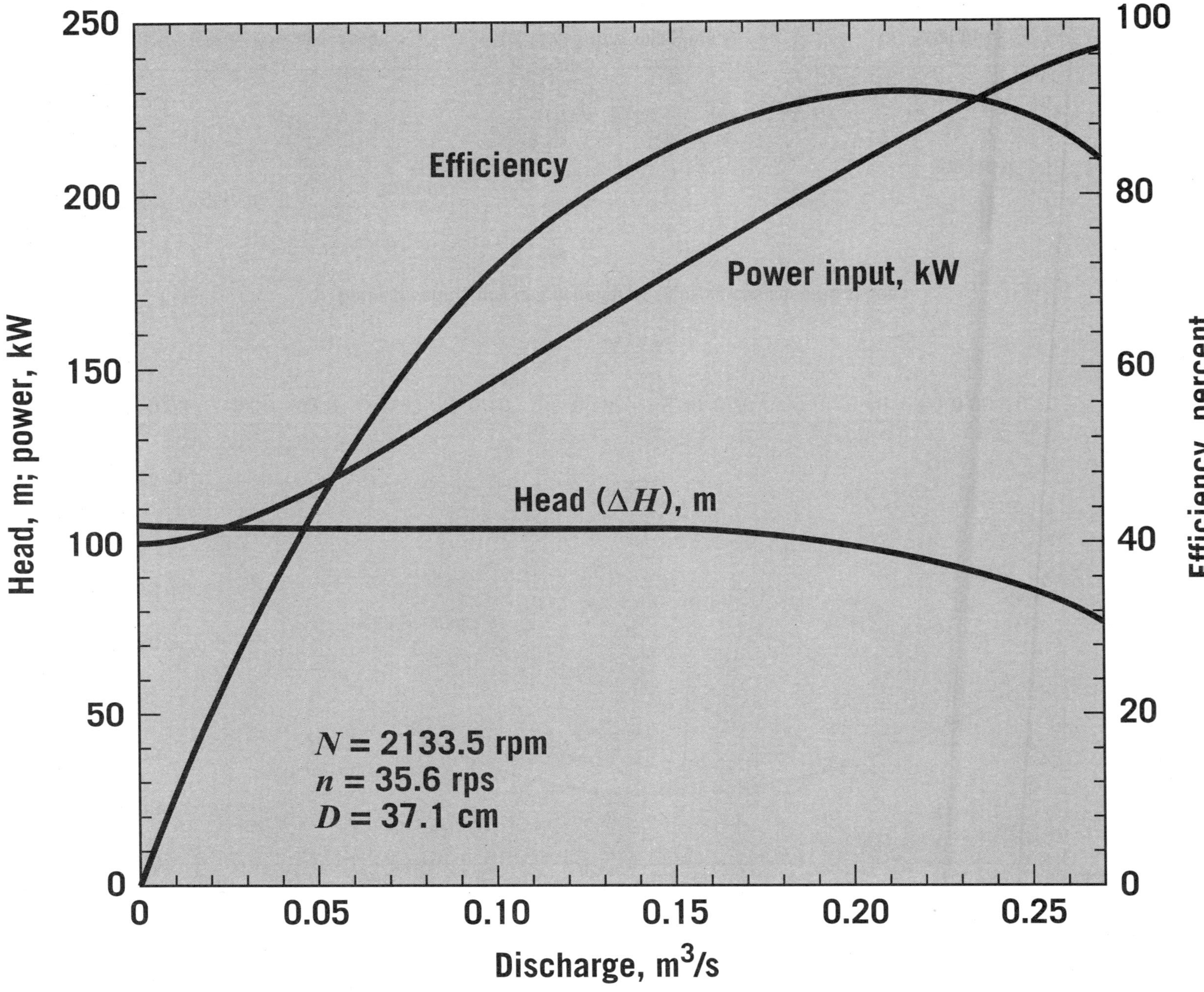

Efficiency
Power input, kW
Head (ΔH), m
N = 2133.5 rpm
n = 35.6 rps
D = 37.1 cm
250
200
150
100
50
0
Head, m; power, kW
100
80
60
40
20
0
Efficiency, percent
0
0.05
0.10
0.15
0.20
0.25
Discharge, m^3/s

FIGURE 14.9

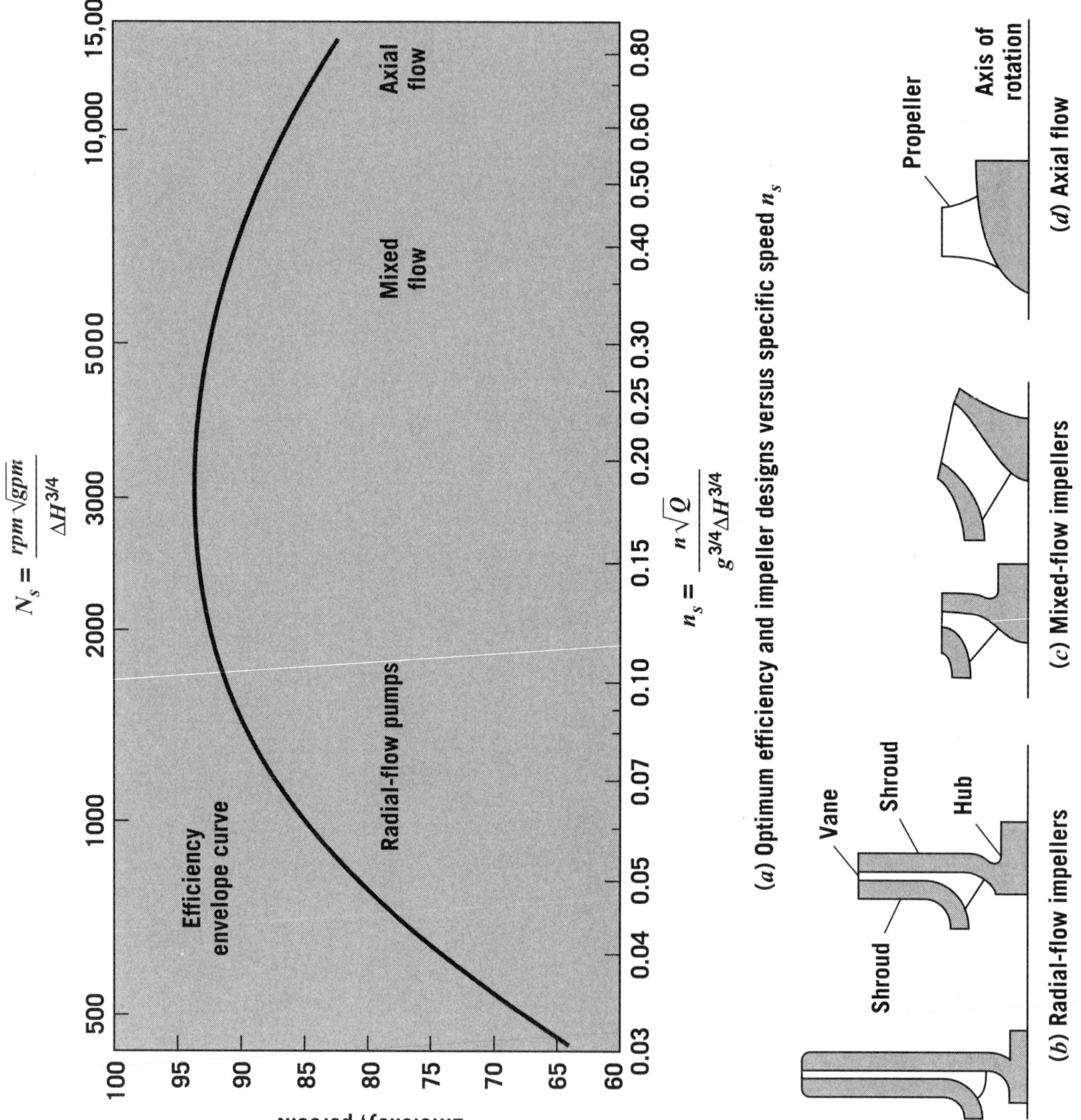

(*a*) Optimum efficiency and impeller designs versus specific speed n_s

(*b*) Radial-flow impellers

(*c*) Mixed-flow impellers

(*d*) Axial flow

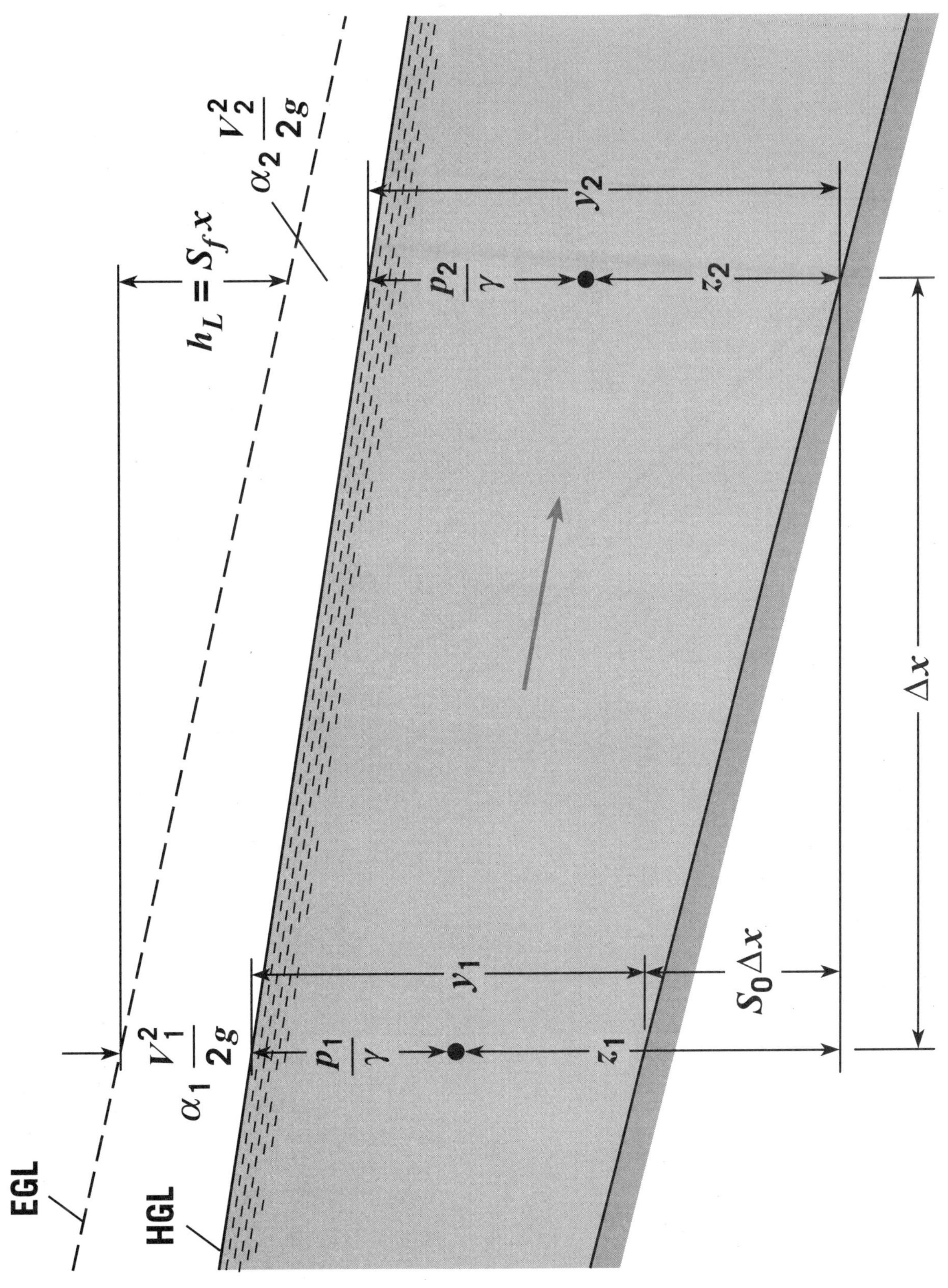
EGL
HGL
$h_L = S_f x$
$\alpha_1 \frac{V_1^2}{2g}$
$\alpha_2 \frac{V_2^2}{2g}$
$\frac{p_1}{\gamma}$
$\frac{p_2}{\gamma}$
y_1
y_2
z_1
z_2
$S_0 \Delta x$
Δx

FIGURE 15.1

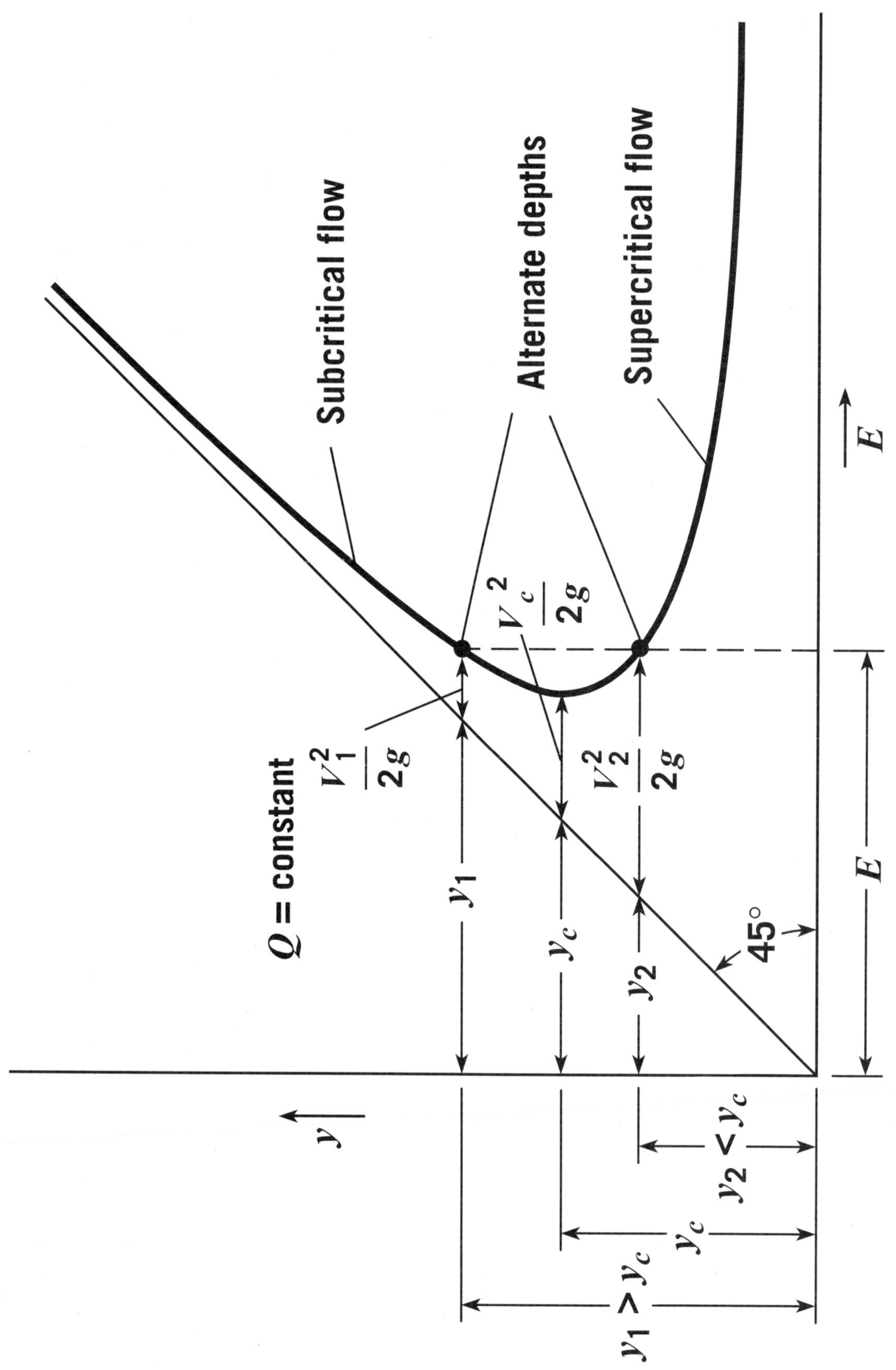
Subcritical flow
Alternate depths
Supercritical flow
Q = constant
$\frac{V_1^2}{2g}$
$\frac{V_c^2}{2g}$
$\frac{V_2^2}{2g}$
y_1
y_c
y_2
45°
E
$y_1 > y_c$
y_c
$y_2 < y_c$
y
E

FIGURE 15.2

TABLE A.2 PHYSICAL PROPERTIES OF GASES AT STANDARD ATMOSPHERIC PRESSURE AND 15°C (59°F)

Gas	Density, kg/m^3 (slugs/ft^3)	Kinematic viscosity, m^2/s (ft^2/s)	R Gas constant, J/kg K (ft-lbf/slug-°R)	c_p $\frac{\text{J}}{\text{kg K}}$ $\left(\frac{\text{Btu}}{\text{lbm-°R}}\right)$	$k = \frac{c_p}{c_v}$
Air	1.22 (0.00237)	1.46×10^{-5} (1.58×10^{-4})	287 (1716)	1004 (0.240)	1.40
Carbon dioxide	1.85 (0.0036)	7.84×10^{-6} (8.48×10^{-5})	189 (1130)	841 (0.201)	1.30
Helium	0.169 (0.00033)	1.14×10^{-4} (1.22×10^{-3})	2077 (12,419)	5187 (1.24)	1.66
Hydrogen	0.0851 (0.00017)	1.01×10^{-4} (1.09×10^{-3})	4127 (24,677)	14,223 (3.40)	1.41
Methane (natural gas)	0.678 (0.0013)	1.59×10^{-5} (1.72×10^{-4})	518 (3098)	2208 (0.528)	1.31
Nitrogen	1.18 (0.0023)	1.45×10^{-5} (1.56×10^{-4})	297 (1776)	1041 (0.249)	1.40
Oxygen	1.35 (0.0026)	1.50×10^{-5} (1.61×10^{-4})	260 (1555)	916 (0.219)	1.40

SOURCES: V. L. Streeter (ed.), *Handbook of Fluid Dynamics,* McGraw-Hill Book Company, New York, 1961; also R. E. Bolz and G. L. Tuve, *Handbook of Tables for Applied Engineering Science,* CRC Press, Inc., Cleveland, 1973; and *Handbook of Chemistry and Physics,* Chemical Rubber Company, 1951.

TABLE A.3 MECHANICAL PROPERTIES OF AIR AT STANDARD ATMOSPHERIC PRESSURE

Temperature	Density	Specific weight	Dynamic viscosity	Kinematic viscosity
	kg/m^3	N/m^3	$N \cdot s/m^2$	m^2/s
−20°C	1.40	13.7	1.61×10^{-5}	1.16×10^{-5}
−10°C	1.34	13.2	1.67×10^{-5}	1.24×10^{-5}
0°C	1.29	12.7	1.72×10^{-5}	1.33×10^{-5}
10°C	1.25	12.2	1.76×10^{-5}	1.41×10^{-5}
20°C	1.20	11.8	1.81×10^{-5}	1.51×10^{-5}
30°C	1.17	11.4	1.86×10^{-5}	1.60×10^{-5}
40°C	1.13	11.1	1.91×10^{-5}	1.69×10^{-5}
50°C	1.09	10.7	1.95×10^{-5}	1.79×10^{-5}
60°C	1.06	10.4	2.00×10^{-5}	1.89×10^{-5}
70°C	1.03	10.1	2.04×10^{-5}	1.99×10^{-5}
80°C	1.00	9.81	2.09×10^{-5}	2.09×10^{-5}
90°C	0.97	9.54	2.13×10^{-5}	2.19×10^{-5}
100°C	0.95	9.28	2.17×10^{-5}	2.29×10^{-5}
120°C	0.90	8.82	2.26×10^{-5}	2.51×10^{-5}
140°C	0.85	8.38	2.34×10^{-5}	2.74×10^{-5}
160°C	0.81	7.99	2.42×10^{-5}	2.97×10^{-5}
180°C	0.78	7.65	2.50×10^{-5}	3.20×10^{-5}
200°C	0.75	7.32	2.57×10^{-5}	3.44×10^{-5}
	$slugs/ft^3$	lbf/ft^3	$lbf\text{-}s/ft^2$	ft^2/s
0°F	0.00269	0.0866	3.39×10^{-7}	1.26×10^{-4}
20°F	0.00257	0.0828	3.51×10^{-7}	1.37×10^{-4}
40°F	0.00247	0.0794	3.63×10^{-7}	1.47×10^{-4}
60°F	0.00237	0.0764	3.74×10^{-7}	1.58×10^{-4}
80°F	0.00228	0.0735	3.85×10^{-7}	1.69×10^{-4}
100°F	0.00220	0.0709	3.96×10^{-7}	1.80×10^{-4}
120°F	0.00213	0.0685	4.07×10^{-7}	1.91×10^{-4}
150°F	0.00202	0.0651	4.23×10^{-7}	2.09×10^{-4}
200°F	0.00187	0.0601	4.48×10^{-7}	2.40×10^{-4}
300°F	0.00162	0.0522	4.96×10^{-7}	3.05×10^{-4}
400°F	0.00143	0.0462	5.40×10^{-7}	3.77×10^{-4}

SOURCE: Reprinted with permission from R. E. Bolz and G. L. Tuve, *Handbook of Tables for Applied Engineering Science,* CRC Press, Inc., Cleveland, 1973. Copyright © 1973 by The Chemical Rubber Co., CRC Press, Inc.

TABLE A.4 APPROXIMATE PHYSICAL PROPERTIES OF COMMON LIQUIDS AT ATMOSPHERIC PRESSURE

Liquid and temperature	Density kg/m^3 ($slugs/ft^3$)	Specific gravity (S) water at 4°C is ref.	Specific weight, N/m^3 (lbf/ft^3)	Dynamic viscosity, $N \cdot s/m^2$ ($lbf\text{-}s/ft^2$)	Kinematic viscosity, m^2/s (ft^2/s)	Surface tension, N/m* (lbf/ft)
Ethyl alcohol[3][1]	799	0.79	7,850	1.2×10^{-3}	1.5×10^{-6}	2.2×10^{-2}
20°C (68°F)	(1.55)		(50.0)	(2.5×10^{-5})	(1.6×10^{-5})	(1.5×10^{-3})
Carbon tetrachloride[3]	1,590	1.59	15,600	9.6×10^{-4}	6.0×10^{-7}	2.6×10^{-2}
20°C (68°F)	(3.09)		(99.5)	(2.0×10^{-5})	(6.5×10^{-6})	(1.8×10^{-3})
Glycerine[3]	1,260	1.26	12,300	6.2×10^{-1}	5.1×10^{-4}	6.3×10^{-2}
20°C (68°F)	(2.45)		(78.5)	(1.3×10^{-2})	(5.3×10^{-3})	(4.3×10^{-3})
Kerosene[2][1]	814	0.81	8,010	1.9×10^{-3}	2.37×10^{-6}	2.9×10^{-2}
20°C (68°F)	(1.58)		(51)	(4×10^{-5})	(2.55×10^{-5})	(2.0×10^{-3})
Mercury[3][1]	13,550	13.55	133,000	1.5×10^{-3}	1.2×10^{-7}	4.8×10^{-1}
20°C (68°F)	(26.3)		(847)	(3.2×10^{-5})	(1.3×10^{-6})	(3.3×10^{-2})
Sea water 10°C	1,026	1.03	10,070	1.4×10^{-3}	1.4×10^{-6}	
at 3.3% salinity	(1.99)		(64.1)	(3×10^{-5})	(1.5×10^{-5})	
Oils — 38°C (100°F)						
SAE 10W[4]	870	0.87	8,530	3.6×10^{-2}	4.1×10^{-5}	
	(1.69)		(54.4)	(7.4×10^{-4})	(4.4×10^{-4})	
SAE 10W-30[4]	880	0.88	8,630	6.7×10^{-2}	7.6×10^{-5}	
	(1.71)		(55.1)	(1.4×10^{-3})	(8.2×10^{-4})	
SAE 30[4]	880	0.88	8,630	1.0×10^{-1}	1.1×10^{-4}	
	(1.71)		(55.1)	(2.0×10^{-3})	(1.2×10^{-3})	

* Liquid–air surface tension values.

SOURCES: (1) V. L. Streeter, *Handbook of Fluid Dynamics,* McGraw-Hill Book Company, New York, 1961; (2) V. L. Streeter, *Fluid Mechanics,* 4th ed., McGraw-Hill Book Company, New York, 1966; (3) J. Vennard, *Elementary Fluid Mechanics,* 4th ed., John Wiley & Sons, Inc., New York, 1961; (4) R. E. Bolz and G. L. Tuve, *Handbook of Tables for Applied Engineering Sciences,* CRC Press, Inc., Cleveland, 1973.